Analog Circuits and Signal Processing

The Analog Circuits and Signal Processing book series, formerly known as the Kluwer International Series in Engineering and Computer Science, is a high level academic and professional series publishing research on the design and applications of analog integrated circuits and signal processing circuits and systems. Typically per year we publish between 5–15 research monographs, professional books, handbooks, edited volumes and textbooks with worldwide distribution to engineers, researchers, educators, and libraries.

The book series promotes and expedites the dissemination of new research results and tutorial views in the analog field. There is an exciting and large volume of research activity in the field worldwide. Researchers are striving to bridge the gap between classical analog work and recent advances in very large scale integration (VLSI) technologies with improved analog capabilities. Analog VLSI has been recognized as a major technology for future information processing. Analog work is showing signs of dramatic changes with emphasis on interdisciplinary research efforts combining device/circuit/technology issues. Consequently, new design concepts, strategies and design tools are being unveiled.

Topics of interest include:

Analog Interface Circuits and Systems;

Data converters;

Active-RC, switched-capacitor and continuous-time integrated filters;

Mixed analog/digital VLSI;

Simulation and modeling, mixed-mode simulation;

Analog nonlinear and computational circuits and signal processing;

Analog Artificial Neural Networks/Artificial Intelligence;

Current-mode Signal Processing;

Computer-Aided Design (CAD) tools;

Analog Design in emerging technologies (Scalable CMOS, BiCMOS, GaAs, heterojunction and floating gate technologies, etc.);

Analog Design for Test;

Integrated sensors and actuators;

Analog Design Automation/Knowledge-based Systems;

Analog VLSI cell libraries;

Analog product development;

RF Front ends, Wireless communications and Microwave Circuits;

Analog behavioral modeling, Analog HDL.

More information about this series at http://www.springer.com/series/7381

Karun Rawat • Patrick Roblin • Shiban Kishen Koul

Bandwidth and Efficiency Enhancement in Radio Frequency Power Amplifiers for Wireless Transmitters

Springer

Karun Rawat
Indian Institute of Technology Roorkee
Roorkee, India

Patrick Roblin
The Ohio State University
Columbus, OH, USA

Shiban Kishen Koul
Indian Institute of Technology Delhi
New Delhi, India

ISSN 1872-082X ISSN 2197-1854 (electronic)
Analog Circuits and Signal Processing
ISBN 978-3-030-38865-2 ISBN 978-3-030-38866-9 (eBook)
https://doi.org/10.1007/978-3-030-38866-9

This Springer imprint is published by the registered company Springer Nature Switzerland AG.
The registered company address is: Gewerbestrasse 11, 6330 Cham, Switzerland

We would like to dedicate our efforts to our families who inspired and supported us through everything.
We would also like to acknowledge our students and colleagues for their support.

Contents

Chapter 1
Introduction to RF Power Amplifier Design and Architecture

1.1 Introduction

Radio frequency (RF) power amplifier (PA) is an essential component of transmitter, which boosts the power of signal being transmitted over a long distance. The PA operates in large signal operation region where the inherent device nonlinearity will produce several distortions which reduces the spectral efficiency of transmission. In addition, the DC to RF conversion efficiency must be high to minimize the power dissipation. In order to maximize this efficiency, appropriate harmonic terminations must be presented to the transistor to appropriately shape the voltage and current waveform. Due to upcoming high-speed requirements in the wireless transmitters, the PA bandwidth must be enhanced to handle such transmission. In order to design wideband PA, the conventional architectures must be modified. This becomes very challenging in cases where appropriate harmonic terminations are required for high-efficiency operation due to which the operation is inherently narrow band. Moreover, the modern wireless communication uses spectrum-efficient modulated signals, for example, high-order quadrature amplitude modulation (QAM) signals, resulting into an envelope varying RF signals. These signals have high peak to average power ratio (PAPR), which operates the PA inefficiently at average power backed-off from its saturation. Therefore, such applications require enhancement of PA efficiency in the back-off region. Load modulation is one of the popular schemes to enhance the PA efficiency in back-off region. Doherty PA (DPA) and Chireix outphasing PA are two popular schemes based on the principle of load modulation. In such case an appropriate load combiner is required to enable load modulation. This load combiner is inherently narrowband and therefore restricts the bandwidth of such PAs. In order to enhance the bandwidth of these PAs, recently, several efforts have been carried out in modifying the architecture of these load combiners which will be discussed in this book.

K. Rawat et al., *Bandwidth and Efficiency Enhancement in Radio Frequency Power Amplifiers for Wireless Transmitters*, Analog Circuits and Signal Processing,
https://doi.org/10.1007/978-3-030-38866-9_1

This chapter discusses basic PA design parameters along with high-efficiency PA designs based on switch-mode operation and harmonic terminations. In addition, continuum of some classes of operation is also discussed for bandwidth enhancement in PAs. In order to develop basic understanding of Doherty and Chireix outphasing PA, a brief discussion on their architecture is also presented in this chapter. This chapter also introduces various linearization schemes which will be further discussed in detail in the later chapters. The chapter also introduces the basic architecture of delta-sigma modulation (DSM)-based transmitter which will be comprehensively discussed in later chapters.

1.2 Power Amplifier Design Parameters

There are several design parameters involved in any PA design based on its application. Each application targets key features such as high efficiency, high gain, and low distortion. The applications involving modulated signals will demand further features such as high average efficiency and low in-band and out-of-band distortions. This section describes these design parameters.

1.2.1 Power Amplifier Design Specifications

The PA in general can be seen as an RF component amplifying the RF power by consuming power from the DC power supply. Figure 1.1 shows the symbolic representation of PA, describing the DC power conversion to RF power at the output.

Due to this conversion, a DC to RF conversion efficiency is defined as one of the design parameters for the PA. This efficiency is often called drain efficiency (DE) in case the PA is designed using field effect transistor (FET) devices. If $P_{\text{out,RF}}$ represents output RF power and P_{DC} represents DC power consumed from the power supply, the DE represented as η_{D} is given by,

$$\eta_{\text{D}} = 100\frac{P_{\text{out,RF}}}{P_{\text{DC}}} \tag{1.1}$$

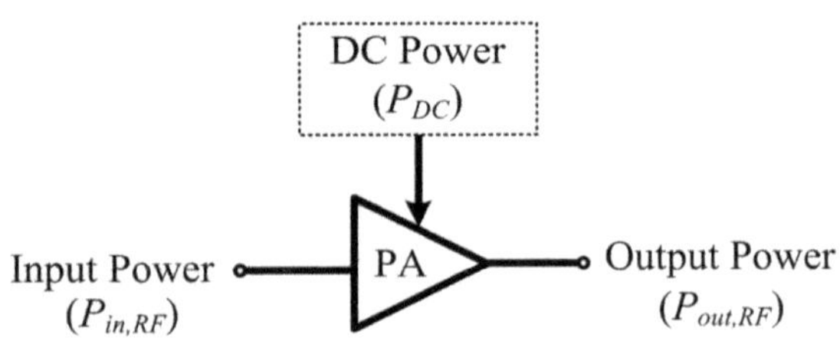

Fig. 1.1 Energy conversion in power amplifier (PA)

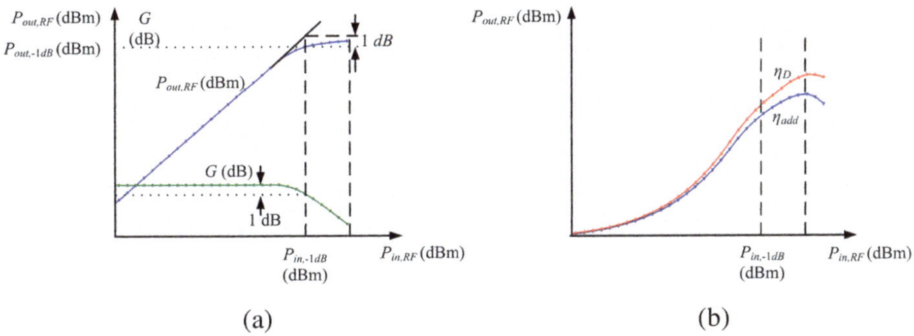

Fig. 1.2 PA RF characteristics (**a**) output power and gain versus input power drive (**b**) drain efficiency and power added efficiency versus input power drive

where η_D is expressed in percentage. The high efficiency corresponds to reduced power supply requirement which is essential for portable systems. In case of RF applications, the PA gain is also an important design parameter. In general, the gain reduces as frequency increases; therefore, it is important to consider the contribution of input power drive while operating at RF frequency. The efficiency corresponding to this added power is defined as power added efficiency (PAE) represented as η_{add} and given by,

$$\eta_{\text{add}} = 100\frac{\left(P_{\text{out,RF}} - P_{\text{in,RF}}\right)}{P_{\text{DC}}} = \eta_{\text{D}}\left(1 - \frac{1}{G}\right) \tag{1.2}$$

Figure 1.2 shows the characteristics of RF PA such as output power, gain, efficiency, and their variation with the input power drive.

One can see from Fig. 1.2a that the gain starts compressing after certain input power drive as the active device enters the large signal region. This gain compression is one of the reasons behind nonlinear operation in PA. The 1 dB gain compression point in such case is the power level at which gain compresses by 1 dB from its small signal value. This is one of the design parameters which is specified by the designer to identify the input power drive above which PA enters the nonlinear region. In practice, the PA is operated well above the P1-dB point where gain compression may be 2–3 dB. Due to this gain compression, the output power saturates after certain input power drive as shown in Fig. 1.2a. Similarly, the drain efficiency and PAE also saturates beyond certain input drive level as shown in Fig. 1.2b.

1.2.2 Power Amplifier Nonlinear Distortion Parameters

The nonlinearity in PA is caused by nonlinear transfer characteristic of the transistor. Therefore, the current produced with the input voltage drive has nonlinear components in case of FET device. Assuming a constant resistive load across the drain terminal of the device, the input and output voltages can be represented by a power series as,

$$v_0 = a_1 v_{\text{in}} + a_1 v_{\text{in}}^2 + a_3 v_{\text{in}}^3 + \cdots + a_n v_{\text{in}}^n \tag{1.3}$$

This simplified Taylor's power series can be truncated to lower order depending upon the extent of nonlinearity. For example, third-order approximation is used in case of cubic nonlinearity.

Based on the input excitation, different nonlinear terms appear at the output. In case of single-tone excitation, one can assume $v_{\text{in}} = V_{\text{S}}\cos(\omega_0 t)$, where V_{S} is the amplitude of the input signal and therefore the input power is given as $P_{\text{in}} = V_{\text{S}}^2/2$. Using this single-tone input signal in Eq. (1.3) truncated up to third harmonic, one can get,

$$\begin{aligned} v_0(t) = a_2\frac{V_{\text{S}}^2}{2} + \left(a_1 + \frac{3}{4}a_3 V_{\text{S}}^2\right) V_{\text{S}} \cos(\omega_0 t) \\ + a_2\frac{V_{\text{S}}^2}{2}\cos(2\omega_0 t) + a_3\frac{V_{\text{S}}^3}{4}\cos(3\omega_0 t) \end{aligned} \tag{1.4}$$

where the first term in Eq. (1.4) represents DC component, second term represents fundamental component, and third and fourth terms represent second and third harmonic components, respectively. The power gain can be written from Eq. (1.4) as,

$$G = \frac{P_{\text{out,RF}}}{P_{\text{in,RF}}} = G_{\text{SSG}}\left(1 + \frac{3}{2}\frac{a_3}{a_1}P_{\text{in,RF}}\right)^2 \tag{1.5}$$

where $G_{\text{SSG}} = a_1^2$ is the small signal gain assuming linear operation (with only first-order term in Eq. (1.3)). One can see that large signal gain as given in Eq. (1.5) depends on small signal gain and a term dependent on the input power drive. If the ratio a_3/a_1 is negative, this term reduces with input power drive resulting in gain compression characteristics. The output power at second and third harmonics are proportional to P_{in}^2 and P_{in}^3, respectively. Similarly, the power at nth harmonic is proportional to P_{in}^n. In general, harmonic components must be suppressed to achieve good dynamic range of the transmitter. The figure of merit which quantifies this suppression is called harmonic distortion. The harmonic distortion for nth harmonic is defined as ratio of output power at harmonic frequency and fundamental frequency. The total harmonic distortion is sum of all harmonic distortions and defined as,

$$\text{THD} = \sum_{k=2}^{n} \frac{P_{\text{out},kf_0}}{P_{\text{out},f_0}} \tag{1.6}$$

where P_{out,kf_0} and P_{out,f_0} are the output power at kth harmonic and fundamental frequency, respectively.

The harmonic distortion is of main concern in case of single-tone excitation. Since harmonic distortion occurs at frequencies which are integral multiple of fundamental frequency, they are far apart and can be suppressed by appropriate filtering. However, in case of multitone excitation, the multiple tones also interact with each other to produce several intermodulation products in addition to the harmonics. These closely placed multiple tones represents a modulated signal with envelope variation. A simple case is two-tone test which are often carried out in PA for nonlinear characterization. In such case the input drive is represented as $v_{\text{in}} = V_{\text{S1}}\cos(\omega_1 t) + V_{\text{S2}}\cos(\omega_2 t)$, where ω_1 and ω_2 are the two close frequencies. Using this input excitation in Eq. (1.3), one can expect the distorted output of PA as,

$$\begin{aligned} v_0(t) &= a_2\frac{V_{\text{S1}}^2}{2} + a_2\frac{V_{\text{S2}}^2}{2} + \left(a_1 + \frac{3}{4}a_3V_{\text{S1}}^2 + \frac{3}{2}a_3V_{\text{S2}}^2\right)V_{\text{S1}}\cos(\omega_1 t) \\ &+ \left(a_1 + \frac{3}{4}a_3V_{\text{S2}}^2 + \frac{3}{2}a_3V_{\text{S1}}^2\right)V_{\text{S2}}\cos(\omega_2 t) \\ &+ a_2\frac{V_{\text{S1}}^2}{2}\cos(2\omega_1 t) + a_2\frac{V_{\text{S2}}^2}{2}\cos(2\omega_2 t) \\ &+ a_2V_{\text{S1}}V_{\text{S2}}\left[\cos(\omega_2-\omega_1)t + \cos(\omega_2+\omega_1)t\right] \\ &+ a_3\frac{V_{\text{S1}}^3}{4}\cos(3\omega_1 t) + a_3\frac{V_{\text{S2}}^3}{4}\cos(3\omega_2 t) \\ &+ a_3\frac{3}{4}V_{\text{S1}}^2V_{\text{S2}}\left[\cos(2\omega_1+\omega_2)t + \cos(2\omega_1-\omega_2)t\right] \\ &+ a_3\frac{3}{4}V_{\text{S2}}^2V_{\text{S1}}\left[\cos(2\omega_2+\omega_1)t + \cos(2\omega_2-\omega_1)t\right] \end{aligned} \quad (1.7)$$

The first term is DC component. The second term represents the two tones amplified. One can see the amplitude of each amplified tone depends on amplitude of both the input tones. The amplitude terms V_{S1}^3 and V_{S2}^3 represent compression of tones at ω_1 and ω_2 frequencies, respectively. However, the amplitude terms $V_{\text{S1}}V_{\text{S2}}^2$ and $V_{\text{S2}}V_{\text{S1}}^2$ represent suppression of tones at ω_1 and ω_2 frequencies, respectively. The third and fifth terms in Eq. (1.7) represent second and third harmonic components of each tone. The fourth, sixth, and seventh terms however represents intermodulation products. These components appear at $n\omega_1 \pm m\omega_2$ frequencies, where $n+m$ define the order of the intermodulation product. Therefore, the fourth term appearing in Eq. (1.7) represents second-order intermodulation components. Similarly, sixth and seventh term in Eq. (1.7) represent third-order intermodulation components. One can see the third-order components $2\omega_1-\omega_2$ and $2\omega_2-\omega_1$ appear very close of the fundamental tones at ω_1 and ω_2 frequencies and therefore difficult to be filtered out. This requires an appropriate scheme such as predistortion to mitigate these unwanted components. The figure of merit that defines the suppression of these unwanted third-order intermodulation components is carrier to third-order intermodulation suppression ratio (C/IMD3) generally expressed in dBc.

1.2.3 Modulated Measurement and Characterization of Power Amplifiers

In Sect. 1.2.2, the nonlinear characteristic represented by Eq. (1.3) denotes a memoryless system. However, in practice, the PA is a dynamic system with memory where the nonlinear behavior also affects the phase of the output signal. In general, this dynamic behavior is more prominent for modulated signals with high bandwidth where carrier is modulated with fast changing envelope. In such case, if the input signal is represented by complex modulated signal $v_{in}(t) = V_S(t)\cos(\omega t + \varphi(t))$, the output depicts nonlinearity in both the amplitude and phase and is given as,

$$v_{out}(t) = G\{V_S(t)\}\cos[\omega t + \varphi(t) + \Psi\{V_S(t)\}] \tag{1.8}$$

where $G\{\}$ and $\Psi\{\}$ represents gain and phase as function of amplitude variation of input signal defined as amplitude modulation to amplitude modulation (AM/AM) conversion and amplitude modulation to phase modulation (AM/PM) conversion, respectively. Figure 1.3 shows the AM/AM and AM/PM plots of a Doherty PA tested with the Wideband Code Division Multiple Access (WCDMA) signal with 5 MHz bandwidth. One can observe from the Fig. 1.3 that the PA output is a function of several instances of the input signal for both AM/AM and AM/PM modulations. This many-to-one mapping seen in the PA characteristics is mainly due to memory effect [1]. An effective behavioral model should imitate the AM/AM and AM/PM characteristics accurately. Due to the nonlinear characteristics of PA in terms of gain (AM/AM) as well as phase (AM/PM), the signal is distorted and transmission error is introduced. The signal transmission errors can be categorized as the in-band and out-of-band error.

As shown in Sect. 1.2.2, out-of-band error is due to intermodulation distortion (IMD) terms appearing at the frequencies which are outside the band of the signal.

This spectrum leakage must be minimized to avoid interfering with the adjacent channel. The figure of merit to measure out-of-band distortion is adjacent channel power ratio (ACPR) given as [1],

$$\text{ACPR}_{\text{TOT}} \triangleq \frac{P_{\text{in-band}}}{P_{\text{adjacent-channels}}} \tag{1.9}$$

where $P_{\text{in-band}}$ is the in-band signal power and $P_{\text{adjacent-channels}}$ is the power measured in the adjacent channel. Equation (1.9) can be defined using power spectral density of the output signal as,

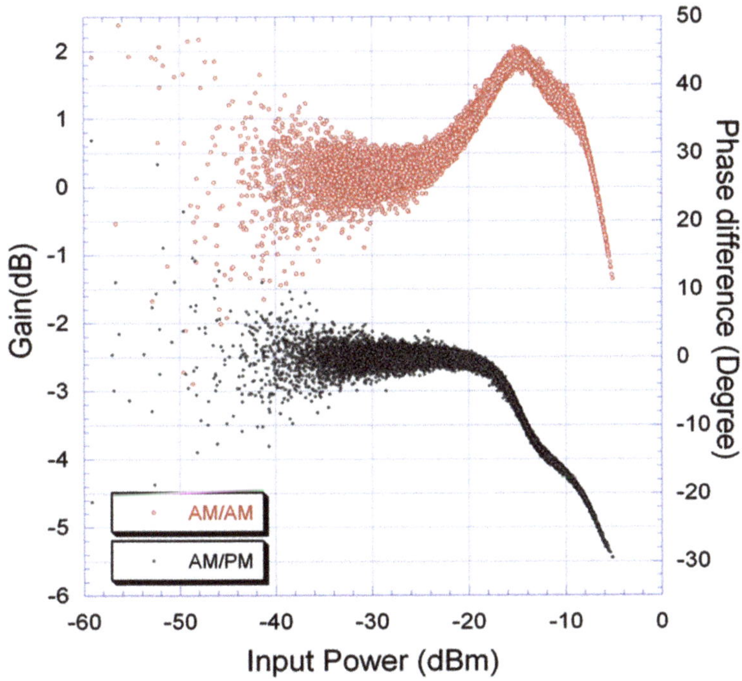

Fig. 1.3 Amplitude/amplitude modulation (AM/AM) and amplitude/phase modulation (AM/PM) characteristics for Doherty PA ([2], reprinted with permission from IEEE)

$$\mathrm{ACPR_{TOT}} = \frac{\int\limits_{\mathrm{B}} P_{\mathrm{out}}(f)\mathrm{d}f}{\int\limits_{\mathrm{LS}} P_{\mathrm{out}}(f)\mathrm{d}f + \int\limits_{\mathrm{US}} P_{\mathrm{out}}(f)\mathrm{d}f} \tag{1.10}$$

where $P_{\mathrm{out}}(f)$ is the power spectrum density of the output signal in dBm/Hz, B represents the band of the signal, LS and US represent lower-side and upper-side adjacent frequency bands, respectively. In addition to the out-of-band distortion in the form of spectrum leakage, the nonlinear characteristics of the PA will also result in in-band distortion/error. The figure of merit for the measurement of in-band error can be given as normalized mean square error (NMSE) [2, 3].

$$\mathrm{NMSE} \triangleq 10\ \log_{10}\left(\frac{\sum_{n=1}^{N} |y_{\mathrm{I}}(n) - y(n)|^2}{\sum_{n=1}^{N} |y_{\mathrm{I}}(n)|^2}\right) \tag{1.11}$$

where $y(n)$ is the output baseband signal, and $y_{\mathrm{I}}(n)$ is the ideal baseband signal/transmitted signal. NMSE provides in-band error information using baseband signal. However, another figure of merit for the in-band error is the error vector magnitude (EVM) that measures the modulation and demodulation accuracy. The

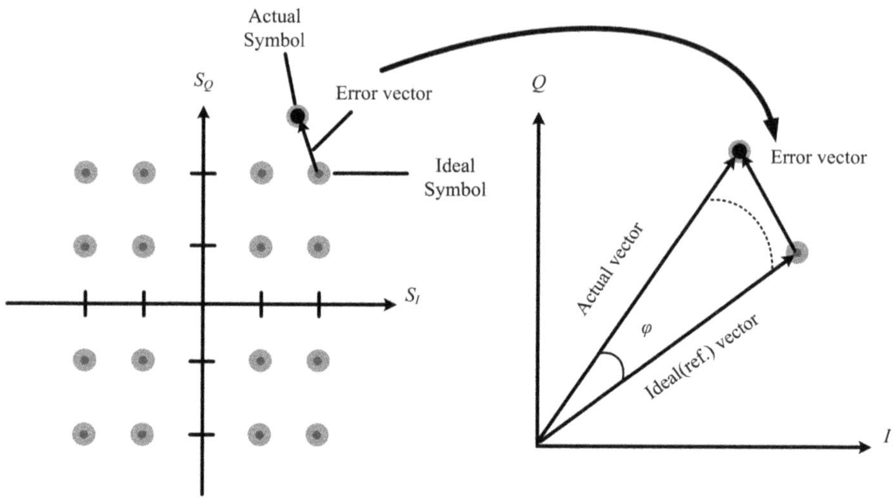

Fig. 1.4 Biasing arrangement of a PA

EVM can be calculated by observing the shift of symbols in constellation diagram as shown in Fig. 1.4. In terms of deviation of in-phase component (I) and quadrature-phase component (Q), EVM is defined as,

$$\mathrm{EVM} = \frac{\sqrt{(I - I_0)^2 + (Q - Q_0)^2}}{|P_0|} \tag{1.12}$$

where $P_0 = I_0 + jQ_0$ represents the original position of a symbol and $P = I + jQ$ denotes the measured position of demodulated symbol in the constellation diagram. EVM is also known as relative constellation error (RCE). EVM can also be related to NMSE while utilizing baseband information as,

$$\mathrm{EVM} \triangleq \sqrt{\frac{\sum_{n=1}^{N} |y_{\mathrm{I}}(n) - y(n)|^2}{\sum_{n=1}^{N} |y_{\mathrm{I}}(n)|^2}} \tag{1.13}$$

where $y(n)$ is the output baseband signal, and $y_{\mathrm{I}}(n)$ is the ideal baseband signal/transmitted signal. Figure 1.5 shows the power spectrum density of sent and received data from a low-cost software-defined radio communication system. Power spectrum density of error is also shown in the Fig. 1.5. One can perceive that the ACPR is approximately −42 dBc. The NMSE of the baseband signal is calculated using Eq. (1.11) as −25.9 dB. It can be observed from Fig. 1.5 that difference of average power between output signal and error signal within band is approximately −25 dB, which justifies the use of NMSE as an in-band performance metric. Therefore, if power spectrum density of signal and error signal is available, NMSE

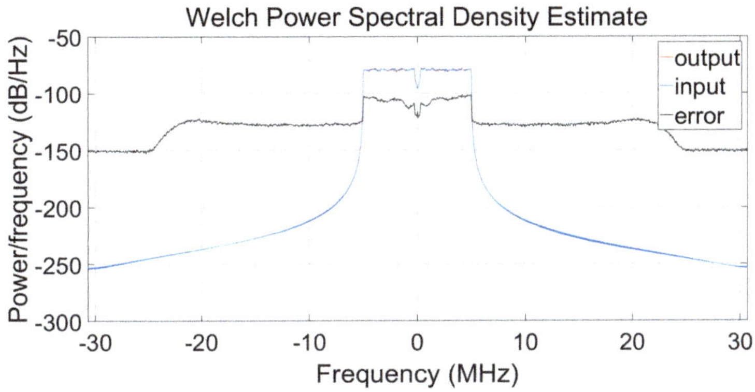

Fig. 1.5 Power spectrum density for in-band and out-of-band error estimation

of the system can be estimated by observing the power level difference between error signal and output signal in the band of interest.

Based on the definitions for ACPR in Eq. (1.10) and NMSE in Eq. (1.11), ACPR is recognized as "frequency-domain figure of merit," while NMSE is recognized as "time-domain figure of merit" for PA nonlinearity characterization.

1.3 High-Efficiency Power Amplifier Design

This section introduces basic classes of operation based on conduction angle. Later, the high-efficiency modes of operation such as switch-mode PAs will be introduced along with their basic theory of operation. This further discusses harmonic manipulation in class F mode to obtain high efficiency. The continuum of some classes is also introduced as potential solution for bandwidth enhancement along with appropriate harmonic termination. Later, the scheme of harmonic injection will be discussed to relax the constraints over passive matching networks in harmonic terminated PAs.

1.3.1 Classes of Operation: Based on Conduction Angle Mode

PAs are generally divided on the basis of their classes of operation and this usually is determined by the conduction angle of PA. Device output conduction angle refers to the fraction of RF signal period when the current flowing through the transistor is nonzero. In general, device conduction angle is set by the biasing point of the transistor, and therefore it can be used in classifying various classes of operation in PA. Transistor in any PA can be biased in one of the four biasing classes: A,

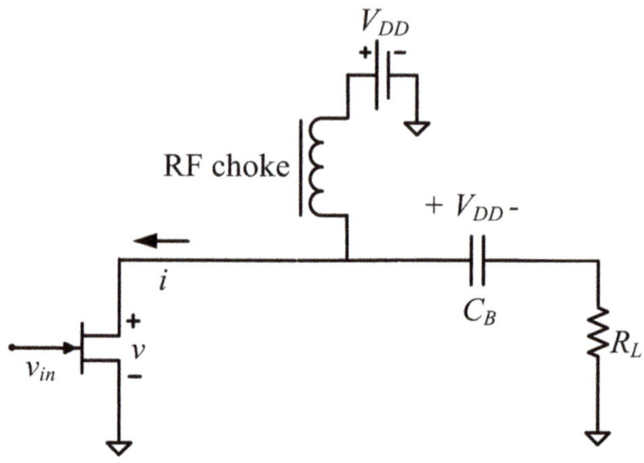

Fig. 1.6 Simplified circuit of PA with biasing arrangement

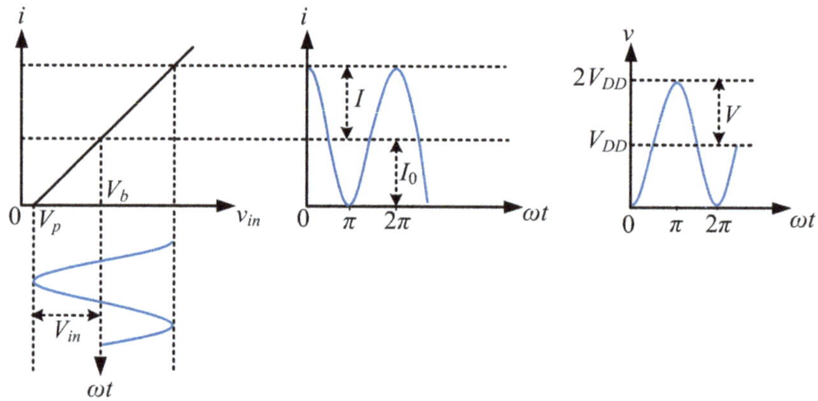

Fig. 1.7 Voltage–current waveforms for class A bias

AB, B, and C and the operating principle of these four classes are described in this section. For the analysis purpose, the transistor (e.g., FET) is assumed as an ideal voltage-controlled current source having zero output resistance in its saturation region. Moreover, the transconductance characteristic relating the output current and input voltage is assumed to be piece-wise linear in this analysis.

Figure 1.6 shows simplified PA circuit consisting of bias circuit and a load. The transistor is biased with an inductor often called RF choke and a capacitor termed as blocking capacitor. The RF choke prevents RF to leak toward DC supply, whereas blocking capacitor C_B blocks the DC to pass to the output load R_L.

For class A operation, the quiescent point of PA is set at the middle of the transistor transfer characteristics as shown in Fig. 1.7.

If V_p is the pinch-off voltage of the device, the gate bias and input drive ensure that V_{in} is always above pinch-off and the device is on all the time. Therefore, output current will be,

$$i = I_0 + I \cos \omega t \tag{1.14}$$

where I_0 is the quiescent current and second term is a cosine term which is produced due to input driving signal. The output voltage v across the drain terminal of the device represents a sum of the DC supply voltage V_{DD} and sinusoidal voltage across the load resistance R_L. Consequently, it represents a linear relationship between current and voltage, that is, with increasing output current, the output voltage across the load resistance R_L increases, and thus the output voltage v across the device output terminal decreases. Therefore, for a purely real load impedance, the output voltage v is shifted by 180° relative to the input voltage v_{in} and can be represented as,

$$v = V_{DD} - V \cos \omega t \tag{1.15}$$

Substituting Eq. (1.14) into Eq. (1.15) gives,

$$v = V_{DD} - I R_L \cos \omega t = V_{DD} - (i - I_0) R_L \tag{1.16}$$

Rearranging Eq. (1.16) gives,

$$i = \left(I_0 + \frac{V_{DD}}{R} \right) - \frac{v}{R_L} \tag{1.17}$$

Equation (1.17) represents the linear relationship between drain current and drain to source voltage and corresponds to the class A mode of operation. The drain efficiency (η) can be calculated as,

$$\eta = \frac{P_{out}}{P_{DC}} = \frac{1}{2} \left(\frac{I}{I_0} \right) \left(\frac{V}{V_{DD}} \right) \cdot 100 \tag{1.18}$$

where $P_{out} = \frac{1}{2} I V$ is the RF output power, whcre I and V are respective peak values of output current and voltage swings as shown in Fig. 1.7. P_{DC} is the DC power given by $P_{DC} = I_0 V_{DD}$, where I_0 and V_{DD} are the DC drain supply current and voltages, respectively, in a typical FET.

One can see from Eq. (1.18) that in order to maximize the efficiency, V/V_{DD} should reach to its maximum value, that is, 1. However, due to the knee voltage appearing in the output characteristic of FET device, the ratio V/V_{DD} will never reach to 1 in practice. Yet, the maximum theoretical efficiency in class A operating mode is 50% if $I/I_0 = 1$ and effect of knee voltage is neglected.

In case $I/I_0 > 1$, the drain current further increases; however, this will change the cosine current waveform into non-cosine waveform (e.g., half-cosine in case of class B operation). This results into the operation of PA in active region as well as in pinch-off region. This, in turn, introduces the higher order terms (i.e., harmonics) in the frequency spectrum. Therefore, PA circuit as shown in Fig. 1.6

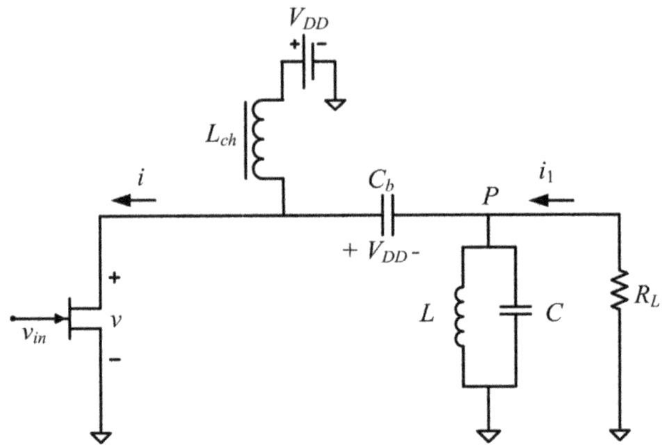

Fig. 1.8 Tuned load circuit of PA for class of operations with reduced conduction angle

Table 1.1 Summary of conduction angle and bias currents for various classes with reduced conduction angle

Class of operation	Conduction angle (2θ)	θ	$\cos\theta$	Bias current (I_0)
Class B	=180°	=90°	=0	=0
Class AB	>180°	>90°	<0	>0
Class C	<180°	<90°	>0	<0

must be modified to suppress these harmonics from reaching load R_L. This requires a frequency selective resonant circuit to be introduced at the output side of PA circuit as shown in Fig. 1.8.

This parallel LC tank circuit is used as a resonant circuit with resonant frequency equal to fundamental frequency. One can obtain the relation between conduction angles and current by considering the instance at which output current given in Eq. (1.14) takes a zero value represents half of current conduction angle (θ). This results in the following expression,

$$\cos\theta = -\frac{I_0}{I} \tag{1.19}$$

where the conduction angle is defined as 2θ. For class B biasing condition, the active device is biased at pinch-off such that the quiescent current $I_0 = 0$. Applying $I_0 = 0$ condition in Eq. (1.19), $\cos\theta = 0°$, and therefore $\theta = 90°$. The definition for various class of operation in terms of half conduction angle is given in Table 1.1.

Considering a conduction angle θ for any reduced class of operation other than class A, the input voltage and output current characteristics are shown in Fig. 1.9.

The output voltage across the load resistor R_L is sinusoidal assuming the LC resonator short all the harmonics currents at node P, allowing only fundamental current to flow in load resistor R_L. Since the output current is not sinusoidal, it can be represented using Fourier series expansion as,

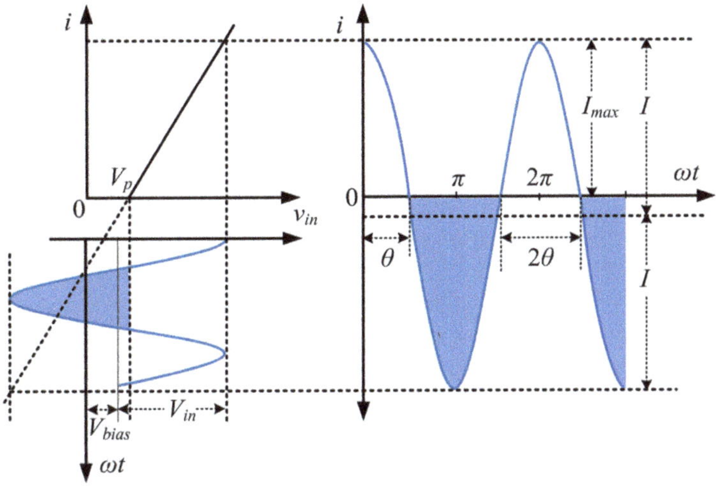

Fig. 1.9 Input voltage drive and output current waveforms for any conduction angle θ in case of reduced conduction angle

$$i = I_0 + I_1 \cos \ \omega t + I_2 \cos \ 2\omega t + \cdots \tag{1.20}$$

The value of the DC, fundamental frequency, and harmonic components can be calculated by obtaining Fourier series coefficient as,

$$I_0 = \frac{1}{2\pi}\int_{-\theta}^{\theta} I\left(\cos\omega t - \cos\theta\right) = \frac{I}{\pi}\left(\sin\theta - \theta\cos\theta\right) \tag{1.21}$$

$$I\left(f_0\right) = \frac{1}{2\pi}\int_{-\theta}^{\theta} I\left(\cos \ \omega t - \cos\theta\right)\cos\left(\omega t\right)\mathrm{d}\left(\omega t\right) = \frac{I}{\pi}\left(\theta - \frac{\sin 2\theta}{2}\right) \tag{1.22}$$

$$\begin{aligned} I\left(nf_0\right) &= \frac{1}{2\pi}\int_{-\theta}^{\theta} I\left(\cos \ \omega t - \cos\theta\right)\cos\left(n\omega t\right)\mathrm{d}\left(\omega t\right) \\ &= \frac{I}{\pi}\left\{\frac{\sin\left(n-1\right)\theta}{n\left(n-1\right)\theta} - \frac{\sin\left(n+1\right)\theta}{n\left(n+1\right)\theta}\right\} \end{aligned} \tag{1.23}$$

where $I = g_{\mathrm{m}} V_{\mathrm{in}}$ is obtained by assuming piece-wise linear transconductance (or transfer characteristics) of transistor. The DC component I_0 comprises bias current as well as average value added due to non-sinusoidal current waveform. Therefore, the DC power consumed is $P_{\mathrm{DC}} = V_{\mathrm{DD}}\, I_0$ and RF output power is given by $P_{\mathrm{out}} =$

$\frac{1}{2}I(f_0)V$. Since the maximum output voltage is V_{DD}, the efficiency of reduced conduction angle case is given by,

$$\eta = \frac{P_{out}}{P_{DC}} = \frac{1}{2}\left(\frac{\theta - \sin\theta \ \cos\theta}{\sin\theta - \theta \ \cos\theta}\right)\left(\frac{V}{V_{DD}}\right) \times 100 \tag{1.24}$$

If the knee effect is neglected, the ratio V/V_{DD} in Eq. (1.18) is set to 1. In such case, if $\theta = 90°$ which represents class B operation, the theoretical maximum efficiency is 78.54%. Similarly, for class C operation if $\theta = 60°$, $\eta \approx 90\%$. In case of class AB operation, where θ is selected as 120°, $\eta \approx 66\%$. One can conclude that with reduction in conduction angle, the efficiency improves which is in agreement with the fact that power dissipation across the device is reduced when it is conducting for small intervals of time. However, the device gain is reduced for small conduction angle in class C operation resulting in poor PAE. In case of tuned load circuit as shown in Fig. 1.8, the optimum load resistance R_L can be obtained as $R_L = V^2/2P_{out}$, where V is the peak voltage swing which is same as V_{DD} if knee voltage is neglected. However, for various conduction angles, the peak value of fundamental current is different if this load resistance is fixed according to the bias voltage and required output power.

1.3.2 Switch-Mode Power Amplifiers

In switch-mode PAs, the transistor is operated as a switch that is either in saturation region or pinch-off region, so as to maximize its efficiency. When switch is turned *on*, high current flows through the transistor and voltage across it would be zero and hence has low resistance. When switch is turned *off*, current flowing through the transistor would be zero and high voltage exists across it. Hence, switch mode results in nonoverlapping voltage-current waveforms and can ideally achieve no power dissipation across the device. This along with appropriate compensation of harmonics may result in 100% efficiency. The switch-mode PA exhibits high nonlinearity since the transistor is operated as an *on-off* switch. However, while dealing with constant envelope modulated signals, switch-mode PAs nonlinearity would have minimal effect. A wide variety of switch-mode PAs have been developed such as class D, class E, and class S which are discussed in the following sections.

Class D Power Amplifier

Class D mode was introduced by Baxandall in 1959 [4]. Class D PA is a switch-mode PA where transistor operates as an *on-off* switch. In general, class D PA is operated in push-pull configuration where two transistors are used. These transistors are driven with input signals with 180° phase shift such that only one transistor will be *on* for each cycle. Class D PAs broadly operate with two types of switching, one

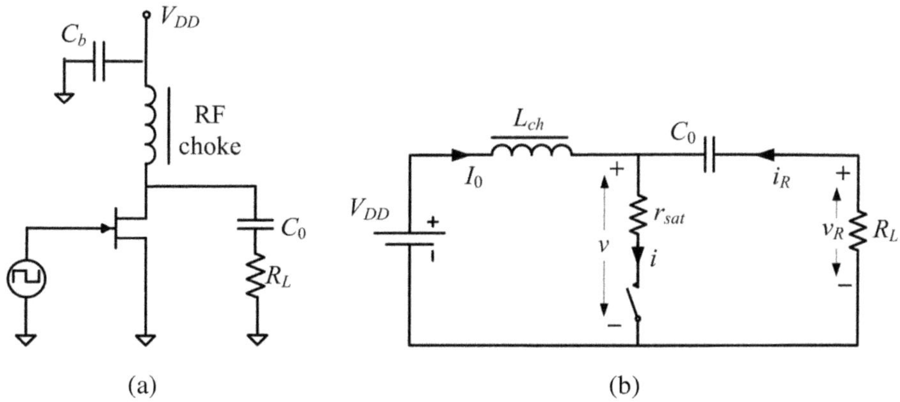

Fig. 1.10 (**a**) Basic architecture, (**b**) simplified equivalent circuit of single ended class D PA

is voltage switching and the other is current switching. In voltage switching class D PA, the intrinsic current is half sinusoidal and voltage is square wave. While in current switching class D PA, intrinsic current is square wave and voltage is half sinusoidal. Hence, class D PA exhibits theoretical efficiency of 100%. The basic architecture of single ended class D PA is shown in Fig. 1.10a. The simplified equivalent circuit, where the device acts as a switch is shown in Fig. 1.10b.

The RF choke as shown in Fig. 1.10 is used to prevent any leakage of RF into DC power supply. The blocking capacitor C_0 is used to avoid DC to reach the load R_{L}. The load current i_R is assumed to be sinusoidal. In the equivalent circuit, the saturation drain to source resistance is represented as r_{sat} and load resistance is represented as R_{L}, DC current is I_0, peak drain current is $I_{\max}$, and maximum drain voltage is $V_{\max}$. The device is driven in such a way that it will switch between its saturation (*on*) and cut-off (*off*) condition with an assumption of 50% duty cycle [5]. When switch is *on* for $0 \leq \omega t < \pi$, the current through the device can be obtained by using KVL in Fig. 1.10b and given by,

$$i\left(\omega t\right) = I_{\max} = \frac{V_{\mathrm{DD}} + R_{\mathrm{L}} I_0}{R_{\mathrm{L}} + r_{\mathrm{sat}}} \tag{1.25}$$

The corresponding voltage across the device is given by,

$$v\left(\omega t\right) = V_{\mathrm{sat}} = r_{\mathrm{sat}} I_{\max} \tag{1.26}$$

Similarly, when switch is *off* during $\pi \leq \omega t < 2\pi$, the current through the device is $i(\omega t) = 0$, whereas the voltage across the device is,

$$v\left(\omega t\right) = V_{\max} = V_{\mathrm{DD}} + R_{\mathrm{L}} I_0 \tag{1.27}$$

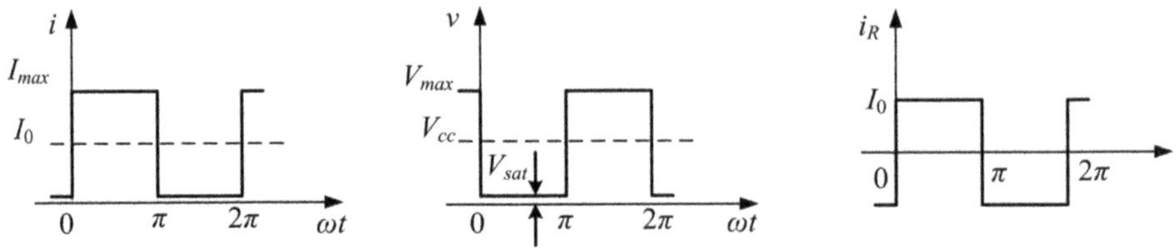

Fig. 1.11 Voltage and current waveforms of single-ended class D PA

The rectangular drain voltage, drain current, and load current waveforms are shown in Fig. 1.11. It shows that when there is maximum drain voltage, then drain current will be minimum (ideally zero), and when drain current is maximum, then drain voltage is minimum. The drain voltage can only attain the minimum value of V_{sat} and cannot be zero like drain current; this is because the active device is not a perfect switch and there will be some voltage drop across its drain and source terminals. This is modeled as saturation resistance (r_{sat}) between drain and source of the active device when operating in saturation region.

Figure 1.11 represents switch-mode operation of the active device which is the basic requirement for class D operation [5]. By taking the Fourier transform of Eq. (1.25), one can obtain DC and fundamental components of current as follows,

$$I_0 = \frac{1}{2\pi}\int_0^{\pi} i\,(\omega t)\,\mathrm{d}\,(\omega t) = \frac{V_{\text{DD}}}{R_{\text{L}} + 2r_{\text{sat}}} \tag{1.28}$$

Similarly, the fundamental component of current amplitude I can be obtained from,

$$I\,(f_0) = \frac{1}{\pi}\int_0^{\pi} i\,(\omega t)\sin(\omega t)\,\mathrm{d}\,(\omega t) = \frac{4}{\pi} I_0 \tag{1.29}$$

The output power for fundamental-frequency can be calculated by,

$$P = \frac{1}{2} I^2\,(f_0)\,R_{\text{L}} = \frac{8}{\pi^2}\frac{R_{\text{L}} V_{\text{DD}}^2}{(R_{\text{L}} + 2r_{\text{sat}})^2} \tag{1.30}$$

In push-pull configuration, two transistors are used and driven 180° out of phase. By giving two out of phase inputs it is ensured that only one transistor is operating for half cycle and another one for the next half cycle. The analysis for push-pull configuration is similar to the single ended class D configuration discussed above. But the maximum drain voltage reaches up to $2V_{\text{DD}}$ [5].

Class E Power Amplifier

Class E is a nonlinear PA where the transistor is operated as a switch instead of current source in order to achieve slow switching voltage-current characteristics [6–8]. This results in minimization of voltage-current overlapping and hence achieving 100% drain efficiency theoretically. The basic schematic diagram of class E PA is shown in Fig. 1.12.

The shunt capacitor C represents the intrinsic output capacitance of device and the external capacitance added by the load network. If the series resonant circuit L_S–C_S acts as an ideal filter at the fundamental frequency, it can be considered short at this frequency. Similarly, the DC feed inductor which is considered as an ideal choke can be considered open circuit. This results in simplified equivalent circuit as shown in Fig. 1.13. This topology of class E PA is known as tuned class E PA with shunt capacitance.

It is worth mentioning that for analysis purpose, it is assumed that transistor has infinite off-resistance, zero saturation resistance, and it has instantaneous and lossless switching action. Moreover, it is assumed that the only lossy component in Fig. 1.13 is resistant R. Since the device behaves as a switch in class E mode, in *on* condition current flows through the device with zero voltage drop across it while in *off* condition the current flows through the capacitor C. During *on* to *off* transition, the current will instantaneously flow into the C with no power dissipation while from *off* to *on* transition, charge stored in C will be discharged through device with soft switching, which may result in power loss. Hence, class E mode should be

Fig. 1.12 Schematic diagram of class E PA

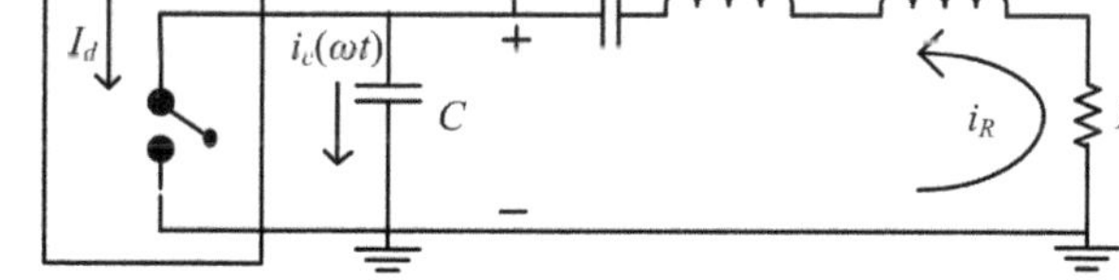

Fig. 1.13 Basic topology of tuned class E PA with a shunt capacitance mode at fundamental frequency [9]

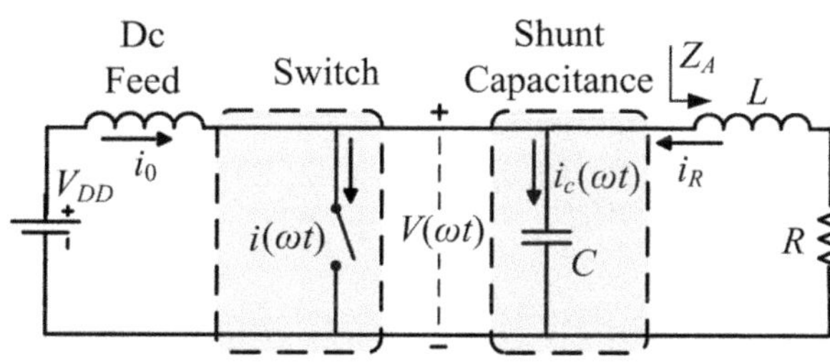

able to turn the voltage *off* once current turns *on*. The following conditions ensures the ideal class E operation: (a) zero voltage across the device when current flows through it, (b) zero current through the device when voltage exists across it, (c) minimize the switching time of the device, (d) voltage delay when switch turns *off*, (e) voltage returns to zero when switch is turned *on* (ZVS), and (f) zero voltage slope when switch is turned *on* (ZVDS) [8]. The last two conditions ensure the lossless operation mode when voltage is applied across the switch just before it turns *on* at the moment $t = 2\pi$. These can be expressed as,

$$\text{ZVS}: \; v(\omega t)|_{\omega t=2\pi} = 0 \tag{1.31}$$

$$\text{ZVDS}: \; \left.\frac{\mathrm{d}v(\omega t)}{\mathrm{d}\omega t}\right|_{\omega t=2\pi} = 0 \tag{1.32}$$

Due to ideal series L_S–C_S, the current flowing through the load i_R can be considered as sinusoidal with ϕ as the initial phase shift. Therefore, it can be expressed as,

$$i_R(\omega t) = I_R \sin(\omega t + \phi) \tag{1.33}$$

When switch is *on* for $0 \leq \omega t < \pi$, the voltage across the device $v(\omega t)$ and the current through the capacitor is $i_C(\omega t) = \omega C \frac{\mathrm{d}v(\omega t)}{\mathrm{d}\omega t}$ are zero. Therefore, current through the device can be written as,

$$i(\omega t) = I_0 + I_R \sin(\omega t + \phi) \tag{1.34}$$

where I_0 is the DC current and since under the initial on-state condition $i(0) = 0$, I_0 can be expressed as,

$$I_0 = -I_R \sin\phi \tag{1.35}$$

When switch is *off* for $\pi \leq \omega t < 2\pi$, the current across the device $i(\omega t) = 0$ and the current across the capacitor is given by,

$$i_c(\omega t) = I_0 + I_R \sin(\omega t + \phi) \tag{1.36}$$

Using Eqs. (1.35) and (1.36) one can obtain the voltage across the capacitor as,

$$v(\omega t) = \frac{1}{\omega C}\int_{\pi}^{\omega t} i_c(\omega t)\,\mathrm{d}\omega t = \frac{-I_R}{\omega C}\left[\cos(\omega t + \phi) + \cos\phi + (\omega t - \pi)\sin\phi\right] \tag{1.37}$$

After applying the ZVS as given in Eq. (1.31) at $\omega t = 2\pi$, one can obtain,

$$\tan\phi = \frac{-2}{\pi} \tag{1.38}$$

The above equation can further be expressed into two trigonometric relations as,

$$\sin\phi = \frac{-2}{\sqrt{\pi^2+4}}; \quad \cos\phi = \frac{\pi}{\sqrt{\pi^2+4}} \tag{1.39}$$

Expanding cosine term in Eq. (1.37) and using Eqs. (1.35) and (1.39), one can write,

$$v(\omega t) = \frac{I_0}{\omega C}\left[\omega t - \frac{3\pi}{2} - \frac{\pi}{2}\cos\omega t - \sin\omega t\right] \tag{1.40}$$

Accordingly, the DC component of Fourier analysis of Eq. (1.40) will determine the supply voltage V_{DD} as,

$$V_{DD} = \frac{1}{2\pi}\int_0^{2\pi}\frac{I_0}{\omega C}\left[\left(\omega t - \frac{3\pi}{2} - \frac{\pi}{2}\cos\omega t - \sin\omega t\right)\right] d\omega t = \frac{I_0}{\pi\omega C} \tag{1.41}$$

Using Eq. (1.41) in Eq. (1.40), one can write normalized steady-state voltage across the device for $\pi \leq \omega t < 2\pi$ as,

$$v_{norm}(\omega t) = \frac{v(\omega t)}{V_{DD}} = \pi\left(\omega t - \frac{3\pi}{2} - \frac{\pi}{2}\cos\omega t - \sin\omega t\right) \tag{1.42}$$

In order to obtain current $i(\omega t)$ when switch is *on* for $0 \leq \omega t < \pi$, the sine term in Eq. (1.34) is expanded and Eq. (1.35) is used to express the normalized current waveform as,

$$i_{norm}(\omega t) = \frac{i(\omega t)}{I_0} = 1 + \frac{\pi}{2}\sin\omega t - \cos\omega t \tag{1.43}$$

The normalized ideal voltage and current waveforms of class E mode according to Eqs. (1.42) and (1.43) are shown in Fig. 1.14.

It can be observed that the voltage is zero and current is sinusoidal when transistor is turned *on*. This current flows through the parallel capacitor C when transistor is turned *off*. Hence, there are nonoverlapping voltage-current waveforms which results in 100% efficiency theoretically. Since the efficiency is theoretically 100% for class E mode, the DC power supplied to the device must be equal to the power delivered to the load at the fundamental frequency. This results in,

$$I_0 V_{DD} = \frac{I_R^2}{2}R \tag{1.44}$$

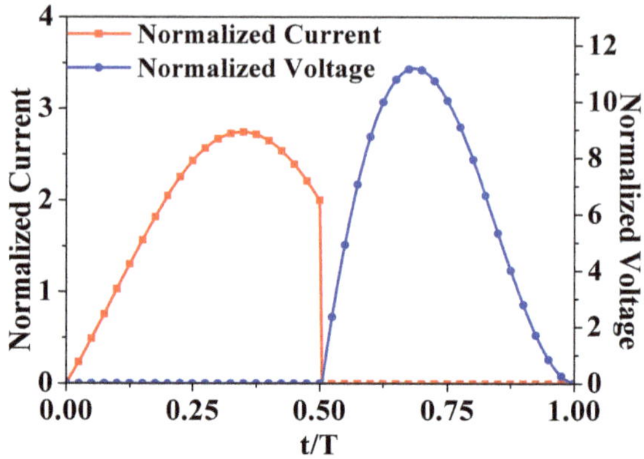

Fig. 1.14 Normalized switch current and voltage waveforms (time is normalized with period T)

Substituting I_R in Eq. (1.44) from Eq. (1.35) and using value of sinϕ from Eq. (1.39), one can get,

$$I_0 = 0.577\frac{V_{\mathrm{DD}}}{R} \tag{1.45}$$

Similarly, the output voltage across resistance R is $V_R = I_R R$. Further, using I_R from Eq. (1.35) with value of sinϕ obtained from Eq. (1.39), one can get,

$$V_R = 1.074V_{\mathrm{DD}} \tag{1.46}$$

At fundamental frequency, the voltage across the switch during $\pi \leq \omega t < 2\pi$ has two quadrature components, one developed across resistor R and the other across L. The amplitude of these voltages can be obtained by Fourier series coefficient as,

$$\begin{aligned} V_R &= -\frac{1}{\pi}\int_0^{2\pi} \pi V_{\mathrm{DD}}\left(\omega t - \frac{3\pi}{2} - \frac{\pi}{2}\cos\omega t - \sin\omega t\right)\sin(\omega t + \phi)\,\mathrm{d}\omega t \\ &= \frac{I_R}{\pi\omega C}\left(\frac{\pi}{2}\sin 2\phi + 2\cos 2\phi\right) \end{aligned} \tag{1.47}$$

$$\begin{aligned} V_L &= -\frac{1}{\pi}\int_0^{2\pi} \pi V_{\mathrm{DD}}\left(\omega t - \frac{3\pi}{2} - \frac{\pi}{2}\cos\omega t - \sin\omega t\right)\cos(\omega t + \phi)\,\mathrm{d}\omega t \\ &= -\frac{I_R}{\pi\omega C}\left(\frac{\pi}{2} + \pi\ \sin^2\phi + 2\sin 2\phi\right) \end{aligned} \tag{1.48}$$

For optimum angle ϕ as obtained from Eq. (1.38), the Eq. (1.47) will simplify to Eq. (1.46). Therefore, the ratio of inductive reactance and resistance $\omega L/R = V_L/V_R$ has value 1.1525 for the optimum angle ϕ from Eqs. (1.47) and (1.48). Similarly, the ratio of resistance and capacitive reactance $\omega CR = \omega CV_R/I_R = 0.1836$ can directly be obtained from Eq. (1.47). Therefore, the series inductance and shunt capacitance can be calculated as,

$$L = 1.1525\frac{R}{\omega} \tag{1.49}$$

$$C = 0.1836\frac{1}{\omega R} \tag{1.50}$$

where R is the optimum load resistance obtained by using Eqs. (1.44) and (1.45) as,

$$R = 0.5768\frac{V_{\mathrm{DD}}^2}{P_{\mathrm{out}}} \tag{1.51}$$

where P_{out} is average power given by $I_R^2 R/2$ and V_{DD} is the supply voltage.

In practice, while realizing class E PA with transmission line, the load including the inductor and resistor while excluding the capacitance is calculated. This load is named as Z_{A} in Fig. 1.13 which is given by,

$$Z_{\mathrm{A}}(n\omega) = \begin{cases} R\,(1 + j\omega L/R) & n = 1 \\ \infty & n = 2, 3, \ldots, \infty \end{cases} \tag{1.52}$$

where $\omega L/R$ can be obtained by Eq. (1.49). A transmission line-based matching network can easily realize the above load while transforming from 50 Ω impedance. A good approximation to class E mode can be obtained by considering matching up to second harmonics [8]. Moreover, the corresponding value of capacitance C is obtained from Eq. (1.50), which can be realized by the intrinsic device capacitance along with some shunt capacitance if required. For example, a 10 W Cree GaN HEMT device CGH40010F is used to design class E PA in [9]. In this case, $R = 42.04\ \Omega$ is obtained for $V_{\mathrm{DD}} = 27$ V and corresponding values of L and C are 2.2 nH and 0.19 pF, respectively. Thus, the value of fundamental load represented by Eq. (1.52) is calculated as $42.04 + j48.38\ \Omega$. The harmonic impedances are set to be strongly reactive. These values are indeed set at intrinsic device plane and later projected to the package plane for appropriate matching [9]. Figure 1.15 shows the schematic of class E PA [9]. One can see an additional capacitance is used in shunt at device drain terminal at package plane to obtain appropriate capacitance for class E mode. The stabilization is performed by shunt R-C network at the input.

Figure 1.16 shows the photograph of class E PA fabricated on Rogers RO4350 substrate. One can see the second harmonic matching is kept highly reactive using the drain bias circuitry. Figure 1.17 shows the measured and simulated results for the class E PA designed to operate at 3.6 GHz. The fabricated class E PA is tested with continuous wave (CW) signal from 3.55 to 3.65 GHz. The simulated and measured results show maximum DE of 70% and 67.42%, respectively. The maximum PAE

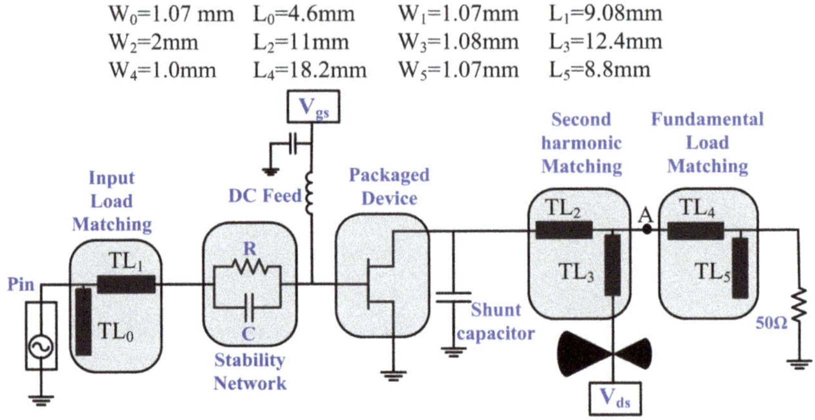

Fig. 1.15 Block diagram of class E PA designed at 3.6 GHz ([9], reprinted with permission from IEEE)

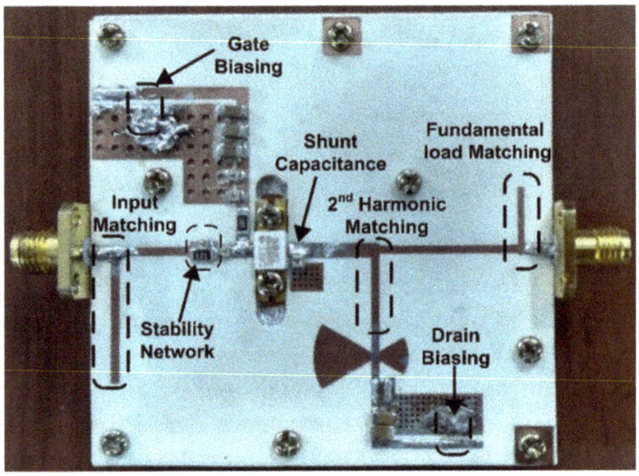

Fig. 1.16 Photograph of class E PA designed at 3.6 GHz ([9], reprinted with permission from IEEE)

is 66.5% in simulation and 63.21% in measurement. The simulation and measured gain at saturation are 12.8 and 12.04 dB, respectively.

The analysis of conventional class E topology with shunt capacitor as shown in Fig. 1.13 assumes an infinite RF choke as DC feed. But in practice, it is impossible to realize the RF choke with infinite impedance. Hence, a new topology is presented which is considered the finite inductance of a DC feed [8]. This topology is called parallel circuit class E PA. Therefore, the proposed class E topology [8] consists of a parallel capacitor, parallel inductor, and a series resonant circuit for fundamental along with load as shown in Fig. 1.18a–c [8].

Since the series combination of L_S and C_S acts as an ideal filter at the fundamental frequency, it can be considered short at this frequency. Therefore, Fig. 1.18b is further modified to Fig. 1.18c for analysis purpose. The parallel circuit class E has advantages like high load resistance, low DC feed inductance, high

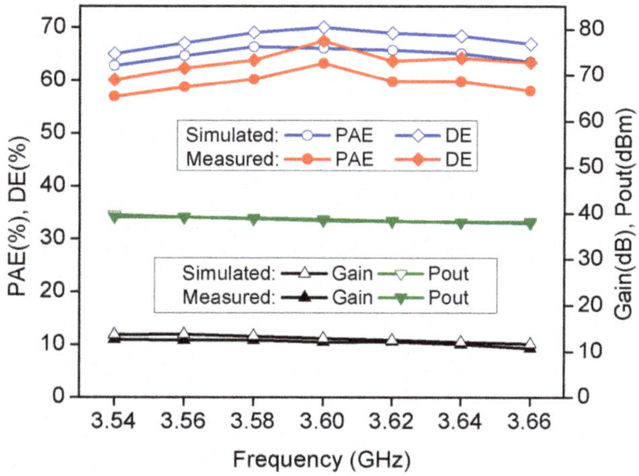

Fig. 1.17 Measured and simulated results of class E PA over frequency ([9], reprinted with permission from IEEE)

operating frequency, and even broadband characteristics with reactive compensation which will be discussed in Chap. 4.

Referring to Fig. 1.18c, the output current flowing through load is assumed to be sinusoidal with ϕ as initial phase shift, that is, $i_R(\omega t) = I_R \sin(\omega t + \phi)$. When switch is *on* for $0 \leq \omega t < \pi$, the voltage $v(\omega t)$ across the switch and the current across the capacitance C is zero. Substituting these conditions on KVL equation of left mesh, one can write $V_{DD} = v_L(\omega t)$, where $v_L(\omega t)$ is the voltage across inductor. Therefore, the current through inductor at steady-state condition is given by,

$$i_L(\omega t) = \frac{1}{\omega L}\int_0^{\omega t} v_L(\omega t)\,\mathrm{d}(\omega t) = \frac{V_{DD}}{\omega L}\omega t \tag{1.53}$$

The current across the switch is given as,

$$i(\omega t) = i_R(\omega t) + i_L(\omega t) = \frac{1}{\omega L}\int_0^{\omega t} V_{DD}\mathrm{d}(\omega t) + i_L(0) + I_R \sin(\omega t + \phi) \tag{1.54}$$

At initial *on*-state condition, $i(0) = 0$, which implies $i_L(0) = -i_R(\omega t)$. Therefore, Eq. (1.54) can be rewritten as,

$$i(\omega t) = \frac{V_{DD}}{\omega L}\omega t + I_R[\sin(\omega t + \phi) - \sin\phi] \tag{1.55}$$

When the switch is *off* for $\pi \leq \omega t < 2\pi$, the current across the switch $i(\omega t) = 0$ and the current flowing through the capacitance is given by $i_C(\omega t) = i_L(\omega t) + i_R(\omega t)$. In such a case, the voltage across inductor is $v_L(\omega t) = V_{DD} - v(\omega t)$, where $v(\omega t)$ is

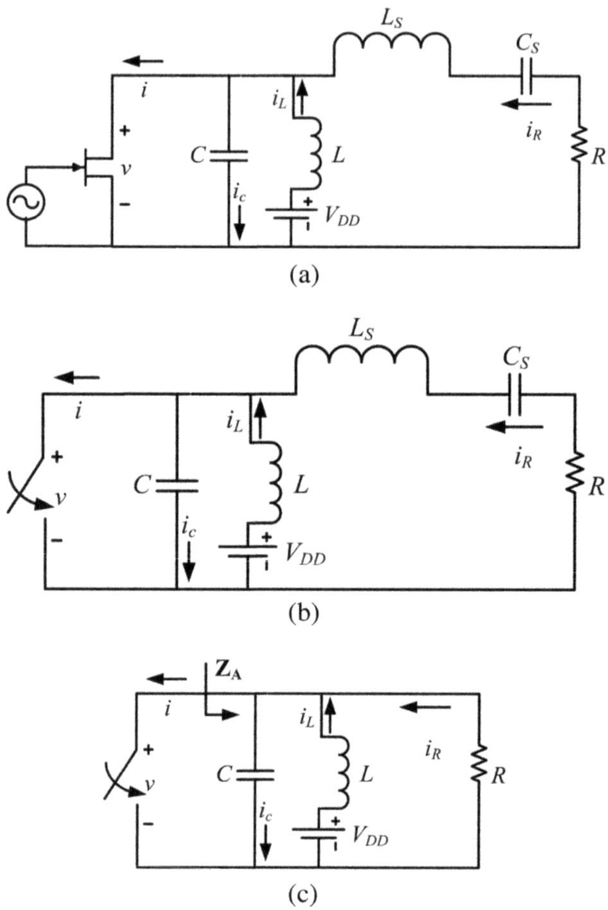

Fig. 1.18 (**a**) Basic architecture, (**b**) equivalent circuit, and (**c**) equivalent circuit at fundamental frequency of parallel circuit class E PA

the voltage across the switch in steady-state condition. Therefore, the current across the capacitor is given by,

$$i_C(\omega t) = \frac{1}{\omega L}\int_{\pi}^{\omega t} [V_{\mathrm{DD}} - v(\omega t)]\,\mathrm{d}(\omega t) + i_L(\pi) + I_R \sin(\omega t + \phi) \tag{1.56}$$

where $i_L(\pi)$ is the initial off-state condition current through the inductor. At this initial state condition, $v(\pi) = 0$; therefore, one can write,

$$i_L(\pi) = i(\pi) - i_R(\pi) = \frac{\pi V_{\mathrm{DD}}}{\omega L} - I_R \sin\phi \tag{1.57}$$

Using Eq. (1.57) in Eq. (1.56), one can get,

$$i_C(\omega t) = \omega C \frac{\mathrm{d}v(\omega t)}{\mathrm{d}(\omega t)} = \frac{1}{\omega L}\int_{\pi}^{\omega t} [V_{\mathrm{DD}} - v(\omega t)]\,\mathrm{d}(\omega t) + \frac{\pi V_{\mathrm{DD}}}{\omega L} - I_R \sin\phi + I_R \sin(\omega t + \phi) \tag{1.58}$$

Differentiating both sides of Eq. (1.58) one can write,

$$\omega^2 LC \frac{\mathrm{d}^2 v(\omega t)}{\mathrm{d}(\omega t)^2} + v(\omega t) - \omega L I_R \cos(\omega t + \phi) = V_{\mathrm{DD}} \tag{1.59}$$

Equation (1.59) is a linear nonhomogeneous second-order differential equation which has general solution given by,

$$v(\omega t) = A\ \cos(k\omega t) + B\ \sin(k\omega t) + V_{\mathrm{DD}} - \frac{k^2}{1-k^2}\omega L I_R \cos(\omega t + \phi) \tag{1.60}$$

where $k = 1/\omega\sqrt{LC}$ and coefficients A and B are determined from initial conditions, that is, $v(\pi) = v(2\pi) = 0$. There are three unknowns in Eq. (1.60); therefore, three conditions will be used to solve for these unknowns. Out of these three, two conditions can be ZVS and ZVDS as mentioned in Eqs. (1.31) and (1.32), rest of the condition is obtained by defining the DC voltage component in terms of Fourier series expansion of voltage across switch, that is, $v(\omega t)$ as $V_{\mathrm{DD}} = \frac{1}{2\pi}\int_0^{2\pi} v(\omega t)\,\mathrm{d}\omega t$. Using these conditions, the exact values of unknown parameters can be numerically obtained as,

$$k = 1.412 \tag{1.61}$$

$$\phi = 15.155^{\circ} \tag{1.62}$$

$$\omega L I_R = 1.21 V_{\mathrm{DD}} \tag{1.63}$$

Similarly, the DC supply current can be obtained using Fourier series coefficient of $i(\omega t)$ given by Eq. (1.55) as,

$$I_0 = \frac{1}{2\pi}\int_0^{2\pi} i(\omega t)\,\mathrm{d}\omega t = \frac{I_R}{2\pi}\left(\frac{V_{\mathrm{DD}}}{\omega L I_R}\frac{\pi^2}{2} + 2\cos\phi - \pi\ \sin\phi\right) \tag{1.64}$$

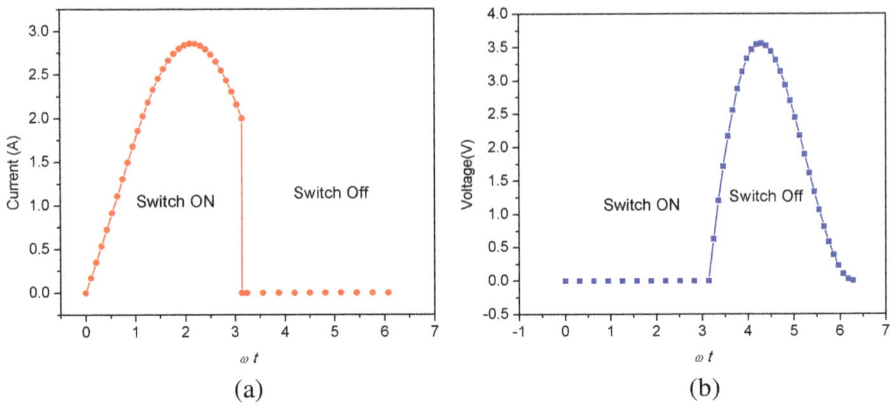

Fig. 1.19 Normalized drain (**a**) current and (**b**) voltage waveforms for ideal parallel class E PA

Using Eqs. (1.61)–(1.63) in Eq. (1.64), one can get,

$$I_0 = 0.826 I_R \tag{1.65}$$

Figure 1.19a, b shows the voltage and current waveforms normalized with respect to bias voltage and current, respectively, for class E mode as given in Eqs. (1.55) and (1.60). One can see that there are nonoverlapping voltage-current waveforms which result in 100% efficiency theoretically.

At fundamental frequency, the current flowing through the switch during $0 \leq \omega t < \pi$ has two quadrature components, namely, I_R and I_X flowing through resistor and reactance (parallel combination of L and C), respectively. The amplitudes of these currents can be obtained by Fourier series coefficient of $i(\omega t)$ in Eq. (1.55) as,

$$\begin{aligned} I_R &= \frac{1}{\pi} \int_0^{2\pi} i\,(\omega t) \sin\,(\omega t + \phi)\,\mathrm{d}\,(\omega t) \\ &= \frac{I_R}{\pi} \left[\frac{V_{\mathrm{DD}}}{\omega L I_R} (\pi \quad \cos\phi - 2\sin\phi) + \frac{\pi}{2} - \sin 2\phi \right] \end{aligned} \tag{1.66}$$

$$\begin{aligned} I_X &= -\frac{1}{\pi} \int_0^{2\pi} i\,(\omega t) \cos\,(\omega t + \phi)\,\mathrm{d}\,(\omega t) \\ &= \frac{I_R}{\pi} \left[\frac{V_{\mathrm{DD}}}{\omega L I_R} (\pi \quad \sin\phi + 2\cos\phi) - 2\sin^2\phi \right] \end{aligned} \tag{1.67}$$

Referring to Fig. 1.18c, the angle ψ between the fundamental voltage and current applied to the switch is given by,

$$\tan\psi = \frac{I_X}{I_R} = \frac{R}{\omega L} - \omega RC = 0.6807 \tag{1.68}$$

Therefore, the normalized load network parameters can be obtained by,

$$\frac{\omega L}{R} = \frac{k^2 - 1}{k^2 \tan\psi} \tag{1.69}$$

$$\omega RC = \frac{\tan\psi}{k^2 - 1} \tag{1.70}$$

Therefore, using Eqs. (1.61) and (1.68), optimum shunt inductance and capacitance can be calculated as,

$$L = 0.732\frac{R}{\omega} \tag{1.71}$$

$$C = \frac{0.685}{\omega R} \tag{1.72}$$

where R is the optimum load resistance for a given value of supply voltage V_{DD} and output power P_{out} can be obtained by using Eq. (1.63) and from the fact that $R = \frac{V_R^2}{2P_{out}}$ as,

$$R = 1.365\frac{V_{DD}^2}{P_{out}} \tag{1.73}$$

In practice, while realizing class E PA with transmission line, the shunt inductor is replaced by a short circuit transmission line with small electrical length θ with characteristic impedance Z_0 related to required inductance L at fundamental frequency as $Z_0 \tan\theta = \omega L$. This line can be shorted by a bypass capacitor and can be used in bias circuitry to replace RF choke. Moreover, the series resonant circuit L_S and C_S as shown in Fig. 1.18a is realized using a matching network. With an appropriate selection of transmission line realizing this matching network one can also realize high impedance for harmonic frequencies.

Class S Power Amplifier

Class S is also the switch-mode PA based on class D architecture. Class S circuit utilizes bi-level constant envelope signals such as obtained from delta-sigma

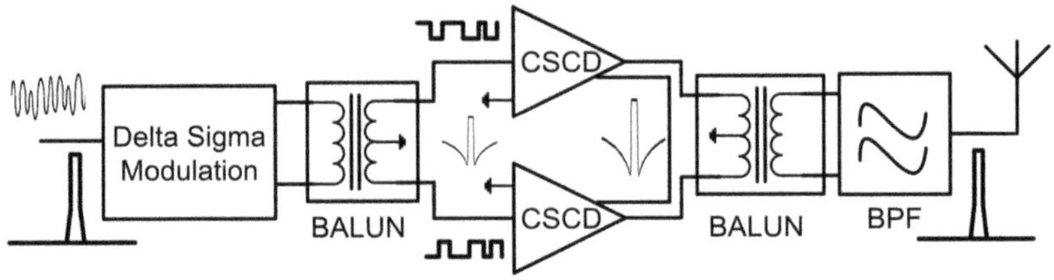

Fig. 1.20 General architecture of class S PA

Fig. 1.21 Output spectrum of 5 MHz LTE signal after amplification from class S PA ([15], reprinted with permission from IEEE)

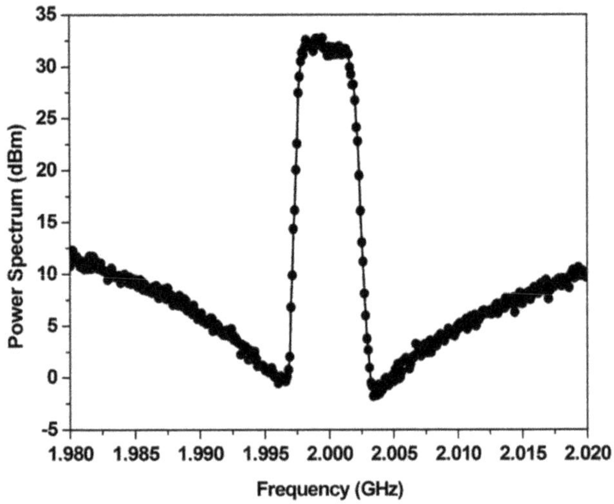

modulation (DSM) or any other signal such as pulse density modulation. In such signals, the information is encoded in the density of pulses [10–14]. The typical architecture of class S is shown in Fig. 1.20 [15]. One can see from Fig. 1.20 that the modulator can be DSM, where the current switching class D PAs are used in push-pull configuration. Two baluns are used, one at the input and other at the output. The input balun splits the input signal into 180° out of phase which is later amplified and combined using balun at the output.

A band-pass filter following the output balun is used to regain the original envelope of the signal back. Figure 1.21 shows the spectrum of 5 MHz LTE signal amplified using class S PA.

One can see from Fig. 1.21, that the 5 MHz LTE signal is placed at 2 GHz and a band-pass filter will suppress the out of band noise [15]. The in-band noise is already reduced by an appropriate noise transfer function of DSM modulation.

Class F Power Amplifiers

Class F PAs consists of multi-harmonic resonator in the load network for waveform shaping in order to achieve high efficiency. This was first invented by Tyler in 1919 [16]. Multi-resonators shape the waveform such that the drain current is allowed to flow when drain-to-source voltage is low, and when the drain current is low the drain-to-source voltage is high. This reduces the power dissipation in the transistor and hence efficiency is improved. Lumped element resonators based class F PAs have been widely used in high-power AM broadcast radio transmitters in low-frequency, medium-frequency, and high-frequency range [17]. However, at very high frequency (VHF), Tyler proposed the class F PA with quarter wavelength transmission line which controls all the odd harmonics. These were used in VHF FM broadcast radio transmitters.

In general, the drain to source voltage and drain current waveforms with all the harmonic components can be represented as [18],

$$\begin{aligned} V_{\mathrm{d}}(\theta) &= V_{\mathrm{DD}} + V_{a1}\cos\theta - V_{b1}\sin\theta + V_{a2}\cos 2\theta \\ &\quad - V_{b2}\sin 2\theta + V_{a3}\cos 3\theta - V_{b3}\sin 3\theta \ldots \end{aligned} \tag{1.74}$$

$$\begin{aligned} I_{\mathrm{d}}(\theta) &= I_0 + I_{a1}\cos\theta - I_{b1}\sin\theta + I_{a2}\cos 2\theta \\ &\quad - I_{b2}\sin 2\theta + I_{a3}\cos 3\theta - I_{b3}\sin 3\theta \ldots \end{aligned} \tag{1.75}$$

where V_{ai} and I_{ai} are the cosine terms in voltage and current waveform, respectively. Here i is an integer defining the harmonic term. Similarly, V_{bi} and I_{bi} are the Sine terms in the voltage and current waveform, respectively. In ideal class F mode of operation, load network should present a short circuit at even harmonics and an open circuit at odd harmonics. This results into the voltage waveform to be a square-wave and current waveform to be a half sinusoid. The basic circuit of class F PA with multiple resonant circuits is shown in Fig. 1.22. The ideal drain-to-source voltage and drain current for class F mode is represented by normalized waveform as shown in Fig. 1.23.

In order to maintain perfect square voltage wave, an infinite number of odd-harmonic resonator is required, which is impossible in practical design point of view. Therefore, for practical implementation of class F PA, appropriate number of

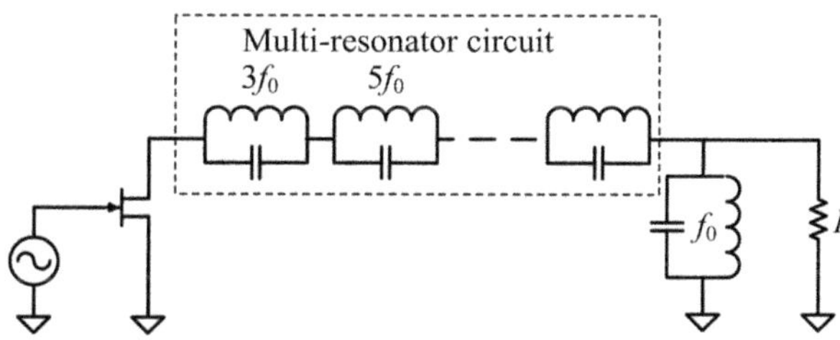

Fig. 1.22 Class F PA with multi-resonator circuit

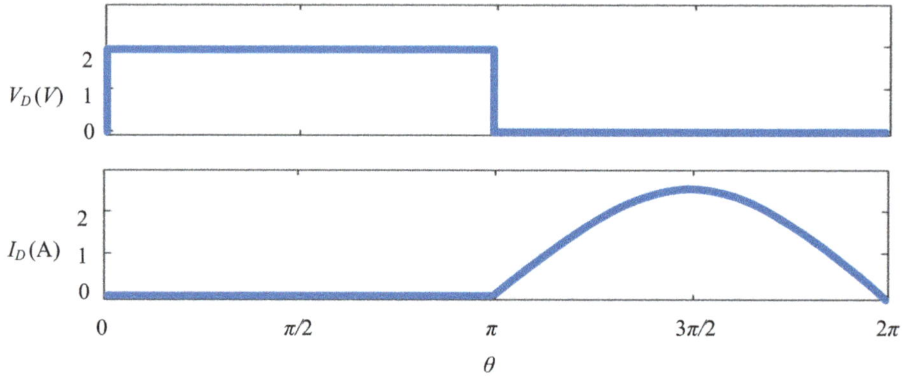

Fig. 1.23 Ideal waveform of class F PA

harmonic components of voltage and current waveform are considered depending upon the desired efficiency and output power capability. The analysis of class F mode with finite harmonics was presented by Raab in two different categories. First category is class F PAs with maximally flat waveforms [19] and second is class F PA with maximum power/efficiency waveforms [20]. The design of class F PAs with maximally flat waveforms is discussed in this section in detail and the design of class F PAs with maximum power/efficiency can be found in [18, 20]. Considering the harmonics condition for class F mode of operation, drain-to-source voltage and drain current, can be represented as [19],

$$V_{\mathrm{d}}(\theta) = V_{\mathrm{DD}} + V_{b1}\sin\theta + V_{b3}\sin 3\theta + V_{b5}\sin 5\theta \ldots \tag{1.76}$$

$$I_{\mathrm{d}}(\theta) = I_0 - I_{b1}\sin\theta - I_{a2}\cos 2\theta - I_{a4}\cos 4\theta \ldots \tag{1.77}$$

The maximally flat voltage waveforms are shown in Fig. 1.24. One can see in Fig. 1.24 that flatness of voltage waveform increases as the number of harmonic component increases. The voltage waveforms arrive its maximum and minimum values at $\theta = \pi/2$ and $\theta = 3\pi/2$, respectively. Since the odd-order derivative of voltage waveform are zero at $\theta = \pi/2$ and $\theta = 3\pi/2$, the even-order derivative is used to find the coefficients. As a result, the second and fourth derivatives of voltage waveform up to fifth harmonic are given as,

$$\frac{\mathrm{d}^2 V_{\mathrm{d}}}{\mathrm{d}\theta^2} = -V_{b1}\sin\theta - 9V_{b3}\sin 3\theta - 25V_{b5}\sin 5\theta \tag{1.78}$$

$$\frac{\mathrm{d}^4 V_{\mathrm{d}}}{\mathrm{d}\theta^4} = V_{b1}\sin\theta + 81V_{b3}\sin 3\theta + 625V_{b5}\sin 5\theta \tag{1.79}$$

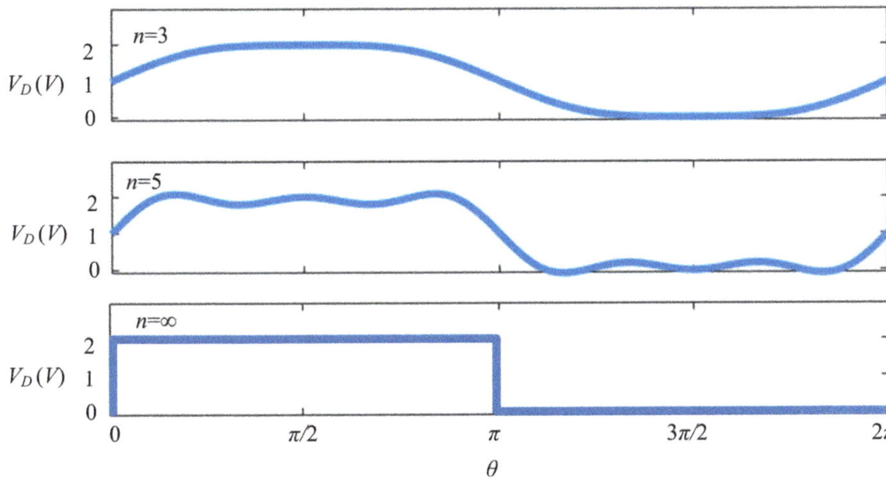

Fig. 1.24 Voltage waveform for n harmonic peaking

For maximum flatness at $\theta = 3\pi/2$ where voltage is minimum, the above derivatives should be set to zero. Considering this, one can rewrite Eqs. (1.78) and (1.79) for $\theta = 3\pi/2$ as,

$$V_{b1} - 9V_{b3} + 25V_{b5} = 0 \tag{1.80}$$

$$V_{b1} - 81V_{b3} + 625V_{b5} = 0 \tag{1.81}$$

Considering the case with $n = 3$ in Fig. 1.24, only third harmonic is peaking; therefore, considering $V_{b5} = 0$ in Eq. (1.80) gives,

$$V_{b3} = \frac{1}{9}V_{b1} \tag{1.82}$$

At $\theta = 3\pi/2$, voltage minima occur; therefore, $V_{\mathrm{d}}(\theta)$ in Eq. (1.76) becomes zero. Enforcing this condition in Eq. (1.76) along with Eq. (1.82) gives,

$$V_{b1} = \frac{9}{8}V_{\mathrm{DD}} \quad \text{and} \quad V_{b3} = \frac{1}{8}V_{\mathrm{DD}} \tag{1.83}$$

For the case with $n = 5$, the third and fifth harmonic components are present. Solving Eqs. (1.80) and (1.81) simultaneously gives,

$$V_{b3} = \frac{1}{6}V_{b1} \quad \text{and} \quad V_{b5} = \frac{1}{50}V_{b1} \tag{1.84}$$

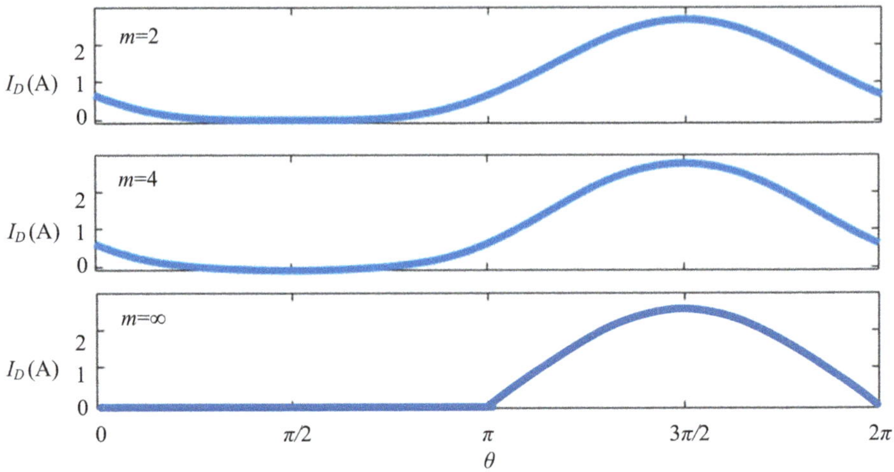

Fig. 1.25 Current waveform for m harmonic peaking

Using the condition that drain voltage in Eq. (1.76) becomes zero at $\theta = 3\pi/2$ and Eq. (1.84) gives,

$$V_{b1} = \frac{75}{64} V_{\mathrm{DD}}, \quad V_{b3} = \frac{25}{128} V_{\mathrm{DD}} \quad \text{and} \quad V_{b5} = \frac{3}{128} V_{\mathrm{DD}} \tag{1.85}$$

When $n =$ infinite as shown in Fig. 1.24, voltage waveform contains all the odd harmonic component and the resultant waveform is a square-wave. The coefficient in this case can be obtained by Fourier analysis and $V_{\mathrm{d}}(\theta)$ can be represented as [8],

$$V_{\mathrm{d}}(\theta) = V_{\mathrm{DD}} \left(1 + \frac{4}{\pi} \sin\theta + \frac{4}{\pi} \sum_{n=3,5,7\ldots}^{N} \frac{\sin\ n\omega t}{n} \right) \tag{1.86}$$

The maximally flat current waveforms attain its maximum and minimum values at $\theta = 3\pi/2$ and $\theta = \pi/2$, respectively, as shown in Fig. 1.25. At these values of θ, the odd-order derivatives of current waveform as given in Eq. (1.77) are inherently zero. Hence the even-order derivative is used to find the coefficients. Moreover, one can also see in Fig. 1.25 that flatness of current waveform increases as the number of harmonic component increases. As a result, the second and fourth derivatives of current waveform up to fifth harmonic are given as,

$$\frac{\mathrm{d}^2 I_{\mathrm{d}}}{\mathrm{d}\theta^2} = I_{b1} \sin\theta + 4 I_{a2} \cos 2\theta + 16 I_{a4} \cos 4\theta \tag{1.87}$$

$$\frac{\mathrm{d}^4 I_{\mathrm{d}}}{\mathrm{d}\theta^4} = -I_{b1} \sin\theta - 16 I_{a2} \cos 2\theta - 256 I_{a4} \cos 4\theta \tag{1.88}$$

For maximum flatness at $\theta = \pi/2$ where current is minimum in Fig. 1.25, the above derivatives of current waveform are set to zero. Therefore substituting $\theta = \pi/2$ in Eqs. (1.87) and (1.88)) and setting them to zero gives,

$$I_{b1} - 4I_{a2} + 16I_{a4} = 0 \tag{1.89}$$

$$I_{b1} - 16I_{a2} + 256I_{a4} = 0 \tag{1.90}$$

For the case with $m = 2$ in Fig. 1.25, only second harmonic is peaking; therefore, considering $I_{a4} = 0$ in Eq. (1.89) gives,

$$I_{a2} = \frac{1}{4} I_{b1} \tag{1.91}$$

At $\theta = \pi/2$, current takes minimum value; therefore, $I_{\mathrm{d}}(\theta)$ in Eq. (1.77) becomes zero. Enforcing this condition in Eq. (1.77) along with Eq. (1.91) gives,

$$I_{b1} = \frac{4}{3} I_0 \quad \text{and} \quad I_{a2} = \frac{1}{3} I_0 \tag{1.92}$$

Taking the case with $m = 4$ in Fig. 1.25, the second and fourth harmonic components are present. Solving Eqs. (1.89) and (1.90) simultaneously gives,

$$I_{a2} = \frac{5}{16} I_{b1} \quad \text{and} \quad I_{a4} = \frac{1}{64} I_{b1} \tag{1.93}$$

Substituting Eq. (1.93) in Eq. (1.77) at $\theta = \pi/2$ gives,

$$I_{b1} = \frac{64}{45} I_0, \quad I_{a2} = \frac{4}{9} I_0 \quad \text{and} \quad I_{a4} = \frac{1}{45} I_0 \tag{1.94}$$

For the last case shown in Fig. 1.25 where infinite even harmonics are present, the current waveform becomes half-sinusoid. The coefficient of all the frequency components can be obtained by Fourier analysis of $I_{\mathrm{d}}(\theta)$ and is given as [8],

$$I_d(\theta) = I_0 \left(1 - \frac{\pi}{2} \sin\theta - 2 \sum_{m=2,4,6\ldots}^{M} \frac{\cos\ m\omega t}{m^2 - 1} \right) \tag{1.95}$$

In ideal class F mode, drain to source voltage and drain current is shaped such that they provide zero overlapping. In such condition, the voltage and current waveform are represented by Eqs. (1.86) and (1.95), respectively. The theoretical efficiency can be calculated as,

$$\eta\% = \frac{P_1}{P_{\mathrm{DC}}} = \frac{\left(\frac{V_1 I_1}{2}\right)}{(V_{\mathrm{DD}} I_0)} = \frac{\frac{1}{2}\left(\frac{4}{\pi} V_{\mathrm{DD}} \times \frac{\pi}{2} I_0\right)}{(V_{\mathrm{DD}} I_0)} = 100\% \tag{1.96}$$

Table 1.2 Maximum efficiency (%) of class F PA for different combinations of harmonics [19]

Order of harmonics	$n = 1$	$n = 3$	$n = 5$	$n = \infty$
$m = 1$	$\eta = 50\%$	$\eta = 57.7\%$	$\eta = 60.3\%$	$\eta = 63.7\%$
$m = 2$	$\eta = 70.7\%$	$\eta = 81.7\%$	$\eta = 85.3\%$	$\eta = 90.0\%$
$m = 4$	$\eta = 75\%$	$\eta = 86.6\%$	$\eta = 90.5\%$	$\eta = 95.5\%$
$m = \infty$	$\eta = 78.5\%$	$\eta = 90.7\%$	$\eta = 94.8\%$	$\eta = 100\%$

For ideal class F mode, the harmonic impedances seen by device can be determined using Eqs. (1.86) and (1.95) as [8, 21],

$$Z_{\text{Fundamental}} = \frac{V_1}{I_1} = \frac{8}{\pi^2}\frac{V_{DD}}{I_0} \tag{1.97}$$

$$Z_{\text{even}} = 0 \tag{1.98}$$

$$Z_{\text{odd}} = \infty \tag{1.99}$$

It is seen from Eq. (1.96) that the maximum efficiency for ideal class F case is 100%. However, in practical PA design, the maximum efficiency depends on the number of voltage harmonics and current harmonics considered during the design of PA. Raab investigated the performance of PAs for various combinations of harmonics in [19] and resultant efficiencies are presented in Table 1.2. The m and n denote maximum order of harmonics in drain current and voltage, respectively.

One can see from Table 1.2 that maximum efficiency increases from 50% (similar to class A) where only fundamental frequency component is used to 100% where infinite harmonic component is considered.

1.3.3 Continuous Mode of Operation

Conventional PA operates in class A, AB, B, and C biasing conditions. These biasing conditions give unique solution for optimum load corresponding to high efficiency and output for each class. These loads are in general obtained using load line and defined at current generator reference plane (CRP) of a typical FET transistor. This unique solution may translate to widespread trajectory after device parasitic which may not be possible to match over the wide bandwidth. Carrubba first demonstrated that for any class of operation, it is possible to have family of solutions for optimum loads which can provide same performance as expected from this particular class. This will present a wide design space of optimum loads that the designer can choose over a wide frequency band [22]. Starting point of these broadband/continuous modes was the class J mode of operation which was presented by Cripps [23]. It was further discovered by Carrubba that by using only one parameter in voltage or current equation one can find infinite number of solutions between class B and class J conditions and each of these solutions maintains high power and high efficiency. The authors have exploited similar broadband results for class A, class B, class F, and inverse class F modes [22]. In each of these modes, output power and efficiency is maintained at a constant level over a broad range of frequencies. If the new parameter is added to the expression of voltage of any particular mode X, the resultant continuous modes will be termed as continuous class XV and vice versa. Generally, the added function is in the form of $(1 - \delta \sin \theta)$ where δ is the angle parameter to vary one waveform for maintaining constant output.

The following two equations describe the general formulation for the overall continuous mode voltage and current waveforms when first three harmonics are taken into consideration,

$$v(\theta) = (1 - \alpha \cos\theta - \beta \cos 2\theta - \gamma \cos 3\theta)(1 - \delta \sin\theta) \tag{1.100}$$

$$i(\theta) = (1 + x \cos\theta + y \cos 2\theta + z \cos 3\theta)(1 - \xi \sin\theta) \tag{1.101}$$

where α, β, γ, and δ are voltage waveform describing parameters and x, y, z, and ξ are the parameters describing current waveform. Substituting different values for these parameters one can obtain corresponding sets of possible solutions for nonoverlapping waveforms. The solution set can be termed as "design space." The parameter values are given in Table 1.3 as illustrated in [22].

Depending on these different values one can obtain different continuum modes. Continuous class B modes have been used more due to high-efficiency performance of class B biased PAs.

Design Strategy for Continuous Class J Power Amplifiers

The continuous class J or B/J is the extension of conventional class B mode and was first introduced in literature by Cripps [24]. Class B mode requires real value of impedance at fundamental frequency with all other higher harmonics shorted at intrinsic current generator reference plane of the transistor. If the harmonic and fundamental impedances are considered reactive, it gives another optimum response point resulting in a new mode named class J. In between these two extremes several similar high-efficiency points can be found by varying voltage/current parameters in the predefined equations. Therefore, voltage-current waveforms of a class J PA starts from that of a class B. The current waveform is kept constant to a half-wave rectified sinusoid as class B and is defined as,

$$I_{\mathrm{d}}(\theta) = I_0 \left(\frac{1}{\pi} + \frac{1}{2} \cos\theta + \frac{2}{3\pi} \cos 2\theta + \frac{2}{15\pi} \cos 4\theta + \cdots \right) \tag{1.102}$$

Table 1.3 Current and voltage parameters for different continuous modes [22]

PA mode of operation	Current parameters				Voltage parameters				Load impedances		
	x	y	z	ξ	α	β	γ	δ	Z_F	Z_{2F}	Z_{3F}
Class A	1	0	0	0	1	0	0	0	R_{opt}	–	–
Continuous class AV	1	0	0	0	1	0	0	−1 to 1	Varies	∞	–
Continuous class AI	1	0	0	−1 to 1	1	0	0	0	Varies	0	–
Class B	$\sqrt{2}$	1/2	0	0	1	0	0	0	R_{opt}	0	–
Continuous class BV	$\sqrt{2}$	1/2	0	0	1	0	0	−1 to 1	Varies	Varies	
Continuous class BI	$\sqrt{2}$	1/2	0	−1 to 1	1	0	0	0	Varies	0	0
Class F^{-1}	$2/\sqrt{3}$	0	$-1/3\sqrt{3}$	0	$\sqrt{2}$	1/2	0	0	R_{opt}	∞	0
Continuous class F^{-1}V	$2/\sqrt{3}$	0	$-1/3\sqrt{3}$	0	$\sqrt{2}$	1/2	0	−1 to 1	Varies	∞	Varies
Continuous class F^{-1}I	$2/\sqrt{3}$	0	$-1/3\sqrt{3}$	−1 to 1	$\sqrt{2}$	1/2	0	0	Varies	Varies	0
Class F	$\sqrt{2}$	1/2	0	0	$2/\sqrt{3}$	0	$1/3\sqrt{3}$	0	R_{opt}	0	∞
Continuous class FV	$\sqrt{2}$	1/2	0	0	$2/\sqrt{3}$	0	$1/3\sqrt{3}$	−1 to 1	Varies	Varies	∞
Continuous class FI	$\sqrt{2}$	1/2	0	−1 to 1	$2/\sqrt{3}$	0	$1/3\sqrt{3}$	0	Varies	0	Varies

The voltage waveform is described by Eq. (1.100). Substituting different values for parameters α, β, γ, and δ, one can obtain corresponding sets of possible solutions for nonoverlapping waveforms. In order to maintain linear behavior $v(\theta) \geq 0$. The value of these coefficients can be determined by applying the limiting condition for this waveform, that is, there is at least one value of θ where

$$v(\theta) = 0; \tag{1.103}$$

$$\frac{\mathrm{d}v(\theta)}{\mathrm{d}\theta} = 0 \tag{1.104}$$

From the basic theory of class J operation, one restricts only up to second harmonic terms in the voltage waveform of Eq. (1.100), that is,

$$v(\theta) = (1 - \alpha \ \cos\theta)(1 - \delta \ \sin\theta) \tag{1.105}$$

where cosine and sine terms represent real and reactive components, respectively. Solving Eqs. (1.103)–(1.105) one can get multiple solutions for the coefficients, all of which specify particular relationships between the harmonics. The formal procedure for solving Eq. (1.105) using Eqs. (1.103) and (1.104) is cumbersome due to the presence of multiple minima in these equations. Thus, one can start the solution assuming limiting cases for class B. In general, for Class B mode the voltage waveform has only real fundamental component. Therefore, one can write,

$$v_{1R} = 1; \tag{1.106}$$

$$v_{1Q} = 0; \tag{1.107}$$

$$v_{nQ} = 0 \tag{1.108}$$

where v_{1R} and v_{1Q} are real and reactive components of fundamental voltage and v_{nQ} is the reactive component of nth harmonics. For Class B mode, using Eq. (1.106) in Eq. (1.105), one can get,

$$\alpha = 1 \tag{1.109}$$

Considering Eqs. (1.107) and (1.105), the δ should be zero in case of Class B mode. On the other hand, Class J mode demands substantial second harmonic reactive component along with simultaneous fundamental reactive component in the voltage waveform. For class J mode, δ is equal to 1 in order to keep the voltage waveform above zero [23]. Therefore, in order to explore the class B/J continuum,

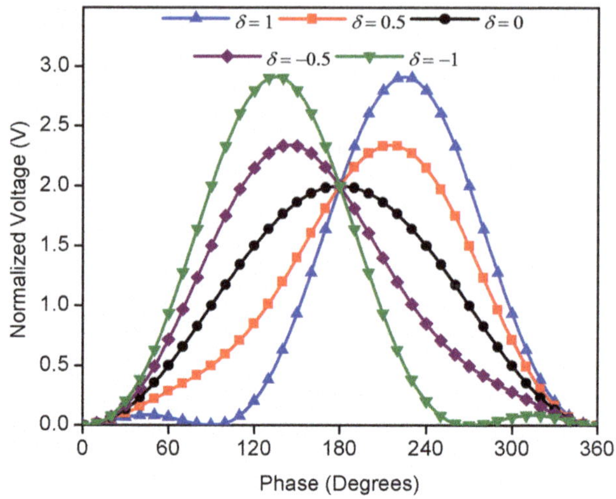

Fig. 1.26 Voltage waveforms for an ideal continuous class B/J mode with δ step size of 0.5

α is set to 1 but δ should be selected over the range $0 \leq \delta \leq 1$ and Eq. (1.105) becomes,

$$v(\theta) = (1 - \cos\theta)(1 - \delta \sin\theta) \tag{1.110}$$

To include class J^{-1}, the δ should be selected over the range $-1 \leq \delta \leq 1$.

Figure 1.26 represents the voltage waveforms for continuous class B/J mode including class J^{-1} plotted using Eq. (1.110). One can see that there are number of solutions available for voltages which can give high-efficiency output.

In designing a continuous class B/J PA, one of the main key points is that these waveforms need to be verified at intrinsic current generator reference plane of the device. One of the easy ways to ensure this is to keep the impedances at the reference plane corresponding to these waveforms. Representing Eqs. (1.102) and (1.110) in phasor forms, one can derive the voltage components as [22],

$$V_{DD} = 1; \tag{1.111}$$

$$V_1 = -1 + j\delta \tag{1.112}$$

$$V_2 = -j\frac{\delta}{2} \tag{1.113}$$

where V_{DD}, V_1, and V_2 are the DC, fundamental, and second harmonic components of output voltage, respectively. Dividing Eqs. (1.112) and (1.113) by corresponding value of currents at fundamental and harmonic frequencies, the impedances at CRP are derived as,

$$Z_F = R_{opt} - j\delta R_{opt} \tag{1.114}$$

$$Z_{2F} = j\delta \frac{3\pi}{8} R_{opt} \tag{1.115}$$

where R_{opt} is the optimum resistance obtained from class B load line.

Theory of Continuous Class F Power Amplifier

The continuous class F mode of operation is the extension of class F mode and was first demonstrated by Carrubba et al. [25–27]. The ideal class F mode requires a square voltage waveform and a half-rectified current waveform at its output current-generator plane as discussed in Sect. 1.3.2. These waveforms are typically described by Eqs. (1.86) and (1.95). In such a case, the even and odd harmonics are terminated with short and open respectively at the current generator reference plane of the transistor. This leads to ideal efficiency of 100% while considering infinite number of harmonics. When voltage harmonics are reduced to three, the maximum achievable efficiency is reduced to 90.7% [23]. However, one of the major limitations of class F PAs is that they present a very narrow band of operation. To extend the operational bandwidth, it was investigated that constant open-circuited third harmonic impedance and short-circuited second harmonic component are not a unique solution for achieving maximum efficiency and output power [25]. Starting from standard class F mode and varying the fundamental and second harmonic impedances while keeping third harmonic open, a continuum of solutions can be obtained for high efficiency and output power. The voltage waveform for continuous class F mode can be derived from generalized continuous mode Eq. (1.100) which can also be expressed as,

$$\begin{aligned} v_d(\theta) = V_{DD} - a_1 \cos\theta + b_1 \sin\theta - a_2 \cos 2\theta + b_2 \sin 2\theta - a_3 \cos 3\theta \\ + b_3 \sin 3\theta - a_4 \cos 4\theta + b_4 \sin 4\theta + \cdots \end{aligned} \tag{1.116}$$

where a_1, a_2, a_3, and a_4 represent the real part of fundamental and harmonic components. Similarly, coefficients b_1, b_2, b_3, and b_4 represent the imaginary part of fundamental and harmonic components of voltage waveform. The voltage components in Eq. (1.116) can be expressed in terms of α, β, γ, and δ described in Eq. (1.100) as [25],

$$V_{DD} = 1 + \frac{1}{2}\alpha^2 - \alpha\beta, \tag{1.117}$$

$$a_1 = 2\alpha - \beta - \frac{3}{4}\alpha^2\beta, \quad a_2 = \alpha\beta - \frac{1}{2}\alpha^2, \quad a_3 = -\frac{1}{4}\alpha^2\beta, \tag{1.118}$$

$$b_1 = \left(\frac{1}{2}\alpha\beta - 1 - \frac{1}{4}\alpha^2\right)\delta, \quad b_2 = \left(\alpha - \frac{1}{2}\beta - \frac{1}{4}\alpha^2\beta\right)\delta,$$
$$b_3 = \left(\frac{1}{2}\alpha\beta - \frac{1}{4}\alpha^2\right)\delta, \quad b_4 = -\frac{1}{8}\alpha^2\beta\delta \tag{1.119}$$

To obtain maximum theoretical efficiency, the even harmonics are short circuited in class F mode of operation [19]. However, the reactive part of even harmonic is introduced in continuous class F with the introduction of new parameter δ. Therefore, the real part of second harmonic, that is, a_2, must be zero. This gives the condition,

$$\beta = \frac{\alpha}{2} \tag{1.120}$$

To obtain the maximum amplitude of fundamental voltage, one can substitute the value of β in the real part of the fundamental component and differentiate as a function of α,

$$\frac{\mathrm{d}a_1}{\mathrm{d}\alpha} = \frac{\mathrm{d}}{\mathrm{d}\alpha}\left(\frac{3}{2}\alpha - \frac{3}{8}\alpha^3\right) = \frac{3}{2} - \frac{9}{8}\alpha^2 \tag{1.121}$$

Substituting $\mathrm{d}a_1/\mathrm{d}\alpha = 0$ in Eq. (1.121) gives,

$$\alpha = \frac{2}{\sqrt{3}} \tag{1.122}$$

Substituting value of α in Eq. (1.120) gives,

$$\beta = \frac{1}{\sqrt{3}} \tag{1.123}$$

In a continuous class F mode, current waveform shape is constant and only voltage waveform is varied with the parameter δ while keeping α and β as constant. Parameter δ is varied over the range $-1 \le \delta \le 1$. Substituting α and β in Eq. (1.100) gives,

$$V_\mathrm{d}(\theta) = \left(1 - \frac{2}{\sqrt{3}}\cos\theta\right)^2\left(1 + \frac{1}{\sqrt{3}}\cos\theta\right)(1 - \delta \ \sin\theta) \tag{1.124}$$

For $\delta = 0$, the waveform is defined as conventional class F mode. Figure 1.27 shows the voltage waveform for theoretical continuous class F mode.

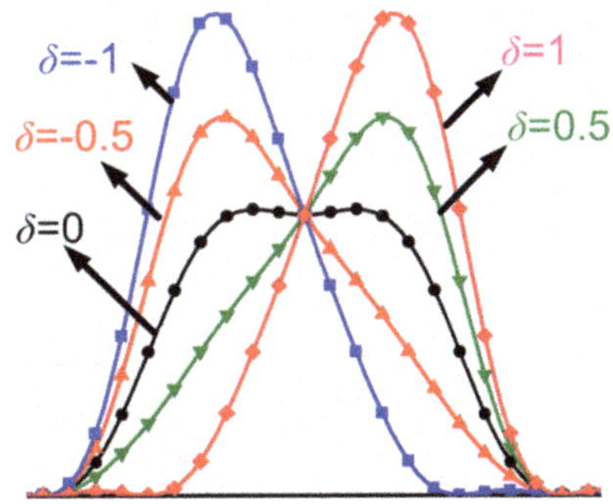

Fig. 1.27 Theoretical continuous class F voltage waveform with δ step size of 0.5 [28]

This represents the continuous family of solutions, where the constant output power and efficiency performance equivalent to standard class F mode can be maintained. The required harmonic impedance for continuous class F mode can be obtained using,

$$Z_n = -\frac{V_n}{I_n} \tag{1.125}$$

where n represents the order of harmonic component. Using Eqs. (1.102) and (1.124), the fundamental and harmonic impedances for continuous class F mode at CRP is calculated as [29, 30],

$$Z_{\mathrm{CRP},1} = \left(\frac{2}{\sqrt{3}} + j\delta\right) R_{\mathrm{opt}} \tag{1.126}$$

$$Z_{\mathrm{CRP},2} = -j\frac{7\sqrt{3}\pi}{24}\delta R_{\mathrm{opt}} \tag{1.127}$$

$$Z_{\mathrm{CRP},3} = \infty \tag{1.128}$$

where $Z_{\mathrm{CRP},1}$, $Z_{\mathrm{CRP},2}$, and $Z_{\mathrm{CRP},3}$ represent the fundamental, second, and third harmonic impedances at current generator reference plane, respectively. R_{opt} represents the optimum impedance of ideal class B mode. The parameter δ used in the above equations of load impedances provides the increased flexibility to the PA designers for maintaining the optimum device performance over wider band of operation [28, 31, 32]. These set of loads are plotted on Smith chart as shown in Fig. 1.28.

The fundamental load impedance moves on constant resistance circle and second harmonic load impedance simultaneously lies around the edge of Smith chart for the range of δ from -1 to 1. The third harmonic impedance is maintained at constant open circuit. The requirement of fourth harmonic component can be neglected as its omission only results in the efficiency degradation of 4.5% in the worst case

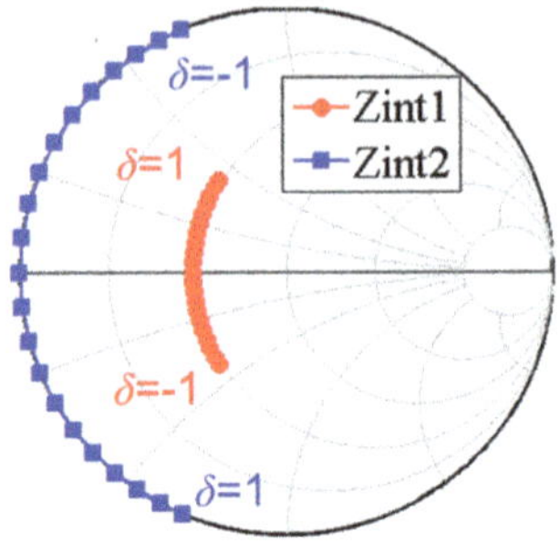

Fig. 1.28 Theoretical continuous class F load impedances [28]

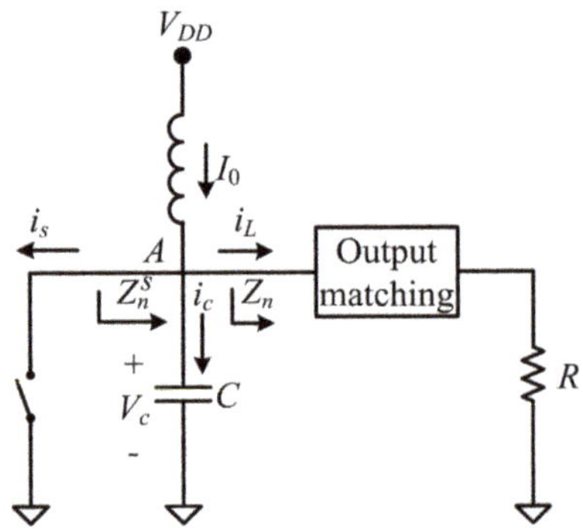

Fig. 1.29 Equivalent circuit of continuous class E PA

[30]. The different values of fundamental and second harmonic load impedances can be chosen to maintain the constant output power and theoretical efficiency performance.

Class E Continuum

The ZVS and ZVDS conditions defined in section "Class E Power Amplifier" for class E mode can also be achieved if the second harmonic switch impedance (Z_2^S) is arbitrarily selected [33]. These arbitrary Z_2^S enables robust shaping of switched waveforms and gives the possibility to obtain continuum of class E solutions. Figure 1.29 shows equivalent circuit of continuous class E PA. This is similar to class E topology with shunt capacitor except, the harmonic impedances Z_n^S are considered. Specifically, the second harmonic is considered during the analysis. Therefore, the load matching network consists of purely reactive lossless passive component with a current i_L flowing into it having component of both fundamental and second harmonic frequency. It will be shown later that conventional class E mode with shunt capacitor is one of the subsets of this class E continuum. The RF choke is ideal one presenting open circuit at fundamental as well as harmonic frequency and hence not considered in the analysis. Further, it has been assumed that the switch is lossless and operates with 50% duty cycle. The resistance R terminates the load matching network.

For successful class E operation, the load matching network should provide fundamental impedance Z_1, second harmonic impedance Z_2, and higher harmonic impedance Z_n as given below,

$$Z_1 = R\left(1 + jX_1\right) \tag{1.129}$$

$$Z_2 = jX_2R \tag{1.130}$$

$$Z_n = \infty, \quad n > 2 \tag{1.131}$$

where X_1 is reactance at fundamental frequency, whereas X_2 is the reactive load presented at second harmonic which can be obtained independent of X_1. When switch is *on* during $0 \le \omega t < \pi$, the current through capacitor $i_C(\omega t) = 0$, whereas voltage across the switch (device) and therefore capacitor $v_C(\omega t) = 0$. As discussed earlier, considering the current through the load having both fundamental and harmonic component of amplitude I_1 and I_2, one can write,

$$i_L\left(\omega t\right) = I_1 \cos\left(\omega t + \phi_1\right) + I_2 \cos\left(2\omega t + \phi_2\right) \tag{1.132}$$

Considering Kirchhoff's current law (KCL) at node A in Fig. 1.29, the current through the switch considering $i_C(\omega t) = 0$ during $0 \le \omega t < \pi$ can be written as,

$$i_S\left(\omega t\right) = I_0 - i_L = I_0 - I_1 \cos\left(\omega t + \phi_1\right) - I_2 \cos\left(2\omega t + \phi_2\right) \tag{1.133}$$

where I_0 is the DC current and ϕ_1 and ϕ_2 are the initial phase shift in fundamental and harmonic currents, respectively.

When switch is *off* during $\pi \le \omega t < 2\pi$, the current starts flowing through the capacitor and it can be found by applying KCL at node A as,

$$i_C\left(\omega t\right) = I_0 - i_L = I_0 - I_1 \cos\left(\omega t + \phi_1\right) - I_2 \cos\left(2\omega t + \phi_2\right) \tag{1.134}$$

The corresponding voltage across the capacitor is given by,

$$\begin{aligned} v_C\left(\omega t\right) &= \frac{1}{\omega C}\int_{\pi}^{\omega t} i_C\left(\omega t\right) \mathrm{d}\left(\omega t\right) \\ &= -\frac{1}{\omega_0 C}\left[I_2 \sin\left(\omega t\right)\cos\left(\omega t + \phi_2\right) + I_1\left(\sin\left(\omega t + \phi_1\right) + \sin\phi_1\right) + I_0\left(\pi - \omega t\right)\right] \end{aligned} \tag{1.135}$$

By solving the above equation along with ZVS and ZVDS conditions mentioned in Eqs. (1.31) and (1.32) of section "Class E Power Amplifier" at $\omega t = 2\pi$, the following two equations are obtained,

$$2I_1 \sin\phi_1 - \pi I_0 = 0 \tag{1.136}$$

$$I_1 \cos\phi_1 + I_2 \cos\phi_2 - I_0 = 0 \tag{1.137}$$

Solving Eqs. (1.136) and (1.137) further gives the relation for I_0 and I_1 in terms of I_2 as,

$$I_0 = \frac{2I_2 \sin\phi_1 \cos\phi_2}{2\sin\phi_1 - \pi\cos\phi_1} \tag{1.138}$$

$$I_1 = \frac{\pi I_2 \cos\phi_2}{2\sin\phi_1 - \pi\cos\phi_1} \tag{1.139}$$

Considering Eqs. (1.132) and (1.135) as the current and voltage conditions for load matching network, one can obtain load impedance Z_n as the ratio of these voltages and currents for fundamental as well as harmonic frequencies. This requires computation of fundamental and second harmonic current and voltage from Fourier analysis of Eqs. (1.132) and (1.135). The amplitude of second harmonic component of voltage can be obtained as Fourier coefficient of v_C given in Eq. (1.135) as,

$$V_{C,2} = \frac{1}{2\pi}\int_{\pi}^{2\pi} v_C(\omega t)\,\mathrm{e}^{-j2\omega t}\,\mathrm{d}\omega t = \frac{j\left(6I_0\pi + 3\mathrm{e}^{j\phi_2} I_2\pi + 8jI_1\cos\phi_1 - 16I_1\sin\phi_1\right)}{24\omega_0 C\pi} \tag{1.140}$$

Similarly, the amplitude of second harmonic component of current can be obtained as Fourier coefficient of i_L given in Eq. (1.132) as,

$$i_{L,2} = \frac{1}{2\pi}\int_{0}^{\pi} i_L(\omega t)\,\mathrm{e}^{-j2\omega t}\,\mathrm{d}\omega t = \frac{I_2}{2}\mathrm{e}^{j\phi_2} \tag{1.141}$$

One can use Eqs. (1.140) and (1.141) to find impedance provided by the load matching network to second harmonic frequency as,

$$Z_2 = \frac{V_{C,2}}{i_{L,2}} = \frac{\mathrm{e}^{-2j\phi_2}(4 + j2\ \tan\phi_1) - j4\ \tan\phi_1 + j3\pi + 4}{12\omega_0 C(\pi - 2\tan\phi_1)} \tag{1.142}$$

If the relation between ϕ_1 and ϕ_2 is obtained, the design equations can be defined with only one design variable. This is possible by setting the real part of Z_2 in

Eq. (1.142) as zero since the harmonic load is already assumed to be purely reactive as shown in Eq. (1.130). Therefore, from Eq. (1.142), the relation between ϕ_1 and ϕ_2 can be given as,

$$\phi_2 = -\tan^{-1}(2\cot\phi_1) \tag{1.143}$$

Similarly, the Fourier coefficient of voltage and current can be calculated for fundamental frequency from Eqs. (1.135) and (1.132), respectively, as,

$$V_{C,1} = V_{\mathrm{DD}} \frac{j16\pi\cot\phi_1 + \left(32 - 3\pi^2\right)(\tan\phi_1 - j) - 4\pi}{6\pi^2\cot\phi_1} \tag{1.144}$$

$$I_{L,1} = \frac{V_{\mathrm{DD}}}{R}\frac{4}{\pi}e^{j\phi_1}\sin\phi_1 \tag{1.145}$$

The expression for reactance X_1 can be obtained as,

$$X_1 = \frac{\mathrm{Im}(Z_1)}{\mathrm{Re}(Z_1)} = \frac{\left(16\pi\ \cos^2\phi_1\cot\phi_1 + 2\pi\ \sin 2\phi_1 + 3\pi^2 - 32\right)}{12\pi\ \cos^2\phi_1} \tag{1.146}$$

where Re(Z_1) and Im(Z_1) are real and imaginary part of impedance of load matching network presented at fundamental frequency and calculated as $Z_1 = V_{C1}/I_{L1}$, where V_{C1} and I_{L1} are given by Eqs. (1.144) and (1.145), respectively. Similarly, the reactance X_2 is given by,

$$X_2 = \frac{Z_2}{jR} = \frac{\pi\ \sec^2\phi_1}{24(\pi - 2\tan\phi_1)} \left[4\sin\left(2\tan^{-1}(2\cot\phi_1)\right) + 3\pi + 2\left(\cos\left(2\tan^{-1}(2\cot\phi_1)\right) - 2\right)\tan\phi_1\right] \tag{1.147}$$

where Z_2 is load presented by load matching network at second harmonic frequency as given by Eq. (1.142).

From Eqs. (1.146) and (1.147), one can see that reactive part of fundamental and second harmonic load, that is, X_1 and X_2 are functions of single variable ϕ_1. Varying ϕ_1 results in the continuum of class E with robust switched waveforms. The power levels similar to conventional class E is obtained if ϕ_1 is selected in between 43° and 78° [33].

Since, the theoretical efficiency of class E mode is 100%, one can equate DC power and fundamental RF power as,

$$I_0 V_{\mathrm{DD}} = \frac{1}{2} I_1^2 R \tag{1.148}$$

The expression for V_{DD} can further be found from DC component of Fourier series expansion of v_C as,

$$V_{DD} = \frac{2\pi}{\omega_0} \int_0^{\frac{2\pi}{\omega_0}} v_C(t) dt = \frac{\pi \left(I_0 \pi - 2I_1 \sin \phi_1 + I_2 \sin \phi_2 \right) + 4I_1 \cos \phi_1}{4\omega_0 C \pi} \tag{1.149}$$

Solving Eqs. (1.138), (1.139), (1.143), and (1.148) in (Eq. 1.149), one can get the expression for switching capacitance C as follows,

$$C = \frac{2}{\pi \omega_0 R} \cos^2 \phi_1 \tag{1.150}$$

The respective voltage and current equations across switch can be obtained from Eqs. (1.133), (1.135), (1.138), (1.139), and (1.148), as,

$$v_C(t) = V_{DD} \tan \phi_1 \begin{bmatrix} 4\omega_0 t \quad \tan \phi_1 + (4\pi \quad \cot \phi_1 - 8) \sin^2 \omega_0 t - 2\pi \quad \sec \phi_1 \sin (\omega_0 t + \phi_1) \\ + (\pi - 2 \tan \phi_1) \sin 2\omega_0 t - 6\pi \quad \tan \phi_1 \end{bmatrix} \tag{1.151}$$

$$i_S(t) = \frac{V_{DD}}{R} \frac{4}{\pi^2} \begin{bmatrix} \cos \phi_1 \left(\pi \quad \cos \begin{pmatrix} 2\omega_0 t \\ - \tan^{-1} (2 \cot \phi_1) \end{pmatrix} \times \sqrt{1 + 4\cot^2 \phi_1} - 4 \sin 2\omega_0 t \right) \\ - \pi \quad \cos (\omega_0 t + \phi_1) + 4 \sin \phi_1 \sin^2 \omega_0 t \end{bmatrix} \sin \phi_1 \tag{1.152}$$

The RF output power can be obtained as,

$$P_{out} = V_{DD} I_0 = V_{DD} \frac{1}{2\pi} \int_0^{2\pi} i_S (\omega t) \, d (\omega t) = \frac{V_{DD}^2}{R} \frac{8 \sin^2 (\phi_1)}{\pi^2} \tag{1.153}$$

The above equations correspond to operation similar to class E if $\phi_1 \in [43°, 78°]$. The $\phi_1 = 57.5°$ corresponds to conventional class E mode, $\phi_1 = 43.3°$ corresponds to class E/F$_2$ mode and class EF$_2$ mode. One can see that all the above equations are function of ϕ_1. According to Eqs. (1.151) and (1.152), the normalized voltage and current waveforms for switch can be plotted for $\phi_1 = 43.3°$, 57.5°, and 78° as shown in Fig. 1.30. Hence by the virtue of single independent variable ϕ_1, the waveform engineering can be done to obtain continuous class E modes.

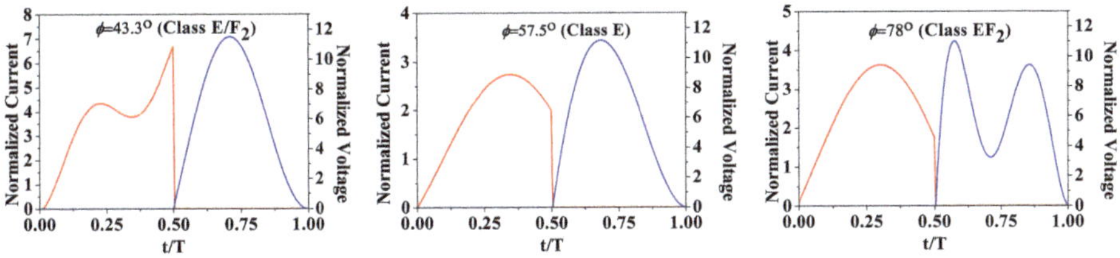

(a) (b) (c)

Fig. 1.30 Normalized switch waveforms for (**a**) $\phi = 43.3°$, (**b**) $\phi = 57.5°$, and (**c**) $\phi = 78°$

1.3.4 Harmonic Injection-Based Power Amplifiers

The PA topology where the harmonics are injected externally is referred as the harmonic injection-based PAs (HI-PA). The harmonic injection can be either at input, output, or both of the active devices. The HI-PA where harmonics are injected at the output of the transistor will enhance the efficiency while the harmonic injection at the input of the transistor will help in the linearity enhancement [34]. This section mainly discusses the HI-PA concept for efficiency enhancement with harmonics injected at the output of the transistor.

In 1986, Zivkovic et al. have designed high-power triode based high-frequency (HF) amplifier with increased efficiency by injecting second harmonic signal into grid (input) and plate (output) [35]. The efficiency enhancement of HF amplifier with third harmonic injection similar to second harmonic injection was presented in [36]. Later in 1992, the concept of HI for high-efficiency PA was patented by Willems et al. where the small portion of input signal was sampled, frequency multiplied, power amplified, and passed through the phase shifter before injecting at the drain terminal of transistor as shown in Fig. 1.31 [37]. This input signal at the fundamental is used to create the harmonic signal using frequency multipliers (e.g., doubler, tripler) and injected at the output of PA with an optimum amplitude and phase. The phase shifters and attenuators are used to adjust the amplitude and phase of the harmonic signal.

In [34], the theoretical and experimental study of linearization scheme for RF PAs based on HI at the input is discussed. In this method, the predistortion circuits were used to generate second harmonic signal and fed to the input simultaneously with the base-band signal for PA linearization.

Concept of Active Harmonic Injection

In conventional class B and class J mode PAs, the transistors are biased to provide half wave sinusoidal current waveform. This provides the fundamental signal similar to sinusoidal waveform, but advantageously reduces the DC component by $(2/\pi)$ and hence the efficiency increases by $(\pi/2)$. In class B mode, the voltage waveform

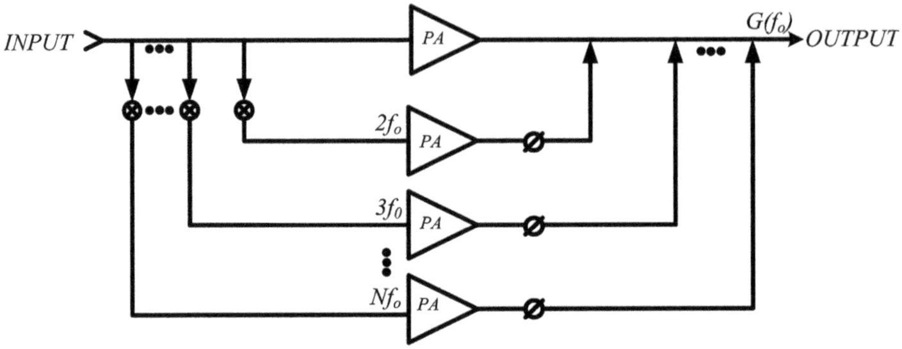

Fig. 1.31 Block diagram of harmonic injection-based PA (HI-PA) with harmonic injection at the output

is shaped to sinusoidal by short circuiting the second harmonic while in class J mode, the voltage waveform is shaped into a half wave sinusoidal by appropriate reactive termination at fundamental and second harmonic. Due to the passive harmonic termination in class J mode, there is a constraint in the reduction of voltage-current overlap, thereby limiting the efficiency [38]. Alternatively, these harmonic terminations can be realized by injecting the harmonics to the PA output in HI-PA. Any amplification in auxiliary path while injecting the harmonics shall consume DC power; therefore, one should include this power consumption while estimating efficiency of HI-PA. The DE of HI-PA is represented as $\eta_{\text{HI-PA}}$ given by [38],

$$\eta_{\text{HI-PA}} = \frac{P_{\text{out}}}{P_{\text{DC}} + \frac{P_{\text{inj}}}{\eta_{\text{inj}}}} \tag{1.154}$$

where P_{inj} and η_{inj} are the DC power and *DE* of harmonic injection path, respectively. The general circuit diagram of HI-PA is shown in Fig. 1.32 [39].

One can see the HI-PA consists of two paths: the main path consists of the PA designed at fundamental frequency while the second harmonic signal is injected from auxiliary path at the drain of main PA to control the overlap between voltage-current waveforms. This injection is provided through diplexer. The diplexer should be carefully designed such that the main PA sees only constant load at fundamental frequency while the active harmonic injection path should see the output impedance of the main PA at second harmonic. The topology shown in Fig. 1.33 can fulfil the function of diplexer [38].

If the diplexer is considered as a three-port network as shown in Fig. 1.32, then its *S* parameter matrix can be written as [39],

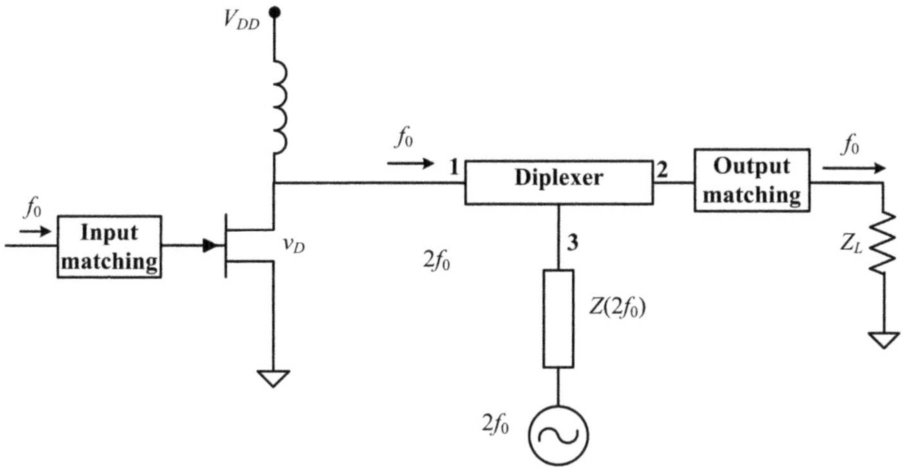

Fig. 1.32 General circuit diagram of HI-PA

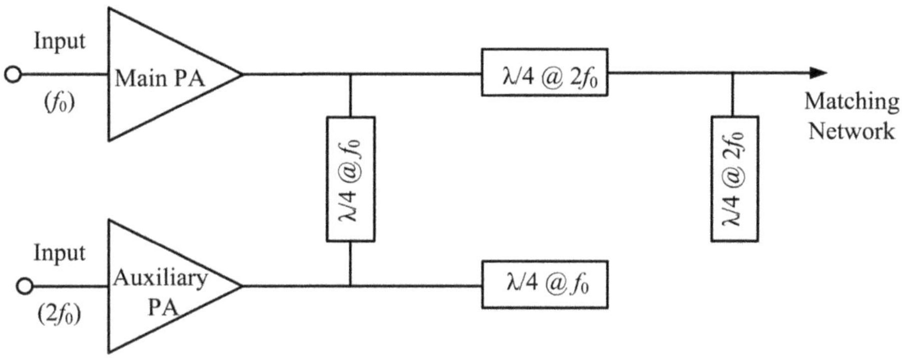

Fig. 1.33 Diplexer topology

$$S(f_0, 2f_0) = \begin{bmatrix} 0, 0 & e^{j\phi_{21}(f_0)}, 0 & 0, e^{j\phi_{31}(2f_0)} \\ e^{j\phi_{21}(f_0)}, 0 & 0, e^{j\phi_{22}(2f_0)} & 0, 0 \\ 0, e^{j\phi_{31}(2f_0)} & 0, 0 & e^{j\phi_{33}(f_0)}, 0 \end{bmatrix} \tag{1.155}$$

The harmonic signal can be generated externally and injected to the main PA using auxiliary path. The amplitude and phase of the harmonics can be controlled externally in digital domain [39–41] or using attenuator and phase shifter in auxiliary path of HI-PA [38, 42]. Alternatively, these harmonics can also be generated using analog multipliers along with the HI-PA circuit [43, 44]. This topology is shown in Fig. 1.34.

One can see that same input source is used to drive main and doubler circuit in the auxiliary path. The doubler circuit generates second harmonic of the input signal. The analog phase shifter which may or may not be a variable will set an

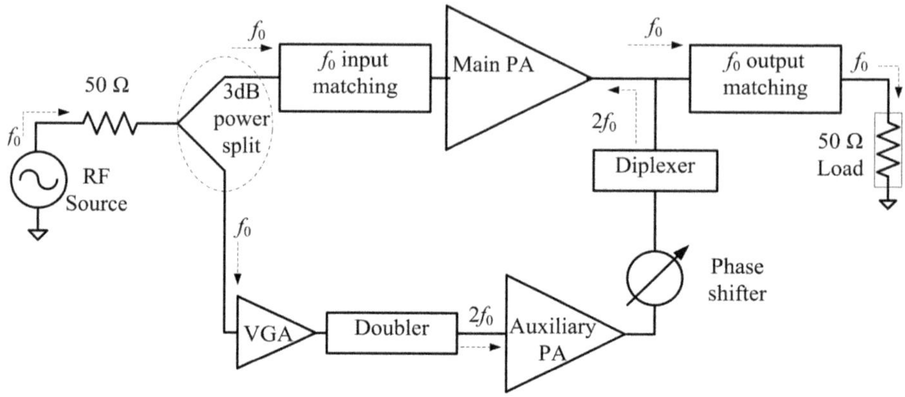

Fig. 1.34 General block diagram of HI-PA

appropriate phase of second harmonic before injection at the output of the main PA. The variable gain amplifier (VGA) is used to control the amplitude of injected harmonic signal. The HI-PA topology where harmonics are generated externally [38–42] has dual-input configuration which is difficult to be used in conventional transmitters. Therefore, the HI-PA topology with on-board multiplier/doubler can be a preferable choice for the designers.

Fourier Analysis of Harmonic Injected Power Amplifier

The normalized voltage and current waveforms with harmonic injection can be expressed using Fourier series as [39, 41],

$$\overline{v(\theta)} = \overline{V_{\mathrm{DD}}} + \sqrt{2}\sin\theta + a_2\cos(2\theta) \tag{1.156}$$

$$\overline{i(\theta)} = \overline{I_0} - \sqrt{2}\sin\theta + a_2\cos(2\theta) \tag{1.157}$$

These equations can be obtained by considering the sine term for fundamental frequency in generalized equations given by Eqs. (1.74) and (1.75) [18]. The voltage and current waveforms are considered as half sinusoidal with 180° out of phase. Moreover, only second harmonic component is considered in these expressions for harmonic injection, where a_2 represents the amplitude of this component. An optimum value of a_2 that maximizes the efficiency can be found by differentiating Eq. (1.156) with respect to θ and equating to zero as,

$$\sin\left(\theta_{\mathrm{critical},v}\right) = \frac{\sqrt{2}}{4a_2} \tag{1.158}$$

$$\theta_{\text{critical},v} = \frac{\pi}{2} + n\pi, \quad n = 0, 1, \ldots, \infty \tag{1.159}$$

where $\theta_{\text{critical},v}$ corresponds to a point at which the first derivative of Eq. (1.156) is equal to zero. Considering the critical points of drain voltage and current waveforms, the optimum values of a_2 that maximizes the efficiency can be given as [41],when

$$\overline{v}_{\text{D}}\left(\theta_{\text{critical},v}\right) \text{ is minimum}, \quad a_2 > +\frac{\sqrt{2}}{4} \tag{1.160}$$

when

$$\overline{v}_{\text{D}}\left(\theta_{\text{critical},v}\right) \text{ is maximum}, \quad a_2 < -\frac{\sqrt{2}}{4} \tag{1.161}$$

The normalized DC power consumed by the main PA and HI can be given as [41],

$$\overline{P_{\text{DC}}} = \overline{V_{\text{DD}}}\,\overline{I_0} + \frac{a_2^2}{2\eta_{\text{inj}}} = \left(\overline{V_{\text{DD}}}\right)^2 + \frac{a_2^2}{2\eta_{\text{inj}}} \tag{1.162}$$

$$\overline{P_{\text{DC}}} = \begin{cases} \left(\frac{1+4a_2^2}{4a_2^2}\right) + \frac{a_2^2}{2\eta_{\text{inj}}}, & a_2 \leq \frac{-\sqrt{2}}{4} \\ \left(\sqrt{2} + a_2\right)^2 + \frac{a_2^2}{2\eta_{\text{inj}}}, & a_2 > \frac{-\sqrt{2}}{4} \end{cases} \tag{1.163}$$

The DE of HI-PA can now be written as [41],

$$\eta_{\text{HI-PA}} = \begin{cases} \frac{8\sqrt{2\left(2+\frac{1}{\eta_{\text{inj}}}\right)}}{\left(2+\sqrt{2\left(2+\frac{1}{\eta_{\text{inj}}}\right)}\right)^2+2}, & a_2 \leq \frac{-\sqrt{2}}{4} \\ \eta_{\text{inj}}\left(\sqrt{\frac{8}{\eta_{\text{inj}}} + 16} - 4\right), & a_2 > \frac{-\sqrt{2}}{4} \end{cases} \tag{1.164}$$

From Eq. (1.164), the maximum value of DE of HI-PA is 89.9% when the η_{inj} is 100%. In practice, η_{inj} cannot be 100%; therefore, $\eta_{\text{HI-PA}}$ decreases with degradation in η_{inj}. It is obvious that the DE of HI-PA decreases and behaves like class A mode as η_{inj} decreases.

1.4 Load Modulation for Average Efficiency Enhancement

1.4.1 Doherty Power Amplifier

The DPA is a back-off efficiency enhancement technique which works on the principle of load modulation [45–55]. This method was first introduced by W. H. Doherty in 1936 [45]. The idea was to improve efficiency at an average power, backed-off from saturation in order to handle signals with high crest factors. The basic idea is to modulate load seen by an amplifier in order to operate it at voltage saturation even at low input drive level. This load modulation is carried out actively using an auxiliary PA and the PA which operates at voltage saturation is termed as the main PA. Figure 1.35a shows the architecture of DPA comprising main PA and an auxiliary PA combined using a quarter-wave transformer. The combined node is terminated with a common load R_L. Sometimes an additional quarter-wave transformer is further used to transform R_L to standard load of 50 Ω. Since the phase of output current of two PAs at load terminal should be equal, a phase compensation network is inserted at the input of auxiliary PA as shown in Fig. 1.35a. This network compensates for the phase shift that is generated by quarter-wave transformer, Z_T. Figure 1.35b shows the equivalent circuit diagram of conventional two-stage DPA. Here, each PA is represented by voltage controlled current source (VCCS) with subscript M and A used for main PA and auxiliary PA, respectively.

The quarter-wave transformer Z_T converts the main current source, $VCCS_M$ into a voltage controlled voltage source ($VCVS_M$) as shown in Fig. 1.35b. To analyze the configuration of two-stage DPA shown in Fig. 1.35b, the ABCD parameters of quarter-wave transformer of characteristic impedance Z_T used in the load combiner can be written as,

$$\begin{bmatrix} V_M \\ I_M \end{bmatrix} = \begin{bmatrix} 0 & jZ_T \\ j/Z_T & 0 \end{bmatrix} \begin{bmatrix} V_A \\ I_{M1} \end{bmatrix} \tag{1.165}$$

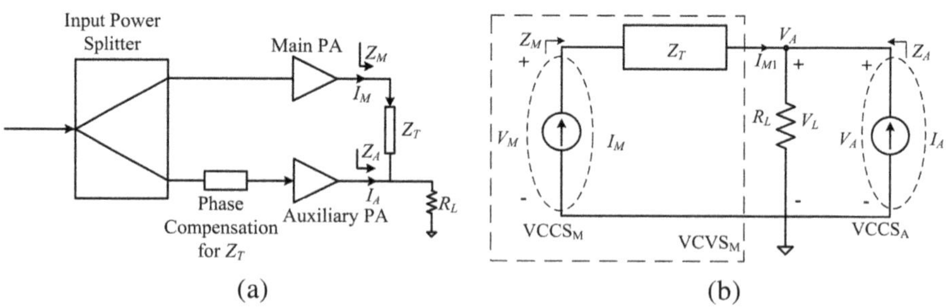

Fig. 1.35 Conventional two-stage DPA, (**a**) architecture and (**b**) equivalent circuit diagram

where Z_T is the characteristic impedance of the transformer. The Eq. (1.165) can be expanded as,

$$V_M = jZ_T I_{M1} \tag{1.166}$$

$$I_M = \left(\frac{j}{Z_T}\right) V_A \tag{1.167}$$

From Fig. 1.35b, one can observe that load voltage V_L is equal to the voltage of auxiliary PA, that is, V_A and can be written as,

$$V_A = (I_{M1} + I_A)\, R_L = -jZ_T I_M \tag{1.168}$$

Solving for I_{M1} from Eq. (1.168) and putting its value in Eq. (1.166), the voltage across main PA, V_M is derived as,

$$V_M = \frac{Z_T^2}{R_L} I_M - jZ_T I_A \tag{1.169}$$

The impedance seen by each PA is obtained by taking ratio of voltage across each PA to the current drawn by each PA, respectively. The main PA and auxiliary PA impedance is given by [56],

$$Z_M = \left(\frac{Z_T}{R_L} - \frac{I_A}{I_M}\right) Z_T \tag{1.170}$$

$$Z_A = \left(\frac{I_M}{I_A}\right) Z_T \tag{1.171}$$

where Z_M is the impedance seen by the main PA and Z_A is the impedance seen by auxiliary PA. It should be noted that the currents I_{M1} and I_A must be in phase. Therefore, the current of auxiliary PA must lag current of main PA by 90°. This phase relation $I_A = -jI_M$ has already been considered while obtaining Eqs. (1.170) and (1.171). One can see from Eq. (1.170) that current of auxiliary PA modulates the main PA impedance, Z_M. In order to obtain the proper load modulation in DPA, the current profiles of each PA is given by,

$$I_M = \frac{V_{in}}{V_{in,max}} I_{max} \quad 0 \leq \frac{V_{in}}{V_{in,max}} \leq 1 \tag{1.172}$$

$$I_A = \begin{cases} 0 & 0 \leq \frac{V_{in}}{V_{in,max}} \leq \frac{1}{\beta} \\ \left(\frac{V_{in}}{V_{in,max}} - \frac{1}{\beta}\right) \frac{I_{A,max}}{\left(1-\frac{1}{\beta}\right)} & \frac{1}{\beta} < \frac{V_{in}}{V_{in,max}} \leq 1 \end{cases} \tag{1.173}$$

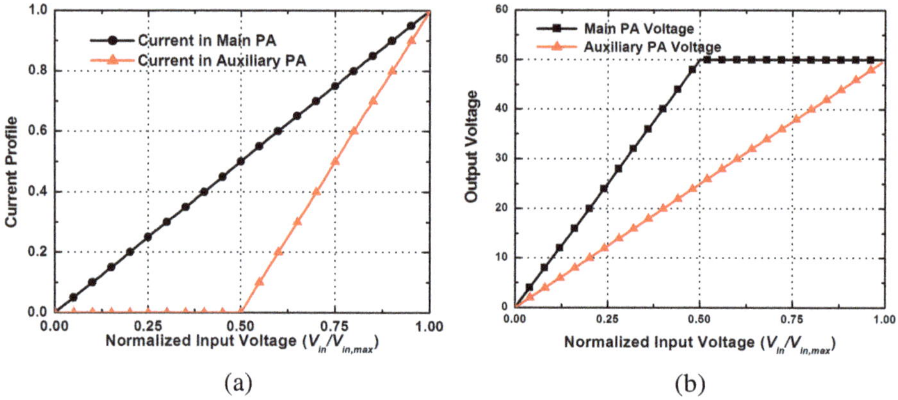

Fig. 1.36 Current and voltage profile of main and auxiliary PA with respect to normalized input voltage, (**a**) current profile and (**b**) voltage profile

where β is the load modulation factor related to input back-off level (X dB) from its maximum value, where high efficiency is required and is given as,

$$\beta = 10^{(X/20)} \tag{1.174}$$

For conventional DPA, X is 6 dB, for which β is equal to 2. The current profile of main and auxiliary PAs is obtained using Eqs. (1.172) and (1.173) and is plotted in Fig. 1.36a. It can be observed from Fig. 1.36a that the main PA current is assumed to increases linearly with normalized input voltage which depicts class B behavior. One can also see that the auxiliary PA current is initially OFF and turns ON at higher input voltage imitating class C behavior. It is also assumed that there is one-to-one relation between current and input voltage in this simplified DPA model. This implies that if the load modulation factor is β, the auxiliary PA starts conducting at $V_{\text{in,max}}/\beta$, where the main PA current at this moment is $I_{\text{M,max}}/\beta$. One can see this in Fig. 1.36a, where β is chosen as 2. Using Eqs. (1.172) and (1.173) in Eqs. (1.170) and (1.171), the impedance seen by main and auxiliary PA can be rewritten as,

$$Z_{\text{M}} = \begin{cases} \dfrac{Z_{\text{T}}^2}{R_{\text{L}}} & 0 \leq \dfrac{V_{\text{in}}}{V_{\text{in,max}}} \leq \dfrac{1}{\beta} \\ \left(\dfrac{Z_{\text{T}}}{R_{\text{L}}} - \dfrac{I_{\text{A}}}{I_{\text{M}}}\right) Z_{\text{T}} & \dfrac{1}{\beta} < \dfrac{V_{\text{in}}}{V_{\text{in,max}}} \leq 1 \end{cases} \tag{1.175}$$

$$Z_{\text{A}} = \begin{cases} \infty & 0 \leq \dfrac{V_{\text{in}}}{V_{\text{in,max}}} \leq \dfrac{1}{\beta} \\ \left(\dfrac{I_{\text{M}}}{I_{\text{A}}}\right) Z_{\text{T}} & \dfrac{1}{\beta} < \dfrac{V_{\text{in}}}{V_{\text{in,max}}} \leq 1 \end{cases} \tag{1.176}$$

From Eq. (1.175), one can conclude that initially the load impedance of main PA is constant, and after auxiliary PA turns ON, the load impedance changes in accordance with current of auxiliary PA. Similarly, the impedance of auxiliary PA is modulated from very high impedance (open circuit) to an optimum value as can be seen from Eq. (1.176).

There are three design parameters of DPA, that is, Z_T, R_L, and current ratio α, which is defined as the ratio of auxiliary PA current with respect to main PA current at saturation. In order to obtain these design parameters (α, Z_T, and R_L) of two-stage DPA for a given β, the proper voltage conditions have to be assumed. First condition is that the output voltage of main PA is same for back-off till saturation, that is, $V_{M,BO} = V_{M,SAT}$ as shown in Fig. 1.36b. Second condition is that the voltage of auxiliary PA is equal to voltage of main PA at saturation, that is, $V_{A,SAT} = V_{M,SAT}$. The voltage of main PA at back-off and saturation is obtained using Eq. (1.169) and is given by,

$$V_{M,BO} = \frac{Z_T^2}{R_L} I_{M,BO} = \frac{Z_T^2}{R_L} \frac{I_{M,Sat}}{\beta} \tag{1.177}$$

$$V_{M,Sat} = \left(\frac{Z_T^2}{R_L} - Z_T \alpha \right) I_{M,Sat} \tag{1.178}$$

Equating (1.177) and (1.178) for load modulation, the following relation is obtained,

$$\alpha = \frac{Z_T}{R_L} \left(\frac{\beta - 1}{\beta} \right) \tag{1.179}$$

The voltage of auxiliary PA at saturation is obtained by multiplying its current at saturation with its impedance given in Eq. (1.171) and is expressed as,

$$V_{A,Sat} = I_{M,sat} Z_T \tag{1.180}$$

Assuming the impedance seen by main PA and auxiliary PA at saturation is $R_{opt,M}$ and $R_{opt,A}$, respectively. Therefore, Eqs. (1.178) and (1.180) can be rearranged in terms of $R_{opt,M}$ and $R_{opt,A}$, respectively, as,

$$R_{opt,M} = \frac{Z_T^2}{R_L} - Z_T \alpha \tag{1.181}$$

$$R_{opt,A} = \frac{Z_T}{\alpha} \tag{1.182}$$

Substituting the value of Z_T from Eq. (1.182) in Eq. (1.179), the load impedance R_L can be defined as,

$$R_{\mathrm{L}} = R_{\mathrm{opt,A}} \left(\frac{\beta - 1}{\beta} \right) \tag{1.183}$$

The relation between α and β can further be obtained by solving Eqs. (1.181)–(1.183) and eliminating Z_{T} and R_{L} in these equations to obtain,

$$\beta = 1 + \alpha^2 \left(\frac{R_{\mathrm{opt,A}}}{R_{\mathrm{opt,M}}} \right) \tag{1.184}$$

For similar devices having same optimum impedance at saturation, $R_{\mathrm{opt,M}} = R_{\mathrm{opt,A}} = R_{\mathrm{opt}}$, Eqs. (1.182) and (1.183) can be rewritten as,

$$\alpha = \sqrt{\beta - 1} \tag{1.185}$$

$$Z_{\mathrm{T}} = \alpha R_{\mathrm{opt}} \tag{1.186}$$

$$R_{\mathrm{L}} = R_{\mathrm{opt}} \left(\frac{\beta - 1}{\beta} \right) \tag{1.187}$$

Thus, for enhancing efficiency at $X = 6$ dB back-off level, $\beta = 2$ from Eq. (1.174), which gives $\alpha = 1$ from Eq. (1.185). Using $\alpha = 1$, one can use Eq. (1.186) to obtain $Z_{\mathrm{T}} = R_{\mathrm{opt}}$. Similarly, $R_{\mathrm{L}} = R_{\mathrm{opt}}/2$ can be obtained from Eq. (1.187). Figure 1.37a shows the load profile seen by main and auxiliary PA for the case, where R_{opt} is assumed as 50 Ω. Consequently, Z_{T} and R_{L} are obtained as 50 and 25 Ω and used to plot Fig. 1.37a. One can see from Fig. 1.37a that the load seen by main PA is constant before back-off and is equal to $2R_{\mathrm{opt}}$, that is, 100 Ω. From the point where auxiliary PA turns ON, the load seen by main PA is modulated and changed from $2R_{\mathrm{opt}}$ (100 Ω) at back-off to R_{opt} (50 Ω) at saturation. This variation in impedance drives main PA to maximum voltage at back-off and thus the efficiency of DPA increases at back-off. Figure 1.37b shows the efficiency profile of DPA along with ideal class B PAs combined in balanced mode. One can see that the efficiency of DPA is better than ideal balanced mode class B PA at 6 dB back-off.

From the above discussion, one can conclude that the DPA has a simple architecture and does not require any extra signal processing circuitry for its implementation as compared to Chireix outphasing PA. Therefore, load modulation using DPA is widely accepted for research and commercial purpose and can easily replace existing PAs in wireless systems without any major modification.

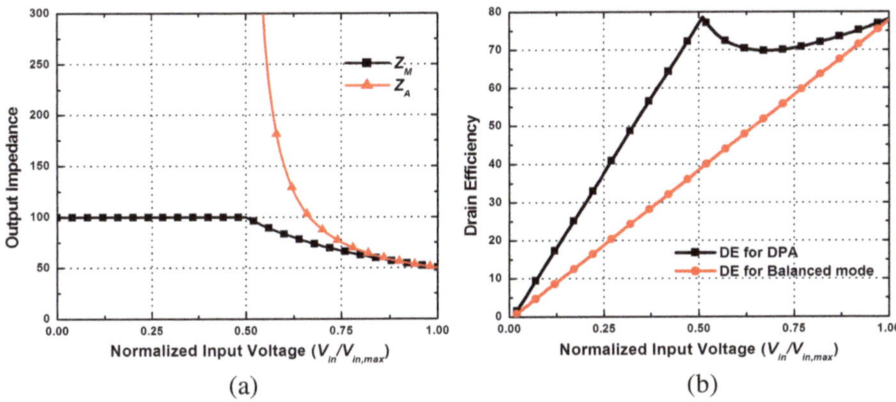

Fig. 1.37 Two-stage DPA with symmetric devices: (**a**) impedance profile of main and auxiliary PA with respect to normalized input voltage, (**b**) efficiency profile in comparison with balanced mode of operation

1.4.2 Multistage Doherty Power Amplifier

Based on the concept of load modulation as discussed in the Sect. 1.4.1, the DPA is initially proposed for the average efficiency improvement at 6 dB back-off [45]. In order to enhance this high-efficiency region to further back-off, multistage DPA architectures have been proposed which provides efficient power amplification at multiple back-off points. There are two kinds of three-stage DPA architectures. Figure 1.38a shows one widely known topology where one DPA is used as a main PA with an additional auxiliary PA. Initially, the first auxiliary PA modulates the load of the main PA similar to any conventional DPA. Later, the second auxiliary PA modulates the load of the previous DPA stage at a higher power [57–60]. The second topology as in Fig. 1.38b has been reported by NXP Semiconductors and later analyzed by several researchers [45, 61–66]. In contrary to the first topology of Fig. 1.38a, this topology is a parallel combination of one main PA and one DPA used as an auxiliary PA.

Three-Stage Doherty Power Amplifier: Topology 1

The topology illustrated in Fig. 1.38a consists of two-quarter-wave transformers Z_{01} and Z_{02} as a part of load combiner, whereas the third quarter-wave transformer Z_T is used to transform common load R_L to 50 Ω.

The topology has three PAs, where each PA behaves as a VCCS when turn ON sequentially one after another. The main PA is ON for low-power levels, and as the input power level increases the first-auxiliary PA followed by second starts conducting one after the other. The main PA saturates at $v_{in} = 1/\beta_1$. At this point, the first-auxiliary PA turns ON and modulates the load impedance of the main PA.

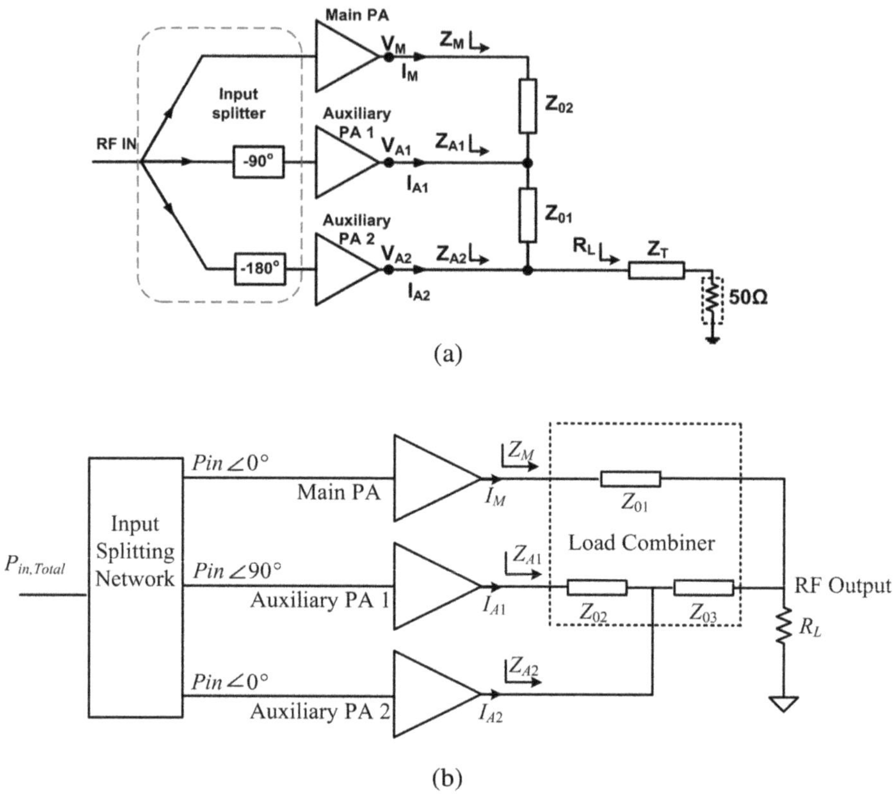

Fig. 1.38 Schematic of three-stage DPA: (**a**) topology 1 ([66], reprinted with permission from IEEE) and (**b**) topology 2

Therefore, the main and first-auxiliary PA behaves as first-DPA pair. At $v_{\text{in}} = 1/\beta_2$, the second auxiliary PA starts conducting and modulates the load impedance of first-DPA pair. At this point, first-auxiliary PA has already reached to saturation. The load modulation factors β_1 and β_2 decide the two back-off points where efficiency enhancement is targeted. The generic expression can be given as,

$$\beta_i = 10^{X_i/20} \tag{1.188}$$

where $i = 1$ and 2 is used for first and second back-offs, respectively. The variable X_i is back-off level expressed in dB. Thus, at input drive of $v_{\text{in}} = 1/\beta_1$, $1/\beta_2$ and at saturation, the three-stage DPA achieves maximum efficiency. Therefore, the first and second auxiliary PA turn ON at normalized input voltage of $v_{\text{in}} = 1/\beta_1$ and $v_{\text{in}} = 1/\beta_2$, respectively, which corresponds to first and second back-off level. Figures 1.39a, b represents the current and voltage profiles of conventional three-stage DPA. One can see that the first and second auxiliary PA turn ON at normalized

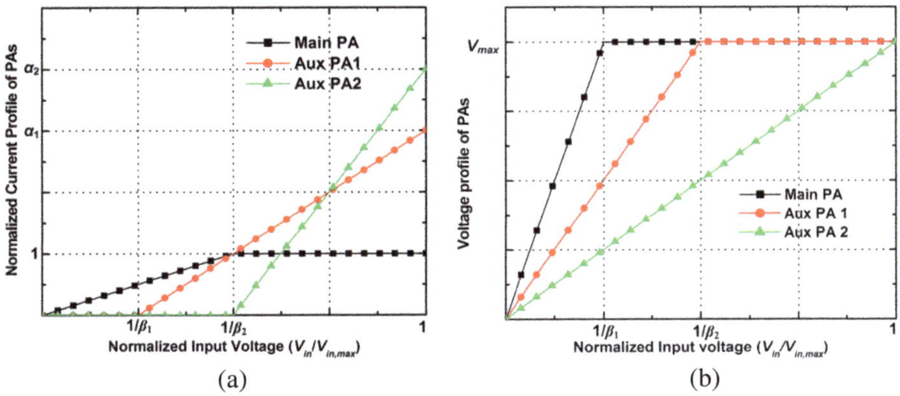

Fig. 1.39 Current and voltage profile of three-stage DPA topology 1: (**a**) current profile ([66], reprinted with permission from IEEE) and (**b**) voltage profile ([66], reprinted with permission from IEEE)

input voltage of $v_{\text{in}} = 1/\beta_1$ and $1/\beta_2$, respectively, which corresponds to first and second back-off level. One can express the current profile in Fig. 1.39a in the form of following equations.

$$I_{\text{M}} = \begin{cases} \beta_2 v_{\text{in}} & \text{for } v_{\text{in}} \leq 1/\beta_2 \\ 1 & \text{for } v_{\text{in}} \geq 1/\beta_2 \end{cases} \tag{1.189}$$

$$I_{\text{A1}} = \alpha_1 \left(\frac{v_{\text{in}} - 1/\beta_1}{1 - 1/\beta_1} \right) \quad \text{for } v_{\text{in}} \geq 1/\beta_1 \tag{1.190}$$

$$I_{\text{A2}} = \alpha_2 \left(\frac{v_{\text{in}} - 1/\beta_2}{1 - 1/\beta_2} \right) \quad \text{for } v_{\text{in}} \geq 1/\beta_2 \tag{1.191}$$

At saturation, α_i is a parameter relating the output currents of auxiliary PAs to the output current of the main PA as,

$$\alpha_i = \frac{I_{\text{A}i,\max}}{I_{\text{M},\max}} \tag{1.192}$$

where i has value 1 for the first auxiliary PA and 2 for the second auxiliary PA.

The ABCD parameters of quarter-wave transformers of characteristic impedances Z_{01} and Z_{02} are used to obtain following load impedances seen by each PA in Fig. 1.38a.

$$Z_{\text{M}} = \frac{Z_{02}^2}{Z_{01}^2} R_{\text{L}} - Z_{02} \left(\frac{I_{\text{A1}}}{I_{\text{M}}} \right) + R_{\text{L}} \frac{Z_{02}}{Z_{01}} \left(\frac{I_{\text{A2}}}{I_{\text{M}}} \right) \tag{1.193}$$

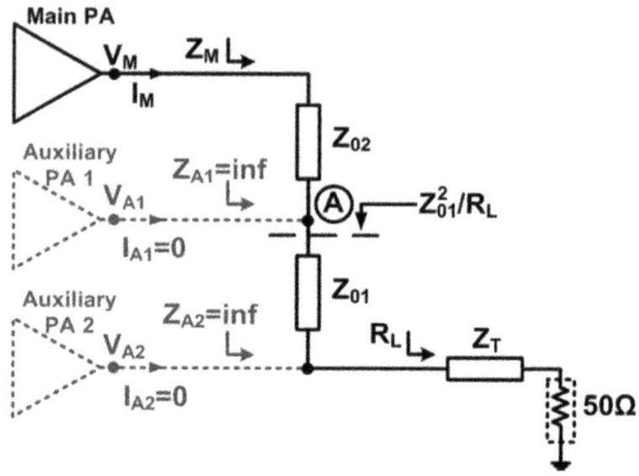

Fig. 1.40 Schematic showing conduction of only main PA before $v_{in} = 1/\beta_1$, that is, before first back-off ([66], reprinted with permission from IEEE)

$$Z_{A1} = Z_{02}\left(\frac{I_M}{I_{A1}}\right) \tag{1.194}$$

$$Z_{A2} = R_L\left(1 + \frac{Z_{02}}{Z_{01}}\frac{I_M}{I_{A2}}\right) \tag{1.195}$$

Here, a 90° phase relationship is assumed between I_M and I_{A1}. Similarly, the phase relationship between I_M and I_{A2} is assumed as 180° in deriving Eqs. (1.193)–(1.195). In practice, an input splitter will realize this required phase relationships. At first back-off, both first and second auxiliary PA are OFF, therefore, Z_{02}, Z_{01}, and R_L appear in series as shown in Fig. 1.40. To achieve maximum bandwidth, the characteristic impedance Z_{02} can be selected as same as input impedance at node A as shown in Fig. 1.40.

Since a transmission line is terminated with the impedance equal to its characteristic impedance, its input impedance is equal to the characteristic impedance of the line irrespective of the operating frequency. Therefore, Z_{02} can be expressed as,

$$Z_{02} = \frac{Z_{01}^2}{R_L} \tag{1.196}$$

Equating the voltage of main PA at first back-off, second back-off, and at saturation, following equations are obtained as,

$$\frac{Z_{02}R_L}{Z_{01}^2} = \frac{\alpha_1\beta_1}{\beta_2(\beta_1 - 1)} \tag{1.197}$$

$$\frac{Z_{01}}{R_L} = \frac{\alpha_2}{\alpha_1}\frac{\beta_2}{\beta_1}\frac{\beta_1 - 1}{\beta_2 - 1} \tag{1.198}$$

Multiplying Eqs. (1.197) and (1.198), the relation between Z_{02} and Z_{01} is determined as,

$$\frac{Z_{02}}{Z_{01}} = \frac{\alpha_2}{\beta_2 - 1} \tag{1.199}$$

In conventional three-stage DPA as proposed in [66], all the three PAs see equal voltage at saturation as shown in Fig. 1.39b. The respective output voltage of main PA, first auxiliary PA, and second auxiliary PA at saturation is determined using expression for current profile as given by Eqs. (1.189)–(1.191) and impedance seen by each PA as given by Eqs. (1.193)–(1.195) as,

$$V_{\mathrm{M,sat}} = \frac{Z_{02}^2}{Z_{01}^2} R_{\mathrm{L}} - \alpha_1 Z_{02} + \alpha_2 \frac{Z_{02}}{Z_{01}} R_{\mathrm{L}} = \frac{\beta_2}{\beta_1} \frac{Z_{02}^2}{Z_{01}^2} R_{\mathrm{L}} \tag{1.200}$$

$$V_{\mathrm{A1,sat}} = Z_{02} \tag{1.201}$$

$$V_{\mathrm{A1,sat}} = \alpha_2 \left(1 + \frac{Z_{02}}{\alpha_2 Z_{01}} \right) R_{\mathrm{L}} \tag{1.202}$$

Equating the voltage of main PA to voltage of second auxiliary PA at saturation, the current ratio, α_2, is obtained as follows,

$$\alpha_2 = \beta_1 (\beta_2 - 1) \tag{1.203}$$

Similarly, the main PA output voltage at saturation is equated to voltage of first auxiliary PA at saturation. The current ratio at saturation between first auxiliary PA and main PA is derived and is given as,

$$\alpha_1 = \beta_1 - 1 \tag{1.204}$$

The load combiner parameters are resolved using Eqs. (1.199)–(1.204) and are expressed as follows,

$$Z_{01} = \beta_2 R_{\mathrm{L}} \tag{1.205}$$

$$Z_{02} = \beta_1 \beta_2 R_{\mathrm{L}} \tag{1.206}$$

The impedance R_{L} is chosen according to the optimum impedance of main PA. If main PA sees an impedance of $R_{\mathrm{opt,M}}$ at saturation, then using Eqs. (1.203)–(1.206) in Eq. (1.200), the load impedance, R_{L}, can be expressed as,

$$R_{\mathrm{L}} = \frac{R_{\mathrm{opt,M}}}{\beta_1 \beta_2} \tag{1.207}$$

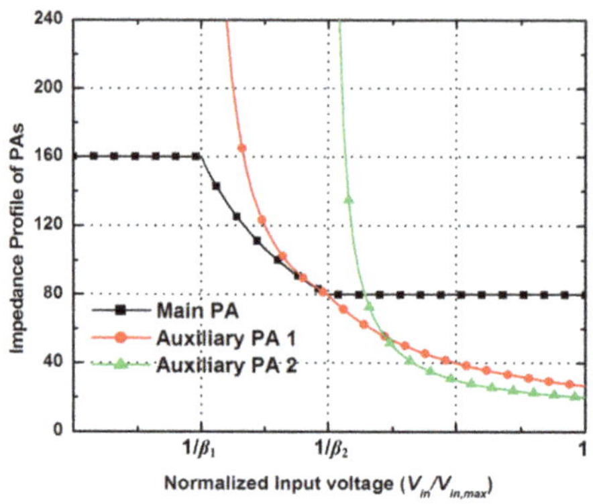

(a)

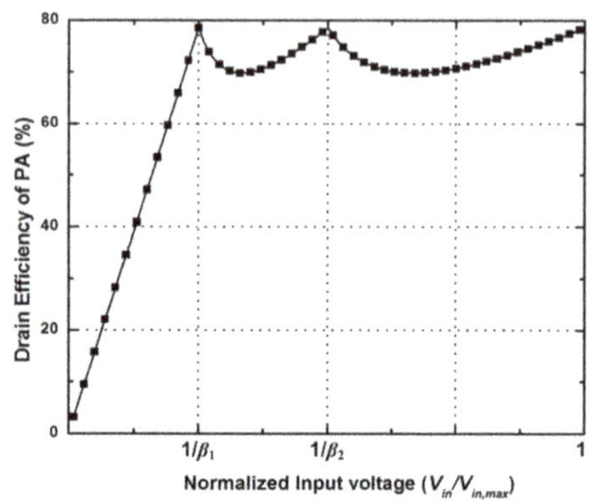

(b)

Fig. 1.41 Load modulation behavior of three-stage DPA (topology 1): (**a**) impedance profile seen by each PA, and (**b**) theoretical efficiency profile

For enhancing efficiency at 12 and 6 dB back-offs, the back-off parameters β_1 and β_2 correspond to 4 and 2, respectively. Using Eqs. (1.203) and (1.204), one can determine the required output current ratios α_1 and α_2 as 3 and 4, respectively. It is worth noticing that the output voltage of all PAs is same at saturation, but the output current of each PA is different at saturation. Therefore, the ratio of output power of main PA, first auxiliary PA and second auxiliary PA is different and is given by 1:3:4. Choosing an optimum impedance of main PA as 80 Ω, R_L is obtained as 10 Ω. The load combiner impedances are determined using Eqs. (1.205) and (1.206). The load modulation characteristics of three-stage DPA (topology 1) are plotted as shown in Fig. 1.41a. Figure 1.41b depicts the efficiency profile of DPA which shows that high efficiency is obtained at two back-off points and saturation.

Three-Stage Doherty Power Amplifier: Topology 2

The topology 2 of the three-stage DPA architecture is shown in Fig. 1.38b. In order to realize this topology in hardware, it can be alternately illustrated as Fig. 1.42a. This consists of one main PA and two auxiliary PAs, as shown in Fig. 1.42a. The load combiner comprises three quarter-wave transformers of characteristic impedances Z_{01}, Z_{02}, and Z_{03}, which combine the output power of all the PAs and deliver it to the common load R_L. The input splitter as shown in Fig. 1.42a splits the input power to the three PAs in a proper quadrature phase relationship necessary for the proper load modulation. In general, each of these PAs can be represented as a VCCS, as shown in Fig. 1.42b. To achieve proper load modulation, the main PA is represented as VCCS and is converted into a VCVS using a quarter-wave transmission line having characteristic impedance Z_{01}. Similarly, the two auxiliary PAs are represented as $\mathrm{VCCS_{Aux1}}$ and $\mathrm{VCCS_{Aux2}}$ with output current I_{A1} and I_{A2}, respectively, which are connected to the load using two quarter-wave transformers

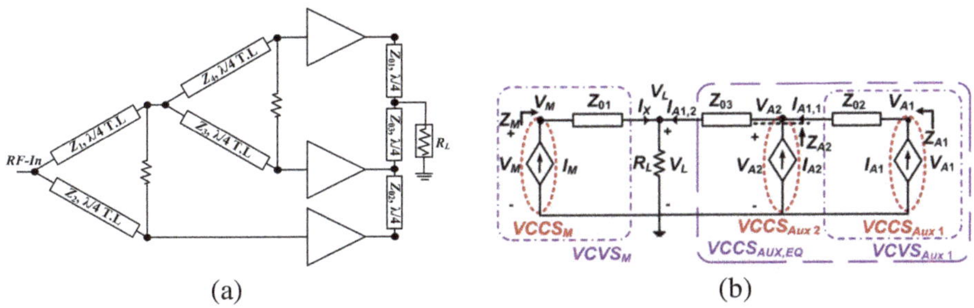

Fig. 1.42 Three-stage DPA (topology 2): (**a**) topology redrawn for circuit design with input splitter, (**b**) equivalent circuit model ([65], reprinted with permission from IEEE)

with characteristic impedances Z_{02} and Z_{03}, as shown in Fig. 1.42b. At higher back-off (first back-off point), the auxiliary PA represented by $\text{VCVS}_{\text{Aux2}}$ is off and current is zero. Therefore, only auxiliary PA1 represented as $\text{VCCS}_{\text{Aux1}}$ is operating, which is further converted back to the current source using a quarter-wave transformer of characteristic impedance Z_{03}. At lower back-off point (second back-off point), the second auxiliary PA turns on and the load seen by first auxiliary PA is modulated by an output current of it. Therefore, the combination of $\text{VCCS}_{\text{Aux1}}$, $\text{VCCS}_{\text{Aux2}}$, and the two quarter-wave transformers represent an equivalent current source represented as $\text{VCCS}_{\text{Aux,EQ}}$ in Fig. 1.42b. This is directly connected to load R_{L} as shown in Fig. 1.42b. This represents a standard DPA equivalent model, where the load is connected with one voltage source VCVS_{M} with current I_{X} and a current source $\text{VCVS}_{\text{AUX,EQ}}$ with the output current $I_{\text{A1,2}}$, as shown in Fig. 1.42b. The load seen by each PA in the symmetric three-stage DPA configuration can be obtained using the following ABCD matrices,

$$\begin{bmatrix} V_{\text{M}} \\ I_{\text{M}} \end{bmatrix} = \begin{bmatrix} 0 & jZ_{01} \\ j/Z_{01} & 0 \end{bmatrix} \begin{bmatrix} V_L \\ I_X \end{bmatrix} \tag{1.208}$$

$$\begin{bmatrix} V_{\text{A2}} \\ I_{\text{A1,1}} + I_{\text{A2}} \end{bmatrix} = \begin{bmatrix} 0 & jZ_{03} \\ j/Z_{03} & 0 \end{bmatrix} \begin{bmatrix} V_L \\ I_{\text{A1,2}} \end{bmatrix} \tag{1.209}$$

$$\begin{bmatrix} V_{\text{A1}} \\ I_{\text{A1}} \end{bmatrix} = \begin{bmatrix} 0 & jZ_{02} \\ j/Z_{02} & 0 \end{bmatrix} \begin{bmatrix} V_{\text{A2}} \\ I_{\text{A1,1}} \end{bmatrix} \tag{1.210}$$

Solving these three equations, the impedances seen by the main, first, and second auxiliary devices can be obtained as,

$$Z_{\text{M}} = \frac{Z_{01}^2}{R_{\text{L}}} - Z_{01}\left(\frac{Z_{02}}{Z_{03}}\right)\left(\frac{I_{\text{A1}}}{I_{\text{M}}}\right) \tag{1.211}$$

$$Z_{A1} = Z_{01}\left(\frac{Z_{02}}{Z_{03}}\right)\left(\frac{I_M}{I_{A1}}\right) - Z_{02}\left(\frac{I_{A2}}{I_{A1}}\right) \tag{1.212}$$

$$Z_{A2} = Z_{02}\left(\frac{I_{A1}}{I_{A2}}\right) \tag{1.213}$$

where Z_M is the output impedance of main PA and defined by the ratio of voltage V_M and current I_M, Z_{A1} is the output impedance of first auxiliary PA and defined by the ratio of voltage V_{A1} and current I_{A1}, and similarly Z_{A2} is the ratio of voltage V_{A2} and current I_{A2} for second auxiliary PA. The above impedance expressions are obtained by further assuming proper quadrature relationship between auxiliary and main PAs current as: $I_{A1} = jI_M$ and $I_{A2} = -jI_{A1}$, that is, $\angle(I_{A1}) - \angle(I_M) = 90°$ and $\angle(I_{A1}) - \angle(I_{A2}) = 90°$. This relation is obtained by input splitting network as shown in Fig. 1.42a. One can see that two Wilkinson Power dividers can realize this splitting configuration.

Thus, from the above impedance expressions one can infer that the impedances seen by the main, first, and second auxiliary PAs depend on the choice of the characteristic impedances of three quarter-wave transformers and the output current of three PAs. Thus, the proper load modulation can be achieved by choosing the relevant characteristic impedances and the current ratios.

Assuming a voltage-current relationship of an ideal class B operation, the required values of load impedances to obtain efficiency enhancement at 6 and 9.54 dB back-off are $\beta_1 R_{opt}$ and $\beta_2 R_{opt}$. Here β_1 and β_2 are 2 and 3.33, respectively. In order to achieve required load modulation characteristics, the current profiles can be assumed as,

$$I_M = I_{M,max}\frac{V_{in}}{V_{in,max}} \tag{1.214}$$

$$I_{A1} = \begin{cases} 0 & 0 \le \frac{V_{in}}{V_{in,max}} \le \frac{1}{\beta_1} \\ \left(\frac{V_{in}}{V_{in,max}} - \frac{1}{\beta_1}\right)\left(\frac{I_{A1,max}}{1-1/\beta_1}\right) & \frac{1}{\beta_1} < \frac{V_{in}}{V_{in,max}} \le 1 \end{cases} \tag{1.215}$$

$$I_{A2} = \begin{cases} 0 & 0 \le \frac{V_{in}}{V_{in,max}} \le \frac{1}{\beta_2} \\ \left(\frac{V_{in}}{V_{in,max}} - \frac{1}{\beta_2}\right)\left(\frac{I_{A2,max}}{1-1/\beta_2}\right) & \frac{1}{\beta_2} < \frac{V_{in}}{V_{in,max}} \le 1 \end{cases} \tag{1.216}$$

where β_1 and β_2 are already defined in Eq. (1.188). The current profiles as described by (Eq. 1.214–1.216) are shown in Fig. 1.43a. The first and second auxiliary PAs are biased in class C which start conducting at the normalized input voltage of value $1/\beta_1$ and $1/\beta_2$, respectively, and the value of output currents of the main PA at these normalized input voltage are $I_{M,max}/\beta_1$ and $I_{M,max}/\beta_2$, respectively. Similarly, α_1 and α_2 relates the output current of auxiliary PAs to the output current of the main PA at saturation as described in Eq. (1.192). Using the above current profiles described in Eq. (1.214–1.216) and the impedance profile as described in

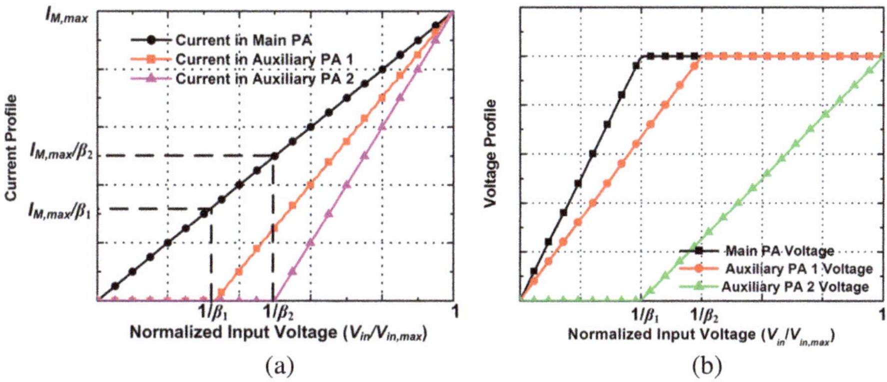

Fig. 1.43 (**a**) General current and voltage profile of main, first, and second auxiliary power amplifier in topology 2 assuming $\alpha_1 = \alpha_2 = 1$: (**a**) current profile ([65], reprinted with permission from IEEE) and (**b**) voltage profile ([65], reprinted with permission from IEEE)

Eqs. (1.211)–(1.213), one can calculate output voltage at the terminals of each PA. This is shown in Fig. 1.43b.

Applying the following three necessary conditions for successful load modulation operation, that is, (i) the voltages at the output of main PA at saturation and both back-offs are equal; (ii) the voltage at the output of the first auxiliary PA is equal at saturation and second back-off; (iii) the voltage of both the auxiliary PAs at saturation is equal, one can obtain,

$$\beta_1 = 1 + \alpha_1 + \alpha_2 \tag{1.217}$$

$$\beta_2 = \frac{\alpha_1 + \alpha_2}{\alpha_2} \tag{1.218}$$

For a conventional symmetric DPA structure, where all the devices deliver same current at saturation, one should choose, $\alpha_1 = \alpha_2 = 1$. For such case, β_1 and β_2 are fixed as 3 and 2, respectively, which corresponds to the back-off of 9.54 and 6 dB, respectively. Therefore, at higher back-off, when both auxiliary PAs are off and the load modulation factor is β_1, the value of load seen by the main PA, that is, Z_M, should be $3R_{opt}$. Using this value in Eq. (1.211), one can obtain characteristic impedance Z_{01} of the quarter-wave transformer in Fig. 1.42a as,

$$Z_{01} = \sqrt{3R_{opt}R_L} \tag{1.219}$$

Equating Z_{A1} and Z_{A2} to R_{opt} at saturation in Eqs. (1.212) and (1.213), respectively, one can obtain the value of characteristic impedances Z_{02} and Z_{03} of the quarter-wave transformers in Fig. 1.42a as,

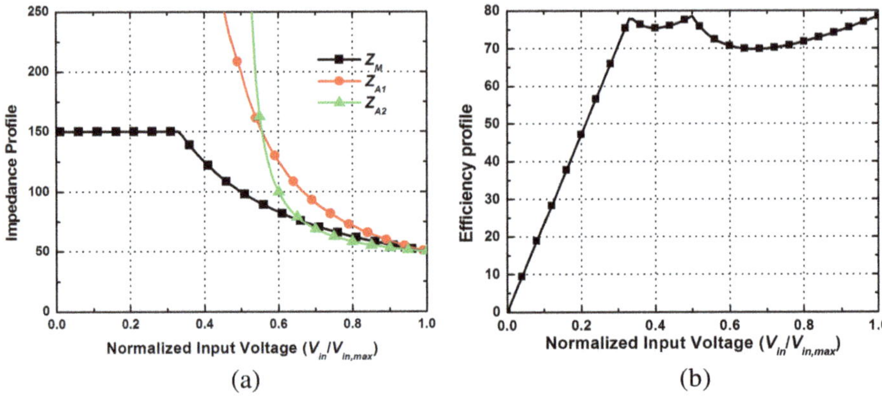

Fig. 1.44 Three-stage DPA design using topology 2 for $R_L = R_{opt} = 50\ \Omega$ and $\beta_1 = 3$, $\beta_2 = 2$, (**a**) impedance seen by main, first, and second auxiliary PAs with respect to normalized input voltage ([65], reprinted with permission from IEEE) and (**b**) efficiency profile

$$Z_{02} = R_{opt} \tag{1.220}$$

$$Z_{03} = \frac{Z_{01}}{2} \tag{1.221}$$

Thus, choice of characteristic impedances in Eqs. (1.219)–(1.221) depends on the load R_L and R_{opt}. Figure 1.44a shows the load modulation characteristic of three-stage DPA using topology 2 with $\alpha_1 = \alpha_2 = 1$. The load combiner characteristic impedances are calculated using Eqs. (1.219)–(1.221) with R_L and R_{opt} assumed as 50 Ω.

Chireix Outphasing Power Amplifier

Chireix outphasing amplifier is an alternative technique based on load modulation for efficiency enhancement at back-off power. It was first proposed by Henri Chireix in 1935 to improve average efficiency and linearity of AM broadcast transmitters [67]. Figure 1.45 shows a simplified block diagram of outphasing PA. One can see that the architecture has a signal separator which separates a complex modulated input signal *S*(*t*) (with amplitude as well as phase modulation) into two phase modulated signals $S_1(t)$ and $S_2(t)$ with constant envelope. These signals at the output of each PA are combined using signal combiner to obtain an amplified version of complex modulated input signal *S*(*t*).

This idea was reintroduced by D. Cox in 1974 as LINC (Linear amplification using nonlinear components) transmitter [68]. LINC architectures are capable of transmitting very wideband signals but the major disadvantage is the power wastage

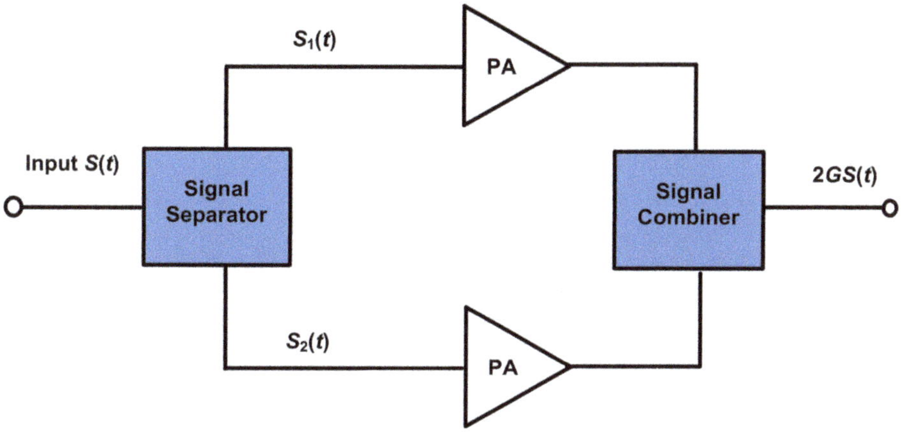

Fig. 1.45 Architecture of outphasing PA

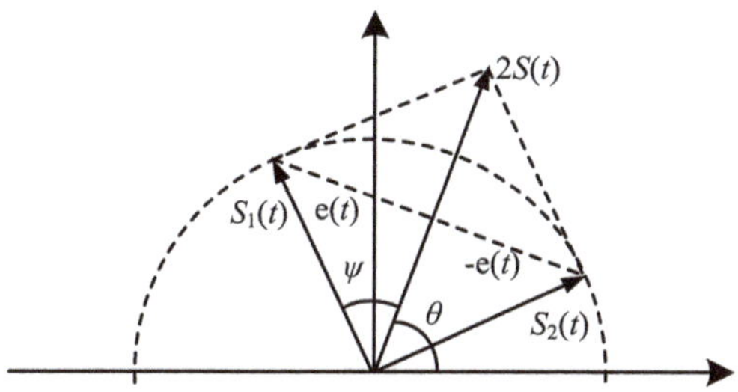

Fig. 1.46 Vector representation of outphasing concept

in the power combiner. LINC architecture uses an isolating power combiner such as a Wilkinson power divider/combiner to avoid signal distortion. These combiners can achieve 100% efficiency at maximum power but even a small phase difference between the signal branches degrades the average efficiency rapidly due to the heat dissipation in isolation resistor [69]. To tackle this problem, non-isolating combiners are used in outphasing architectures for power combining. Chireix proposed a non-isolating combiner which consists of compensating reactive element to enhance the power combining efficiency [67]. Figure 1.46 shows the vector representation of outphasing concept.

The input complex modulated signal S(t) is represented by,

$$S(t) = r(t)\mathrm{e}^{j\theta(t)} \tag{1.222}$$

In order to split the signal into two outphasing constant envelope signals, the envelope $r(t)$ can be expressed as constant envelope $r_{\max}$ with phase modulation $\psi(t)$ in Eq. (1.222). This results in an expression given by,

$$S(t) = r_{\max}\cos\psi(t)\mathrm{e}^{j\theta(t)} = \frac{r_{\max}}{2}\left[\mathrm{e}^{j\psi(t)} + \mathrm{e}^{-j\psi(t)}\right]\mathrm{e}^{j\theta(t)} \tag{1.223}$$

where r_{max} is the maximum value of $r(t)$ and $\psi(t)$ is the outphasing angle. In order to have constant envelope phase modulation, $S(t)$ is decomposed into two constant amplitude signals $S_1(t)$ and $S_2(t)$ as shown in Fig. 1.46.

$$S_1(t) = S_1(t) + e(t) = \frac{r_{max}}{2}\left(e^{j\psi(t)}\right)e^{j\theta(t)} \tag{1.224}$$

$$S_2(t) = S_1(t) - e(t) = \frac{r_{max}}{2}\left(e^{-j\psi(t)}\right)e^{j\theta(t)} \tag{1.225}$$

Subtracting Eq. (1.224) from Eq. (1.225) will give e(t) as,

$$2e(t) = jr_{max}e^{j\theta(t)}\sin\psi(t) \tag{1.226}$$

Rearranging Eq. (1.226) in terms of $S(t)$ will give,

$$e(t) = j0.5S(t)\sqrt{\frac{r_{max}^2}{r^2(t)} - 1} \tag{1.227}$$

These two constant envelope signals $S_1(t)$ and $S_2(t)$ are separately amplified by two independent identical PAs and then recombined. Addition of $S_1(t)$ and $S_2(t)$ will recover original signal $S(t)$. The combiner will give the amplified output signal which is the replica of input amplitude and phase modulated signal.

Chireix Outphasing Combiner

The non-isolated Chireix combiner as mentioned earlier is shown in Fig. 1.47a, b. In addition to the combining operation, this combiner will also provide load modulation. Figure 1.47a, b shows the simplified circuit diagram of conventional Chireix combiner with and without compensation, respectively. One can see that the conventional Chireix combiner consists of quarter-wave transmission lines in both the branches of outphasing PA [70–73]. The amplifiers operating at saturation is represented by constant voltage sources V_1 and V_2 with outphasing angles ψ_1 and ψ_2, respectively. This is a common mode topology of Chireix outphasing which is a more practical representation from implementation point of view than differential mode topology [73]. Y_{01} and Y_{02} in Fig. 1.47a represents the load admittance presented to upper and lower branch PAs, respectively, without any compensation applied. Y_{c1} and Y_{c2} in Fig. 1.47b represent the load admittance presented to upper and lower branch PAs, respectively, with compensation applied.

The input-output relation of the quarter-wave line for upper and lower branch of outphasing shown in Fig. 1.47a can be expressed using ABCD parameters of quarter-wave transformer of characteristic impedance Z_0.

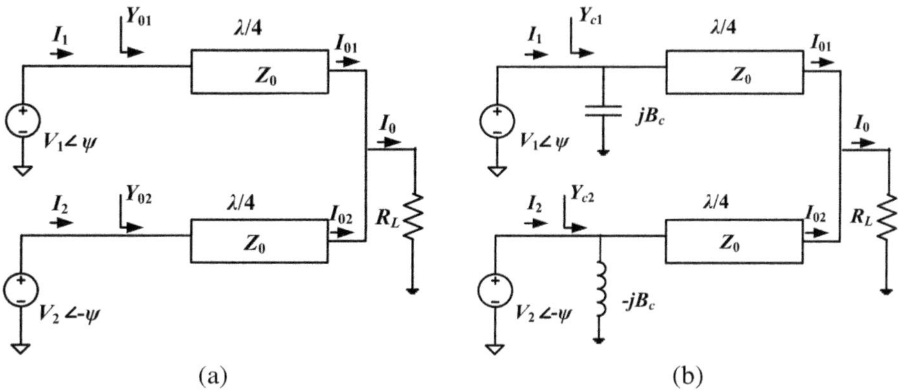

Fig. 1.47 Topology of Chireix outphasing combiner: (**a**) without reactance compensation and (**b**) with reactance compensation

$$\begin{bmatrix} V_1 \\ I_1 \end{bmatrix} = \begin{bmatrix} 0 & jZ_0 \\ \frac{j}{Z_0} & 0 \end{bmatrix} \begin{bmatrix} V_L \\ I_{01} \end{bmatrix} \tag{1.228}$$

$$\begin{bmatrix} V_2 \\ I_2 \end{bmatrix} = \begin{bmatrix} 0 & jZ_0 \\ \frac{j}{Z_0} & 0 \end{bmatrix} \begin{bmatrix} V_L \\ I_{02} \end{bmatrix} \tag{1.229}$$

The load current I_L is the sum of the currents I_{01} and I_{02}. These branch currents can be expressed in terms of voltages V_1 and V_2 using Eqs. (1.228) and (1.229).

$$I_L = \frac{V_1}{jZ_0} + \frac{V_2}{jZ_0} \tag{1.230}$$

If the two PAs are same, the V_1 and V_2 can be expressed as $V_0e^{+j\psi}$ and $V_0e^{-j\psi}$, respectively. In such case Eq. (1.230) can be rewritten as,

$$I_L = \frac{2V_0 \cos\psi}{jZ_0} \tag{1.231}$$

The V_L is the voltage across R_L developed with I_L. Further, using Eqs. (1.228) and (1.229), the current I_1 and I_2 can be expressed as,

$$I_1 = I_2 = \frac{j}{Z_0}V_L = \frac{2V_0 \cos\psi}{Z_0^2}R_L \tag{1.232}$$

Thus, the load admittances Y_{01} and Y_{02} as shown in Fig. 1.47a can be expressed as,

$$Y_{01} = \frac{I_1}{V_1} = \frac{2\cos^2\psi}{R_{L,T}} - j\frac{\sin 2\psi}{R_{L,T}} = G_0 - jB_0 \quad (1.233)$$

$$Y_{02} = \frac{I_2}{V_2} = \frac{2\cos^2\psi}{R_{L,T}} + j\frac{\sin 2\psi}{R_{L,T}} = G_0 + jB_0 \quad (1.234)$$

where $R_{L,T}$ is the intermediate variable given by,

$$R_{L,T} = \frac{Z_0^2}{R_L} \quad (1.235)$$

From Eqs. (1.233) and (1.234), one can see that admittance Y_{01} and Y_{02} are complex values and function of outphasing angle ψ. Thus, the load conductance and susceptance are modulated by outphasing angle. Figure 1.48a, b shows the plot of load conductance and susceptance versus outphasing angle, respectively, for upper and lower branch of outphasing PA.

One can observe that when $\psi = 0°$, loads are purely real since at this point B_0 is zero. At other outphasing angles, imaginary part of admittance causes the efficiency degradation and therefore it should be compensated. Chireix outphasing architecture with reactance compensation network is shown in Fig. 1.48b. By adding two fixed reactive components jB_c and $-jB_c$, susceptance values (jB_0 and $-jB_0$) can be canceled at certain outphasing angle ψ_c. This will lead to efficiency enhancement at back-off power corresponding to outphasing angle ψ_c. The selection of outphasing angle for susceptance cancellation ψ_c is decided according to the desired output power back-off (OPBO) using,

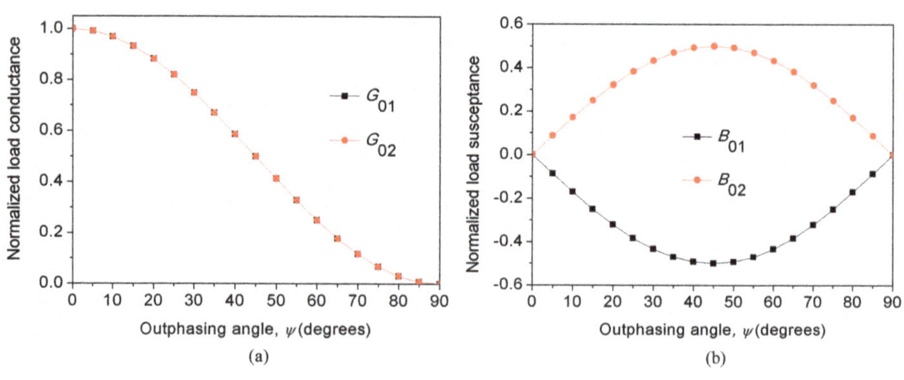

Fig. 1.48 Admittance variation with outphasing angle: (**a**) susceptance versus outphasing angle, (**b**) conductance versus outphasing angle

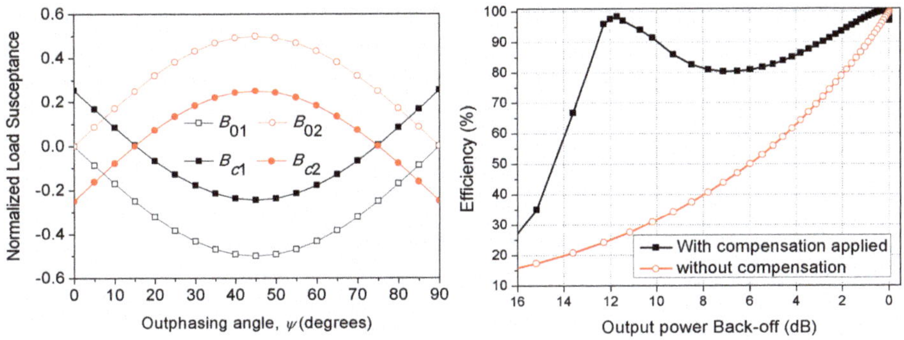

(a) (b)

Fig. 1.49 (**a**) Susceptance versus outphasing angle with and without compensation, (**b**) combiner efficiency with and without compensation

$$\mathrm{OPBO} = -20 \ \log\left(\cos \psi_c\right) \tag{1.236}$$

This can be explained with example shown in Fig. 1.49a, b. Figure 1.49a shows the plot of susceptance versus outphasing angle with compensation at $\psi_c = 75°$. This compensation results in efficiency enhancement up to 11.7 dB OPBO according to Eq. (1.236). The plot of efficiency versus OPBO is shown in Fig. 1.49b.

The methodology discussed above is a conventional outphasing amplifier where both the amplifiers are working in deep saturation. In such case, the analysis assumes that the two branch PAs are driven by constant amplitude signals and are conventionally modeled as ideal voltage sources. Therefore, their efficiency is assumed constant and independent of the variation of the outphasing angle and the overall efficiency is product of PA efficiency and combiner efficiency. In such case, the outphasing angle between these input signals has only influence on power combiner and controls the output power. However, in practice, this assumption is not correct as the instantaneous output voltage across the active device changes with the variation in outphasing angle. Thus, the efficiency of active device does not remain constant with change in outphasing angle and rather degrade in practical implementation of outphasing PA [74]. Furthermore, considering the parasitic and nonlinear active device behavior, this assumption severely limits the performance in hardware development and measurement. This limitation is solved by analyzing Chireix PA with mixed mode of operation where outphasing amplifiers are driven with conditioned input signals [74–78]. In such case, both the phase and amplitude of the branch PA drives are modulated [78]. The relative phase can control the load modulation while the power drives can be typically set to maintain saturation without overdriving the PA [78]. Therefore, mixed-mode operation can offer higher back-off efficiency in comparison to conventional mode of operation and optimum performance over a wide dynamic range [78]. The mixed-mode operation is discussed in later chapters, where a generalized combiner is used which offers direct computation of combiner network parameter from the desired load conditions required for any mode of operation of the PAs in each branch [79]. Once obtaining the network parameters, the combiner can be synthesized using lumped or transmission line elements [80]. The following section describes outphasing PA design using generalized combiner.

Generalized Combiner Synthesis Technique for Class E Chireix Outphasing Amplifier

The idea of analyzing the Chireix outphasing PA by analytically solving class E equations rather than assuming the active devices as ideal voltage source was reported in [79, 81]. A semi-analytical combiner approach was developed in [81] to realize class E-based outphasing PA. However, the technique requires a numerical optimization to obtain circuit elements of combiner. Later, a fully analytical generalized load combiner is presented which can directly realize class E load conditions for designing outphasing PA [79]. In this case, the combiner design starts with a two-port reciprocal lossy network as shown in Fig. 1.50.

Z_n^L and Z_n^R are the impedances seen by the left and right transistor in Fig. 1.50 at nth harmonic. For fundamental frequency $n = 1$. The z-parameters of this combiner is represented as matrix Z_C. The optimum load impedances Z_n^L and Z_n^R can be selected to obtain any operating mode in PA. This network is considered lossy as the load is included in this network for the analysis purpose. The terminal voltages V_1, V_2 and currents I_1 and I_2 in Fig. 1.50 can be expressed as

$$\begin{bmatrix} V_1 \\ I_1 \end{bmatrix} = \begin{bmatrix} Z_{11} & Z_{12} \\ Z_{21} & Z_{22} \end{bmatrix} \begin{bmatrix} I_1 \\ I_2 \end{bmatrix} \tag{1.237}$$

where due to reciprocity one can consider $Z_{12} = Z_{21}$, and therefore three unknown parameters are left to be determined for the above matrix Z_C. Considering the peak occurring at outphasing angle ψ_P, the fundamental load required by any one of the PA (e.g., left side PA) is $Z_1^L(\psi_P)$. Similarly, considering outphasing angle is ψ_B at output power backed-off by γ, the fundamental load required by this PA is $Z_1^L(\psi_B)$. Since these loads must be presented by the two-port network of Fig. 1.50,

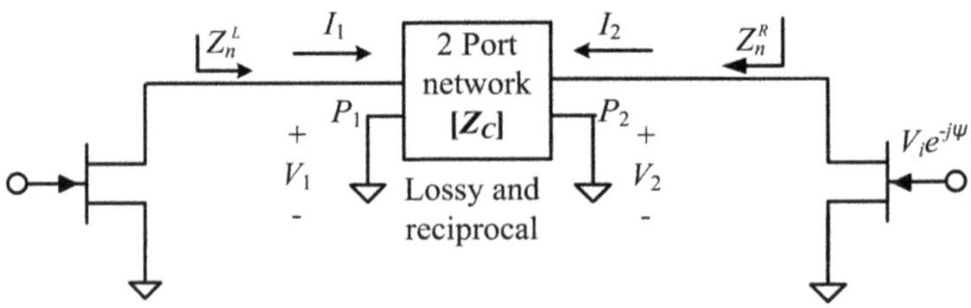

Fig. 1.50 Schematic used for the derivation of the two-port network parameters

$Z_1^L(\psi_P) = \frac{V_1}{I_1}\Big|_{\psi=\psi_P}$ and $Z_1^L(\psi_B) = \frac{V_1}{I_1}\Big|_{\psi=\psi_B}$ can be used in Eq. (1.237) to obtain following z-parameters,

$$Z_{11} = \frac{Z_1^L(\psi_B) - Z_1^L(\psi_P)}{1 + e^{j2\psi_P}} + Z_1^L(\psi_P) \tag{1.238}$$

$$Z_{12} = \frac{1}{2}\left(Z_1^L(\psi_P) - Z_1^L(\psi_B)\right)\sec\psi_P \tag{1.239}$$

$$Z_{22} = \frac{Z_1^L(\psi_P) - Z_1^L(\psi_B)}{1 + e^{j2\psi_P}} + Z_1^L(\psi_B) \tag{1.240}$$

where relation between ψ_P and ψ_B is assumed as,

$$\psi_B = \pi - \psi_P \tag{1.241}$$

Now, in order to realize this two-port network with three-port lossless reciprocal network terminated with a purely resistive load R_L, the two-port lossy network must be reciprocal and following constraint holds,

$$\mathrm{Re}\{Z_{12}\}^2 = \mathrm{Re}\{Z_{11}\}\,\mathrm{Re}\{Z_{22}\} \tag{1.242}$$

For $\psi = \psi_P$, the above condition Eq. (1.242) can be satisfied for four different values of ψ_P, that is, $\pm\psi_{PS}$ and $\pm(\pi - \psi_{PS})$ where ψ_{PS} depends on back-off factor. For a given value of γ, the ψ_{PS} can be obtained numerically using Eqs. (1.238)–(1.240) and (1.242). Figure 1.51 shows the variation of ψ_{PS} with γ. Here ψ_{PS} is represented as θ_{1X} in the reprint figure. Once, ψ_{PS} is known one can obtain all the z-parameters of two-port network of Fig. 1.50 for given Z_n^L and Z_n^R.

The next step is to convert two-port lossy reciprocal network into three-port lossless reciprocal network. This three-port network can be presented as two lossless two-port networks as shown in Fig. 1.52.

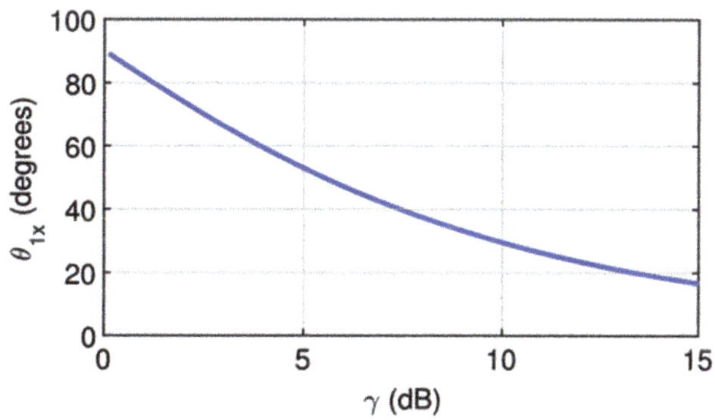

Fig. 1.51 Variation of solution of outphasing angle at back-off (ψ_{PS}) with back-off factor γ ([79], reprinted with permission from IEEE)

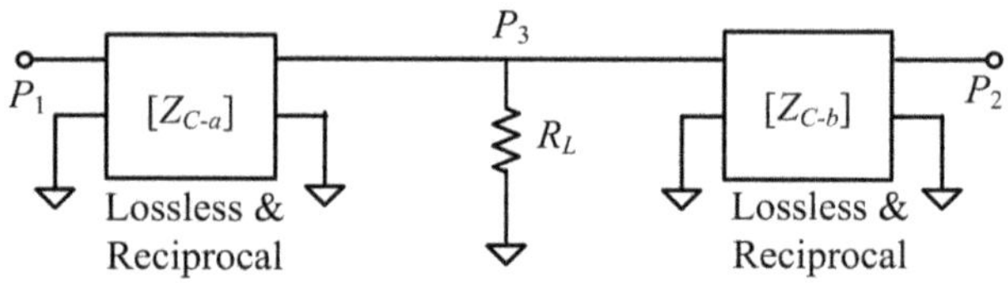

Fig. 1.52 Realization of the combiner as a three-port lossless reciprocal network terminated with a resistive load

For the realization of lossless combiner, the z-parameters of two lossless networks, $Z_{C\text{-}a}$ and $Z_{C\text{-}b}$ are derived in terms of Z_C and R_L. Since the three-port lossless network as shown in Fig. 1.52 comprises three cascaded networks, it is easier to operate in terms of *ABCD* parameters. In such case the *ABCD* parameter of lossy and reciprocal two-port network as shown in Fig. 1.50 should be compared with *ABCD* parameter of lossless and reciprocal network shown in Fig. 1.52. The *ABCD* parameter of lossy and reciprocal two-port network in Fig. 1.50 can be written as,

$$T_{2P} = \begin{bmatrix} A_r + jA_i & B_r + jB_i \\ C_r + jC_i & D_r + jD_i \end{bmatrix} \tag{1.243}$$

This can be obtained by z to *ABCD* network parameter conversion, where the z-parameter Z_C in Fig. 1.50 is already obtained for given Z_n^L and Z_n^R and γ. The *ABCD* parameters of three-port network in Fig. 1.52 can be written as:

$$\mathbf{T_{2P}} = \mathbf{T_{C\text{-}a}} \mathbf{T_R} \mathbf{T_{C\text{-}b}} = \begin{bmatrix} A_a & jB_a \\ jC_a & D_a \end{bmatrix} \begin{bmatrix} 1 & 0 \\ 1/R_L & 1 \end{bmatrix} \begin{bmatrix} A_b & jB_b \\ jC_b & D_b \end{bmatrix} \tag{1.244}$$

where $\mathbf{T_{C\text{-}a}}$ and $\mathbf{T_{C\text{-}b}}$ are *ABCD* parameters of two lossless two-port networks as shown in Fig. 1.52. One can see that the diagonal elements have purely real values, whereas off-diagonal elements have purely imaginary values to guarantee lossless and reciprocal condition in these two-port networks. T_R is the *ABCD* parameter of resistor R_L.

Equating matrix Eqs. (1.243) and (1.244) will give following seven equations

$$B_a = \frac{A_i D_a}{C_r} \tag{1.245}$$

$$C_a = \frac{C_r - C_r A_a D_a}{A_i D_a} \tag{1.246}$$

$$D_a = \pm \frac{C_r \sqrt{R_L} \sqrt{A_i B_i + A_r B_r}}{A_i} \tag{1.247}$$

$$A_b = \frac{C_r R_L}{D_a} \tag{1.248}$$

$$B_b = -\frac{B_r C_r R_L}{A_i D_a} \tag{1.249}$$

$$C_b = \frac{A_a C_r^2 R_L}{A_i D_a} - \frac{A_i D_a}{B_r C_r R_L} + \frac{B_i C_r}{B_r D_a} \tag{1.250}$$

$$D_b = \frac{C_r \left(A_i B_i D_a + A_a B_r C_r R_L\right)}{A_i D_a^2} \tag{1.251}$$

One can see that there are seven Eqs. (1.245)–(1.251) to determine eight unknowns. Therefore, one unknown, that is, A_a, is kept as the free design variable. Once knowing the *ABCD* parameters of each two-port networks of Fig. 1.52, one can easily realize these using T- or Pi-type networks [82]. The above methodology has been used to design an outphasing combiner operating at 900 MHz using class E PA in [79]. In this example, P_{out} is considered as 15 W, V_{DD} is 28 V, and γ is selected as 7.5 dB. Using class E load conditions, one can obtain $\psi_{PS} = 39.5°$. Therefore, Z_C is calculated as [79],

$$Z_C = \begin{bmatrix} 88.16 + j199.68 & -138.32 - j101.01 \\ -138.32 - j101.01 & 217.04 + j23.19 \end{bmatrix} \tag{1.252}$$

Later, the lossless two-port network parameters $T_{C\text{-}a}$ and $T_{C\text{-}b}$ can be obtained from Eqs. (1.245)–(1.251). Using positive root in Eq. (1.247), the resulting lumped element realization of combiner is shown in Fig. 1.53.

The harmonic matching circuitry realizes –j152 Ω at second harmonic as per optimum class E operation [79]. A simple short circuit is realized by an open circuited quarter wavelength stub at $2f_0$. This along with a series transmission line of electric length of 54° and characteristic impedance 50 Ω will realize the harmonic matching. The effect of harmonic matching circuit can be de-embedded from the calculated combiner network parameter to design rest of the section.

The methodology discussed above will give a single set of network parameter which can be used to design a combiner at single operating frequency. For wideband

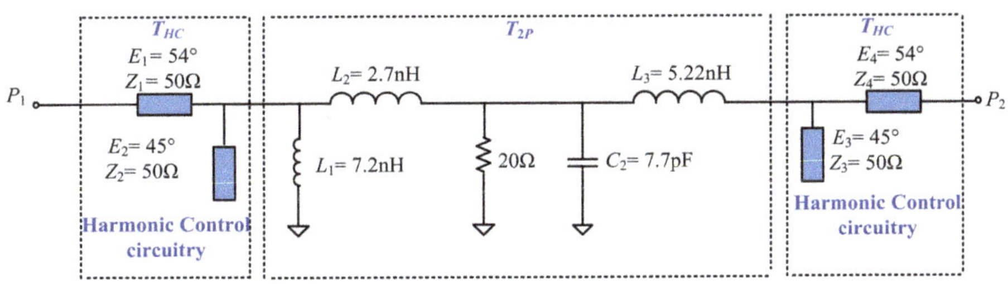

Fig. 1.53 Synthesized lumped element outphasing combiner with harmonic matching

outphasing PA design, the combiner can be synthesized for realizing the family of impedance solutions for continuous class E mode of operation as reported in [33, 83] and discussed in later chapters according to the design example reported in [79].

1.5 Power Amplifier Linearization

1.5.1 Nonlinear Characterization

As established in Sects. 1.2.2 and 1.2.3, due to its nonlinear characteristics, the PA exhibits AM/AM and AM/PM distortions [84]. These distortions are significant when the signal peaks approach the PA's saturation region, therefore PA needs to be driven at a lesser average power level [2]. However, typically, PA efficiency is high at higher power levels near saturation region. It is therefore desirable to extend the linear range of the PA as high as possible toward the saturation point to obtain a reasonable tradeoff between linearity and power efficiency [84]. First step toward design of such linearizers is understanding the nature and effects of nonlinear distortions. PA characteristics can be captured in terms of "behavioral" or "forward" models, which produces same output as the actual PA. Such models can be used to devise proper linearization strategy by porting them into appropriate simulation tools. The "forward modeling" of PA comprises mainly four steps: baseband complex waveform capturing, delay compensation, model generation/characterization, and model validation. These steps are described in the following sections.

Baseband Complex Waveform Capture

This step consists of acquiring baseband waveforms at PA input as well as output. Figure 1.54 shows a typical PA characterization setup [2]. An appropriate training signal is selected as single carrier, two-tone, multitone or broadband signal and its baseband signal is uploaded in vector signal generator (VSG). This signal is converted to analog signal and further up-converted to the RF frequency by VSG. The RF signal is applied to the PA as excitation signal. The attenuator at the output of the PA reduces the signal power to avoid saturating the analog to digital converter (ADC) of the vector signal analyzer (VSA). VSA down-converts the signal to baseband frequency and performs analog to digital conversion.

The VSA and VGS have synchronized baseband clock, which does not allow any phase drift between captured input and output signals. VSA and VSGs are driven at their linear region so that the captured signal may exhibit only distortion due to PA nonlinearity for accurate characterization. The functions of VSA and VSG are controlled via computer using GPIB or LAN. The captured baseband signal at PA output is sent to laptop/computer for further processing in MATLAB/C++ platform.

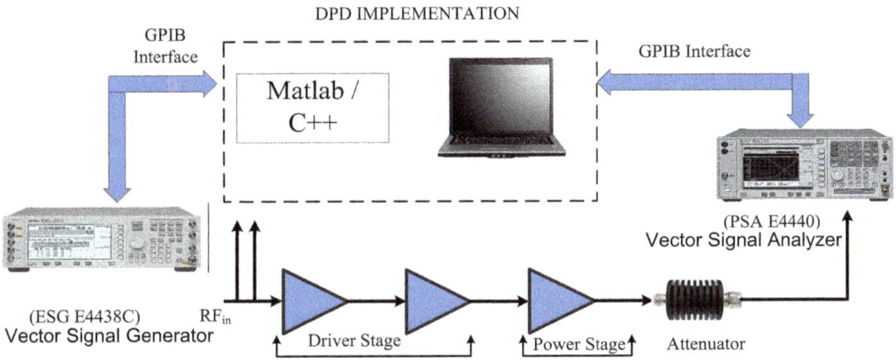

Fig. 1.54 PA characterization setup

Delay Compensation

The captured output baseband signal is ideally delayed version of the input signal. This delay is due to group-delay of transmission path via cable. In addition, VSA may start capturing the signal at the middle of training signal frame which may appear as signal delay. VSA and VSG can have triggers enabled so that the VSA will start capturing data at the instant when the VSG starts transmitting. In the trigger-enabled mode, a constant delay can be observed, which is due to the delay of the characterization path. For PA characterization, this delay can be estimated in either time domain or frequency domain by maximizing the cross-correlation between the two waveforms. It should be noted that cross-correlation is maximized for linear systems and although PA is nonlinear in its saturation region, most of the low-power region represents linear function, which allows the use of cross-correlation-based delay estimation. If input baseband signal sample *u*(*n*) as well as output baseband signal sample *v*(*n*), have mean value of 0, then cross-correlation is called covariance, which is given as:

$$C_{uv} = \begin{cases} \sum_{n=0}^{N-m-1} \left(u\left(n+m\right)-\overline{u}\right)\left(v^{*}(n)-\overline{v}\right) & m \geq 0 \\ \sum_{n=0}^{N+m-1} \left(u^{*}\left(n-m\right)-\overline{u}^{*}\right)\left(v(n)-\overline{v}\right) & m < 0 \end{cases} \tag{1.253}$$

where $\overline{v} = \frac{1}{N}\sum_{i=0}^{N-1} v(i)$ and $\overline{u} = \frac{1}{N}\sum_{i=0}^{N-1} u(i)$. The time delay can be given as,

$$\tau = m_{\max} \times \frac{1}{f_{\mathrm{s}}} \tag{1.254}$$

$m_{\max}$ is the index of output data sample which provides the maximum covariance and f_{s} denotes the sampling rate of the waveform. The output waveform is shifted in time so that the input and output waveforms are time-synchronized. The precision of such method is limited by sampling time-step, therefore, for fine tuning, the Lagrange's interpolation is applied for the segment of waveform near the estimated delay point and cross-covariance is again calculated and more precise delay is calculated [85]. Alternatively, time delay alignment can also be done using frequency domain cross-correlation technique given in [86, 87].

The frequency domain cross-correlation of input and output signals is given as [86],

$$F\left(\vec{u} \times \vec{v}\right) = F\left(\vec{u}\right) \times F\left(\vec{v}\right) \tag{1.255}$$

where F represents the Fourier transform. As power amplifier is linear in low-power ranges, Eq. (1.255) can be rewritten as,

$$F\left(\vec{u} \times \vec{v}\right) = G \times F\left(\vec{u}\right) \times F\left(\vec{u}\right) \mathrm{e}^{j(2\pi f\tau+\theta)} \tag{1.256}$$

where G is the gain, τ is the time delay, and θ is the phase rotation of output signal with reference to the input signal. It is evident that Eq. (1.256) is maximum when $\mathrm{e}^{j(2\pi f\tau + \theta)}$ is equal to 1. In other words, $2\pi f\tau + \theta = 0$. If one draws a line $y = 2\pi f\tau + \theta$ with f as input variable, then τ is slope and θ is intersection with y-axis. Figure 1.55a shows the Fourier transform of input and output waveforms, while Fig. 1.55b shows the frequency versus $(2\pi f\tau + \theta)$. One can observe that a straight line is observed for the frequencies where input-output signal is present. τ and θ can be estimated as slope and intersection of the line. The time adjusted signal is then given as,

$$\overline{v}_{\text{adj}} = \overline{v}\mathrm{e}^{-j(2\pi\tau+\theta)} \tag{1.257}$$

The frequency-domain correlation is a batch mode approach, where processing is applied on a vector of data, which provides fast results. However, when the measurement system has high measurement noise, there may not be a clear line for slope estimation and time-domain method may provide better results.

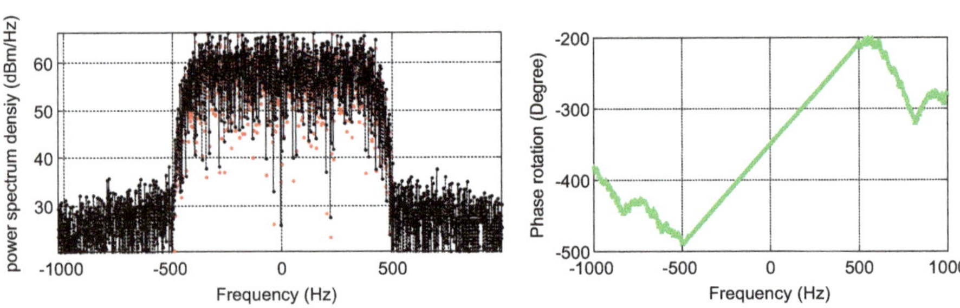

(a) (b)

Fig. 1.55 (**a**) Fourier transform of a modulated signal (**b**) dependence of phase on frequency

Model Characterization

Behavioral models represent the relation between input and output waveform. The simplest model for characterizing nonlinearity of PA is known as look-up-table (LUT) model given as [88],

$$x_{\text{out}}(n) = G\left(|x_{\text{in}}(n)|\right) \times x_{\text{in}}(n) \tag{1.258}$$

where $G(|x_{\text{in}}(n)|)$ represents the instantaneous complex gain of the device under test (DUT). LUT contains the calculated gain values for any input baseband signal, which is mapped linearly to the $|x_{\text{in}}(n)|$. This model is able to predict the nonlinear response of the PA; however, there is no provision for memory effects. In order to cater to the memory effects of the PAs, models such as Weiner model and Hammerstein Model that relies on capturing the memory effects by introducing finite impulse response (FIR) filter before/after memoryless model are used [89]. The linear memory effect is efficiently modeled by Weiner model and Hammerstein model. However, PA memory effects may also be nonlinear, therefore; parallel Weiner-Hammerstein model and augmented Weiner-Hammerstein model were proposed for nonlinear memory effects [90].

To mimic the PAs with nonlinear memory effects, a nonlinear predistorter with dynamic function is required. The basic model for such predistorter is given as Volterra series [91].

$$y(t) = H_1\left[x(t)\right] + H_2\left[x(t)\right] + \cdots + H_n\left[x(t)\right] + \cdots \tag{1.259}$$

$$H_n\left[x(t)\right] = \int_{-\infty}^{\infty} \cdots \int_{-\infty}^{\infty} h_n\left(\tau_1, \ldots, \tau_n\right) x\left(t - \tau_1\right) \ldots x\left(t - \tau_n\right) \mathrm{d}\tau_1 \ldots \mathrm{d}\tau_n \tag{1.260}$$

In order to represent the model, this series requires large number of basis functions. In [92], a special case of this model known as memory polynomial (MP) model has been introduced to reduce the number of basis function while capturing memory effects. The MP model can be represented as [91]:

$$y_{\text{MP}}(n) = \sum_{j=0}^{M} \sum_{i=1}^{N} b_{i,j} x\left(n - j\right) \left|x\left(n - j\right)\right|^{i-1} \tag{1.261}$$

where $x(n)$ is the discrete complex input signal, $y(n)$ is predistorted signal, M is the memory depth, and N is the nonlinearity order.

Equation (1.261) is written in matrix form as,

$$\mathbf{y} = \mathbf{Ua} \tag{1.262}$$

where $\mathbf{a}$ is a vector of coefficients to be extracted, $\mathbf{U}$ is the observation matrix containing the monomials kernels for each i and j in Eq. (1.261) as column vector for an input signal vector containing training data, and $\mathbf{y}$ is a output signal vector of the power amplifier. Then using the Least-Squares (LS) solution, the estimated coefficient vector $\hat{\mathbf{a}}$ is extracted as,

$$\hat{\mathbf{a}} = \left(\mathbf{U}^H\mathbf{U}\right)^{-1}\mathbf{U}^H\mathbf{y} = \mathbf{U}^+\mathbf{y} \tag{1.263}$$

where $\mathbf{U}^H$ denotes hermitian transpose of $\mathbf{U}$ and $\mathbf{U}^+$ is called the pseudo-inverse of the observation matrix.

It should be observed that for $M = 0$, Eq. (1.261) provides a memory less polynomial model, which is comparable to LUT model shown in Eq. (1.258); however, the performance of polynomial and LUT models depends on different factors for two models due to different extraction and implementation process, which has been theoretically analyzed in [93]. Other attempts contain utilizing splines [94], rational functions [95], layered polynomial models [96], and neural networks [2] for behavioral modeling of the PA.

Model Validation

Behavioral models are extracted based on the training input and output baseband signals captured using setup similar to the one shown in Fig. 1.54. However, it is essential to validate the models under diverse conditions to ascertain its usefulness. The PA characteristics are sensitive to bias conditions and measurement conditions such as temperature. In addition, model performance also depends on training signal characteristics such its statistics (Rayleigh, Gaussian, Chi-square, etc.), peak to average power ratio, and bandwidth. Therefore, behavioral model should be validated under different signals and different PA conditions. If the signal properties and PA conditions are similar for test and training data, then the training and test signals are measured at different time instances. The model performance is evaluated in terms of time-domain and frequency-domain waveforms. Figure 1.56 shows the time-domain and frequency-domain validation of the real-valued focused-time-delay neural network (RVFTDNN) models [2]. It can be observed that model output follows the measured output closely.

Model validation is quantified in terms of NMSE for in-band error and adjacent channel error power ratio (ACEPR) for out-of-band modeling performance matrices.

The NMSE in case of modeling is defined as follows [2],

$$\text{NMSE} \triangleq 10\ \log_{10}\left(\frac{\sum_{n=1}^{N}|y_{\text{meas.}}(n) - y_{\text{model}}(n)|^2}{\sum_{n=1}^{N}|y_{\text{meas.}}(n)|^2}\right) \tag{1.264}$$

where $y_{\text{meas.}}(n)$ is measured baseband output signal and $y_{\text{model}}(n)$ is the modeled baseband output signal.

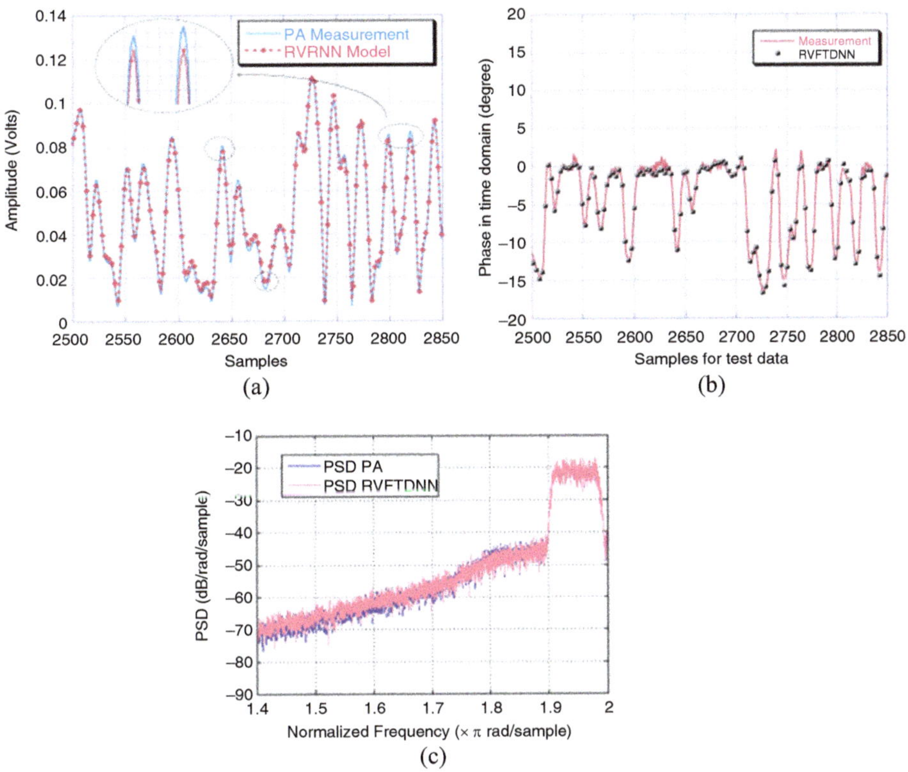

Fig. 1.56 (**a**) Time-domain validation as amplitude of the PA output signal [2], reprinted with permission from IEEE], (**b**) time-domain validation as phase of the PA output signal ([2], reprinted with permission from IEEE), and (**c**) frequency-domain validation as power spectrum density of the PA output signal ([2], reprinted with permission from IEEE)

The ACEPR is defined as,

$$\text{ACEPR} = \frac{\int\limits_{\text{B}} P_{\text{out}}(f)\,\mathrm{d}f}{0.5\left(\int\limits_{\text{LS}} e(f)\,\mathrm{d}f + \int\limits_{\text{US}} e(f)\,\mathrm{d}f\right)} \tag{1.265}$$

$P_{\text{out}}(f)$ is the power spectrum density of the measured output signal in dBm/Hz, $e(f)$ is the power spectrum density of the error signal ($y_{\text{meas.}}(n) - y_{\text{model}}(n)$) in dBm/Hz, B represents the band of the signal, LS and US. The LS and US represent lower-side and upper-side adjacent frequency bands, respectively.

As an example, the NMSE is calculated as −36 dB and ACEPR is calculated as −51 dB for the model performance shown in Fig. 1.56a–c.

1.5.2 Predistortion

Predistortion is conceptually the simplest approach of linearization for an RF PA. The concept of predistortion linearization is shown in Fig. 1.57. To compensate for the nonlinearity of PA, predistortion has inverse AM-AM (gain distortion) and AM-PM (phase distortion) characteristic to that of the nonlinear PA [97–100]. It creates the inverse of the PA transfer function at the input to the amplifier. Predistortion scheme can be divided into two main operation: (a) inverse nonlinearity generator (ING), which provides IMD components to counter the IMD component of the PA. (b) Linear control elements such as attenuation, phase shift, and time-delay shift to apply the generated IMDs of ING with the suitable gain/phase to cancel the IMD components of the PA. Both of these operations can either be implemented in digital (baseband) domain, analog/RF domain, or both (hybrid) domains. The predistortion has few advantages over other linearization techniques, such as (a) the linearizer is implemented using low-power components as the power to linearizer output is provided by PA itself, (b) digital predistortion (DPD) can be adaptive without the need of extra resources, (c) analog predistortion scheme can be adopted

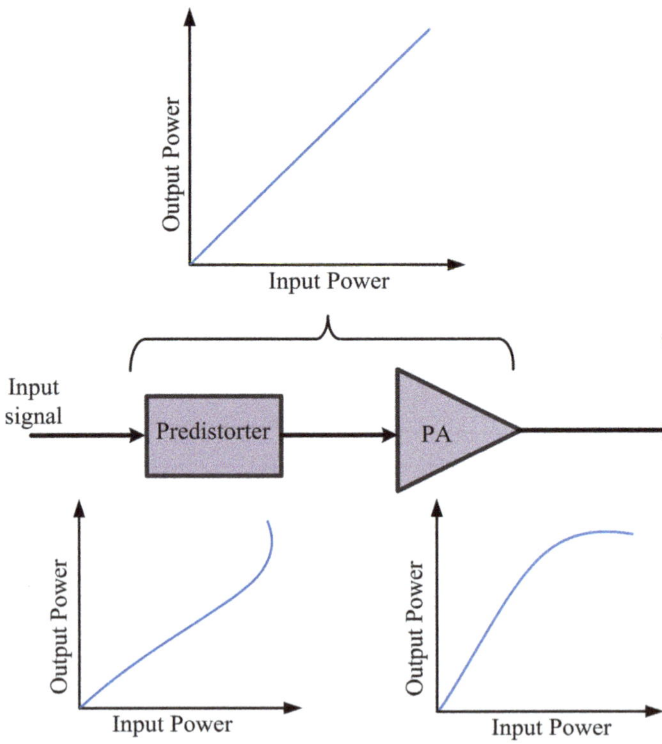

Fig. 1.57 Predistortion for PA linearization

for broadband applications, (d) the predistortion can be implemented in hybrid (digital/analog) domain as per requirement, (e) although predistortion can be applied as an adaptive solution, it is inherently open loop solution leading to a stable solution, and (f) predistortion is less complex and requires less resources as compared to feed-forward technique.

Analog Predistortion

Analog predistorter (APD) utilizes nonlinear components such as diodes or transistors for creating IMD terms, which is applied before PA, to counter the IMD components generated due to nonlinearity of the PA. While DPD is popular due to flexibility and excellent linearization capabilities, APD is a good compromise between bandwidth, complexity, size, and linearity. There are two prominent topologies for the APD, which are shown in Fig. 1.58a, b. Figure 1.58a shows reflective APD, which includes a hybrid coupler, in which the output signal is collected from the isolated port. The through and coupled ports have Schottky diodes attached at their respective ends to introduce nonlinearity. The biasing circuit is used to vary the nonlinear behavior of the diode. This circuit is capable of giving magnitude and phase expansion depending on the input power level [101] which is approximately inverse of the compression characteristics of the PA. The reflective predistorter attempts to create a signal which is inverse to that of the PA characteristics. However, the predistorted signal is created by reflected signals, therefore, though we are able to achieve expansion characteristics, the output of the APD is very small. As a result, power applied to PA is significantly smaller than the intended power. Hence, PA works at a significant back-off and an additional linear PA is required to compensate for the loss of power [102].

Figure 1.58b shows another topology of APD with separate IMD generation path. In this method, instead of relying on diode/transistor characteristics to create expansion characteristics in time domain, the focus is on generating appropriate

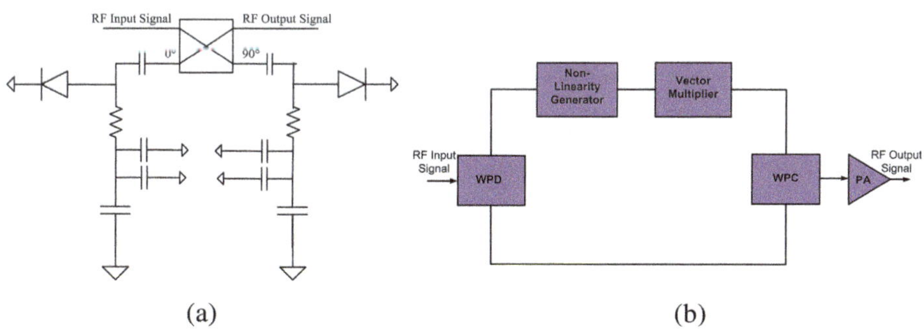

Fig. 1.58 Analog predistorter (APD): (**a**) reflective analog predistorter for PA linearization and (**b**) conventional intermodulation distortion (IMD) separator APD

IMD terms in frequency domain. For application of this APD, the input signal is divided into two paths using a Wilkinson power divider (WPD). One path carries the input signal and other path carries the phase variable IMD components. The second path has IMD generator, which utilizes transistors or diodes along with the provision to cancel the original signal. This path allows only distortion components, that is, IMD, to be propagated. Vector multiplier provides suitable modification in the amplitude and phase of the original fundamental signal to counteract the third-order IMD nonlinear distortion of PA. The output of IM generator is combined with the original fundamental signal using wideband power combiner (WPC) to create predistortion signal for PA [103].

Analog predistortion is preferred over its digital counterpart due to its broadband nature; however, this bandwidth is limited by the operational bandwidth of control elements such as vector multiplier. The details of different APDs and their control schemes will be discussed in Chap. 5.

Digital Predistortion

In the last two decades, DPD technique had been established as an effective technique for linearization of PA. The DPD technique takes advantage of using already available field-programmable gate array (FPGA) or digital processor in the base-station [104, 105]. ING component for the DPD is generated in digital domain using behavioral models discussed in Sect. 1.5.1. The DPD application can be defined in two categories such as (a) indirect learning architecture (b) direct learning architecture.

Indirect learning approach requires performing inverse modeling of the PA as shown in Fig. 1.59a [106], where the complex baseband normalized input signal to PA is provided as training output signal and normalized baseband output signal from PA is provided as training input signal. Once the model is trained to capture this inverse behavior, any incoming data is processed through it first to create inverse

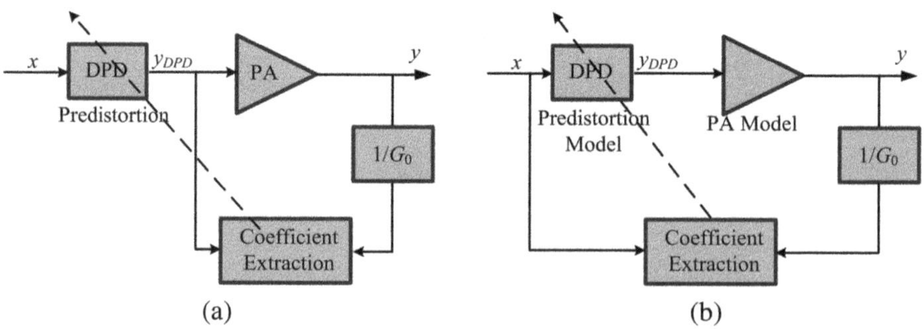

Fig. 1.59 Digital predistorter (DPD), (**a**) indirect learning architecture and (**b**) direct learning architecture

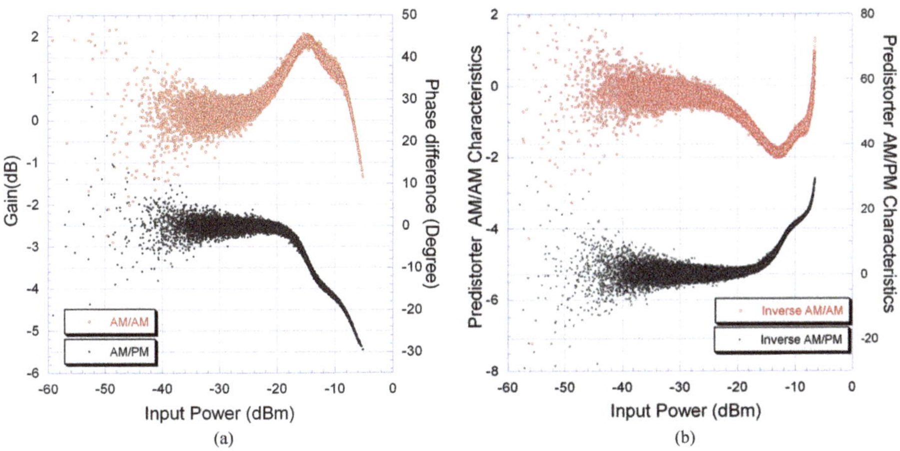

Fig. 1.60 (**a**) PA characteristics ([2], reprinted with permission from IEEE) and (**b**) inverse model characteristics ([2], reprinted with permission from IEEE)

distortion. This predistorted signal is then transmitted via PA. Distortion of the DPD counters the distortion of the PA leading to linear PA output.

Direct learning architecture approach shown in Fig. 1.59b changes the predistorter parameters based on input and output waveforms of the complete linearizer. While designing DPD using direct learning architecture, the PA behavioral model is utilized to observe the output of the linearizer during training phase [107].

Both the techniques utilize the behavioral models discussed in Sect. 1.5.1 as either "PA model" or "inverse model"; therefore, it is essential to create an accurate model. Figure 1.60a, b shows the application of RVFTDNN model as inverse model for DPD application. Figure 1.60a shows the DPA measured characteristics, while Fig. 1.60b shows the modeled inverse model utilizing RVFTDNN model. Figure 1.61a, b shows the DPD performance in terms of power spectrum density of PA output for LDMOS class AB amplifier with 0 dBm peak power. Out-of-band distortion is suppressed and ACPR increases from 35 to 50 dBc. In addition, it is observed that with the inclusion of memory effects, further 5 dBc improvement is observed, providing ACPR of 55 dBc.

The major challenges in the implementation of DPD are wider bandwidth support, multi-band operation, power consumption, and hardware cost of digital processor or FPGA. The MP model is a simple and effective solution as far as complexity is concerned. However, implementation of MP model depends on the property of observation matrix. When PA is excited with wideband signal it exhibits high degree of memory effects. This requires more number of coefficients to implement MP model which will lead to increase in the size of observation (predistorter) matrix of MP model. This in turn has high condition number and dispersion of coefficient leading to numerical unstable solutions in lower-bit digital signal processing (DSP). A better conditioning of the DPD model can help in its

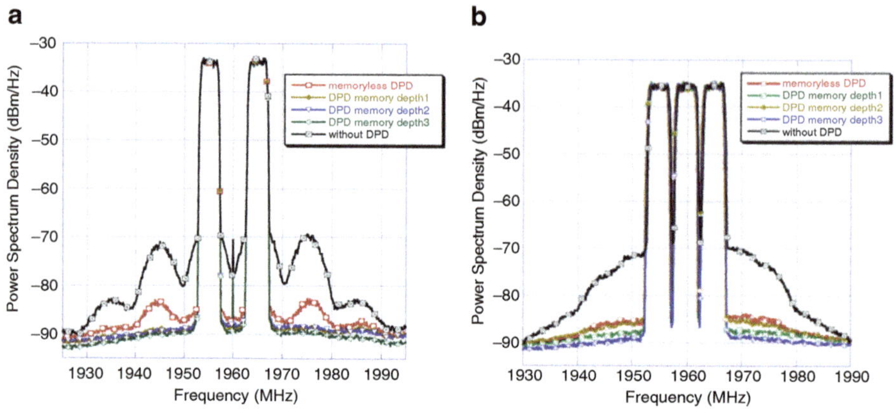

Fig. 1.61 Linearizer output for (**a**) WCDMA 101 signal ([2], reprinted with permission from IEEE) (**b**) WCDMA 111 signal ([2], reprinted with permission from IEEE)

implementation with lower number of bits in FPGA. Later, orthogonal memory polynomial (OMP) model is proposed over MP model which has better numerically stability [97]. The output of this model is obtained by multiplying an upper triangular matrix with MP model's matrix as given by Eq. (1.261). This is given by,

$$y_{\mathrm{OMP}}(n) = \sum_{m=0}^{M} \sum_{k=1}^{K} b_{mk} \psi_k \left(x\left(n-m\right)\right) \tag{1.266}$$

where

$$\psi_k\left(x(n)\right) = \sum_{u=1}^{k} (-1)^{u+k} \frac{(k+u)!}{(u-1)!\,(u+1)!\,(k-u)!} x(n) \times |x(n)|^{u-1} \tag{1.267}$$

where b_{mk} are the coefficients of the OMP model.

The OMP model is numerically stable for known signal probability density functions (PDFs) such as Gaussian, uniform, truncated exponential, and Rayleigh distributions. Although comparatively more stable than the MP model, OMP was implemented in 32-bit and 64-bit floating-point processors [97]. In addition, it is beneficial to reduce overall memory requirement in DSP domain. For implementation of volterra series-based DPD model, an LUT-assisted gain indexing and time-division multiplexing for multiplier sharing have been used to save memory size in FPGA [98].

In [99], a nonlinear autoregressive moving average (NARMA) digital adaptive predistorter (DAPD) is implemented and adapted in real-time FPGA. In [100], the computational complexity of least square (LS)-based model extraction is reduced by using a 1-bit ridge (1-bit RR) method, and errors between the real transmit signal and the training data are reduced by a root mean square (RMS)-based coefficients weighing and averaging for small number of training samples. In [108, 109], principal component analysis (PCA) and independent component analysis were used as pruning technique to reduce the number of coefficients and condition number of observation (predistorter) matrix.

Hybrid Predistortion

While DPD provides accurate characterization and compensation of the PA nonlinearity as compared to the APD, its implementation requires access to the baseband information of input and output signals. Moreover, an additional feedback receiver is required for the digital characterization. Analog predistortion provides a cost-effective, compact linearization solution, which does not require any baseband information and can be applied to any incoming RF signal.

Indeed, digital and analog domains implementation bring their distinct advantage; therefore, there have been efforts to customize the linearization solutions as hybrid solutions as per requirement. Specifically, for broadband signal transmission, the bandwidth of the signal imposes restriction on the speed of digital to analog converter (DAC) and ADC according to Nyquist criterion. As the output of the PA due to nonlinear distortion spreads to the adjacent channels, the DAC and ADC are required to process signal bandwidth that is three to seven times of the original signal bandwidth. This bandwidth requirement leads to very high-speed converters. To circumvent such limitations, two main hybrid techniques have been proposed in the literature, which can be categorized as hybrid RF-DPD and hybrid digitally assisted APD techniques. Figure 1.62a shows the schematic of a hybrid predistorter, where "predistortion function" can be implemented in either digital or analog domain [110]. Figure 1.62b shows the implementation of a hybrid RF-DPD [111]. When the predistortion function is implemented in "digital domain," yet its linear control is in RF domain, it is referred as hybrid RF-DPD. The DPD model can be implemented in LUT form in FPGA, where the entries are indexed with the instantaneous envelope power of the input modulated signals.

The instantaneous envelope power is detected by logarithmic envelope detector (EDET) and later converted into digital signals using ADC as shown in Fig. 1.62b. The digitized value of envelope at any instant will go to the address of RAMs implementing the LUT in FPGA. At the same time, the input-modulated signal is applied to the RF vector multiplier with an appropriate delay. The vector multiplier will adjust the gain and phase of the input signal. The gain and phase of this vector multiplier is controlled by the in-phase (V_I) and quadrature-phase (V_Q) control voltages. These control voltages are generated according to the DPD model stored in LUT. Since the DPD implemented as LUT represents a nonlinear gain and phase indexed according to the envelope power, it is essential to apply the required gain and phase at correct instant of the signal envelope. A delay line is used to ensure this. A delay line shown in Fig. 1.62b provides sufficient group delay to the input

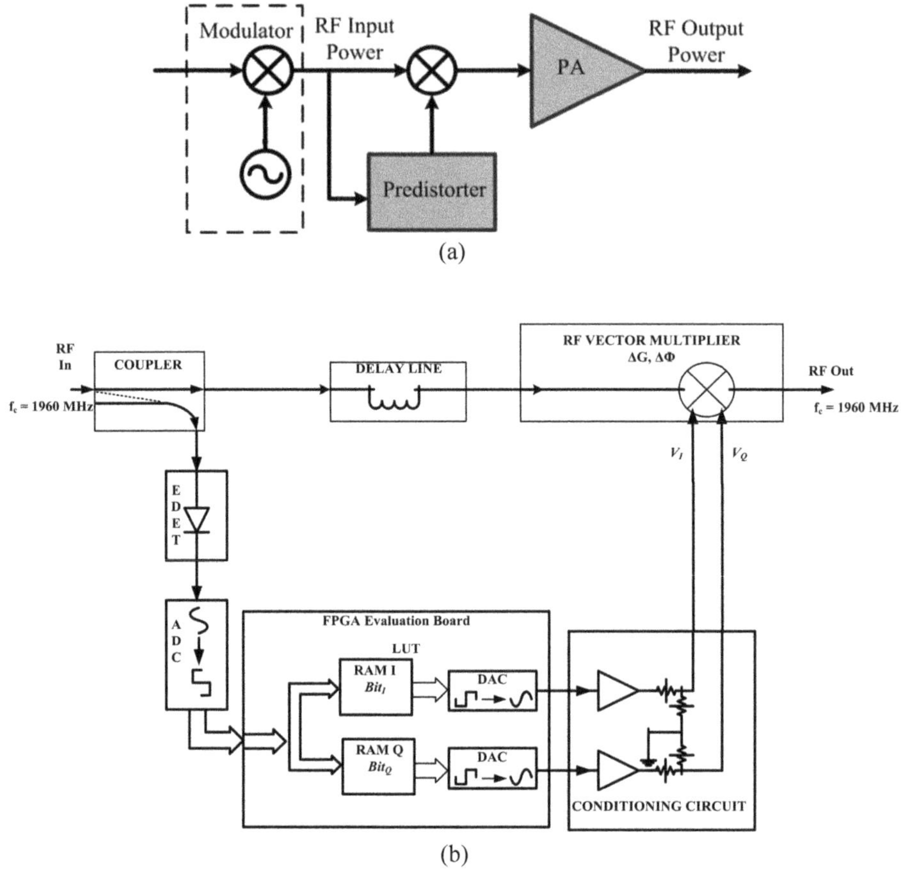

Fig. 1.62 (**a**) Generic schematic of hybrid predistorter, (**b**) hybrid digital predistorter ([111], reprinted with permission from IEEE)

modulated signal such that it reaches the vector multiplier at same instance when the correct control voltages corresponding to its envelope power (as stored in LUT) are applied to the vector multiplier. There are two main steps which are involved in implementation of hybrid RF-DPD system: (1) nonlinear characterization of PA and its predistorter model generation, (2) implementation of predistortion (PD) function as LUT in the FPGA.

The nonlinear characterization and generation of PD model is same as conventional DPD. However, the required complex gain of PD function is first converted in rectangular format, i.e., G_{I} and G_{Q} which are mapped to the respective control voltages V_{I} and V_{Q} of the vector multiplier. This mapping results into the bit values that have to be stored in LUT implemented in an FPGA.

If the gain and phase of the PD is G_{DPD} and Φ_{DPD}, respectively, then one can represent them in rectangular form as:

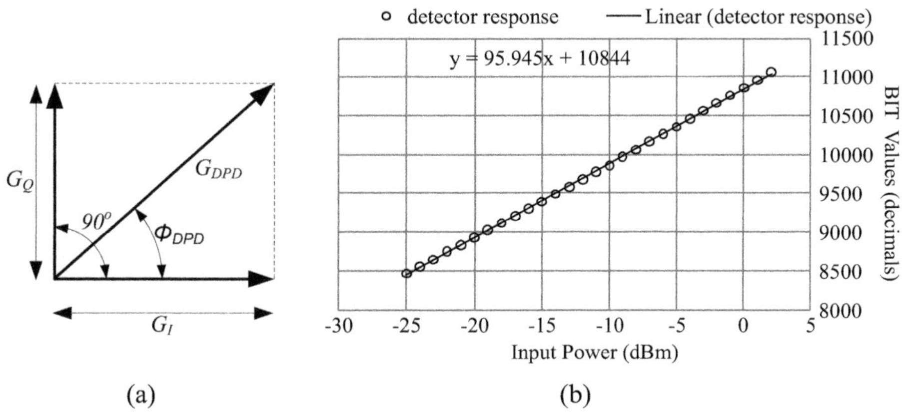

Fig. 1.63 (**a**) Vector decomposition of the predistorter complex gain characteristic for generating LUT entries corresponding to the control voltages of the RF vector multiplier with no I–Q imperfection ([111], reprinted with permission from IEEE). (**b**) Linear response of logarithmic detector based on its characterization ([111], reprinted with permission from IEEE)

$$G_{\mathrm{I}} = G_{\mathrm{DPD}} \cos(\Phi_{\mathrm{DPD}}) \tag{1.268}$$

$$G_{\mathrm{Q}} = G_{\mathrm{DPD}} \sin(\Phi_{\mathrm{DPD}}) \tag{1.269}$$

where G_{I} and G_{Q} are the respective in-phase and quadrature components of the required complex gain characteristics of the PD. This decomposition in rectangular components is shown in Fig. 1.63a.

Assuming a linear relationship between these components and corresponding bits, one can write the following relations.

$$\mathrm{Bit}_{\mathrm{I}} = K_{\mathrm{I}} G_{\mathrm{I}} \tag{1.270}$$

$$\mathrm{Bit}_{\mathrm{Q}} = K_{\mathrm{Q}} G_{\mathrm{Q}} \tag{1.271}$$

where K_{I} and K_{Q} are respective proportionality constants which will be obtained by system calibration. In this calibration, all the entries of both the RAMs corresponding to V_{I} and V_{Q} in Fig. 1.62b are set to same bit value and the gain of RF path is observed. The target is to find bit Bit_{ref} at which the RF path has unity gain. Since equal bits correspond to same V_{I} and V_{Q} at the control pins of the RF vector multiplier, one can assume a phase shift of 45° in RF path with respect to the phase of the system when only $\mathrm{Bit}_{\mathrm{I}}$ is applied and with $\mathrm{Bit}_{\mathrm{Q}}$ set to zero. Therefore, K_{I} and K_{Q} in Eqs. (1.270) and (1.271) can be expressed in terms of $\mathrm{Bit}_{\mathrm{ref}}$ [110, 111],

Table 1.4 Measured ACLR summary of various conditions for one- and three-carrier WCDMA signals [110]

Contents		w/o Hybrid RF-DPD	Hybrid RF-DPD ON
ACPR (dBc)	One-carrier	−37.9/−37.2	−50.3/−52.8
Lower/upper	Three-carrier	−38.5/−39.0	−48.9/−47.3

$$\mathrm{Bit_I} = \frac{\mathrm{Bit_{ref}}G_\mathrm{I}}{\cos\left(45^\circ\right)} = \mathrm{Bit_{ref}}G_\mathrm{I}\sqrt{2} \tag{1.272}$$

$$\mathrm{Bit_Q} = \frac{\mathrm{Bit_{ref}}G_\mathrm{Q}}{\sin\left(45^\circ\right)} = \mathrm{Bit_{ref}}G_\mathrm{Q}\sqrt{2} \tag{1.273}$$

The above equation converts G_I and G_Q in corresponding bit values according to the calibration of hardware system developed. The envelope power corresponding to these G_I and G_Q in an LUT is also converted into bits which correspond to the address of RAMs in FPGA. This conversion is according to the linear relation between the input power (in dBm) to the logarithmic EDET and its digitized output voltage. Figure 1.63b shows this linear characteristic of the logarithmic EDET obtained by varying input power (in dBm) and measuring the corresponding bit values using inbuilt signal tap logic analyzer function in Quartus software of Altera FPGA. A linear fitting of this characteristic can be used to generate a mathematical expression as shown in Fig. 1.63b. This expression will be used to generate address locations for values of G_I and G_Q in RAM according to the envelope power level. The bit values obtained from Eqs. (1.272) and (1.273) are then sorted according the ascending order of address bits which corresponds to the increasing order of the envelope power level.

Table 1.4 shows the ACPR at ±3 MHz offset for WCDMA signal, which is achieved after application of hybrid RF-DPD. It can be observed that for one-carrier signal with 3.84 MHz bandwidth, out-of-band distortion was supressed by 12 dBc, while for three-carrier signal with approximately 15 MHz bandwidth, out-of-band distortion was supressed by 10 dBc.

The LUT-based method demonstrated here compensates for static nonlinearity using indirect learning method based on the open loop characterization of PA. However, the proposed characterization and compensation methodology for hybrid RF-DPD is independent of the dynamic behavior of the PA. If the characteristics of the PA change, the LUT is updated to maintain the performance of predistorter, with no need to compute new mapping coefficients, K_I and K_Q; hence, the proposed compensation remains valid. Here, the DAC does not have to process predistorted signal; therefore, high-speed requirements on DAC are alleviated. However the restrictions on ADC are still similar to DPD as the PA characterization process is the same.

Another hybrid technique is hybrid digitally assisted APD, that takes the advantage of low complexity, yet flexible digital control while applying basic APD design. Instead of using RF delay lines for delay compensation, it is easier to compensate delay digitally. Also gain and phase of a signal can be controlled digitally instead of using vector multiplier [112]. As this technique is more relevant for broadband PA design, Chap. 5 deals with this method in detail.

1.5.3 Feedforward Architecture

The feedforward technique was first proposed by H.S. Black [113, 114], and gained interest in RF application with the work presented by Seidel et.al. [115, 116]. The idea of feedforward technique is based on generating the IMD3 component in an RF section auxiliary to the main PA and adding these components in appropriate phase at the output of the PA. This will cancel the IMD3 components generated by the PA resulting in overall high C/IMD3 ratio at the output of feedforward system [113–118]. Figure 1.64 shows the architecture of feedforward system. This comprises main PA which is primarily responsible for amplifying the signals [117, 118]. The input two-tone signal splits into two parts, one amplified through the main PA path and other is passed through a phase shifter and attenuator in the auxiliary path. This main PA amplifies the two tones while generating unwanted IMD products at its output due to its inherent nonlinearity. A sample of this output from the main PA is fed to the auxiliary section using a sampling coupler (C_1) in the main PA path as shown in Fig. 1.64. This sample signal comprises main tones as well as their IMD products. The main tone fed at the input of auxiliary PA is then subtracted from this sample signal to generate an error signal. This subtraction is performed using a variable phase shifter, a variable attenuator and a combiner in the auxiliary path. This section is often called nulling loop of auxiliary section [118]. This nulling loop controls the phase and amplitude of the main tone fed to the auxiliary section before it combines with the signal sampled from the main PA output. An appropriate nulling loop control will result in suppression of the two fundamental tones in the error signal as shown in Fig. 1.64. This results in an error signal comprising IMD

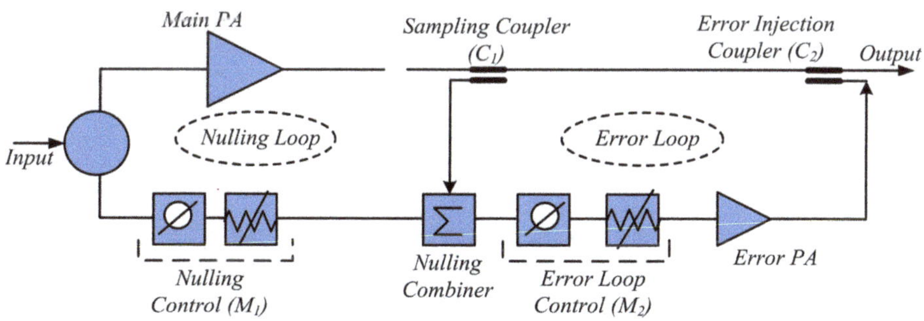

Fig. 1.64 Architecture of feedforward system

products only. This error signal is further processed through the error loop of the auxiliary section [118]. This error loop comprises a variable attenuator, a variable phase shifter, and an error amplifier. The error loop control as shown in Fig. 1.64 will control the amplitude and phase of the error signal whereas an error amplifier later amplifies this error signal to an appropriate level such that it will cancel the IMD products at the output of the PA when injected in the main PA path. This is performed by an error injection coupler (C_2).

The output power of main PA is $P_{\text{out,PA}}$ and the third-order IMD3 level relative to carriers (in case of two-tone test) is C/IMD3 in dBc. The power level of third-order IMD product (per tone) is,

$$P_{\text{IMD3}} = P_{\text{out,PA}} 10^{\frac{\text{C/IMD3}}{10}} = P_{\text{out,PA}} S_{\text{IM}} \tag{1.274}$$

where P_{IMD3} and $P_{\text{out,PA}}$ both can be expressed in Watts. The S_{IM} is the ratio representing C/IMD3 in linear scale. The effective power of this IMD3 product after encountering the loss L_{S2} in linear scale presented by coupler C_2 is given by,

$$P_{\text{out,IMD3,PA}} = P_{\text{IMD3}} L_{\text{S2}} \tag{1.275}$$

The error amplifier in Fig. 1.64 is amplifying error signal comprising only IMD3 components. If the output power of these IMD3 components at the output of error amplifier is $P_{\text{out,EA}}$ and the coupler C_2 has coupling coefficient C_{C2}; then the IMD3 power contribution of error amplifier at the output of feedforward system is,

$$P_{\text{out,IMD3,EA}} = P_{\text{out,EA}} C_{\text{C2}} \tag{1.276}$$

In order to cancel the IMD3 components at the output of feedforward system, the powers given in Eqs. (1.275) and (1.276) should be equal, provided that the phase of IMD3 products in error loop are properly adjusted. This results in the following equation,

$$P_{\text{out,EA}} = \frac{P_{\text{out,PA}} S_{\text{IM}} L_{\text{S2}}}{C_{\text{C2}}} \tag{1.277}$$

The overall efficiency of feedforward system is given by,

$$\eta_{\text{F}} = \frac{P_{\text{out,FF}}}{P_{\text{DC,PA}} + P_{\text{DC,EA}}} = \frac{P_{\text{out,PA}} L_{\text{S2}}}{\frac{P_{\text{out,PA}}}{\eta_{\text{PA}}} + \frac{P_{\text{out,EA}}}{\eta_{\text{EA}}}} \tag{1.278}$$

where η_{PA} and η_{EA} are the efficiencies of main PA and error amplifier, respectively. The loss and the coupling coefficient of coupler C_2 is given as $L_{\text{S2}} = 1 - C_{\text{C2}}$. Using this relation and Eq. (1.277) in Eq. (1.278), one can express overall efficiency of feedforward system as [117],

$$\eta_{\text{F}} = \frac{\eta_{\text{PA}}\eta_{\text{EA}}C_{\text{C2}}\left(1 - C_{\text{C2}}\right)}{\eta_{\text{EA}}C_{\text{C2}} + \eta_{\text{PA}}S_{\text{IM}}\left(1 - C_{\text{C2}}\right)} \tag{1.279}$$

Once can see that η_{F} given in Eq. (1.279) depends on coupling coefficient C_{C2} and one can obtain its optimum value for maximizing the overall efficiency. This can be obtained by equating first-order differentiation of η_{F} with respect to coupling coefficient C_{C2} as zero. This results in a quadratic equation in terms of coupling coefficient C_{C2} which has roots given by [118],

$$C_{\text{C2}} = \frac{K \pm \sqrt{K}}{K - 1} \tag{1.280}$$

where ratio K is given by $\eta_{\text{PA}}S_{\text{IM}}/\eta_{\text{EA}}$. The positive sign in Eq. (1.280) results in coupling coefficient greater than 1 and hence rejected. The negative sign will provide the optimum value of coupling coefficient resulting in optimum efficiency given as [117, 118],

$$\eta_{\text{F,opt}} = \frac{\eta_{\text{PA}}}{\left(1 - \sqrt{K}\right)^2} \tag{1.281}$$

In case of ideal feedforward system, the IMD3 will be perfectly canceled at the output. However, in practice, the IMD3 components will never cancel completely. This cancellation is limited by imperfect cancellation due to imprecision of nulling and error loop control. In addition, the nonlinearity of error amplifier may introduce extra modulation products which may limit the C/IMD3 to low value at the output of feedforward system. These effects can be studied independently and modeled to predict the linearity correction in practical feedforward system [118].

1.6 Delta-Sigma Modulation-Based Transmitters

The modern wireless transmitters often encounter modulated signals having high PAPR. These signals drive PA in an inefficient region at average output power level backed-off from its saturation. Load modulation as discussed earlier is one of the techniques to efficiently operate PA with such envelope varying high PAPR signals. Alternatively, the signal can be shaped such that all the information is encoded in phase while keeping the envelope constant. In such case, highly efficient switch-mode PAs can be used without operating them in power back-off. However, the signal shaping must be performed such that envelope of the signal can be reconstructed at the output of PA without losing any information. DSM is one such technique where the envelope varying signal is encoded into bi-level constant envelope signal using a quantizer. This quantization however introduces significant quantization noise to the signal. An oversampling distributes this quantization noise over the wide frequency band beyond the signal bandwidth reducing the in-band noise [119, 120]. An oversampling ratio (OSR) of the baseband modulated signal is defined as the ratio of DSM sampling frequency and bandwidth of baseband signal. High OSR results in distribution of quantization noise to relatively large frequency and thereby further decreasing the in-band quantization noise. However, this requires high clock speed for oversampling. The feedback loop later shapes the quantization noise generated at the output of DSM such that it falls outside the signal band and therefore can be easily filtered using an RF filter. In general, this method is one type of pulse density modulation where the information is encoded in the density of pulses [10, 14, 121]. In spite of the noise shaping in DSM, some part of quantization noise still exists within band and is amplified through PA. Therefore, two figure of merits, that is, signal to noise and distortion ratio (SNDR) and coding efficiency (CE) are generally used for DSM-based transmitter. The SNDR is used to evaluate the performance of DSM transmitter for the signal quality and in-band noise. This SNDR is given by [121],

$$\mathrm{SNDR} = 10\ \log\left(\frac{\mathrm{Signal\ Power}}{\mathrm{In-Band\ noise\ and\ distortion\ power}}\right) \tag{1.282}$$

where SNDR is expressed in dB and depends on oversampling ratio as well as order of DSM. However, the CE measures the effect of the quantization noise on the transmitter efficiency. The CE is given by [121],

$$\eta_{\mathrm{CE}} = \frac{\mathrm{Signal\ Power}}{\mathrm{Signal\ power} + \mathrm{quantization\ noise\ power}} \tag{1.283}$$

Figure 1.65 shows the DSM-based transmitter architecture, where time and frequency domain signals are shown. The DSM transmitter architecture consists of DSM modulator, an up converter, PA, and a band-pass filter. DSM converts envelope varying input baseband signal into constant envelope pulse shaped signal and fed this signal to PA to operate at saturation to get the maximum efficiency. The amplified output is passed through band-pass filter to filter the unwanted quantization noise before transmission [121].

In general, there are three topologies of DSM, that is, low-pass, band-pass, and high-pass that can be used in wireless transmitter. Among these, the low-pass DSM is quite popular [10, 11, 121–123]. The basic first-order low-pass DSM is shown in Fig. 1.66.

The topology consists of an integrator and quantization block in forward path. The output of quantizer is feedback to input to the integrator. If the quantizer is modeled with an adder that introduces a quantization error $e(n)$ to the signal, one can write the transfer function of first-order low-pass DSM as,

$$Y(z) = z^{-1}X(z) + \left(1 - z^{-1}\right)E(z) \tag{1.284}$$

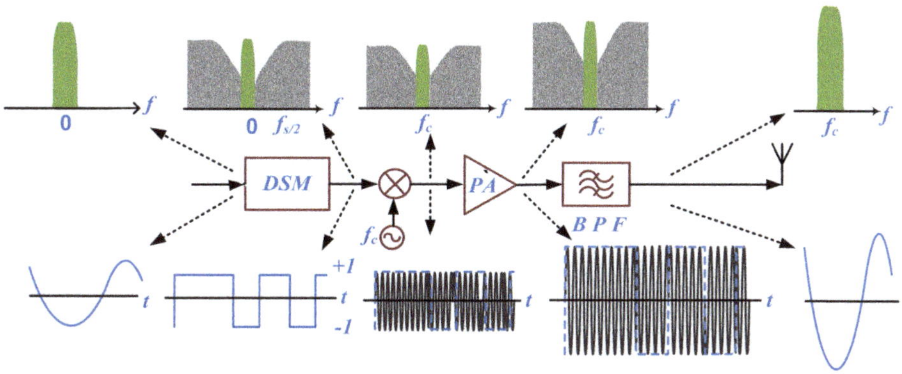

Fig. 1.65 A simplified diagram of delta-sigma modulation (DSM)-based transmitter in time and frequency domain

Fig. 1.66 Topology of first-order low-pass DSM

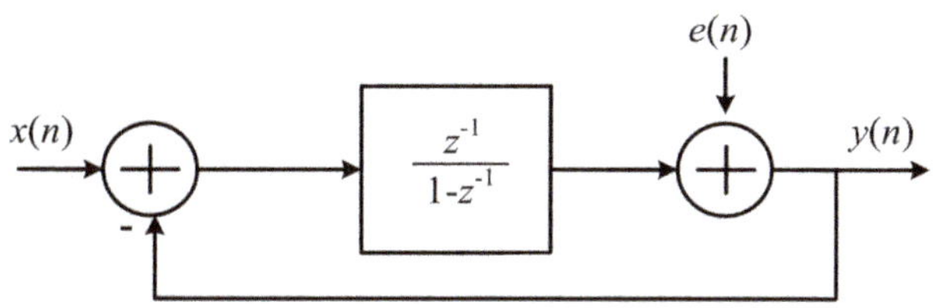

where $X(z)$ and $Y(z)$ are the z-transforms of input and output signals, respectively. $E(z)$ is the z-transform of quantization error $e(n)$. One can see from Eq. (1.284) that the signal is delayed by one clock, whereas the quantization noise is multiplied by $(1 - z^{-1})$ representing a high-pass filter. The former is called signal transfer function (STF), whereas later is termed as noise transfer function (NTF). The order of DSM is defined from the order of NTF. The signal quality improves with the order of DSM with a risk of instability [119]. However, second- and third-order modulator can be used with an optimal balance between signal quality and stability [119, 121]. The DSM transmitter topology using low-pass DSM followed by the up-converter is analogous to the direct conversion transmitter [124, 125]. Therefore, this topology offers several advantages such as image noise and intermediate frequency (IF) suppression etc. inherently present in direct conversion architecture [124]. At the same time, this topology has issues similar to direct conversion transmitters such as quadrature imbalance, local oscillator (LO) leakage, DC offset etc. which deteriorates the linearity of the transmitter [124, 126].

The band-pass DSM and high-pass DSM are the alternate topologies where the signal is located at IF. Usually this IF can be at quarter and half of the sampling frequency for band-pass DSM and high-pass DSM, respectively. The output of these DSMs may directly drive PA. However, in such case the required clock speed will be four-times the carrier frequency in case of band-pass DSM and two-times the carrier frequency in the case of high-pass DSM which is unfeasible. Therefore, an alternative solution where an IQ modulator or a digital multiplexer is used

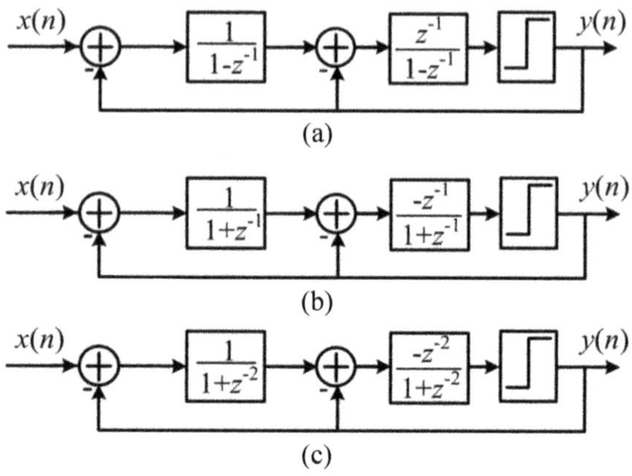

Fig. 1.67 Second-order DSM topology: (**a**) low-pass, (**b**) high-pass, and (**c**) band-pass

as up-converter after DSM is proposed as low-IF architecture [127]. This low-IF architecture is free from distortions that are present in direct conversion architecture with the expense of being comparatively more complex and costly [124].

The transfer functions of band-pass DSM and high-pass DSM can be obtained from low-pass DSM by applying following mapping functions [119, 121].

$$z^{-1} \xrightarrow{\text{Low-pass to Band-pass}} -z^{-2} \tag{1.285}$$

$$z^{-1} \xrightarrow{\text{Low-pass to high-pass}} -z^{-1} \tag{1.286}$$

Figure 1.67a–c shows the second-order low-pass DSM and corresponding band-pass and high-pass DSM architectures, respectively, which are obtained using Eqs. (1.285) and (1.286).

As discussed earlier, the SNDR of DSM improves with the order of DSM. The following equations show the NTFs of each topology for *K*th order [121].

$$\text{NTF}_{\text{low-pass}} = \left(1 - z^{-1}\right)^K \tag{1.287}$$

$$\text{NTF}_{\text{high-pass}} = \left(1 + z^{-1}\right)^K \tag{1.288}$$

$$\text{NTF}_{\text{band-pass}} = \left(1 + z^{-2}\right)^K \tag{1.289}$$

In order to simplify the analysis of the quantization noise, $e(n)$ is assumed as a sample sequence of a stationary random process and it is uncorrelated with the input sequence $x(n)$ in Fig. 1.66. For a zero mean $e(n)$, its variance and power spectral density is given as [119]:

$$\sigma_e^2 = \frac{\Delta^2}{12} \tag{1.290}$$

$$\mathrm{PSD_e}(f) = \frac{\Delta^2}{6 f_\mathrm{S}} \tag{1.291}$$

where Δ is the step size of the quantizer and σ_e^2 is the mean square value or noise power. The in-band noise power at the output of DSM modulator is calculated using the NTF and the $\mathrm{PSD_e}(f)$ over the frequency range $[-f_\mathrm{B}, f_\mathrm{B}]$ which is given as,

$$P_\mathrm{e} = \int_{-f_\mathrm{B}}^{f_\mathrm{B}} \mathrm{PSD_e}(f)|\mathrm{NTF}|^2 \mathrm{d}f \tag{1.292}$$

Considering the Kth order low-pass NTF, the frequency response can be obtained by replacing z with $\mathrm{e}^{j2\pi f/f_\mathrm{S}}$ in Eq. (1.292). This results in $|\mathrm{NTF_{low\text{-}pass}}|^2 = \left|2\sin\left(\pi\frac{f}{f_\mathrm{S}}\right)\right|^{2K}$. If the OSR of modulator is high enough (e.g., more than 10), the ratio f/f_S is small enough and the trigonometric function can be approximated as first-order Taylor series as $|\mathrm{NTF_{low\text{-}pass}}|^2 = \left[2\left(\pi\frac{f}{f_\mathrm{S}}\right)\right]^{2K}$. Using this in Eq. (1.292) along with $\mathrm{PSD_e}(f)$ given in Eq. (1.291), one can obtain,

$$P_\mathrm{e} = \frac{\pi^{2K}}{2K+1}\frac{1}{(\mathrm{OSR})^{2K+1}} e_\mathrm{rms}^2 \tag{1.293}$$

If the signal $x(n)$ is also considered as a zero mean random process and its mean square value or power is given as σ_x^2, the SNDR for Kth order low-pass DSM in dB is given as,

$$P_\mathrm{e} = 10\ \log\left(\sigma_x^2\right) - 10\ \log\left(\sigma_e^2\right) - 10\ \log\left(\frac{\pi^{2K}}{2K+1}\right) + (2K+1)\,10\ \log(\mathrm{OSR}) \tag{1.294}$$

One can see from Eq. (1.294), for first-order low-pass DSM, doubling the oversampling ratio, the SNDR improves by 9 dB. As discussed above, the low-pass DSM is used in direct conversion transmitter. Figure 1.68 shows the direct conversion architecture of low-pass DSM with spectrum at the output of various stages. In this structure modulated signals at baseband are directly converted to RF carrier frequency. The quadrature modulator used for this up-conversion will receive bi-level signals from low-pass DSM of level +1 and −1. These signals are directly up-converted to RF carrier frequency and therefore its output is constant envelope signal with four different phases 45°, 135°, 225°, 315°, similar to quadrature phase shift keying (QPSK) modulator [128].

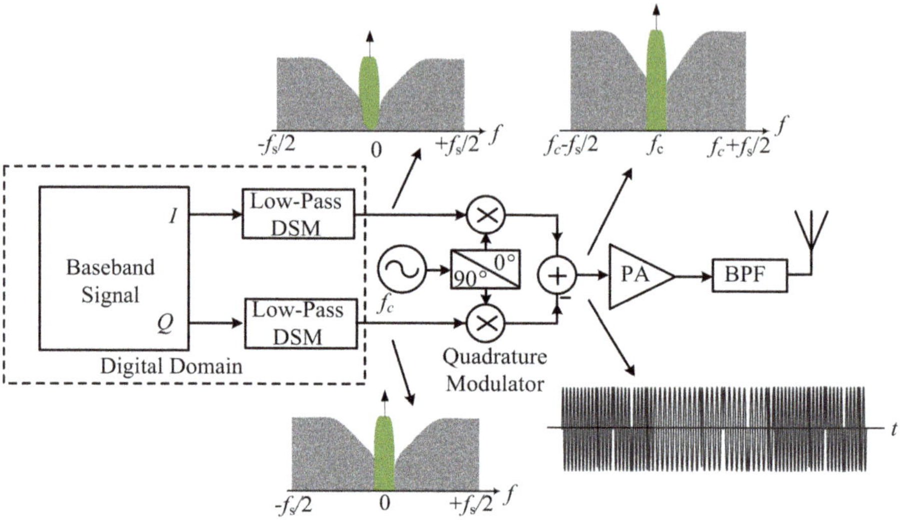

Fig. 1.68 Direct conversion transmitter architecture with low-pass DSM

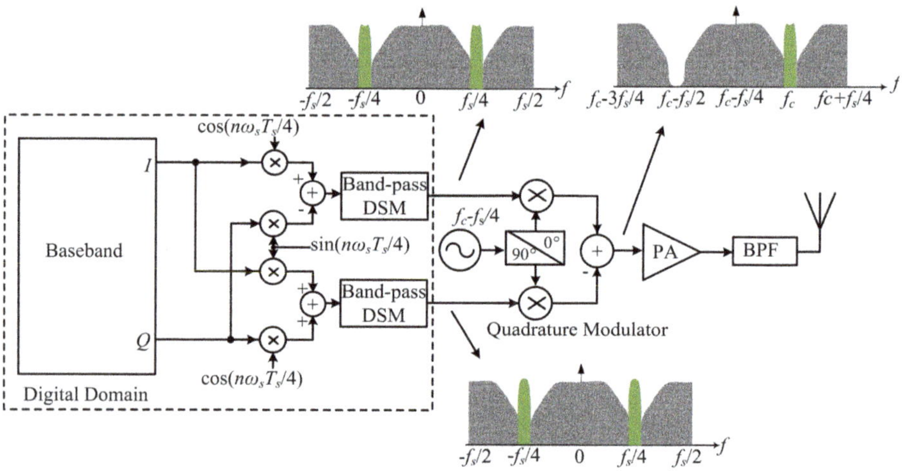

Fig. 1.69 Low IF transmitter architecture with band-pass DSM

Figure 1.69 shows the low-IF transmitter architecture using band-pass DSM. One can see from this figure that the architecture comprises two digital quadrature up-converters and two band-pass DSMs, and one analog quadrature up-converter. The output signal from this analog quadrature up-converter will be further amplified with PA. A band-pass filter will then suppress out of band noise at the output of PA before transmission. The digital quadrature up-converter transfers baseband I and Q signals to $f_S/4$, where f_S is sampling frequency. These digitally up-converted signals are then used by band-pass DSM to generate constant envelope bi-level signals.

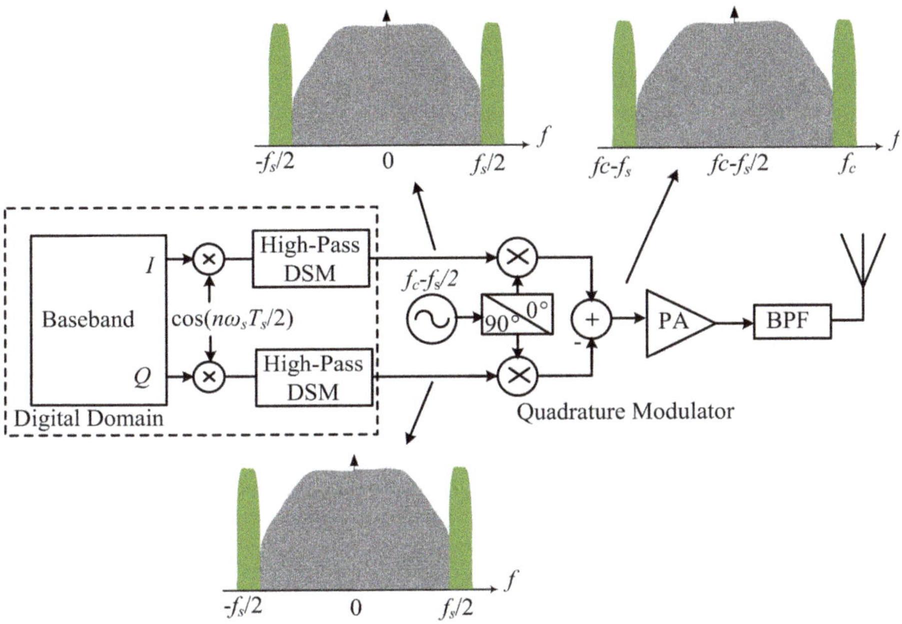

Fig. 1.70 Low IF transmitter architecture with high-pass DSM

Since I and Q signals are up-converted in digital domain and image at $-f_S/4$ also appears as shown in Fig. 1.69. Any image rejection filters between DSMs and PA as used in conventional low IF architecture may affect the amplitude which must be kept constant [129]. Therefore, such filter is avoided and instead the image is suppressed using analog quadrature modulator as shown in Fig. 1.69.

The high-pass DSM transmitter architecture is also low-IF transmitter similar to the band-pass DSM transmitter. However, in such case the digital up-conversions of I and Q data are carried out by simple mixers unlike band-pass DSM transmitter where quadrature digital up-conversion is used. This is clear from Fig. 1.70, where the image at the output of high-pass DSM appears at $-f_S/2$.

Later the signal is up-converted using analog quadrature modulator before PA. The LO frequency of this analog quadrature modulator is $f_C-f_S/2$ which translates the modulated signals to carrier frequency f_C. However, an image is still present at the output of analog quadrature modulator at f_C-f_S which is far from the carrier frequency f_C and can be suppressed by final band-pass filter at the output of PA as shown in Fig. 1.70.

References

1. Martins, P., Cabral, P. M., Carvalho, N. B., & Pedro, J. C. (2006). A metric for the quantification of memory effects in power amplifiers. *IEEE Transactions on Microwave Theory and Techniques, 54*(12), 4432–4439.
2. Rawat, M., Rawat, K., & Ghannouchi, F. M. (2010). Adaptive digital predistortion of wireless power amplifiers/transmitters using dynamic real-valued focused time-delay line neural networks. *IEEE Transactions on Microwave Theory and Techniques, 58*(1), 95–104.
3. Landin, P. N., Isaksson, M., & Handel, P. (2008). Comparison of evaluation criteria for power amplifier behavioral modeling. In *IEEE MTT-S International Microwave Symposium Digest* (pp. 1441–1444). Washington, DC: IEEE.
4. Baxandall, P. J. (1959). Transistor sine-wave LC oscillators. Some general considerations and new developments. *Proceedings of the IEE-Part B: Electronic and Communication Engineering, 106*(16S), 748–758.
5. Grebennikov, A., Sokal, N. O., & Franco, M. J. (2012). *Switch mode RF and microwave power amplifiers*. Cambridge: Academic Press.
6. Sokal, N. O., & Sokal, A. D. (1975). Class E a new class of high-efficiency tuned single-ended switching power amplifiers. *IEEE Journal of Solid-State Circuits, 10*(3), 168–176.
7. Raab, F. (1977). Idealized operation of the Class E tuned power amplifier. *IEEE Transactions on Circuits and Systems, 24*(12), 725–735.
8. Grebennikov, A. (2005). *RF and microwave power amplifier design*. New York: McGraw-Hill.
9. Kakkad, P., Aggrawal, E., & Rawat, K. (2017). De-embedded model based Class-E power amplifier using waveform engineering. In *2017 8th International Conference on Computing, Communication and Networking Technologies (ICCCNT)* (pp. 1–4). Washington, DC: IEEE.
10. Gustavsson, U., Eriksson, T., Nemati, H. M., Singerl, P., & Fager, C. (2011). An RF carrier bursting system using partial quantization noise cancellation. *IEEE Transactions on Circuits and Systems I: Regular Papers, 59*(3), 515–528.
11. Helaoui, M., Hatami, S., Negra, R., & Ghannouchi, F. M. (2008). A novel architecture of delta-sigma modulator enabling all-digital multiband multistandard RF transmitters design. *IEEE Transactions on Circuits and Systems II: Express Briefs, 55*(11), 1129–1133.
12. Ebrahimi, M. M., Helaoui, M., & Ghannouchi, F. M. (2011). Time-interleaved delta-sigma modulator for wideband digital GHz transmitters design and SDR applications. *Progress in Electromagnetics Research B, 34*, 263–281.
13. Jayaraman, A., Chen, P. F., Hanington, G., Larson, L., & Asbeck, P. (1998). Linear high-efficiency microwave power amplifier using bandpass delta-sigma modulators. *IEEE Microwave and Guided Wave Letters, 8*(3), 121–123.
14. Johnson, T., & Stapleton, S. P. (2006). RF class-D amplification with bandpass sigma–delta modulator drive signals. *IEEE Transactions on Circuits and Systems I: Regular Papers, 53*(12), 2507–2520.
15. Kumar, N., Poonia, J. D., & Rawat, K. (2017). Class S power amplifier based on CSCD with delta-sigma modulation. In *2017 IEEE Applied Electromagnetics Conference (AEMC)* (pp. 1–2). Washington, DC: IEEE.
16. Tyler, V. J. (1958). A new high-efficiency high power amplifier. *Marconi Review, 21*(130), 96–109.
17. Bowers, D. F. (1982). HEAD—A high efficiency amplitude-modulation system for broadcasting transmitters. *Communication and Broadcasting, 7*(2), 15–23.
18. Raab, H. (2001). Class E, class C, and class F power amplifiers based upon a finite number of harmonics. *IEEE Transactions on Microwave Theory and Techniques, 49*(8), 1462–1468.
19. Raab, F. H. (1997). Class F power amplifiers with maximally flat waveforms. *IEEE Transactions on Microwave Theory and Techniques, 45*, 2007–2012.
20. Raab, F. H. (2001). Maximum efficiency and output of class F power amplifiers. *IEEE Transactions on Microwave Theory and Techniques, 47*, 1162–1166.

21. Grebennikov, A. V. (2000). Circuit design technique for high efficiency class F amplifiers. In *2000 IEEE MTT-S International Microwave Symposium Digest (Cat. No. 00CH37017)* (Vol. 2, pp. 771–774). Washington, DC: IEEE.
22. Carrubba, V. (2012). *Novel highly efficient broadband continuous power amplifier modes* (PhD thesis). Cardiff University
23. Cripps, S. C., Tasker, P. J., Clarke, A. L., Lees, J., & Benedikt, J. (2009). On the continuity of high efficiency modes in linear RF power amplifiers. *IEEE Microwave and Wireless Components Letters, 19*(10), 665–667.
24. Steve, C., & Cripps, R. F. (2006). *Power amplifiers for wireless communications* (2nd ed.). Norwood, MA, ISBN: 0-89006-989-1: Artech House Publishers Inc..
25. Carrubba, V., et al. (2011). On the extension of the continuous class-F mode power amplifier. *IEEE Transactions on Microwave Theory and Techniques, 59*(5), 1294–1303.
26. Carrubba, V., Lees, J., Benedikt, J., Tasker, P. J., & Cripps, S. C. (2011). A novel highly efficient broadband continuous class-F RFPA delivering 74% average efficiency for an octave bandwidth. In *2011 IEEE MTT-S International Microwave Symposium* (pp. 1–4). Washington, DC: IEEE.
27. Carrubba, V., Clarke, A. L., Akmal, M., Lees, J., Benedikt, J., Tasker, P. J., & Cripps, S. C. (2010). The continuous class-F mode power amplifier. In *The 40th European Microwave Conference* (pp. 1675–1677). Washington, DC: IEEE.
28. Aggrawal, E., Rawat, K., & Roblin, P. (2017). Investigating continuous class-F power amplifier using nonlinear embedding model. *IEEE Microwave and Wireless Components Letters, 27*(6), 593–595.
29. Chen, J., He, S., You, F., Tong, R., & Peng, R. (2014). Design of broadband high efficiency power amplifiers based on series of continuous modes. *IEEE Microwave and Wireless Components Letters, 24*(9), 631–633.
30. Tuffy, N., Guan, L., Zhu, A., & Brazil, T. J. (2012). A simplified broadband design methodology for linearized high-efficiency continuous class-F power amplifiers. *IEEE Transactions on Microwave Theory and Techniques, 60*(6), 1952–1963.
31. Aggrawal, E., Saxena, S., & Rawat, K. (2016). Broadband power amplifier design by exploring design space of continuous class-F mode. In *2016 Asia-Pacific Microwave Conference (APMC) 5th–9th December, 2016, New Delhi, India* (pp. 1–4). Washington, DC: IEEE.
32. Saxena, S., Rawat, K., & Roblin, P. (2017). Continuous Class-B/J power amplifier using nonlinear embedding technique. *IEEE Transactions on Circuits and Systems II: Express Briefs, 64*(7), 837–841.
33. Özen, M., Jos, R., & Fager, C. (2012). Continuous class-E power amplifier modes. *IEEE Transactions on Circuits and Systems II: Express Briefs, 59*(11), 731–735.
34. Fan, C.-W., & Cheng, K.-K. (2002). Theoretical and experimental study of amplifier linearization based on harmonic and baseband signal injection technique. *IEEE Transactions on Microwave Theory and Techniques, 50*(7), 1801–1806.
35. Zivkovic, Z., & Markovic, A. (1986). Increasing the efficiency of the high-power triode HF amplifier -why not with the second harmonic? *IEEE Transactions on Broadcasting, 32*(1), 5–10.
36. Zivkovic-Dzunja, Z., & Markovic, A. (1989). Plate and grid modulated HF high-power tuned amplifier with increased efficiency. *IEEE Transactions on Broadcasting, 35*, 97–107.
37. Willems, D. A., Griffin, E. L., Bahl, I. J., & Pollman, M. D. (1992). *High efficiency harmonic injection power amplifier*. U.S. Patent No. 5,172,072.
38. AlMuhaisen, A., Wright, P., Lees, J., Tasker, P. J., Cripps, S. C., & Benedikt, J. (2010). Novel wide band high-efficiency active harmonic injection power amplifier concept. In *2010 IEEE MTT-S International Microwave Symposium* (pp. 664–667). Washington, DC: IEEE.
39. Dani, A., Roberg, M., & Popovic, Z. (2012). Efficiency and linearity of power amplifiers with external harmonic injection. In *2012 IEEE/MTT-S International Microwave Symposium Digest* (pp. 1–3). Washington, DC: IEEE.

40. AlMuhaisen, A., Lees, J., Cripps, S. C., Tasker, P. J., & Benedikt, J. (2011). Wide band high-efficiency power amplifier design. In *2011 6th European Microwave Integrated Circuit Conference* (pp. 184–187). Washington, DC: IEEE.
41. Dani, A., Michael, R., & Popovic, Z. (2012). PA efficiency and linearity enhancement using external harmonic injection. *IEEE Transactions on Microwave Theory and Techniques, 60*(12), 4097–4106.
42. Seo, M., Lee, H., Gu, J., Kim, H., Ham, J., Choi, W., & Yang, Y. (2014). High-efficiency power amplifier using an active second-harmonic injection technique under optimized third-harmonic termination. *IEEE Transactions on Circuits and Systems II: Express Briefs, 61*(8), 549–553.
43. Alizadeh, A., Yaghoobi, M., & Medi, A. (2017). Class-J2 power amplifiers. *IEEE Transactions on Circuits and Systems I: Regular Papers, 64*(8), 1989–2002.
44. Asha Latha, Y. M., Rawat, K., Helaoui, M., & Ghannouchi, F. M. (2019). Broadband continuous mode power amplifier with on-board harmonic injection. *IET Microwaves, Antenna & Propagation, 13*, 1402–1407. https://doi.org/10.1049/iet-map.2018.5885.
45. Doherty, W. H. (1936). A new high efficiency power amplifier for modulated waves. *Proceedings of the Institute of Radio Engineers, 24*(9), 1163–1182.
46. Raab, F. H. (1987). Efficiency of Doherty RF power-amplifier systems. *IEEE Transactions on Broadcasting, 33*(3), 77–83.
47. Kim, B., Kim, J., Kim, I., & Cha, J. (2006). The Doherty power amplifier. *IEEE Microwave Magazine, 7*(5), 42–50.
48. Jang, H., Roblin, P., Quindroit, C., Lin, Y., & Pond, R. D. (2014). Asymmetric Doherty power amplifier designed using model-based nonlinear embedding. *IEEE Transactions on Microwave Theory and Techniques, 62*(12), 3436–3451.
49. Raab, F. H., et al. (2002). Power amplifiers and transmitters for RF and microwave. *IEEE Transactions on Microwave Theory and Techniques, 50*(3), 814–826.
50. Wu, D. Y. T., & Boumaiza, S. (2012). A modified Doherty configuration for broadband amplification using symmetrical devices. *IEEE Transactions on Microwave Theory and Techniques, 60*(10), 3201–3213.
51. Sweeney, R. (2008). Practical magic. *IEEE Microwave Magazine, 9*(2), 73–82.
52. Bousnina, S. (2009). Analysis and design of high-efficiency variable conduction angle doherty amplifier. *IET Microwaves, Antennas & Propagation, 3*(3), 416–425.
53. Rawat, K., Hashmi, M. S., & Ghannouchi, F. M. (2012). Double the band and optimize. *IEEE Microwave Magazine, 13*(2), 69–82.
54. Iwamoto, M., Williams, A., Chen, P.-F., Metzger, A. G., Larson, L. E., & Asbeck, P. M. (2001). An extended Doherty amplifier with high efficiency over a wide power range. *IEEE Transactions on Microwave Theory and Techniques, 49*(12), 2472–2479.
55. Kim, B., Kim, I., & Moon, J. (2010). Advanced Doherty architecture. *IEEE Microwave Magazine, 11*(5), 72–86.
56. Ghosh, S., & Rawat, K. (2018). A wideband two-stage Doherty power amplifier at high back-off by exploring feasible design space. *International Journal of RF and Microwave Computer-Aided Engineering, 28*(9), E21528.
57. Srirattana, N., Raghavan, A., Heo, D., Allen, P. E., & Laskar, J. (2009). Analysis and design of a high-efficiency multistage Doherty power amplifier for wireless communications. *IEEE Transactions on Microwave Theory and Techniques, 53*(3), 852–859.
58. Neo, W. C. E., Qureshi, J., Pelk, M. J., Gajadharsing, J. R., Pengelly, R. S., & de Vreede, L. C. N. (2007). A mixed signal approach towards linear and efficient-way Doherty amplifiers. *IEEE Transactions on Microwave Theory and Techniques, 55*(5), 866–879.
59. Pelk, M. J., Neo, W. C. E., Gajadharsing, J. R., Pengelly, R. S., & de Vreede, L. C. N. (2008). A high-efficiency 100-W GaN three-way Doherty amplifier for base-station applications. *IEEE Transactions on Microwave Theory and Techniques, 56*(7), 1582–1591.
60. Nghiem, X. A., Guan, J., & Negra, R. (2014). Design of a broadband three-way sequential Doherty power amplifier for modern wireless communications. In *2014 IEEE MTT-S International Microwave Symposium (IMS2014)*. Washington, DC: IEEE.

61. Gajadharsing, J. (2009). Recent advances in Doherty amplifiers for wireless infrastructure. In *IEEE MTT-S International Microwave Symposium Workshop WSC*.
62. Kim, I., Moon, J., Jee, S., & Kim, B. (2010). Optimized design of a highly efficient three-stage Doherty PA using gate adaptation. *IEEE Transactions on Microwave Theory and Techniques, 58*(10), 2562–2574.
63. Barthwal, A., Ajmera, G., Rawat, K., Basu, A., & Koul, S. K. (2014). Design scheme for dual-band three stage Doherty power amplifiers. In *2014 IEEE International Microwave and RF Conference (IMaRC)*. Washington, DC: IEEE.
64. Golestaneh, H., Malekzadeh, F. A., & Boumaiza, S. (2013). An extended bandwidth three-way Doherty power amplifier. *IEEE Transactions on Microwave Theory and Techniques, 61*(9), 3318–3328.
65. Barthwal, A., Rawat, K., & Koul, S. K. (2015). Bandwidth enhancement of three-stage Doherty power amplifier using symmetric devices. *IEEE Transactions on Microwave Theory and Techniques, 63*(8), 2399–2410.
66. Barthwal, A., Rawat, K., & Koul, S. K. (2018). A design strategy for bandwidth enhancement in three-stage Doherty power amplifier with extended dynamic range. *IEEE Transactions on Microwave Theory and Techniques, 66*(2), 1024–1033.
67. Chireix, H. (1935). High power outphasing modulation. *Proceedings of the Institute of Radio Engineers, 23*(11), 1370–1392.
68. Cox, D. (1974). Linear amplification with nonlinear components. *IEEE Transactions on Communications, 22*(12), 1942–1945.
69. Hakala, I., Choi, D. K., Gharavi, L., Kajakine, N., Koskela, J., & Kaunisto, R. (2005). A 2.14-GHz Chireix outphasing transmitter. *IEEE Transactions on Microwave Theory and Techniques, 53*(6), 2129–2138.
70. Beltran, R., Raab, F. H., & Velazquez, A. (2009). HF outphasing transmitter using class-E power amplifiers. In *2009 IEEE MTT-S International Microwave Symposium Digest* (pp. 757–760). Washington, DC: IEEE.
71. Beltran, R., & Raab, F. (2015). Simplified analysis and design of outphasing transmitters using class-E power amplifiers. In *2015 IEEE Topical Conference on Power Amplifiers for Wireless and Radio Applications (PAWR)* (pp. 1–3). Washington, DC: IEEE.
72. Raab, F. H. (1985). Efficiency of outphasing RF power-amplifier systems. *IEEE Transactions on Communications, 33*(10), 1094–1099.
73. Bi, J. (2008). Chireix's/LINC power amplifier for base station applications using GaN Devices with load compensation (MS thesis). Delft University of Technology. Retrieved from http://resolver.tudelft.nl/uuid:80c01881-4210-4979-8820-57ed80b8995c.
74. Qureshi, J. H., et al. (2009). A 90-W peak power GaN outphasing amplifier with optimum input signal conditioning. *IEEE Transactions on Microwave Theory and Techniques, 57*(8), 1925–1935.
75. Barton, T. W., Dawson, J. L., & Perreault, D. J. (2013). Experimental validation of a four-way outphasing combiner for microwave power amplification. *IEEE Microwave and Wireless Components Letters, 23*(1), 28–30.
76. de Falco, P. E., et al. (2017). Load modulation of harmonically tuned amplifiers and application to outphasing systems. *IEEE Transactions on Microwave Theory and Techniques, 65*(10), 3596–3612.
77. Roblin, P., Chang, H.-C., Martinez-Rodriguez, F. J., Xie, C., & Martinez-Lopez, J. I. (2017). On the design of GaN Chireix power amplifiers using an embedding device model. *International Journal of Numerical Modelling: Electronic Networks, Devices and Fields, 30*(1), e2148.
78. Chang, H. C., Hahn, Y., Roblin, P., & Barton, T. W. (2018). new mixed-mode design methodology for high-efficiency outphasing Chireix amplifiers. *IEEE Transactions on Circuits and Systems I: Regular Papers, 66*(4), 1594–1607. https://doi.org/10.1109/TCSI.2018.2882770.
79. Özen, M., van der Heijden, M., Acar, M., Jos, R., & Fager, C. (2017). A generalized combiner synthesis technique for Class-E outphasing transmitters. *IEEE Transactions on Microwave Theory and Techniques, 46*(5), 1126–1139.

80. Özen, M., Andersson, K., & Fager, C. (2016). Symmetrical Doherty power amplifier with extended efficiency range. *IEEE Transactions on Microwave Theory and Techniques, 64*(4), 1273–1284.
81. Zhang, R., et al. (2014). Generalized semi-analytical design methodology of class-E outphasing power amplifier. *IEEE Transactions on Circuits and Systems I: Regular Papers, 61*(10), 2951–2960.
82. Pozar, D. M. (2012). *Microwave engineering* (4th ed.). Hoboken, NJ: Wiley.
83. Özen, M., et al. (2014). Wideband and efficient watt-level SiGe BiCMOS switching mode power amplifier using continuous class-E modes theory. In *2014 IEEE Radio Frequency Integrated Circuits Symposium* (pp. 1–4). Washington, DC: IEEE.
84. Kenington, P. B. (2000). *High-linearity RF amplifier design*. Boston, MA: Artech House.
85. Liu, T., Ye, Y., Zeng, X., & Ghannouchi, F. M. (2008). Accurate time-delay estimation and alignment for RF power amplifier/transmitter characterization. In *2008 4th IEEE International Conference on Circuits and Systems for Communications* (pp. 70–74). Washington, DC: IEEE.
86. Rawat, M., Quindroit, C., Roblin, P., Narharishetti, N., Pond, R., Salam, K., & Xie, C.. (2014). Characterization and modeling scheme for harmonics at power amplifier output. In *83rd ARFTG Microwave Measurement Symposium*, Tampa, FL.
87. Boumaiza, S., Helaoui, M., Hammi, O., Liu, T., & Ghannouchi, F. M. (2007). Systematic and adaptive characterization approach for behavior modeling and correction of dynamic nonlinear transmitters. *IEEE Transactions on Instrumentation and Measurement, 56*(6), 2203–2210.
88. Muhonen, K. J., Kavehrad, M., & Krishnamurthy, R. (2000). Look-up table techniques for adaptive digital predistortion: A development and comparison. *IEEE Transactions on Vehicular Technology, 49*(9), 1995–2002.
89. Melendez-Cano, A., Juárez-Cázares, S. A., Galaviz-Aguilar, J. A., Cárdenas-Valdez, J. R., Garcia-Ortega, M. J., Calvillo-Téllez, A., Roblin, P., & Núñez-Pérez, J. C. (2016). Behavioral modeling for power amplifiers comparing MPM wiener and hammerstein with FPGA-based implementation. In *2016 International Conference on Mechatronics, Electronics and Automotive Engineering (ICMEAE)* (pp. 119–124). Washington, DC: IEEE.
90. Xu, G., Liu, T., Ye, Y., Xu, T., Wen, H., & Zhang, X. (2014). Generalized two-box cascaded nonlinear behavioral model for radio frequency power amplifiers with strong memory effects. *IEEE Transactions on Microwave Theory and Techniques, 62*(12), 2888–2899.
91. Zhu, A., & Pedro, J. C. (2007). Amplifier distortion evaluation of RF power amplifiers using dynamic deviation reduction based volterra series. In *2007 IEEE/MTT-S International Microwave Symposium* (pp. 965–968). Washington, DC: IEEE.
92. Ding, L., Zhou, G. T., Morgan, D. R., Ma, Z., Kenney, J. S., Kim, J., & Giardina, C. R. (2006). A robust digital baseband predistorter constructed using memory polynomial. *IEEE Transactions on Communications, 53*(9), 1468–1479.
93. Barradas, F. M., Cunha, T. R., Lavrador, P. M., & Pedro, J. C. (2014). Polynomials and LUTs in PA behavioral modeling: A fair theoretical comparison. *IEEE Transactions on Microwave Theory and Techniques, 62*(12), 3274–3285.
94. Naraharisetti, N., Roblin, P., Quindroit, C., & Gheitanchi, S. (2015). Efficient least-squares 2-D-cubic spline for concurrent dual-band systems. *IEEE Transactions on Microwave Theory and Techniques, 63*(7), 2199–2210.
95. Rawat, M., Rawat, K., & Ghannouchi, F. M. (2014). Generalized rational function for reduced complexity behavioural modelling and digital predistortion of broadband wireless transmitter. *IEEE Transactions on Instrumentation and Measurement, 63*(2), 485–498.
96. Rawat, M., Rawat, K., & Ghannouchi, F. M. (2013). Three layered biased memory polynomial for dynamic modelling with predistortion of transmitter with memory. *IEEE Transactions on Circuits and Systems I: Regular Papers, 60*(3), 768–777.
97. Raich, R., Qian, H., & Zhou, G. T. (2004). Orthogonal polynomials for power amplifier modeling and predistorter design. *IEEE Transactions on Vehicular Technology, 53*(5), 1468–1479.

98. Guan, L., & Zhu, A. (2010). Low-cost FPGA implementation of Volterra series-based digital predistorter for RF power amplifiers. *IEEE Transactions on Microwave Theory and Techniques, 58*(4), 866–872.
99. Gilabert, P. L., Montoro, G., & Bertran, E. (2011). FPGA implementation of a real-time NARMA-based digital adaptive predistorter. *IEEE Transactions on Circuits and Systems II: Express Briefs, 58*(7), 402–406.
100. Guan, L., & Zhu, A. (2012). Optimized low-complexity implementation of least squares based model extraction for digital predistortion of RF power amplifiers. *IEEE Transactions on Microwave Theory and Techniques, 60*(3), 594–603.
101. Mallet, C., et al. (2017). Analog predistortion for high power amplifier over the Ku-band (13, 75–14, 5 GHz). In *2017 47th European Microwave Conference (EuMC)*. Washington, DC: IEEE.
102. Jeong, H.-Y., Park, S.-K., Ryu, N.-S., Jeong, Y.-C., Yom, I.-B., & Kim, Y. (2005). A design of k-band predistortion linearizer using reflective schottky diode for satellite TWTAS. In *2005 European Microwave Conference*. Washington, DC: IEEE.
103. Lee, Y.-S., Lee, M.-W., Kam, S.-H., & Jeong, Y.-H. (2010). A high linearity wideband power amplifier with cascaded third-order analog predistorters. *IEEE Microwave and Wireless Components Letters, 20*(2), 112–114.
104. Ma, Y., Yamao, Y., Akaiwa, Y., & Ishibashi, K. (2014). Wideband digital predistortion using spectral extrapolation of band-limited feedback signal. *IEEE Transactions on Circuits and Systems I: Regular Papers, 61*(7), 2088–2097.
105. Ding, L., Zhou, G. T., Morgan, D. R., Zhengxiang, M., Kenney, J. S., Jaehyeong, K., & Giardina, C. R. (2004). A robust digital baseband predistorter constructed using memory polynomials. *IEEE Transactions on Communications, 52*(1), 159–165.
106. Eun, C., & Powers, E. J. (1997). A new Volterra predistorter based on the indirect learning architecture. *IEEE Transactions on Signal Processing, 45*(1), 223–227.
107. Jardin, P., & Baudoin, G. (2007). Filter look up table method for power amplifiers linearization. *IEEE Transactions on Vehicular Technology, 56*(3), 1076–1087.
108. Praveen Jaraut, P., & Rawat, M. (2017). Application of principal component analysis based effective digital predistortion technique for low-cost FPGA implementation. *Journal of RF and Microwave Computer-Aided Engineering, 27*(6), e21095. https://doi.org/10.1002/mmce.21095.
109. Jaraut, P., Rawat, M., & Roblin, P. (2019). Digital predistortion technique for low resource consumption using carrier aggregated 4G/5G signals. *IET Microwaves, Antennas & Propagation, 13*(2), 197–207.
110. Rawat, K. (2012). *Multi-band/multi-standard wireless transmitter for software defined radios* (PhD thesis). University of Calgary.
111. Rawat, K., Rawat, M., & Ghannouchi, F. M. (2010). Compensating I–Q imperfections in hybrid RF/digital predistortion with an adapted lookup table implemented in an FPGA. *IEEE Transactions on Circuits and Systems II: Express Briefs, 57*(5), 389–393.
112. Gumber, K., & Rawat, M. (2017). A modified hybrid RF predistorter linearizer for ultra wideband 5G systems. *IEEE Journal on Emerging and Selected Topics in Circuits and Systems, 7*(4), 547–557.
113. Black, H. S. (1928). *Translating system*. U.S. Patent No. 1,686,792.
114. Black, H. S. (1937). U.S. Patent No. 2, 102,671.
115. Seidel, H., Beurrier, H. R., & Friedman, A. N. (1968). Error controlled high power linear amplifiers at VHF. *Bell System Technical Journal, 47*, 651–722.
116. Seidel, H. (1971). A feedforward experiment applied to an L-4 carrier system amplifier. *IEEE Transactions on Communication Technology, 19*(3), 320–325.
117. Kenington, P. B. (1992). Efficiency of feedforward amplifiers. *IEE Proceedings G (Circuits, Devices and Systems), 139*(5), 591–593.
118. Eid, E., Eid, F. M., & Ghannouchi, F. B. (1995). Optimal feedforward linearization system design. *Microwave Journal*, 78–84.

119. Schreier, R., & Temes, G. C. (2005). *Understanding delta-sigma data converters* (Vol. 74). Piscataway, NJ: IEEE Press.
120. Aziz, P. M., & Sorensen, H. V. (1996). An overview of sigma-delta converters. *IEEE Signal Processing Magazine, 13*(1), 61–84.
121. Ebrahimi, M. M., Helaoui, M., & Ghannouchi, F. M. (2013). Delta-sigma-based transmitters: Advantages and disadvantages. *IEEE Microwave Magazine, 14*(1), 68–78.
122. Nielsen, M., & Larsen, T. (2007). A transmitter architecture based on delta-sigma modulation and switch-mode power amplification. *IEEE Transactions on Circuits and Systems II: Express Briefs, 54*(8), 735–739.
123. Thiel, B. T., Ozmert, A., Guan, J., & Negra, R. (2011). Lowpass delta-sigma modulator with digital upconversion for switching-mode power amplifier. In *2011 IEEE MTT-S International Microwave Symposium* (pp. 1–4). Washington, DC: IEEE.
124. Luzzatto, A., & Shirazi, G. (2007). *Wireless transceiver design mastering the design of modern wireless equipment and systems* (1st ed.). New York: Wiley.
125. Jerng, A., & Sodini, C. G. (2007). A wideband delta-sigma digital-rf modulator for high data rate transmitters. *IEEE Journal of Solid-State Circuits, 42*(8), 1710–1722.
126. Anttila, L., Handel, P., & Valkama, M. (2010). Joint mitigation of power amplifier and I/Q modulator impairments in broadband direct conversion transmitters. *IEEE Transactions on Microwave Theory and Techniques, 58*(4), 730–739.
127. Yoon, S., & Park, S. (2009). All-digital transmitter architecture based on bandpass delta-sigma modulator. In *2009 9th International Symposium on Communications and Information Technology* (pp. 703–706). Washington, DC: IEEE.
128. Haykin, S. (1988). *Digital communications*. New York: Wiley.
129. Mak, P. I., Seng-Pan, U., Martins, R. P., & Mak, P. I. (2007). Transceiver architecture selection review, state-of-the-art survey and case study. *IEEE Circuits and Systems Magazine, 7*(2), 6–25.

Chapter 2
Nonlinear Device Characterization and Modeling for Power Amplifiers

2.1 Introduction

In this chapter we shall review the physics, characterization, and modeling of the transistors used in RF power amplifiers (PAs). For this review we will mostly focus on base station PAs realized using LDMOSFETs (lightly doped metal-oxide semiconductor field-effect transistors) and GaN HEMTs (high electron mobility transistors). The dominant technology in base station has long been silicon LDMOSFET. Currently LDMOSFETs, GaN HEMTs, and GaAs MESFETs are each sharing about a third of the market. However, as wireless is expanding, the share of LDMOSFETs is expected to shrink as 5G base stations operate above 3.5 GHz where the LDMOS technology does not appear to be a viable option. Furthermore, the GaN HEMT technology is expected to displace the GaAs technology for 5G and future wireless communication operating at millimeter-wave frequencies.

2.2 Transistors Used in Power Amplifiers

In this section we present two of the leading field-effect transistors (FETs) which are dominating the base station PA market. The first one is the low-cost LDMOS which is mostly used below 2 or 3 GHz [1]. The second one is the GaN HEMT which operates up to mm-waves owing to its high electron mobility. Both devices are able to sustain very large drain-to-source voltage up to 60 V as required for power amplification. Drain voltages up to 90 V are even possibly in GaN devices.

Figure 2.1 shows the cross-section of a single-finger LDMOSFET. The LDMOSFET is a MOSFET enhanced with a lightly doped channel region (n-drain) between the gate and the drain contacts. This lightly doped region can support very large voltages in saturation. Note that the small intrinsic knee voltage V_{on}, typically 2 V

K. Rawat et al., *Bandwidth and Efficiency Enhancement in Radio Frequency Power Amplifiers for Wireless Transmitters*, Analog Circuits and Signal Processing,
https://doi.org/10.1007/978-3-030-38866-9_2

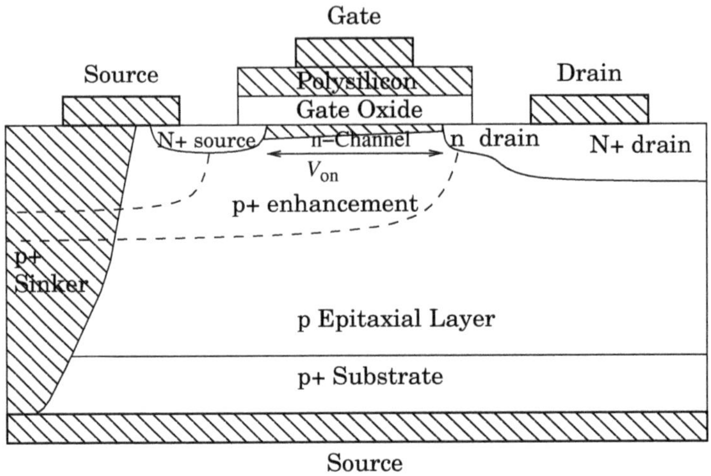

Fig. 2.1 Idealized cross-section of a lightly doped metal-oxide-semiconductor field-effect transistor (LDMOSFET) identifying the location of the inverted n-channel in the p-epitaxial layer

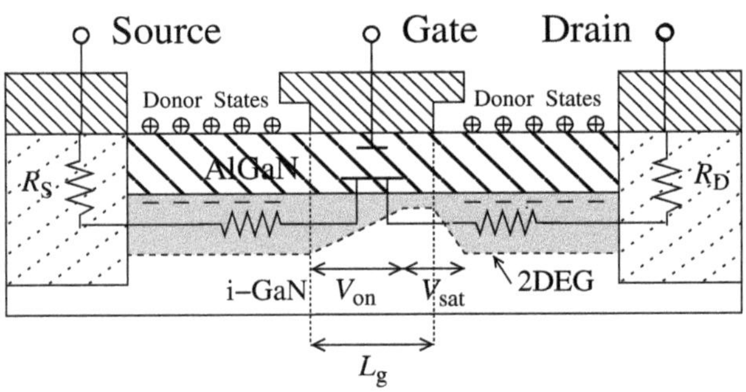

Fig. 2.2 Idealized cross-section of a GaN high electron mobility transistor (HEMT) identifying the location of the two-dimensional electron gas (2DEG) channel

is mostly appearing across the channel below the gate. Also, the source is connected to the substrate via a p+ sinker to achieve a very low contact resistance and to disable the parasitic NPN bipolar transistor arising between source and drain.

Figure 2.2 shows the cross-section from source to drain of a single finger of a GaN HEMT. The AlGaN region which exhibits a very wide bandgap acts as an insulator for the FET. A two-dimensional electron gas (2DEG) forms between source and drain in the intrinsic GaN region near the AlGaN region. This 2DEG charge constituting the HEMT channel originates from the surface states appearing at the top of the AlGaN region which act as donors. Since the channel is undoped, electrons in the 2DEG exhibit a very high mobility, thus justifying calling this FET an HEMT. Note that the small intrinsic knee voltage V_{on}, typically 2–3 V is essentially the intrinsic drain-to-source voltage dropped across the channel below the gate. A fraction of a volt is typically dropped across R_S and R_D and the 2DEG channel access resistance. Most of the drain voltage V_{DS} applied, for example, 60 V is appearing in the pinched-down region which supports the large saturation voltage $V_{sat} \simeq v_{DS} - V_{on}$. In this region the 2DEG is not fully depleted and current saturation results from the electrons traveling quasi-ballistically at an average saturation velocity. The voltage V_{sat} is supported by the small depleted 2DEG charge distribution in this region. Transport in this region is thus referred as space-charge limited in contrast with the channel region sustaining V_{on} which is mostly operating under drift transport.

2.3 Figures-of-Merit

In order to analyze the performance of a device we shall make use of small- and large-signal figures-of-merit.

The most important type of small-signal figures-of-merit are (1) the short-circuit unity current-gain cutoff frequency and (2) the maximum frequency of oscillation. Both of these small-signal figures-of-merit are dependent on the operating point and will thus vary with the DC biases applied.

The short-circuit unity current-gain cutoff frequency is defined in the frequency domain from the current gain

$$G_i(\omega) = \left.\frac{i_2(\omega)}{i_1(\omega)}\right|_{v_2=0} = \left.\frac{i_D(\omega)}{i_G(\omega)}\right|_{v_{DS}=0} = \frac{y_{21}(\omega)}{y_{11}(\omega)}.$$

The short-circuit unity current-gain cutoff frequency $f_T = \omega_T/(2\pi)$ is then the frequency at which the current gain is unity

$$G_i(\omega_T) = 1.$$

The short-circuit unity current-gain cutoff frequency f_T is a broadband figure-of-merit which gives an estimate of the upper frequency at which digital circuits such a ring oscillators will be able to operate. At the short-circuit unity current-gain cutoff frequency f_T, the device is still active and will provide power gain under small-signal operation if the input and output device's impedances are properly terminated by conjugate impedances assuming the device is stable.

The maximum frequency at which a device can provide power gain is the maximum frequency of oscillation f_{max}. The maximum frequency of oscillation f_{max} is defined from the unilateral power gain U. The unilateral power gain U is the maximum available power gain (MAG) of a device which has been unilaterized ($z_{12} = y_{12} = S_{12} = 0$). Because conjugate matching at the input and output is required at the frequency of operation to implement the maximum available power gain, the unilateral power gain is a narrowband figure-of-merit. In terms of the y-parameters of the device (before it is unilaterized) the unilateral power gain is defined as [2]

$$U = \frac{|y_{21} - y_{12}|}{4\,[\mathrm{Re}(y_{11})\mathrm{Re}(y_{22}) - \mathrm{Re}(y_{12})\mathrm{Re}(y_{21})]}.$$

The maximum frequency of oscillation f_{max} is defined as the frequency at which the unilateral power gain is equal to 1

$$U(f_{max}) = 1.$$

The unilateral power gain typically decreases with frequency at the rate of 20 dB per decade of frequencies and can thus be approximated up to f_{max} by

$$U(f) = \left(\frac{f}{f_{max}}\right)^2,$$

where f_{max} is the frequency at which the device switches from active to passive (power gain lower than 1). Typically a power gain of at least 10 dB is desired such that the maximal useful frequency for the device is $f_{max}/3.16$.

Both f_T and f_{max} are a function of the bias and are typically larger in the linear region ($V_{DS} > V_{on}$) and will peak for a gate voltage V_{GS} for which the transconductance is the largest. In the cutoff region f_T and f_{max} rapidly vanish since the device becomes passive.

Large-signal figures-of-merit are of greater importance than small-signal ones when transistors are used to realize power amplifiers. In the design of PAs, the transistor selected must deliver the required output power P_{out} and the maximum RF power the device can deliver is naturally a key figure-of-merit. Consider the PA circuit shown in Fig. 2.3 where the transistor is operating in class B (see Fig. 2.4). The output power P_{out} critically depends on the breakdown voltage V_{max}, the knee voltage V_{on}, and the maximum current I_{max} the device can sustain at the current-source reference planes which are all defined in Fig. 2.4. The RF output power which

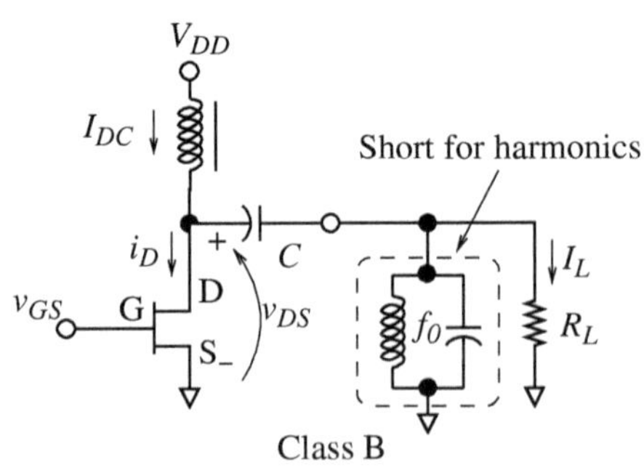

Fig. 2.3 Circuit of a transistor operated in class B

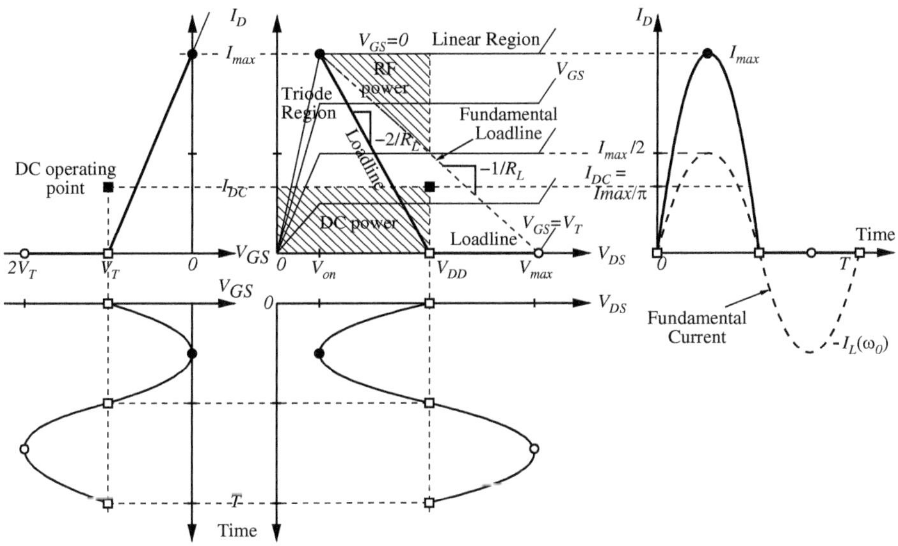

Fig. 2.4 Ideal current–voltage (IV) characteristics and current and voltage waveforms at the current-source reference planes. Also shown are the instantaneous and fundamental load lines and the DC power dissipated and RF power generated

can be graphically calculated from the area of the shaded "RF power" region in Fig. 2.4 is evaluated to be

$$P_{out} = \frac{1}{8}(V_{max} - V_{on})I_{max}.$$

Since PAs are usually the circuit component in a transmitter system dissipating the most power, the optimal efficiency that a device can provide at peak power is of critical importance. The drain efficiency η_D is defined as

$$\eta_D = \frac{P_{out}}{P_{DC}},$$

where P_{DC} is the DC power which is graphically presented by the shaded "DC power" region in Fig. 2.4. The drain efficiency η_D measures the ability of the device in converting the DC power P_{DC} provided by the supplies to the PA into the output RF power P_{out}. At the current-source reference planes, the efficiency in class B of operation in the linear region is a function of the knee-to-maximum voltage ratio V_{on}/V_{max}

$$\eta_D(\text{class-B}) = \frac{\pi}{4}\left(\frac{1 - \frac{V_{on}}{V_{max}} P_{DC}}{1 + \frac{V_{on}}{V_{max}} P_{DC}}\right).$$

The knee-to-maximum voltage V_{on}/V_{max} ratio plays thus a critical role in class B and also in class A and F operations where the transistor also operates mostly in the linear region. In GaN devices, the knee voltage is of the order of 3 V and the maximum voltage of the order of 60 V or more. The efficiency of class B then reduces from 78.5% ($\pi/4$) for zero knee voltage to 71% for $V_{on}/V_{max} = 0.05$. Higher efficiencies can however be attained by pushing the device operation in the triode region ($v_{DS} < V_{on}$) at the price of lower output power and higher nonlinearities.

In class-E operation, the transistor is operating in the triode region and the efficiency then depends critically on the channel conductance $R_{on} \simeq V_{on}/I_{max}$ in that region. An approximate expression provided in [3] gives the following estimate for 50% duty-cycle operation:

$$\eta_D(\text{class-E}) = \frac{R_L}{R_L + 1.365 R_{on}}.$$

A more realistic frequency-dependent drain efficiency has been reported in [4].

Beside providing the desired output power, a PA must also provide power gain. The power added efficiency (PAE) improves on the drain efficiency by accounting for the input power

$$\text{PAE} = \frac{P_{out} - P_{in}}{P_{\text{DC}}}.$$

When transistors are used toward their maximum output power, their RF power gain starts decreasing with increasing output power and the device becomes markedly nonlinear. A gain compression of 1 dB is usually considered to be associated with a reasonable level of nonlinearity. The output power for 1 dB compression usually noted P1dB is thus a good measure of the PA output power characteristics. Nonetheless, sometimes a compression of up to 4 dB is used to maximize the PA efficiency at the price of more severe nonlinearities.

Since PAs are to be used with modulated signals, their ability to amplify broadband signals with a reasonable amount of spectral regrowth is thus of great importance to not perturb adjacent bands. The output-referred third-order intercept IP3 can be used as a figure-of-merit to estimate the device nonlinearity. IP3 is obtained from two-tone measurements ($a(\omega_1)$ and $a(\omega_2)$) as the intercept between the output power ($|b(\omega_1)|^2$ and $|b(\omega_2)|^2$) and the output third-order intermodulation power ($|b(2\omega_1 - \omega_2)|^2$ and $|b(2\omega_2 - \omega_1)|^2$), respectively, linearly extrapolated using 10 dB and 30 dB per decade. Unless linearization is used, the IP3 is typically about 10 dB above P1dB. Given the complex modulations used in wireless communication, it is however preferable to directly rely on the adjacent channel power ratio (ACPR) for signals of various bandwidth and peak-to-backoff power ratios (PAPR) with and without digital predistortion (DPD) linearization.

2.4 Transistor Modeling

Although heterojunction bipolar transistor (HBT) has still important applications in mobile-phone PAs, FETs have become the work-horse of electronics with LDMOSFET, GaAs, and GaN HEMT sharing about equally the base station markets. We shall therefore focus on the FET transistor to introduce the concept of large-signal device modeling at RF frequencies for high-power generation.

In their useful frequency-range of application, FETs can be well-represented at microwave frequencies using a lump equivalent circuit. This is due to the fact that FETs like HBTs are transit-time devices and must be kept small compared to the wavelength of operation for them to provide power gain. This is not that distributed effects are not taking place but rather that they can be well-represented by circuits of lower order. In fact, at higher frequencies when higher-order distributed effects dominate, the transistors become passive and lose their usefulness. In FETs, the distributed effects which are taking place are of three dominant types: (1) the wave propagation from source to drain along the gate length L_g, (2) the wave propagation along the gate width W_g, and (3) the parallel multifinger wave propagation taking place in large power transistors. See Fig. 2.5 for the layout of a single-finger FET defining the gate length L_g and gate width W_g and the terminals' location.

We shall discuss all these three types of propagation in turns as they collectively contribute to the microwave performances and equivalent circuit of power transistors. Of critical importance to the device performance are the reduction of the parasitics contributed by these distributed effects to minimize the lossy dissipation and the energy storage so that the device operates the closest possible to the frequency of operation of small size devices.

The intrinsic device which provides the power amplification defines the primary limitation of the device high-frequency response. The high-frequency response of the ideal intrinsic FET device can be studied by solving the large-signal MOSFET wave-equation within the gradual-channel approximation [5]

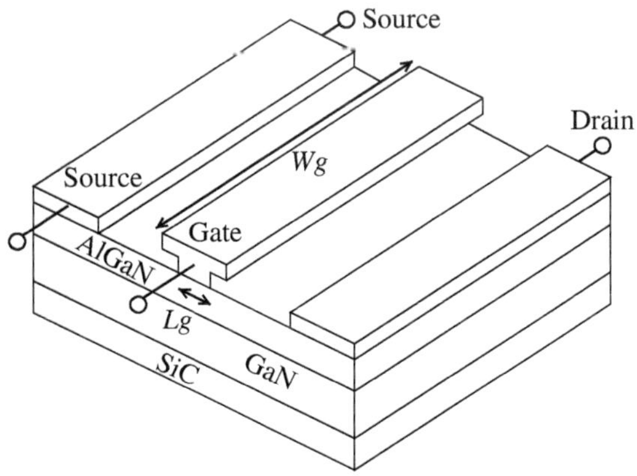

Fig. 2.5 Three-dimensional layout of a single-finger FET defining the gate length and width

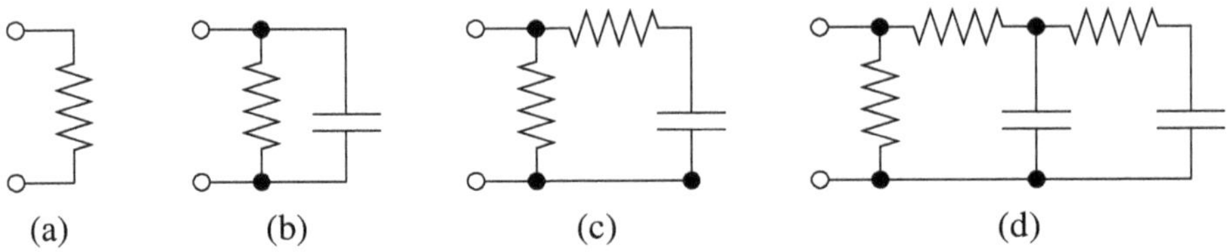

Fig. 2.6 Equivalent circuit for the (**a**) static, (**b**) quasi-static, (**c**) first-order non-quasi-static (NQS), and (**d**) second-order NQS response of y_{11} or y_{22}

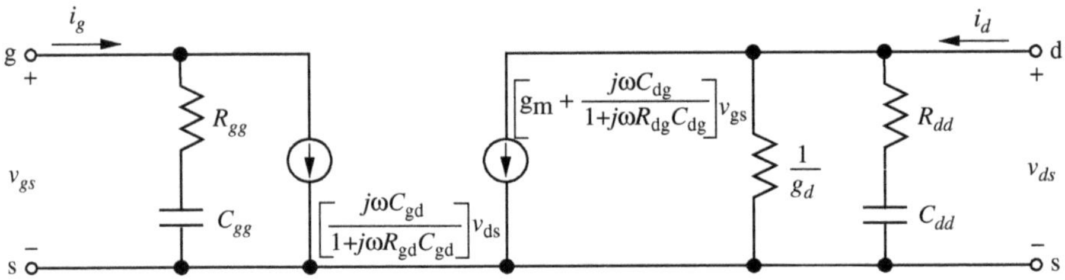

Fig. 2.7 First-order NQS small-signal equivalent circuit for the MOSFET/HEMT

$$\frac{\partial^2}{\partial x^2}\left(v_{GC}^2(x,t)\right) = \frac{2}{\mu}\frac{\partial v_{GC}(x,t)}{\partial t},$$

with μ as the electron mobility. Exact analytic small-signal solutions for both the constant mobility and velocity-saturated wave-equation have been reported in the frequency domain[5, 6]. Equivalent circuits of various orders can be developed to approximate these exact solutions. For example, as shown in Fig. 2.6a–d, the static, quasi-static, first-order non-quasi-static (NQS), and second-order NQS responses can be used, respectively, to represent the DC response, the digital response up to f_T, and the microwave response up to f_{max} and beyond. Second-order NQS has been reported [7] but they are mostly useful when the transistor is operating at frequencies higher than f_{max} where the device is no longer active. It is to be mentioned that the intrinsic f_T and f_{max} are much larger than the extrinsic $f_{T,ext}$ and $f_{max,ext}$ measured for the device given the extrinsic parasitics associated with the contacts, the distributed effects along the gate width and multifinger layout all contribute to decrease the device performance. Thus, a first-order NQS approximation is largely sufficient except when the device is operated in cutoff. In such a case the y-parameters of the intrinsic device can be well-represented by [8]

$$y_{ij}(\omega) = g_{ij} + \frac{j\omega C_{ij}}{1 + j\omega R_{ij}C_{ij}},$$

where each of the g_{ij}, C_{ij}, and $\tau_{ij} = R_{ij}C_{ij}$ is bias-dependent. Note that four NQS time-constants τ_{ij} are defined for the four two-port y-parameters of the intrinsic device. The resulting first-order NQS equivalent circuit is shown in Fig. 2.7.

The short-circuit unity current-gain cutoff frequency previously defined approximately yields for the intrinsic long-channel FET at

$$\omega_T \simeq \frac{g_m}{C_{gs}} = \frac{\mu(V_{GS} - V_T)}{L_g^2}.$$

In short-channel devices, the transconductance becomes approximately constant in a wide range of gate voltages in the linear region and the unity-current gain cutoff frequency reduces now to [8]

$$\omega_T \simeq \frac{g_m}{C_{gs}} = \frac{v_{sat}}{L_g},$$

where v_{sat} is the electron saturation velocity in the drain region. Clearly, for a given semiconductor material, a higher frequency of operation is obtained by shrinking the gate length. However, this typically comes at the price of a reduced breakdown voltage and thus reduced output power.

Having presented an approximate small-signal model for the MOSFET wave-equation we turn now toward its large-signal implementation. If we assume $\tau_G = \tau_{11} = \tau_{12}$ and $\tau_D = \tau_{21} = \tau_{22}$ we can then develop a simplified charge-based NQS large-signal model in which the gate, drain, and source currents are given by

$$\begin{aligned} i_G &= i_{disp,G}, \\ i_D &= I_D(v_{GS}, v_{DS}, T_{dev}) + i_{disp,D}, \\ i_S &= i_D + i_G, \end{aligned} \tag{2.1}$$

where $I_D(v_{GS}, v_{DS}, T_{dev})$ is the intrinsic IV characteristics at the temperature T_{dev} and $i_{disp,G}$ and $i_{disp,D}$ are the displacement currents of the charge elements Q_G and Q_D defined by

$$\begin{aligned} i_{disp,G} &= \frac{dQ_G(v_{GS}, v_{DS}, T_{dev})}{dt} - \frac{d}{dt}\left[\tau_G(v_{GS}, v_{DS}, T_{dev})i_{disp,G}\right], \\ i_{disp,D} &= \frac{dQ_D(v_{GS}, v_{DS}, T_{dev})}{dt} - \frac{d}{dt}\left[\tau_D(v_{GS}, v_{DS}, T_{dev})i_{disp,D}\right]. \end{aligned}$$

Note that τ_G and τ_D are the NQS charge-redistribution time-constants associated with the charge elements Q_G and Q_D, respectively. These equations define a new circuit element consisting of a charge and an associated charging time-constant. It can be verified that this large-signal equivalent circuit conserves charges when the device moves from one operating point to another one. Indeed, when integrating the gate and drain currents, the same charge is retrieved independently of the voltage path of $v_{GS}(t)$ and $v_{DS}(t)$ once the device has relaxed to a DC operating point. This originates from the following three features: (1) the model is defined in terms of charges Q_G and Q_D, (2) the NQS components of the displacement currents are

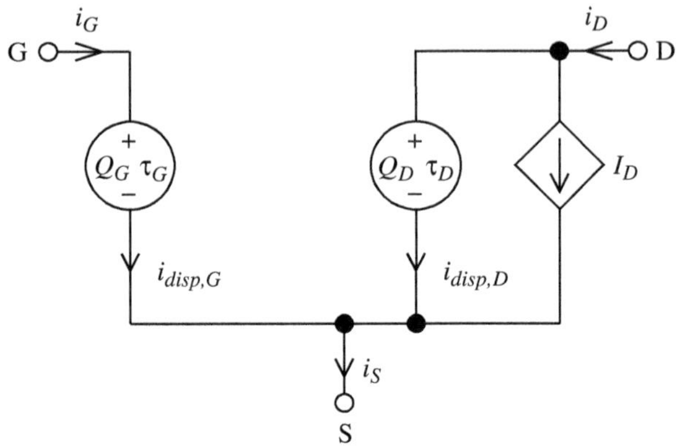

Fig. 2.8 First-order NQS large-signal model for the MOSFET wave-equation

represented by a total derivative, and (3) the displacement currents vanish when the device relaxes to a DC operating point. Figure 2.8 shows a possible topology for implementing this charged-based large-signal model. A more complex model relying on four NQS time-constants can also be devised.

The present model provides a reasonable approximate solution of the large-signal wave-equation as long as the device does not enter the cutoff region. Under large-signal transient operation when the device is switched from cutoff (passive mode) to the triode or linear mode (active modes) higher-order distributed effects can be approximately accounted for by adding a channel propagation delay in series with the intrinsic drain current I_D. In the long-channel case, the drain propagation delay has been verified [9, 10] to be given by

$$\tau_D \simeq 0.38 \frac{L_g^2}{\mu\left(V_{GS}-V_T\right)}.$$

2.5 Distributed Effects Along the Gate Width

The NQS effects described in Sect. 2.2 defines the intrinsic device limitation associated with the gate length. Once the device parasitics such as the source R_s, drain R_d, and gate R_g resistances, the source L_s, drain L_d, and gate L_g inductances, and the fringe capacitances G_{gdf} and G_{gsf} are added to the intrinsic device model as in Fig. 2.9, the performance of the extrinsic device degrades.

The origin of the above device parasitics is investigated now in more detail by considering first distributed effects along the gate width for a single-finger device as shown in Fig. 2.10 before considering multiple-finger FETs. For long gate width the wave-equation needs to be solved along the gate width. The device is then represented by a four-port (six terminals) device as shown in Fig. 2.11.

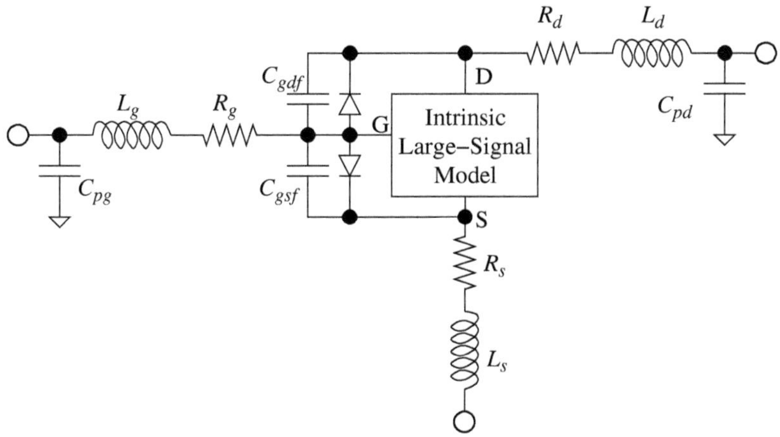

Fig. 2.9 Extrinsic FET model including the intrinsic large-signal model, the fringe capacitances, and the pad parameters

Fig. 2.10 Single-finger FET with its six terminals identified. Two more terminals need to be included if the body/substrate is floating

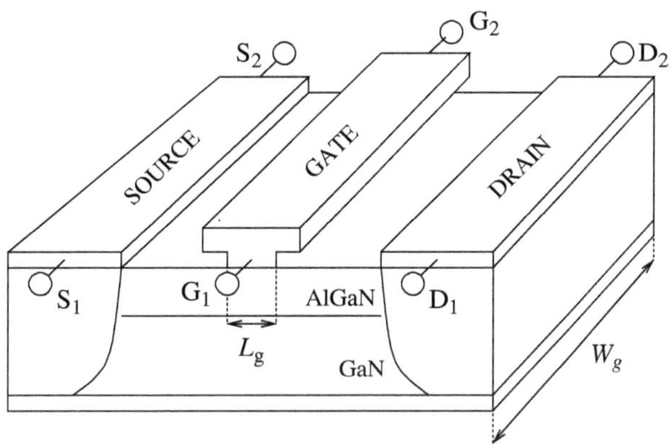

Fig. 2.11 Six-terminal circuit for a single-finger FET

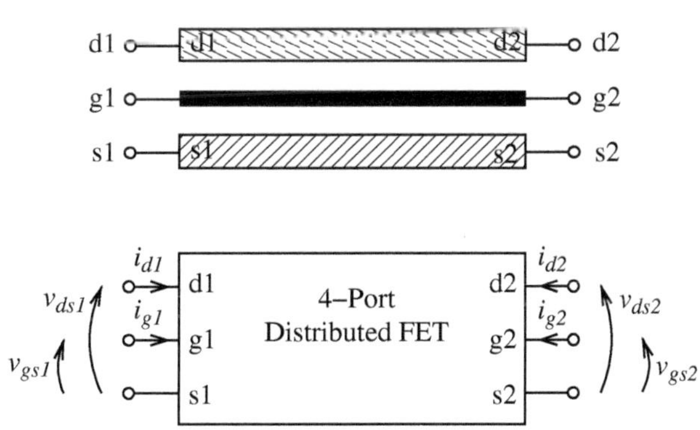

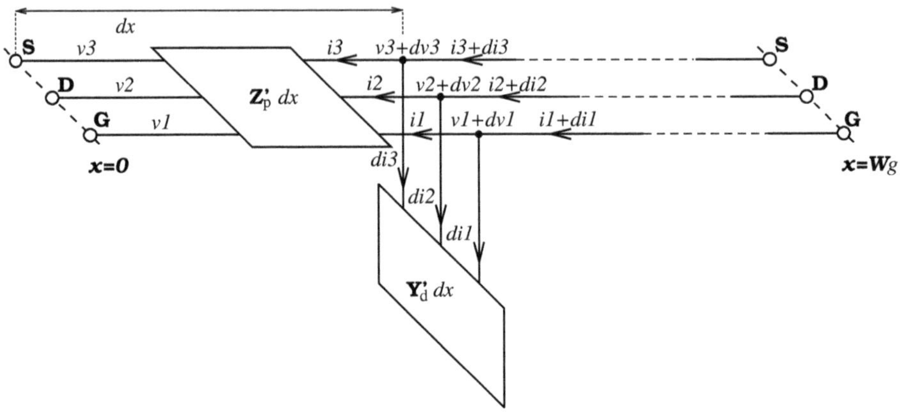

Fig. 2.12 Coupled-line models for four-port FET building block with distributed parasitics and arbitrary gate width

An exact analytic solution has been reported [11] for the small-signal response of the circuit shown in Fig. 2.11. Figure 2.12 gives the four-port equivalent distributed circuit of a general building block for typical transistor layouts. The equivalent circuit in Fig. 2.12 introduces the following matrices for the parasitics $Z_p^{'}$ and the device $Y_d^{'}$:

$$Z_p^{'} = \begin{bmatrix} Z_{go}^{'} & M_{12}^{'} & M_{13}^{'} \\ M_{12}^{'} & Z_{do}^{'} & M_{23}^{'} \\ M_{13}^{'} & M_{23}^{'} & Z_{so}^{'} \end{bmatrix} \quad \text{and} \quad Y_d^{'} = \begin{bmatrix} Y_{11}^{'} & Y_{12}^{'} & Y_{13}^{'} \\ Y_{21}^{'} & Y_{22}^{'} & Y_{23}^{'} \\ Y_{31}^{'} & Y_{32}^{'} & Y_{33}^{'} \end{bmatrix}$$

$$\text{with } Z_{go}^{'} = \frac{Z_{go}}{W_g},\ Z_{do}^{'} = \frac{Z_{do}}{W_g},\ Z_{so}^{'} = \frac{Z_{so}}{W_g},\ M_{ij}^{'} = M_{ji}^{'} = \frac{M_{ij}}{W_g} \text{ and } Y_{ij}^{'} = \frac{Y_{ijo}}{W_g},$$

where $Z_{go}^{'}$, $Z_{so}^{'}$, $Z_{do}^{'}$ are the lateral gate, source, and drain parasitic impedances per unit gate width, $M_{ij}^{'}$ are the lateral mutual impedances per unit gate width, and $Y_{ij}^{'}$ are the vertical FET 3×3 y-parameters per unit gate width. For the derivation of the equivalent circuit, the device admittance matrix $Y_d^{'}$ needs to be first converted into an impedance matrix $Z_d^{'}$

$$Z_d^{'} = Y_d^{'-1} = \begin{bmatrix} Z_{11}^{'} & Z_{12}^{'} & Z_{13}^{'} \\ Z_{21}^{'} & Z_{22}^{'} & Z_{23}^{'} \\ Z_{31}^{'} & Z_{32}^{'} & Z_{33}^{'} \end{bmatrix}.$$

In the limit of small gate width, the 4×4 Z-matrix solution of the wave-equation [11] reduces to

$$Z = \begin{bmatrix} Z_{11} + \frac{Z_g+Z_s}{3} & Z_{12} + \frac{Z_s}{3} & Z_{11} - \frac{Z_g+Z_s}{6} & Z_{12} - \frac{Z_s}{6} \\ Z_{21} + \frac{Z_s}{3} & Z_{22} + \frac{Z_d+Z_s}{3} & Z_{21} - \frac{Z_s}{6} & Z_{22} - \frac{Z_d+Z_s}{6} \\ Z_{11} - \frac{Z_g+Z_s}{6} & Z_{12} - \frac{Z_s}{6} & Z_{11} + \frac{Z_g+Z_s}{3} & Z_{12} + \frac{Z_s}{3} \\ Z_{21} - \frac{Z_s}{6} & Z_{22} - \frac{Z_d+Z_s}{6} & Z_{21} + \frac{Z_s}{3} & Z_{22} + \frac{Z_d+Z_s}{3} \end{bmatrix},$$

where the parameters Z_g, Z_d, Z_s, and Z_{ij} are defined as

$$Z_g = Z'_g W_g, \quad Z_d = Z'_d W_g, \quad Z_s = Z'_s W_g, \qquad \text{and} \;\; Z_{ij} = \frac{Z''_{ij}}{W_g},$$

using the parameters Z'_g, Z'_d, Z'_s and Z''_{ij} themselves defined as,

$$Z'_g = Z'_{go} - M'_{12} - M'_{13} + M'_{23}, \tag{2.2}$$

$$Z'_d = Z'_{do} - M'_{12} + M'_{13} - M'_{23}, \tag{2.3}$$

$$Z'_s = Z'_{so} + M'_{12} - M'_{13} - M'_{23}, \tag{2.4}$$

$$Z''_{ij} = Z'_{ij} - Z'_{3j} - Z'_{i3} + Z'_{33} \qquad i, j = 1, 2. \tag{2.5}$$

This 4 × 4 Z-matrix solution can be represented by the equivalent circuit shown in Fig. 2.13. Note the presence of three negative trans-impedances. The diagonal Z_{ii} parameters however remain positive.

In practical FET layouts, the gate-to-source and drain-source drives are applied on opposite sides on the four-port terminal device and two terminals are left opened as shown in Fig. 2.14. The equivalent circuit reduces then in the limit of small gate width to the two-port equivalent circuit shown in Fig. 2.15. A negative trans-impedance is still present due to excitation of the distributed device on opposite side.

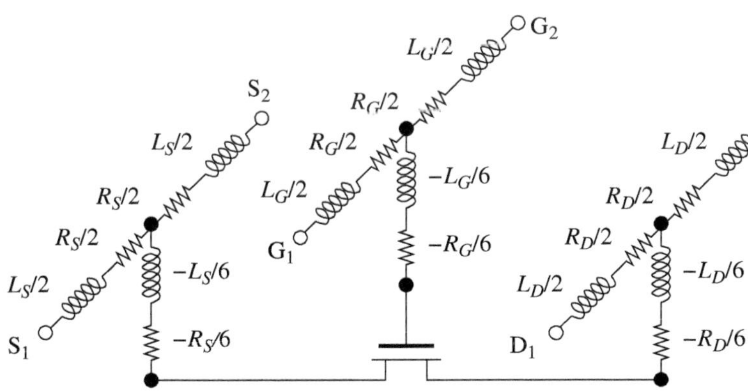

Fig. 2.13 Equivalent circuit model for a six-terminal FET with small gate width

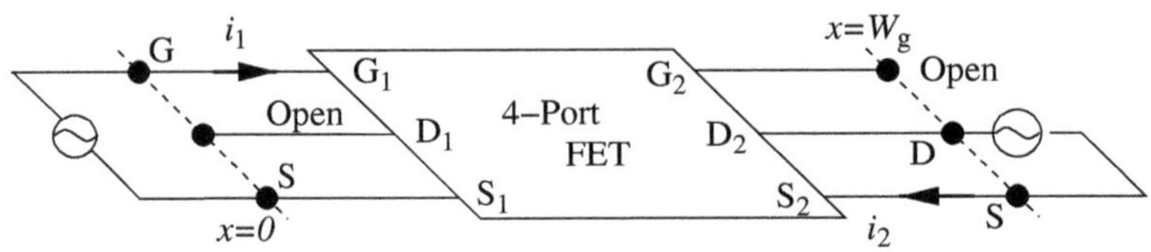

Fig. 2.14 Conventional excitations of the six-terminal FET on opposite sides ([11], reprinted with permission from IEEE)

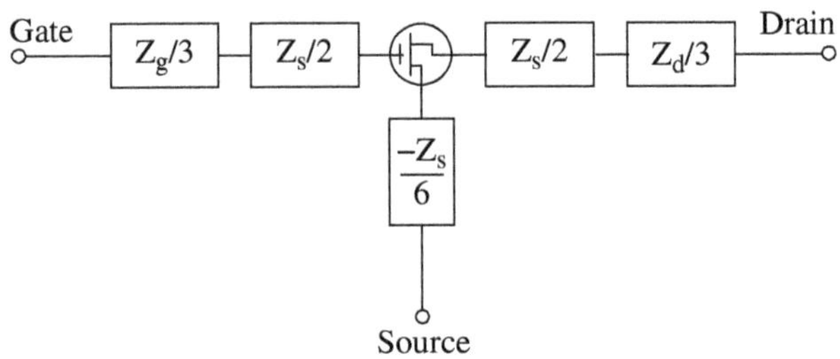

Fig. 2.15 Small gate width two-port equivalent circuits corresponding to the conventional boundary condition ([11], reprinted with permission from IEEE)

However, the contribution of Z_s to the diagonal Z_{11} and Z_{22} impedances which is $Z_s/2 - Z_s/6 = Z_s/3$ remains positive.

As an example, consider a power LDMOSFET with $N_f = 224$ fingers each of a gate width $W_g = 90\,\mu\text{m}$, for a total lateral source resistance of $R_s = R_s' W_g/(3N_f) = 1.3\,\Omega$. Figure 2.16a–c show the MAG/MSG and $|S_{21}|^2$ versus gate width W_g at 1 GHz. Above 40 μm the MAG or maximum stable gain (MSG) is seen to rapidly decrease with frequency as the gate width is increased. Clearly distributed effects along the gate width are seen to be of critical importance. The equivalent circuit and its associated linear scaling of the parasitics used in the short gate width approximation is seen to hold well up to about 50 μm but overestimates the loss for larger gate widths.

We saw in our discussion that there is a diminishing return in terms of power gain when increasing the gate width. Typically, the gate width W_g is kept around a twelfth of the wavelength in the semiconductor. To generate a large output power, transistors must then use a large number N_f of fingers. The total gate periphery is then defined at $N_f W_g$. GaN HEMTs typically exhibit an output power per periphery approaching 10 W/mm in continuous wave (CW) operation and 30 W/mm in pulse operation depending on the process. The addition of a large number of fingers creates an effective synthetic transmission line and additional frequency limitations will arise. Typically, a finite number of fingers can be placed in parallel before lateral distributed effects cripple again the device power gain. An example of a layout with six parallel fingers is shown in Fig. 2.17a. Note that the fingers on the edge contribute larger propagation delays from the input gate to the output drain terminals so that the forward transmitted waves of the various fingers do not perfectly combine in phase. Furthermore, the input and output reflected waves are not themselves balanced, and complex load-pulling may take place between

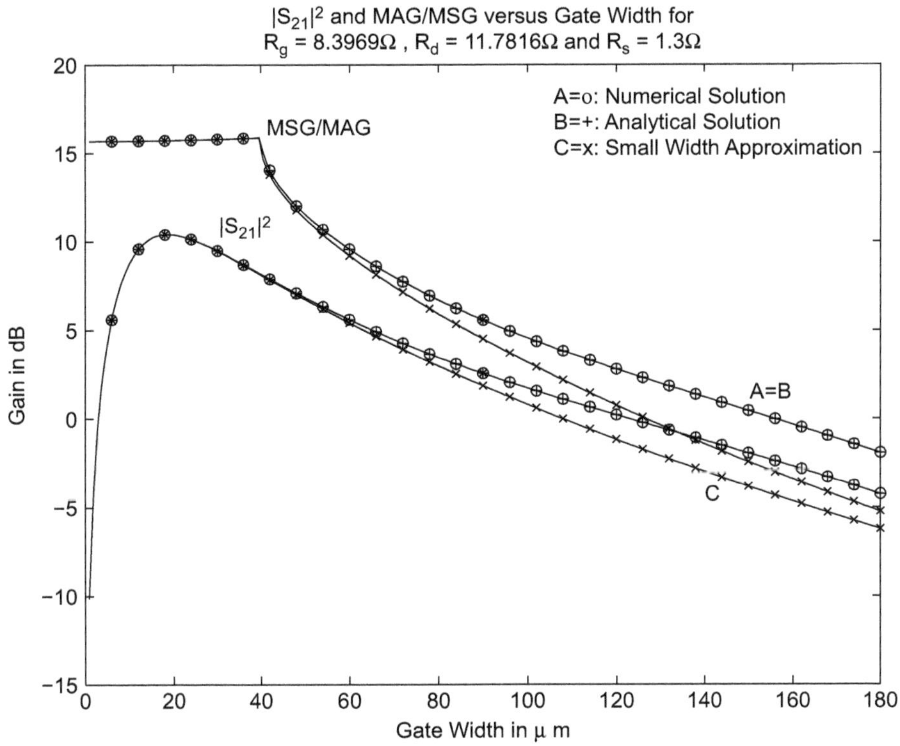

Fig. 2.16 Comparison of (A) the numerical solution, (B) the exact analytic solution, and (C) the small gate width approximation for MSG/MAG and $|S_{21}|^2$ versus gate width ([11], reprinted with permission from IEEE)

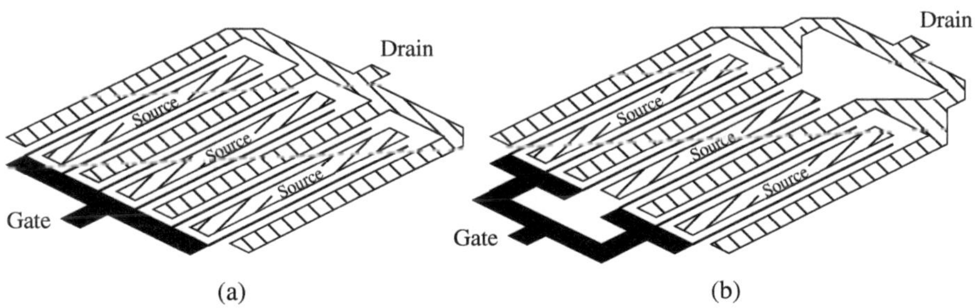

Fig. 2.17 Example of (**a**) symmetrical and (**b**) corporate-tree multifinger layouts

the fingers. A preferred approach for each frequency operation is to rely on the corporate tree layout shown in Fig. 2.17b where the four different branches of the input and output power combiner/splitter are all fully balanced, limiting any load-pulling effect. When combining the power of multiple transistors, a Wilkinson power combiner can also be used to dissipate any residual signals arising from unbalanced device characteristics.

In this layout the source was assumed to be connected to the ground substrate using adjacent via-holes. This is achieved using p-sinkers for the LDMOS technology and using air bridges for the GaN technology. The air bridges can then be also used to shield the gate terminals from the large voltage applied at the drain terminal and help reduce memory effects as will be discussed in Sect. 2.6.

2.6 Characterizing and Modeling Memory Effects in Transistors

In our discussion of the frequency response of microwave transistors we have so far focused on the high-frequency dispersion effects associated with short-term memory effects. In this section we consider in addition the low-frequency dispersion effects associated with long-term memory effects such as self-heating and trapping. Given that microwave transistors are used at very high frequencies, one may wonder on the need to consider low-frequency dispersion effects. As we shall discuss below, the low- and high-frequency responses are linked and must be determined self-consistently. That is, the device microwave response, for example, the RF load line, is dependent on the device temperature or trap occupation and reciprocally the device temperature and trap occupation are dependent on the RF load line.

Let us consider the generic charge-based FET model topology shown in Fig. 2.18. The model topology is divided in multiple layers associated with different reference planes ranging from (a) the current source core, (b) the intrinsic layer, and (c) the extrinsic model.

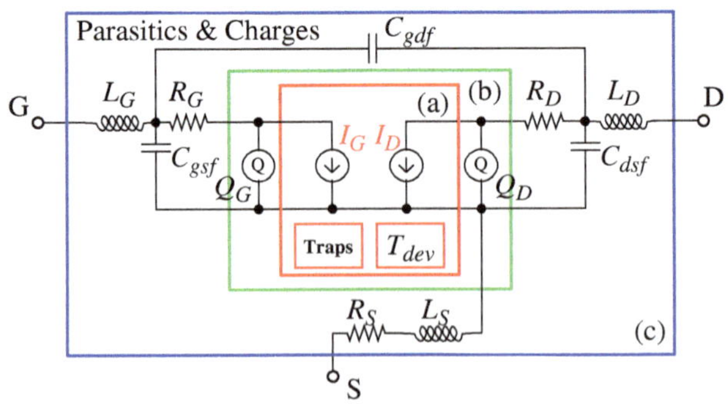

Fig. 2.18 Charge-based model including traps and thermal modeling

The inner core consists of the drain current $I_D(v_{GS}, v_{DS}, T_{dev}, \overline{f}_{T,D})$ and the gate leakage current $I_G(v_{GS}, v_{DS}, T_{dev}, \overline{f}_{T,D})$. Beside the instantaneous gate and drain voltages v_{GS} and v_{DS}, these currents also depend on the operating temperature T_{dev} and the average fractional trap occupation $\overline{f}_{T,D}$.

The circuits calculating the device operating temperature T_{dev} and the trap occupation factor $\overline{f}_{T,D}$ are represented by red boxes in Fig. 2.18. The red frame labeled (a) defines the reference planes for the current source inner core. In an ideal device, the inner core layers should be memoryless. However, trapping and self-heating introduce low-frequency dispersions in the DC characteristic I_G and I_D which modulate the RF device response.

The next layer includes the displacement currents of the nonlinear gate and drain charges

$$i_{G,displ} = \frac{dQ_G(v_{GS}, v_{DS}, T_{dev}, \overline{f}_{T,D})}{dt}, \tag{2.6}$$

$$i_{D,displ} = \frac{dQ_D(v_{GS}, v_{DS}, T_{dev}, \overline{f}_{T,D})}{dt}. \tag{2.7}$$

This layer features charges which could also be dependent on the operating temperature T_{dev} and to a less extent the fractional trap occupation $\overline{f}_{T,D}$. These charges combined with the current sources of the inner core make up the intrinsic device model and its green frame labeled (b) defines then the intrinsic reference planes.

The next layer includes all the lossy and lossless parasitics associated with the physical layout network used to access the intrinsic device. Its blue frame labeled (c) defines then the extrinsic reference planes. The access resistance R_S and R_D are affected by the surface state occupation at the surface of the AlGaN layer on both side of the gate which acts as two virtual gates in GaN HEMTs. Field plates shielding these channel access regions are usually used to reduce the associated memory effects.

Under CW operation, the device temperature T_{dev} is set by the DC power dissipated by the device $P_{diss} = I_D V_{DD}$ according to

$$T_{dev} = R_{th} P_{diss} + T_{sub},$$

where T_{sub} is the substrate temperature and R_{th} the thermal resistance. The device temperature T_{dev} profoundly affects the performance of FETs. The average device temperature principally controls the low-field mobility and the associated channel conductance in the triode region. In the saturation region the electrons are essentially moving ballistically. The average saturation velocity and thus the saturation current are then principally established by space-charge limited transport rather than through the energy relaxation of the electron gas with the lattice.

Figure 2.19a shows the IV characteristics of an LDMOSFET with the average surface temperature measured with an infrared thermometer focused on the die

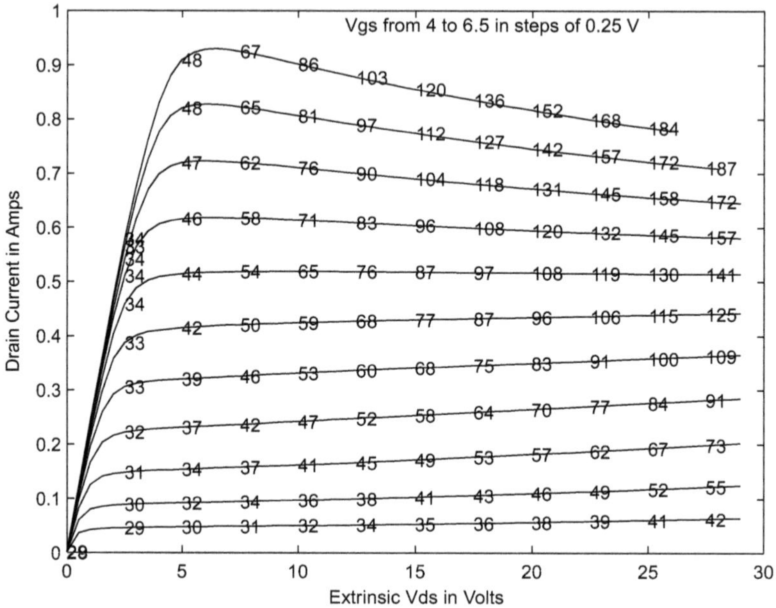

(a)

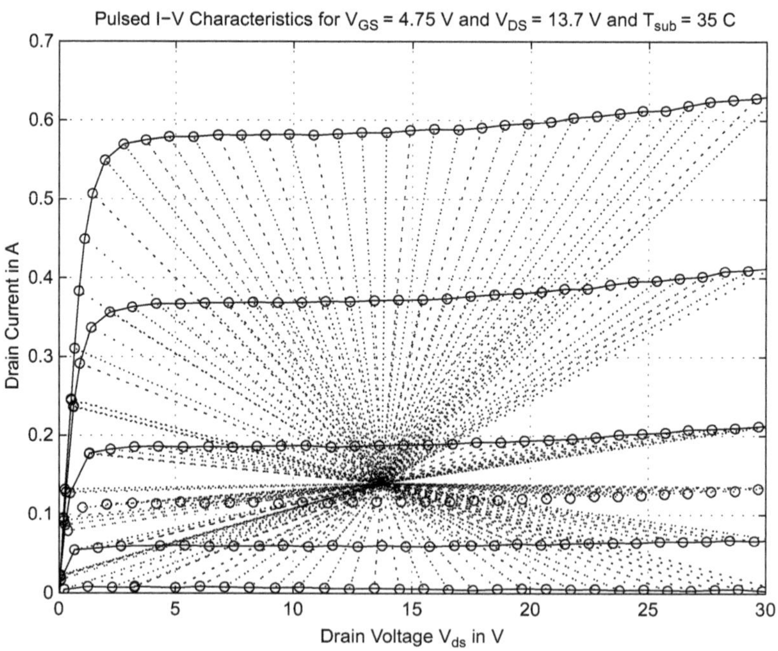

(b)

Fig. 2.19 (**a**) Measured IV characteristics of an LDMOSFET with the average device temperature indicated in degree Celsius. (**b**) Pulsed-IV characteristics for a DC bias of $V_{GS} = 4.75\,\text{V}$ and $V_{DS} = 13.7\,\text{V}$ and substrate temperature $T_{sub} = 35\,°\text{C}$ ([12], reprinted with permission from Cambridge University Press)

[12]. A negative drain conductance is observed at high gate voltages due to the reduction of the saturation velocity with the increase of the device temperature via self-heating at large V_{DS}. Pulsed-IV shown in Fig. 2.19b illustrates the isothermal IV characteristics used by the device in RF operation [12]. IV collapses and knee walkout are also observed to occur as the temperature rises in GaN HEMTs. This effect is exacerbated in GaN devices on sapphire substrate due to their reduced thermal conductivity. Actually, for GaN HEMT on sapphire the knee walkout is dominantly of thermal origin at low drain voltages. This is demonstrated by comparing pulsed-IV pulsed-RF measurements at various device temperatures versus CW RF measurements as shown in Fig. 2.20 [14].

When studying the thermal step response of semiconductors, the thermal time scales are found to vary from a fraction of a microsecond to seconds (see Fig. 2.21). A distributed thermal network with three to five RC time-constants is typically needed to represent accurately the device thermal step response [15]. A three time-constant example of thermal network is shown in Fig. 2.22. Note that nonlinear thermal dynamic can be accounted for by using temperature-dependent thermal resistances and capacitances in Fig. 2.22.

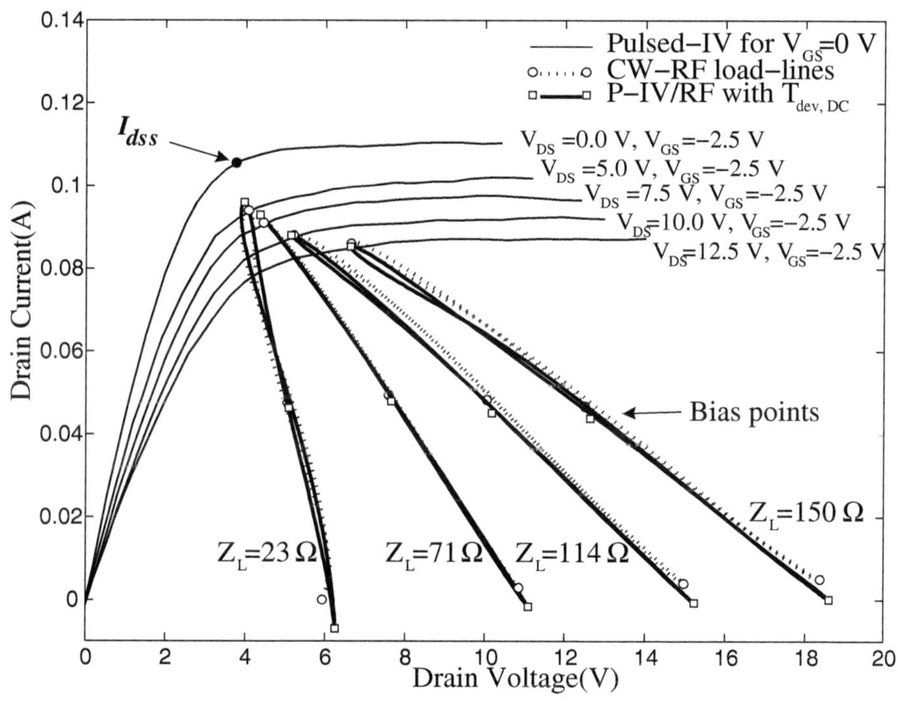

Fig. 2.20 Dynamic load lines (square and bold plain lines) obtained from pulsed-IV/RF measurements with the substrate set to the normal CW device temperature of 58, 72, 100, and 112 °C for V_{DS} of 5.0, 7.5, 10.0, and 12.5 V, respectively. The CW load lines (dashed lines) are also included for reference ([13], reprinted with permission from Cambridge University Press)

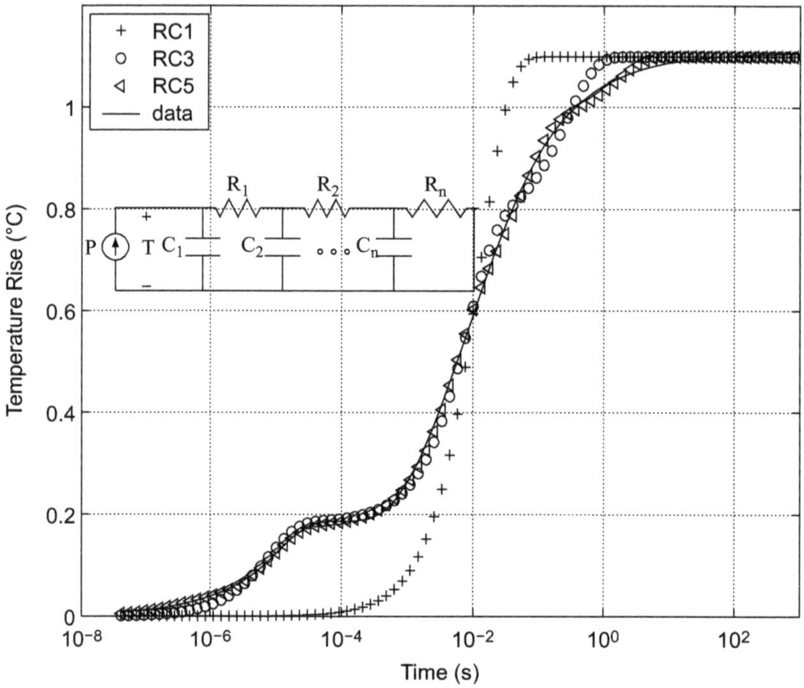

Fig. 2.21 Fit of the thermal step response of a 60 W LDMOSFET using a thermal network with one (RC1), three (RC3), and five (RC5) time-constants ([13], reprinted with permission from Cambridge University Press)

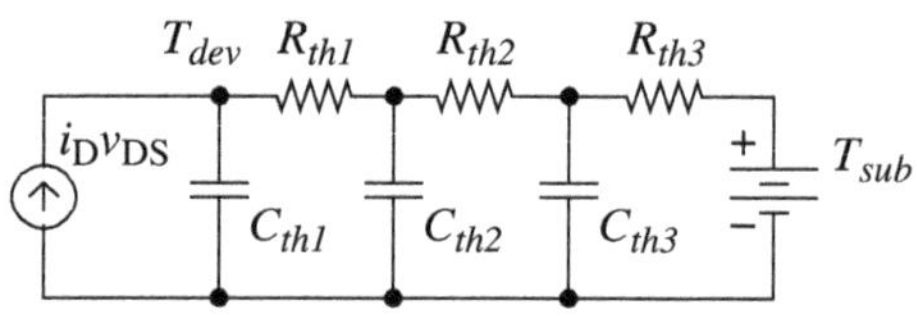

Fig. 2.22 Circuit used for the modeling of the average thermal response with $R_{th} = R_{th1} + R_{th2} + R_{th3}$

Under CW operation a single time-constant RC network is sufficient to model the thermal response of a transistor since the device response is too slow to respond to the instantaneous RF power dissipation. For modulated RF signals with unfrequent short bursts of peak power with duration below a microsecond, it is usually sufficient to rely on the device average temperature. Indeed, given the thermal response is negligible below a microsecond in Fig. 2.21, the instantaneous operating device temperature is not expected to substantially depart from its average value. The main function of the thermal network is then to calculate the average temperature of operation for the device.

Traps can also greatly affect the operation of GaN HEMTs particularly under the presence of large instantaneous drain voltages. For GaN HEMTs on SiC without passivation and field plates, pulsed-RF Cold FET measurements can be used to verify that the drain resistance increases rapidly for large drain voltages [16]. The increase in drain resistance becomes then the leading cause of current collapse and knee walkout in these devices. This effect is commonly referred as the virtual gate effect [17] as the surface donor traps act as a gate controlling the 2DEG population in the ungated region between the gate and drain contacts. In the presence of large drain voltages, the large lateral surface electric field promotes the migration of electrons from the gate to the surface of the ungated regions where they compensate the surface donor traps. In commercialized GaN HEMTs field plates and passivation are used to greatly reduce this effect.

Residual trapping effects are also observed inside the gated region when the devices are excited by high DC or RF bias drain voltages (typically above 50 V). In one possible mechanism, acceptor traps located near the GaN-SiC substrate act as a back gate contact effectively shifting the threshold voltage as they get filled. The interactions of these traps with the RF signals in power amplifiers are also of great interest.

The time constants for the emission and capture of electrons are typically several orders of magnitude larger than the period of the RF signals. Yet it has been observed that under large-signal RF excitations, the average trap population can be strongly affected by the dynamic RF load line trajectory if this one partially extends over a region of high-drain voltages exhibiting a very large capture rate. When the time spent in such a region accumulates to a time comparable to the capture time-constant, the trap occupation will gradually increase. This effect is referred as the cyclostationary effect [18]. It was first observed in silicon MOSFETs where the flicker noise generated by the traps was found to be dependent on the RF load lines. A similar cyclostationary effect was also observed for flicker ($1/f$) noise in GaN HEMTs under large-signal operation [19] and a two trap-level model was proposed [13].

Consider the following state-equation for a single donor trap where the emission τ_e and capture τ_c time-constants are modulated by the RF signals [20] via their bias dependence

$$\frac{df_{T,D}(t)}{dt} + \left(\frac{1}{\tau_e\left(v_{GS}(t), v_{DS}(t), T_{dev}\right)} + \frac{1}{\tau_c\left(v_{GS}(t), v_{DS}(t), T_{dev}\right)}\right) f_{T,D}(t)$$
$$= \frac{1}{\tau_c\left(v_{GS}(t), v_{DS}(t), T_{dev}\right)}. \tag{2.8}$$

It has been determined [20] that for periodical RF excitations, the time average of the capture and emission rates should be used for calculating the average trap population $\langle f_{T,D}(t)\rangle$ and the noise spectral density $S_{W_g \Delta\ell_D n_{T,D}}$ for the channel charge $n_{T,D} = f_{T,D}(t)N_{T,D}$

$$\langle f_{T,D}\rangle = \frac{\langle g_c\rangle}{\langle g_c\rangle + \langle g_e\rangle} = \frac{\langle g_c\rangle}{\langle \lambda_0\rangle}, \tag{2.9}$$

$$S_{W_g \Delta \ell_D n_{T,D}} = N_{T,D} \langle f_{T,D} \rangle \left(1 - \langle f_{T,D} \rangle\right) \frac{2\langle \lambda_0 \rangle}{1 + \langle \lambda_0 \rangle^2 \Delta \omega^2} W_g \Delta \ell_D, \quad (2.10)$$

where we have $\langle \lambda_0 \rangle = \langle g_c \rangle + \langle g_e \rangle$ using the definitions

$$\langle g_c \rangle = \frac{1}{T} \int_0^T \frac{1}{\tau_c \left(v_{GS}(t), v_{DS}(t), T_{dev}\right)} dt, \quad \text{and}$$
$$\langle g_e \rangle = \frac{1}{T} \int_0^T \frac{1}{\tau_e \left(v_{GS}(t), v_{DS}(t), T_{dev}\right)}, dt \quad (2.11)$$

where T is the RF period, W_g the gate width, $\Delta \ell_D$ the width of the pinchoff region in the gate-to-drain spacer, and $N_{T,D}$ the concentration of donor traps. Under CW large signal excitations, the average rates $\langle g_{c/e} \rangle$ will differ from the DC rates $1/\tau_{c/e}(V_{GS,dc}, V_{DS,dc}, T_{dev})$, due to the bias dependence of $\tau_{c/e}$ upon v_{GS} and v_{DS}. The resulting drastic difference in noise spectral density this can produce is referred to as the cyclostationary effect. The cyclostationary effect can also have a strong impact on the device microwave response since the trap occupation can profoundly modify the IV characteristics.

The trapping equations (2.9) and (2.10) hold for CW RF excitations. Let us now discuss their generalization to communication waveforms which exhibit time-varying envelopes. Since the variation of the trap occupation is slow compared to the RF excitations, the envelope of the trap occupation should follow the same rate equation

$$\frac{d\overline{f}_{T,D}(t)}{dt} + \overline{\lambda}_0(t)\overline{f}_{T,D}(t) = \overline{g}_c(t), \quad (2.12)$$

where we now use the following time-moving averages:

$$\overline{g}_c(t) = \frac{1}{T} \int_{t-T}^{t} \frac{1}{\tau_c[v_{GS}(t), v_{DS}(t), T_{dev}]} dt \quad \text{and}$$
$$\overline{g}_e(t) = \frac{1}{T} \int_{t-T}^{t} \frac{1}{\tau_e[v_{GS}(t), v_{DS}(t), T_{dev}]} dt, \quad (2.13)$$

with $\overline{\lambda}_0 = \overline{g}_c + \overline{g}_e$. Figure 2.23 shows a possible circuit implementation of this trap model where the RC time-constant is selected to be larger than the RF period T and smaller than the smallest trap-constant: $T << RC << (\overline{\lambda}_0)^{-1}$. It should be noted that the electron capture and emission processes of deep donor states are facilitated by Fowler–Nordheim tunneling for large applied drain voltages and can be modeled as a field-assisted thermal process (Poole–Frenkel effect) [21].

Having introduced the master equation describing the occupation of deep traps under modulated large-signal RF operation and the resulting cyclostationary effect we discuss now the case of the virtual gate in the gate-to-drain access region. The

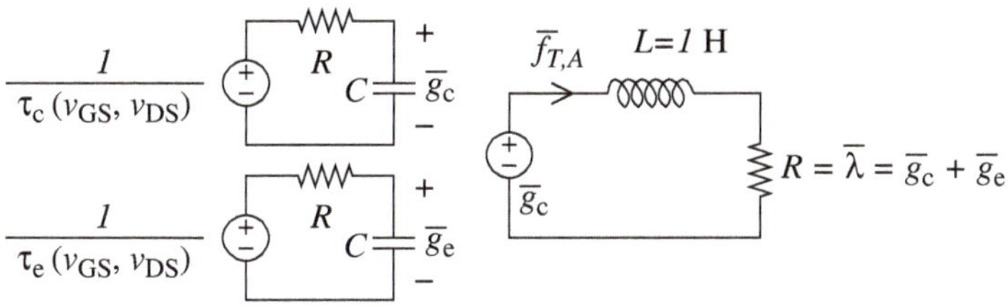

Fig. 2.23 Trap circuit model accounting for the cyclostationary effect

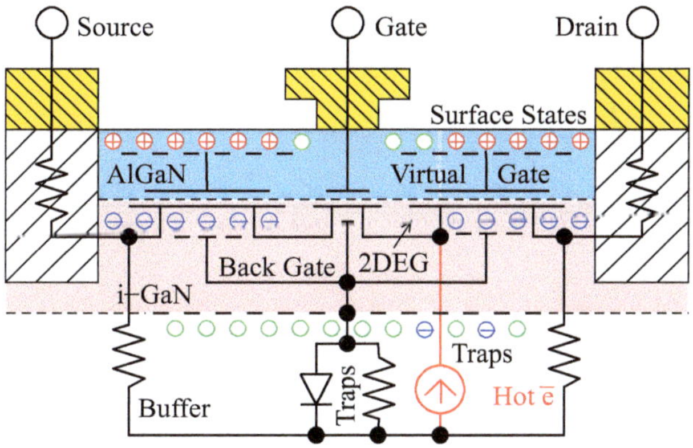

Fig. 2.24 Physical model of a GaN HEMT with various possible trapping mechanism ([16], reprinted with permission from IEEE)

access region shown in Fig. 2.24 can exhibit a large variation in drain resistance R_D when the donor surface states capture electrons at high-drain voltages. Figure 2.25 gives the extracted trap occupation in the depleted region as a function of bias for an unpassivated device [16]. The trap occupation is estimated from the physical modeling of the bias dependence of R_D measured using the pulsed cold-FET technique [16]. The larger the trap occupation, the larger the drain resistance and knee walkout. The bias dependence of the trap occupation also indicates in virtue of (2.9) that the capture rate is higher than the emission rate at high-drain voltages. In practice, passivation and field plates are used to mitigate this effect. Figure 2.26 shows the impact of the load lines for a load varying from 50 to 170 Ω upon the average DC operating point (circles) and average trap occupation factor (squares). The higher the impedances, the higher the instantaneous drain voltage reaches and the higher the trap occupation and knee walkout are, as revealed by the load line shapes.

In this section we have seen how both thermal and trapping memory effects can induce knee voltage walkout degrading the device performance. Pulsed-IV pulsed RF experiments segregating between thermal and trapping relaxations were reported for GaN HEMTs [22]. In general, thermal effects usually dominate for instantaneous drain voltages below 40 V. On the other hand, when the peak instantaneous drain

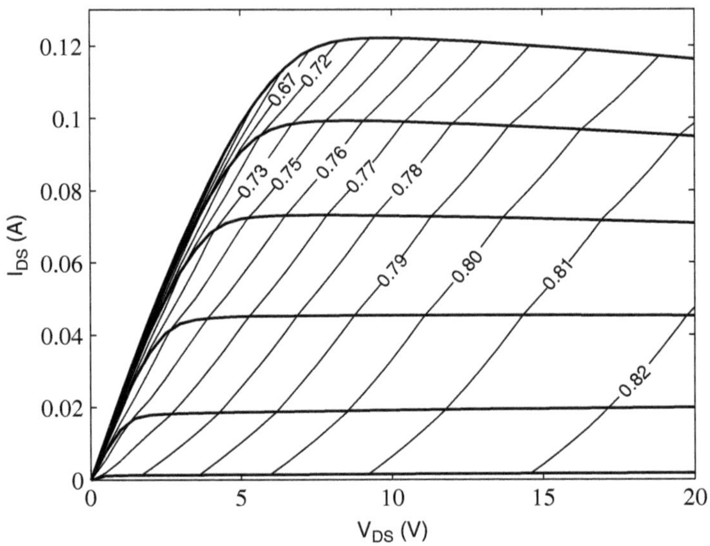

Fig. 2.25 Donor trap occupation in an unpassivated GaN HEMT

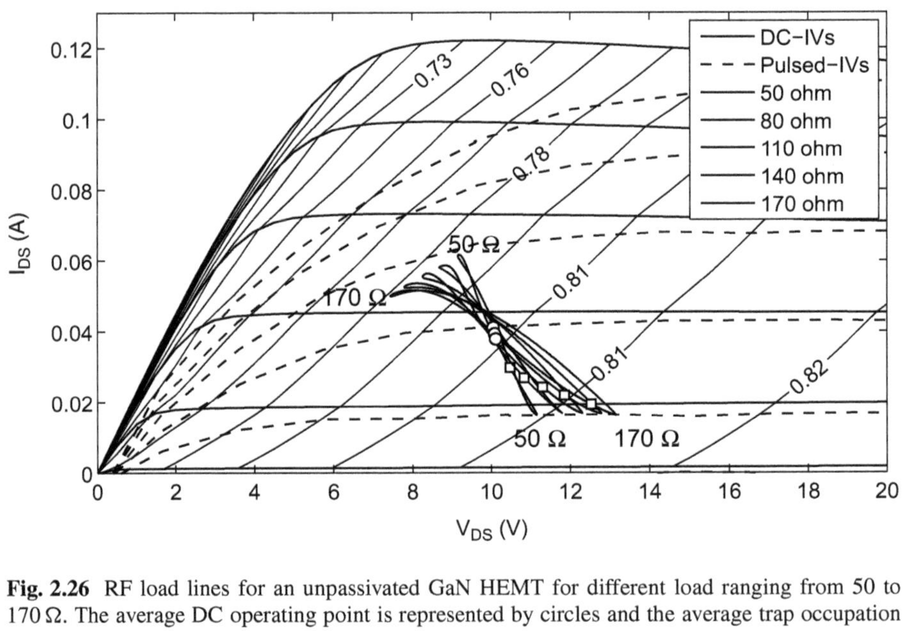

Fig. 2.26 RF load lines for an unpassivated GaN HEMT for different load ranging from 50 to 170 Ω. The average DC operating point is represented by circles and the average trap occupation factor by squares ([19], reprinted with permission from IEEE)

voltage reaches above 50 V under sustained peak power operation, traps will rapidly charge. The charging time was estimated to be around 150 μs in [22] for a device under pulsed class-A operation. Once the device returns to lower power operation, a relaxation time of 300 ms was estimated in the device considered in [22]. During this recovery interval the device exhibits a noticeable degradation in its IV characteristic (knee walkout). The sustained operation of GaN HEMTs at peak power width drain voltage above 50 V is thus not desirable. Luckily for modern modulated signals used in wireless communication, periods of peak power occur at low rate and are not usually sustained for a duration of more than a few microseconds.

2.7 Large Signal Measurement Techniques

The development of large-signal microwave models for FETs has been traditionally conducted using the measurement of the current–voltage characteristics and the measurement of small-signal S-parameters at various bias points. In the last decade, the large-signal vector characterization of transistors at high frequencies has been made possible, thanks to the availability of equipment such as the nonlinear vector network analyzer (NVNA) or the large signal network analyzer (LSNA). Following [23] we shall refer to them as LSVNA.

LSVNAs are pieces of equipment focused on the acquisition of the multi-harmonic response of transistors excited by CW or periodically modulated signals as shown in Fig. 2.27. That is, the amplitude and phase of the incident and reflected waves $a_r(n\omega)$ and $b_r(n\omega)$ at port r are measured for a large number N of harmonic and the waveforms can be reconstituted in the time domain using a truncated Fourier series

$$a_r(t) = \sum_{n=1}^{N} a_r(n\omega)\exp\left[jn\omega t\right],$$

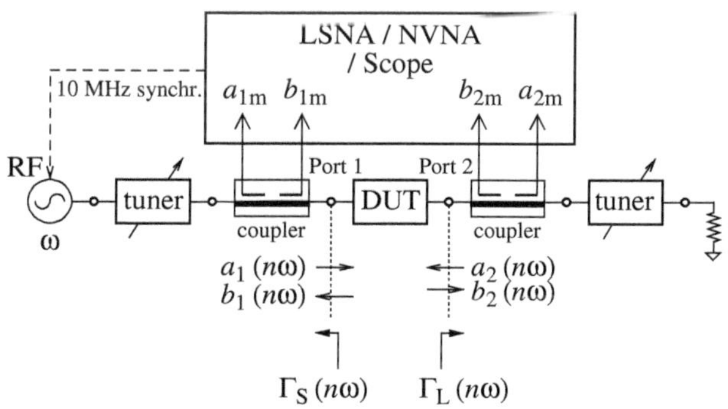

Fig. 2.27 Typical large-signal network analyzer (LSNA) testbed

$$b_r(t) = \sum_{n=1}^{N} b_r(n\omega)\exp\left[jn\omega t\right].$$

When using LSVNAs, the wave normalization for a_r and b_r is such that the instantaneous drain voltages v_r and currents i_r are given by

$$v_r(n\omega) = a_r(n\omega) + b_r(n\omega) \qquad \text{and} \qquad i_r(n\omega) = \frac{a_r(n\omega) - b_r(n\omega)}{50},$$

$$v_r(t) = a_r(t) + b_r(t) \qquad \text{and} \qquad i_r(t) = \frac{a_r(t) - b_r(t)}{50}.$$

Thus, LSVNAs have the ability to acquire the complex incident and reflected waves $a_r(n\omega)$ and $b_r(n\omega)$ at all port r and for a large number of harmonics. This differs from conventional vector network analyzers (VNA) for which only the S-parameter ratios

$$S_{ji}(\omega) = \left.\frac{b_j(\omega)}{a_i(\omega)}\right|_{a_{q\neq i}(\omega)=0}$$

are extracted from the measurements assuming all ports q are matched (terminated by $50\,\Omega$) except the port i where the excitation is applied. Also unlike for conventional VNA, the loads in LSVNA are not limited to $50\,\Omega$ and complex passive or active (RF sources) multi-harmonic source $\Gamma_S(n\omega)$ and load $\Gamma_L(n\omega)$ can be used as shown in Fig. 2.27. However all the RF sources used whether they be at the input or output must be phase-locked to the LSVNA receivers via the shared 10 MHz reference (see Fig. 2.27).

A discussion of the architecture of LSVNAs is beyond the scope of this chapter. However, it can be stated that LSNAs rely on subsampling down-conversion, whereas NVNAs rely on mixer down-conversion. The LSNA thus acquires all the harmonics simultaneously instead of acquiring them one at a time like in the NVNA. However, owing to its small receiver bandwidth, an improved dynamic range is then achieved by the NVNA at the price of longer acquisition times. Note that the calibration of LSVNA requires three types of calibration: (1) relative calibration, (2) power calibration, and (3) harmonic phase reference (HPR). The relative calibration like in VNAs relies on conventional calibration techniques such as SOLT (short-open-load-thru), LRM (load-reflect-match), and TRL (thru-reflect-load). This calibration is performed at all the harmonic frequencies. A power meter is used for the power calibration of the fundamental and harmonics. A comb generator is used for the phase reference calibration between the fundamental and harmonics. Note that the NVNA requires the use of a second HPR on a dedicated channel during the device measurement to determine the phase of the harmonics given they are measured at different times in that system. We explore in Sect. 2.8 the application of such a LSVNA testbed to the quasi-interactive design of PAs.

2.8 Load-Pull and Interactive Load Line-Based PA Design

The LSVNAs can be used to perform *load-pull* and *source-pull* measurements to obtain the optimal load $Z_L(n\omega)$ or source $Z_S(n\omega)$ impedances at the fundamental and harmonics for maximum output power or gain or efficiency. Instead of using passive tuners, an active load-pull (ALP) and source-pull (ASP) configuration can implemented as shown in Fig. 2.28 using various CW multi-harmonic excitations. The phase and power of each fundamental and harmonic tones can then be controlled except for the phase of the input fundamental signal which can take arbitrary values since it is the reference.

Real-time active load-pull (RTALP) and source-pull (RTASP) can also be realized by using an offset frequency Δf to automatically sweep in turns the phase of each harmonic or output fundamental tones. RTALP requires then the acquisition of the modulated multi-harmonic incident and reflected waves of frequencies $nf_0 + p\Delta f$. A single modulated measurement will then provide such a wealth of data that a circuit-based device model can be extracted from a single RTALP measurement [22]. Load-pull measurement is itself greatly accelerated by RTALP.

Let us consider the case of a packaged pHEMT at 2 GHz. As shown in Fig. 2.29 for the second-harmonic RTALP, each measurement yields a closed-trajectory for $\Gamma_L(2\omega)$. The radius of the closed-contour trajectories increases with the injected power level. It is seen that with just 13 different measurements the Smith chart is well covered by the associated load trajectories. Note that each RTALP measurement is performed in 1 ms when selecting the frequency resolution to be

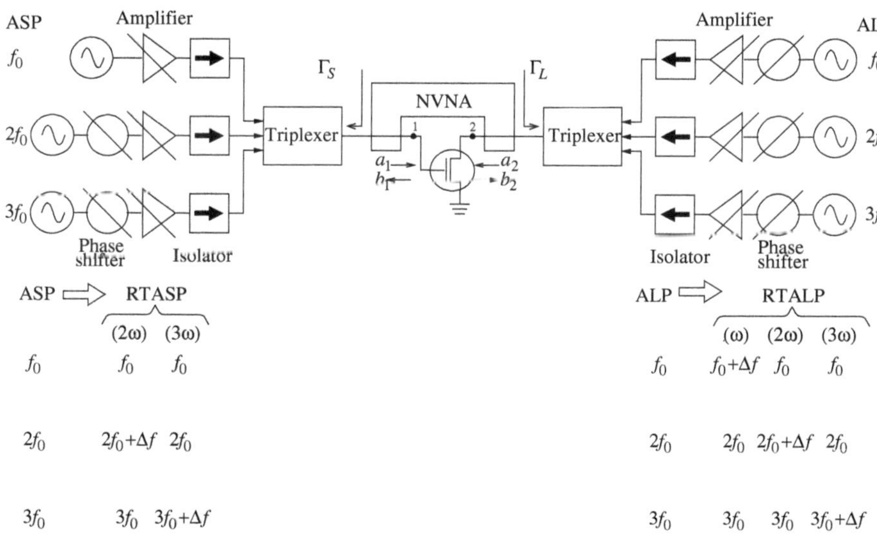

Fig. 2.28 Real-time active load-pull/source-pull (RTALP/RTASP) and active load-pull/source-pull (ALP/ASP) setups for transistor characterization in various modes of operation

Fig. 2.29 Closed-trajectories of $\Gamma_L(2\omega)$ of pHEMT for 13 different power levels

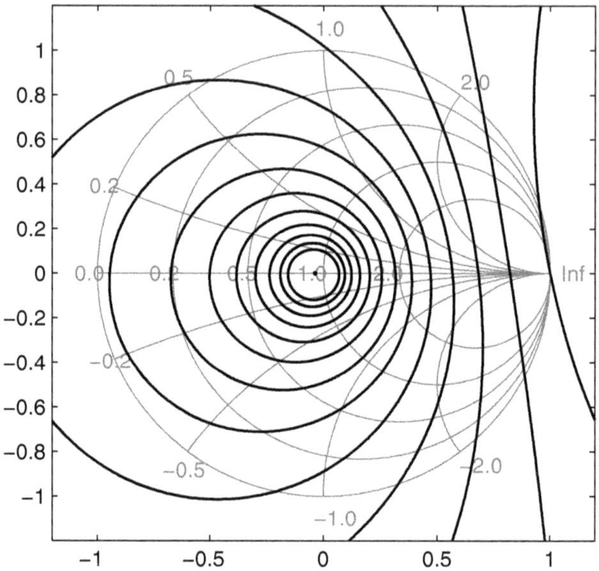

1 kHz. RTALP with the LSNA opens then the possibility for interactive load-pull measurements (Fig. 2.30).

The resulting second-harmonic load-pull contour plot for the PAE is shown in Fig. 2.31. Note that the maximum PAE is taking place for $\Gamma_L(2\omega)$ on the edge of the Smith chart and located near point A. Using ALP this corresponds to about 72% PAE. The associated load lines obtained from the second-harmonic RTALP for this transistor are shown in Fig. 2.31. The benefit of using a LSNA/NVNA is that one can inspect the voltage and current waveforms. Figure 2.32a shows the voltage and current waveforms for $\Gamma_L(2\omega)$ located at the point A in Fig. 2.31. The instantaneous power is also plotted in Fig. 2.32b. It is seen that the instantaneous power dissipated is negative at times. This indicates that the device and its package are storing energy. The average power dissipated in the RF period corresponding to the area under the curve remains, however, positive. The output power, gain, and efficiency are plotted versus the input power in Fig. 2.33.

Although the LSVNA permits to visualize the load lines and the voltage and current waveforms of the transistor at the package reference planes, it is not possible to determine with certainty the mode of operation of the device as this one is defined by the load lines and the voltage and current waveforms at the current-source reference planes (see Fig. 2.34). De-embedding from the package to the current-source reference planes can be performed but this requires the availability of an accurate de-embedding device model (see Chap. 3). Nonetheless, the measured waveforms and load lines acquired by the LSVNA provide the most useful and accurate tools available to verify that a device is operating as expected at the package reference planes when compared to waveforms and load lines obtained in simulations using the device model.

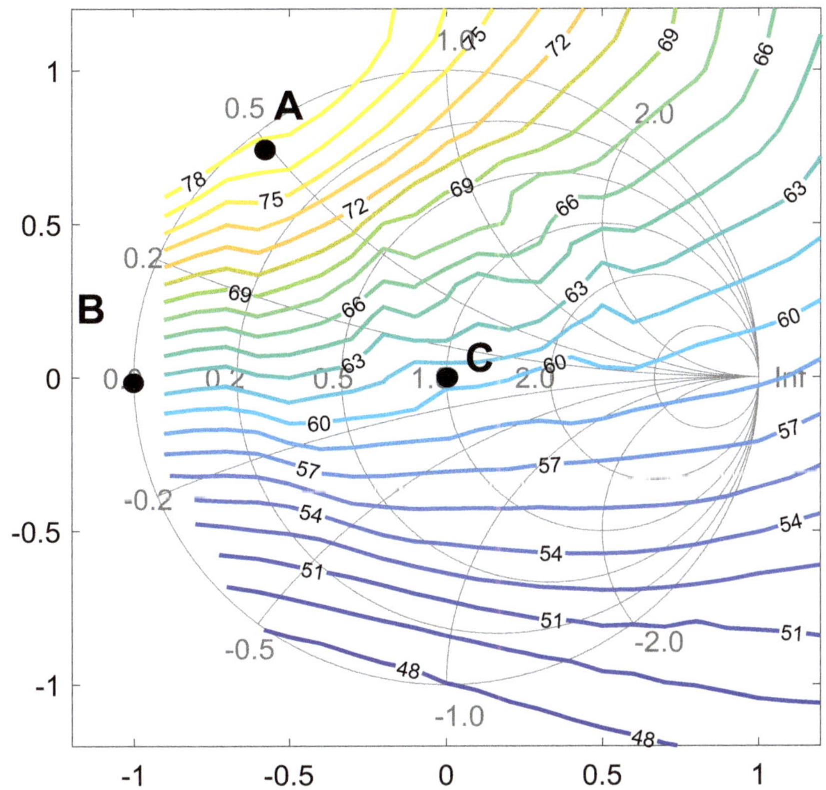

Fig. 2.30 Contour plot for the power added efficiency (PAE) obtained from a few second-harmonic RTALP measurements ([24], reprinted with permission from IEEE)

2.9 Behavioral Modeling

2.9.1 Harmonic Volterra Functions

In this section we discuss the behavioral modeling of nonlinear devices characterized by a LSVNA.

For this introduction to the behavioral modeling of nonlinear devices or circuits we will initially focus on single-input single-output (SISO) devices to simplify the presentation. As shown in Fig. 2.35a–c, a SISO model can be directly applied to a single-port (two-terminal) device as in Fig. 2.35b or to a two-port (four-terminal) device as in Fig. 2.35c when the load $Z_L(\omega)$ at port 2 is assumed to be known for the fundamental and all harmonics.

Assume we now consider injecting a signal consisting of two harmonic-related incident waves $a(\omega_0)$ and $a(2\omega_0)$ at the input of a SISO nonlinear circuit with infinite nonlinearity order. All other incident harmonics $a(n\omega_0)$ with $n > 2$ are

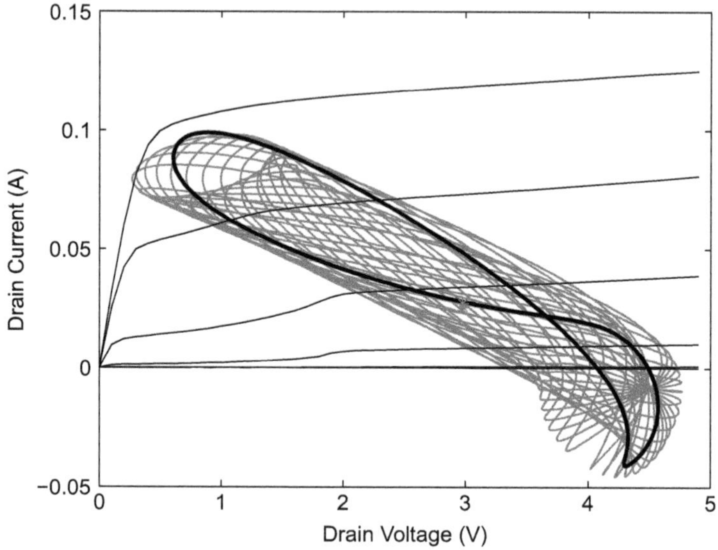

Fig. 2.31 Load lines obtained in the second-harmonic RTALP measurements ([24], reprinted with permission from IEEE)

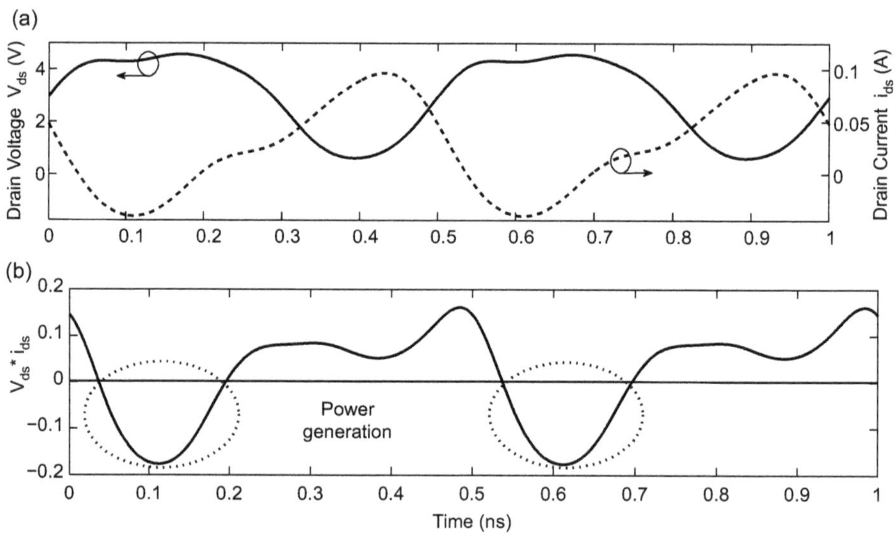

Fig. 2.32 (**a**) Voltage and current waveforms and (**b**) instantaneous power dissipated ([24], reprinted with permission from IEEE)

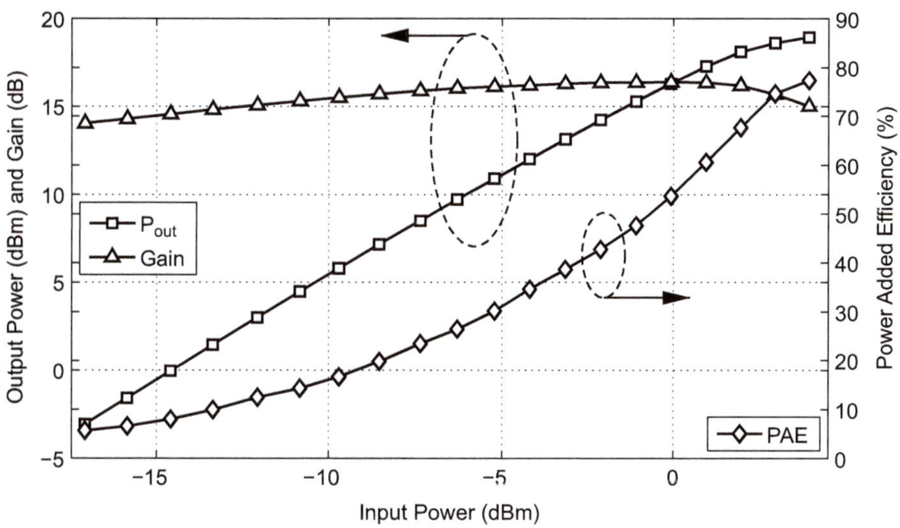

Fig. 2.33 The output power, gain, and efficiency versus the input power ([24], reprinted with permission from IEEE)

Fig. 2.34 Package and current-source reference planes in a device model

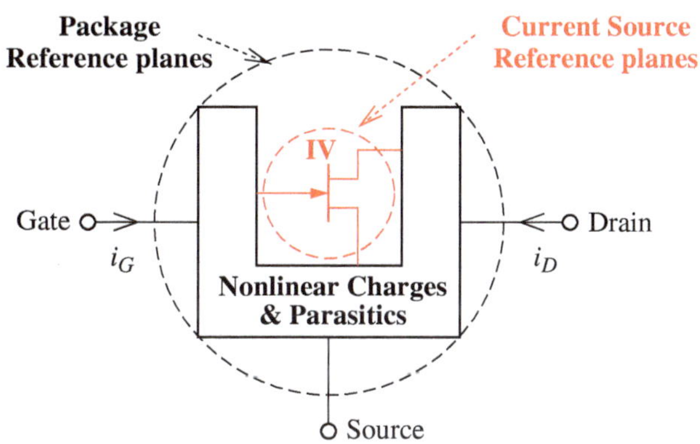

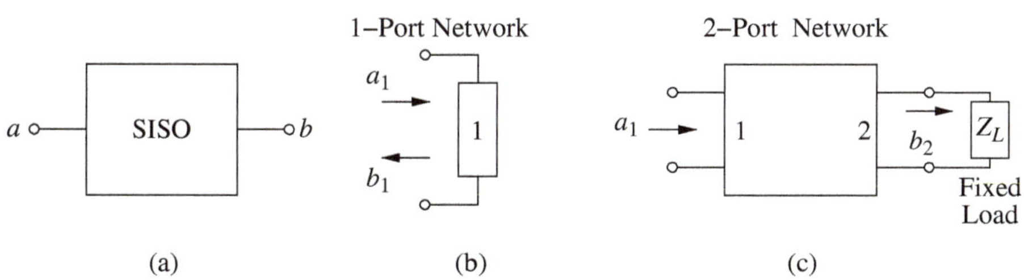

Fig. 2.35 (**a**) Single-input single-output (SISO) model for (**b**) one-port and (**c**) two-port circuits

assumed to be either negligible or terminated by constant harmonic impedances. The system being of an infinite order, an infinite number of harmonics $b(n\omega)$ is generated. Each of these harmonics can then be expressed in terms of a *describing* function B_n of the form

$$b(n\omega) = B_n\left(V_{\mathrm{DC}}, a(\omega_0), a^*(\omega), a(2\omega_0), a^*(2\omega_0)\right).$$

It is seen that the B_n functions are expended in terms of complex variables $a(n\omega_0)$ and their conjugates. This is a requirement for the B_n functions to be holomorphic such that their derivatives in the complex domain be independent of direction and each function B_n be equal to its own Taylor series [25]. Alternately, the B_n functions can be expanded in terms of real numbers (amplitude and phase or real and imaginary components)

$$\begin{aligned} b(n\omega) &= B_n'\left(V_{\mathrm{DC}}, |a(\omega_0)|, \angle a(\omega), |a(2\omega_0)|, \angle a(2\omega_0)\right), \\ &= B_n''\left(V_{\mathrm{DC}}, \mathrm{Re}\left[a(\omega_0)\right], \mathrm{Im}\left[a(\omega)\right], \mathrm{Re}\left[a(2\omega_0)\right], \mathrm{Im}\left[a(2\omega_0)\right]\right). \end{aligned}$$

It is however possible to further detail the nature of the B_n functions. Indeed, it can be verified that the reflected waves can be represented in the frequency domain using the following harmonic Volterra series expansion [13] (only the first three harmonics are shown here),

$$b(0) = V_{0,1}(\mathbf{x}_2), \tag{2.14}$$

$$b(\omega_0) = a(\omega_0)\cdot V_{1,1}(\mathbf{x}_2) + a^*(\omega_0)a(2\omega_0)\cdot V_{1,2}(\mathbf{x}_2), \tag{2.15}$$

$$b(2\omega_0) = a(2\omega_0)\cdot V_{2,1}(\mathbf{x}_2) + a^2(\omega_0)\cdot V_{2,2}(\mathbf{x}_2), \tag{2.16}$$

$$\begin{aligned} b(3\omega_0) &= a^3(\omega_0)\cdot V_{3,1}(\mathbf{x}_2) + a(\omega_0)a(2\omega_0)\cdot V_{3,2}(\mathbf{x}_2) \\ &\quad + a^*(\omega_0)a^2(2\omega_0)\cdot V_{3,3}(\mathbf{x}_2). \end{aligned} \tag{2.17}$$

The functions $V_{n,i}(\mathbf{x}_2)$ for the harmonic n in Eqs. (2.14)–(2.17) are found to be dependent on five DC or baseband terms $\mathbf{x}_2 = [x_{2,0}, x_{2,1}, x_{2,2}, x_{2,3}, x_{2,4}]$ given by

$$\begin{aligned} &x_{2,0} = a(0), \\ &x_{2,1} = a^*(\omega_0)a(\omega_0), \quad x_{2,2} = a^*(2\omega_0)a(2\omega_0), \\ &x_{2,3} = a^*(2\omega_0)a^2(\omega_0),\ x_{2,4} = x_{2,3}^* = a(2\omega_0)a^{*2}(\omega_0). \end{aligned}$$

As an example of application of the *harmonic Volterra function* expansion consider the measured output power contour plot shown in Fig. 2.36 which is obtained from a second-harmonic load-pull performed on a nonlinear negative resistance. The *harmonic Volterra function* is seen to be able to accurately fit the measured output power. Note that the harmonic Volterra function expansion for incident waves with three and four incident harmonics has been reported in [13].

In the time domain the measured incident and transmitted waves are of the form

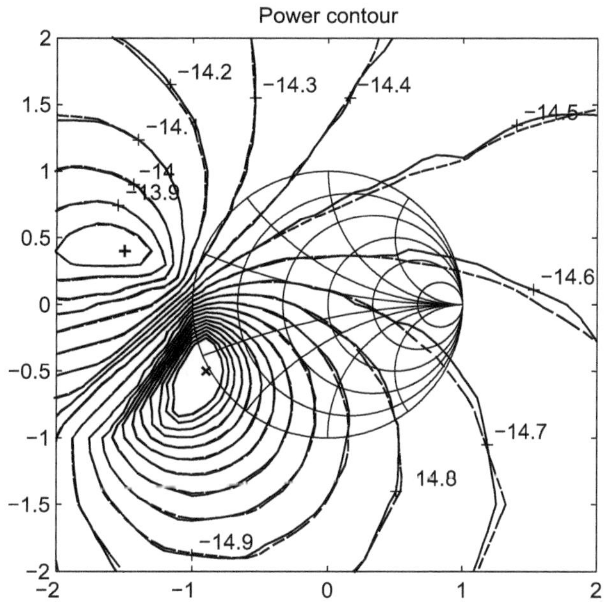

Fig. 2.36 Measured (plain lines) and *harmonic Volterra function* representation (dashed lines) of the output power of a device obtained under second-harmonic load-pull measurement ([13] reprinted with permission from Cambridge University Press)

$$
\begin{aligned}
a(t) &= \mathrm{Re}\left\{\sum_p a(p\omega_0)\exp\left[jp\omega_0 t\right]\right\} \\
&= \mathrm{Re}\left\{\sum_p |a(p\omega_0)|\exp\left[j\left(p\omega_0 t + \angle a(p\omega_0)\right)\right]\right\}, \\
b(t) &= \mathrm{Re}\left\{\sum_p b(p\omega_0)\exp\left[jp\omega_0 t\right]\right\} \\
&= \mathrm{Re}\left\{\sum_p |b(p\omega_0)|\exp\left[j\left(p\omega_0 t + \angle b(p\omega_0)\right)\right]\right\}.
\end{aligned}
$$

It is to be noted that nonlinear measurements are acquired up to an arbitrary phase which is usually taken as the phase of the fundamental incident wave at port 1 $\angle a_1(\omega_0)$ in multi-port devices. We can move the time origin by a delay τ such that we have $a_1(\omega_0) = |a_1(\omega_0)|$ or equivalently $\angle a_1 = 0$. It results that all the harmonic phases at port r are then updated from $\angle a_r(n\omega_0)$ to $\angle a_r(n\omega_0) - n\angle a_1(\omega_0)$ and from $\angle b_r(n\omega_0)$ to $\angle b_r(n\omega_0) - n\angle a_1(\omega_0)$. Such a time-shift operation on the measured data will be assumed to have been performed in the perturbative nonlinear model presented next such that the resulting $b_r(p\omega)$ are now independent of the arbitrary input phase $\angle a_1$ and found to be invariant upon successive measurements.

To simply the behavioral model it is possible to use a perturbative approximation which relies on a simple linear expansion of the harmonic Volterra functions in terms of the harmonics ($h > 1$) of the incident waves $a_p(h\omega_0)$ and $a_p^*(h\omega_0)$ applied at port p. Higher-order expansions are only retained for the incident waves at the fundamental frequency $a_p(\omega_0)$. This is the so-called *poly-harmonic distortion* (PHD) approximation (also referred to as X-parameters) [25]. For a SISO device the transmitted wave from port 1 to port 2 are then given by

$$b_2(n\omega_0) = \sum_{h=1}^{H} S_{2,1;n,h}(V_0, |a_1(\omega_0)|)\ a_1(h\omega_0) + T_{2,1;n,h}(V_0, |a_1(\omega_0)|)\ a_1^*(h\omega_0),$$

using a harmonic expansion up to a number H. Note that no phasor is appearing in this presentation since we have changed the time reference to obtain $\angle a_1 = 0$. The indices 1 and 2 correspond to the input port 1 and output port 2 of the SISO model of the two-port system. For a DIDO (double input double output) model of a two-port device, this equation for the transmitted wave b_r at port $r = 1, 2$ is generalized to

$$\begin{aligned} b_r(n\omega_0) = \sum_{p=1}^{2}\sum_{h=1}^{H} \quad & S_{r,p;n,h}(V_0, |a_1(\omega_0)|, |a_2(\omega_0)|, \angle a_2(\omega_0))\ a_p(h\omega_0) \\ & + T_{r,p;n,h}(V_0, |a_1(\omega_0)|, |a_2(\omega_0)|, \angle a_2(\omega_0))\ a_p^*(h\omega_0). \end{aligned}$$

In the PHD approximation, the multi-harmonic transmitted and reflected waves are only a linear function of the harmonics of the incident waves $a_p(h\omega_0)$ and $a_p^*(h\omega_0)$. It is experienced in practice that most amplifiers operate in a mild nonlinear regime, and the linear expansion in terms of the harmonics used by the PHD model approaches very well the results of the harmonic Volterra function if a sufficiently large number H of incident harmonic waves is used in the expansion. This linearization then simplifies the system identification of the PA behavioral model. For example, the fit of the measured second-harmonic load-pull in Fig. 2.36 can be achieved with a *linear* expansion in terms of only four harmonics ($H = 4$) instead of a *nonlinear* expansion in terms of two harmonics ($H = 2$) using the Harmonic Volterra series.

It also results from the PHD linear approximation for the harmonics that the operating point reduces to the DC biases and the incident power $|a_1(\omega_0)|^2$ and $|a_2(\omega_0)|^2$ for the fundamental frequency. This operating point is referred to as the large-signal operating point (LSOP) since the harmonics are only acting as perturbations. We shall see in Chap. 3 that the tuning of the LSOP is of critical importance for the optimization of the power efficiency in a PA.

References

1. Pritiskutch, J., & Hanson, B. (1996). Understanding LDMOS device fundamentals. *Microwaves & RF, 37*(8), 114–116.
2. Mason, S. J. (1954). Power gain in feedback amplifiers. *IRE Transactions on Circuit Theory, CT-1*(2), 20–25.

3. Raab, F. H., & Sokal, N. O. (1978). Transistor power losses in the Class-E tuned power amplifier. *IEEE Journal of Solid State Circuits, 13*(6), 912–914.
4. Mader, T., Bryerton, E., Markovic, M., Forman, M., & Popovic, Z. (1998). Switched-mode high-efficiency microwave power amplifiers in a free-space power-combiner array. *IEEE Transactions Microwave Theory and Techniques, MTT-46*(10), 1391–1398.
5. Burns, J. R. (1967). High-frequency characteristics of the insulated gate field-effect transistor. *RCA Review, 28*, 385–418.
6. Roblin, P., Kang, S. C., & Morkoc, H. (1990). Analytic solution of the velocity-saturated MOSFET/MODFET wave-equation and its application to the prediction of the MODFET microwave characteristics. *IEEE Transactions on Electron Devices, ED-37*(7), 1608–1623.
7. Kang, S. C., Roblin, P., & Liou, W. R. (1992). Optimal second-order non-quasi-static equivalent circuit for the MOSFET wave equation. *IEEE Transactions on Electron Devices, 39*(8), 1909–1915.
8. Roblin, P., Kang, S. C., & Liou, W. R. (1991). Improved small-signal equivalent circuit model and large-signal state equations for the MOSFET/MODFET wave equation. *IEEE Transactions on Electron Devices, 38*(8), 1706–1718.
9. Burns J. R. (1969). Large-signal transit-time effects in the MOS transistor. *RCA Review, 30*, 15–35.
10. Mancini, P., Turchetti, C., & Masetti G. (1987). A nonquasi-static analysis of the transient behavior of the long-channel MOST valid in all regions of operation. *IEEE Transactions on Electron Devices, ED-34*, 325–344.
11. Lee, S., Roblin, P., & Lopez, O. (2002). Modeling of distributed parasitics in power FETS, *IEEE Transactions on Electron Device, 49*(10), 1799–1806.
12. Roblin, P., & Rohdin, H. (2002). *High-speed heterostructure devices* (688 pp.). Cambridge: Cambridge University Press.
13. Roblin, P. (2011). *Nonlinear RF circuits and nonlinear vector network analyzers: Interactive measurement and design techniques* (283 pp.). Cambridge: Cambridge University Press.
14. Doo, S. J. (2008). *New Pulsed-IV Pulsed-RF Measurement Techniques for Characterizing Power FETs for Pulsed-RF Power Amplifier Design*. The Ohio State University, Dissertation, Doctor of Philosophy.
15. Dai, W. (2004). *Large-Signal Electro-Thermal LDMOSFET Modeling and the Thermal Memory Effects in RF Power Amplifiers*. The Ohio State University, Dissertation, Doctor of Philosophy.
16. Yang, C. K., Roblin, P., De Groote, F., Ringel, S., Rajan, S., Teyessier, J.-P., et al. (2010). Pulsed-IV pulsed-RF cold-FET parasitic extraction of biased AlGaN/GaN HEMTs using large signal network analyzer. *IEEE Transactions on Microwave Theory and Techniques, 58*(5), 1077–1088.
17. Vetury, R., Zhang, N. Q., Keller, S., & Mishra U. K. (2001). The impact of surface states on the DC and RF characteristics of AlGaN/GaN HFETs. *IEEE Transactions Electron Devices, 48*, 560–566.
18. Miguez, M., & Arnaud, A. (2008). A study of flicker noise in MOS transistor under switched bias condition. *Journal Integrated Circuits and Systems, 3*(2), 63–68.
19. Suh, I., Roblin, P., Ko, Y., Yang, C.-K., Malonis, A., Arehart, A., et al. (2009). Additive phase noise measurements of AlGaN/GaN HEMTs using a large signal network analyzer and a tunable monochromatic light source. In *Microwave Measurement Symposium, 74th ARFTG*, pp. 1–5.
20. Roy, A. S., & Enz, C. C. (2007). Analytical modeling of large-signal cyclo-stationary low-frequency noise with arbitrary periodic input. *IEEE Transactions Electron Devices, 54*(9), 2537–2545.
21. Quéré, R., Jardel, O., Sommet, R., Piotrowicz, S., & Teyssier, J. P. (2015). Self-heating and trap characterization and simulation for large signal GaN transistors modeling. In *Workshop on "Direct Extraction of FET Circuit Models from Microwave and Baseband Large-Signal Measurements for Model-Based Microwave Power Amplifier Design," International Microwave Symposium*, May 18, 2015.

22. Ko, Y., Roblin, P., Yang, C. K., Jang, H., & Poling, B. (2012). New thermometry and trap relaxation characterization techniques for AlGaN/GaN HEMTs using pulsed-RF excitations. In *IEEE Microwave Symposium Digest (MTT)*, pp. 1–3.
23. Angelov, I., Thorsell, M., Andersson, K., Rorsman, N., & Zirath, H. (2012). Recent results on using LSVNA for Compact modeling of GaN FET devices. In *IEEE International Microwave Symposium Workshop (WMB): On "Device Model Extraction form Large-Signal Measurements"*, Montreal, June 2012.
24. Cui, X., Doo, S. J., Roblin, P., Strahler, J., & Rojas-Teran, R. G. (2008). High efficiency RF power amplifier designed with harmonic real-time active load-pull. *IEEE Microwave and Wireless Components Letter, 18*(4), 266–268.
25. Root, D. E., Verspecht, J., Sharrit, D., Wood, J., & Cognata, A. (2005). Broad-band polyharmonic distortion (PHD) behavioral models from fast automated simulations and large-signal scattering functions. *IEEE Transactions on Microwave Theory and Techniques, 53*(11), 3656–3664.

Chapter 3
Power Amplifier Design Using Nonlinear Embedding

3.1 Introduction to the Nonlinear Embedding Device Model

The design of power amplifiers (PAs) involves complex optimizations to deliver the required output power with high efficiency and acceptable linearity over the targeted bandwidth. For narrowband applications the waveforms at the current-source reference planes providing the best efficiency such as class B, class F, inverse class F, class E are well established. In support of the recent interest in developing power-efficient broadband PAs, there is also a sustained effort in developing new waveforms at the current-source reference planes to optimize the power-efficiency over a wide range of frequencies. Both these classic and new waveforms are provided at the current-source reference planes where the device characteristics are accurately represented by its memoryless IV characteristics. Given that in normal transistor operation the IV core is solely responsible for converting DC into RF power, PA design theory is rightly focused on the optimization of the voltage and current waveforms at the current-source reference planes (see Fig. 3.1).

When a specific optimal internal mode of operation is targeted at the device current-source reference planes, the PA designer must then determine the required waveforms at the package reference planes to sustain that internal mode of operation. Two processes are possible for this purpose: nonlinear de-embedding and nonlinear embedding. Both are conceptually presented in Fig. 3.1. Nonlinear de-embedding is the most current approach. When the PA is designed with a simulator, the PA designer just needs to monitor the load lines and waveforms at the current-source reference planes. Alternatively when experimental load-pull measurements have been performed at the package reference planes, a nonlinear de-embedding model must be used to calculate the voltages and currents at the current-source reference planes given the measured voltages and currents at the package reference planes. In both cases, a device model must be available to check if the desired internal mode of operation is approached.

K. Rawat et al., *Bandwidth and Efficiency Enhancement in Radio Frequency Power Amplifiers for Wireless Transmitters*, Analog Circuits and Signal Processing,
https://doi.org/10.1007/978-3-030-38866-9_3

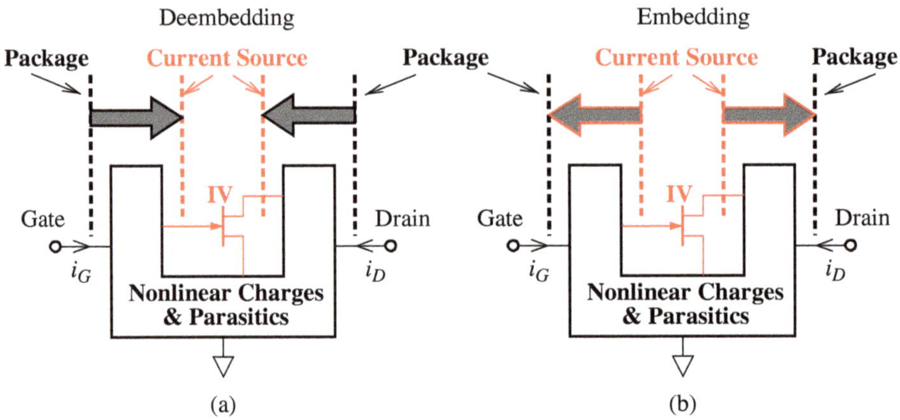

Fig. 3.1 Field effect transistor (FET) models with arrows representing the (**a**) de-embedding and (**b**) embedding processes used to move from the package to current-source reference planes and vice versa

In the de-embedding process just outlined, a very wide design space must be explored using simultaneous multi-harmonic load-pull and source-pull simulations at the package reference planes to determine the multi-harmonic load and source impedance terminations delivering the targeted voltage and current waveforms at the current-source reference planes. These multi-harmonic load-pull and source-pull are affecting each other and must be performed self-consistently for reliable results. As a result, time-consuming simulations are required to obtain an accurate solution even when just considering the first three harmonics.

The alternate design of PAs using nonlinear embedding has been proposed [1, 2] to bypass this blind search. The original approach relied on experimental measurements at low-frequencies, for example, 10 MHz, to select the desired intrinsic device load lines while accounting for memory effects. Nonlinear embedding of the RF contributions of the gate, drain charges, and parasitics was then used to obtain the RF waveforms at the package reference planes. Alternatively, the nonlinear embedding process can be performed automatically in a circuit simulator with the help of a nonlinear embedding device model [3]. The PA designer can then, in a single simulation for each power level, predict from the desired mode of operation at the current-source reference planes the multi-harmonic waveforms required at the transistor package reference planes.

Conceptually nonlinear embedding can be realized with the help of an anti-circuit transfer model relying on negative resistances, negative capacitances, negative inductances, and negative nonlinear charges to cancel the parasitics so as provide access to the intrinsic gate and drain current sources at the current-source reference planes. Such an example is shown in Fig. 3.2 for an extrinsic (package) device model. The resulting nonlinear embedding model provides at the outer reference planes (red line) a direct access to the IV characteristics at the current-source reference planes while at the same time calculating the voltages and currents at the

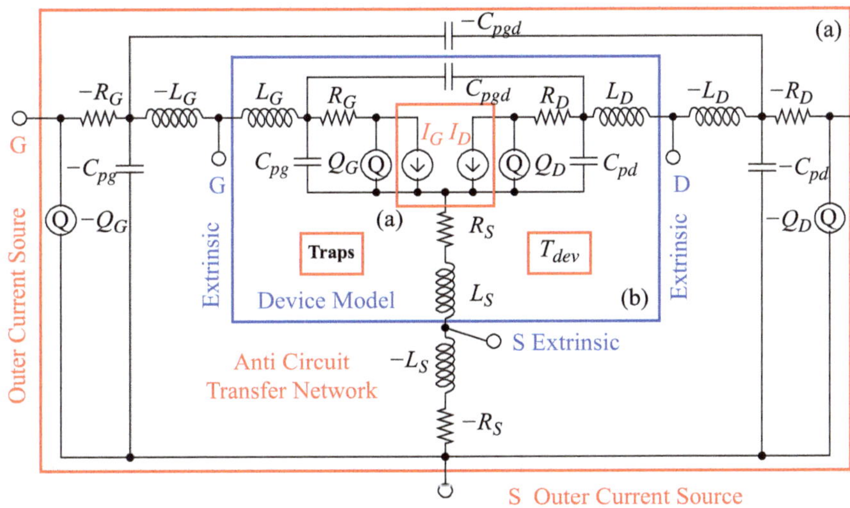

Fig. 3.2 Conceptual embedding device model moving from (**a**) the current-source reference planes (outer red line) to (**b**) the extrinsic reference planes (blue line)

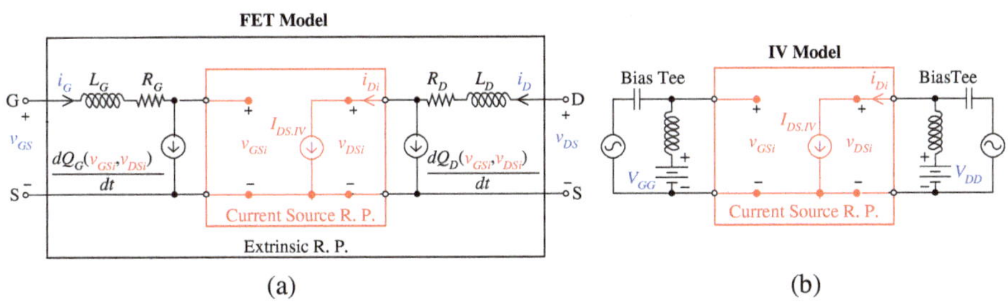

Fig. 3.3 (**a**) Quasi-static FET model and (**b**) IV model circuit defining the desired mode of operation at the current-source reference planes

extrinsic (or package) reference planes (b) (blue line) which are required to sustain the operation selected at the current-source reference planes.

In practice, the implementation of an embedding model in a circuit simulator does not require the use of negative resistances, capacitances, or inductances. In Veriloga, it is simply implemented using the device model equation. Consider the quasi-static field effect transistor (FET) model example presented in Fig. 3.3a. The desired mode of operation at the current-source reference planes is established in the circuit shown in Fig 3.3b such that v_{GSi}, v_{DSi}, and $I_{IV}(v_{GSi}, v_{DSi})$ are known. The calculation of the voltages and currents at the package reference planes is then simply given by the equations

$$i_G = \frac{dQ_G(v_{GSi}, v_{DSi})}{dt},$$

$$
\begin{aligned}
v_{GS} &= v_{GSi} + R_G i_G + L_G \frac{di_G}{dt} \\
i_D &= I_D(v_{GSi}, v_{DSi}) + \frac{dQ_D(v_{GSi}, v_{DSi})}{dt}, \\
v_{DS} &= v_{DSi} + R_D i_D + L_D \frac{di_D}{dt}.
\end{aligned}
$$

An alternate implementation can also be developed in schematic using symbolically defined devices (SDD). Examples of such embedding device circuits were reported in [3] for a GaN HEMT Angelov device model [4] and in [5] for a silicon-on-insulator (SOI) metal-oxide-semiconductor field-effect transistor (MOSFET) using an artificial neural network [6]. In all cases when implemented in a circuit simulator, the embedding device model calculates the exact waveforms required at the package reference planes up to numerical precision. Note that no parameter extraction is required for developing a nonlinear embedding device model since it relies on the same exact model equations and model parameters as the original model. The FET model and its associated nonlinear embedding model are symbolically represented in Fig. 3.4a, b.

To experimentally investigate the design of a PA using *nonlinear embedding*, we will first focus on the synthesis of a PA operating in class-B modes for different fundamental loads. A commercial packaged GaN device with 15 W peak power modeled using an Angelov model is used for this demonstration.

The drain voltage (solid lines ←, red) and drain current (solid lines →, blue) waveforms generated during the embedding process at each of the reference planes of interest are shown in Fig. 3.5a–f for a GaN HEMT operating in class B. Figure 3.5a and e shows the selected intrinsic load lines and waveforms, respectively, obtained in simulation for six different power levels. The same exact

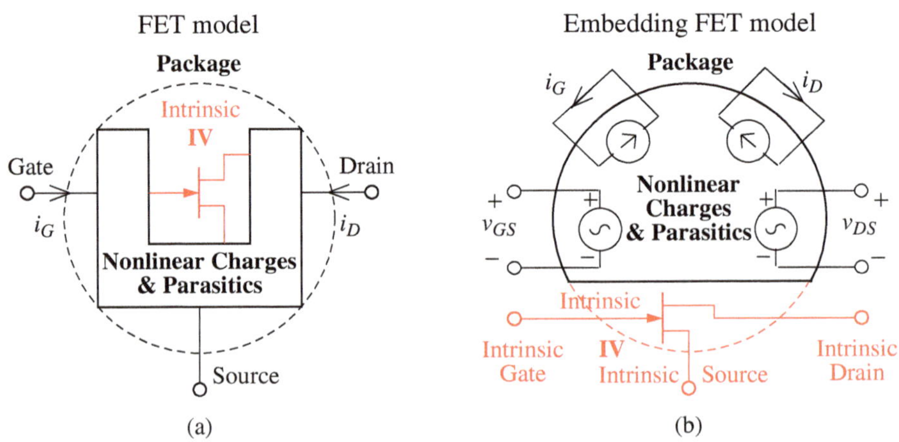

Fig. 3.4 Circuit symbol for (**a**) the *device model* and (**b**) the *embedding device model* ([7], reprinted with permission from IEEE)

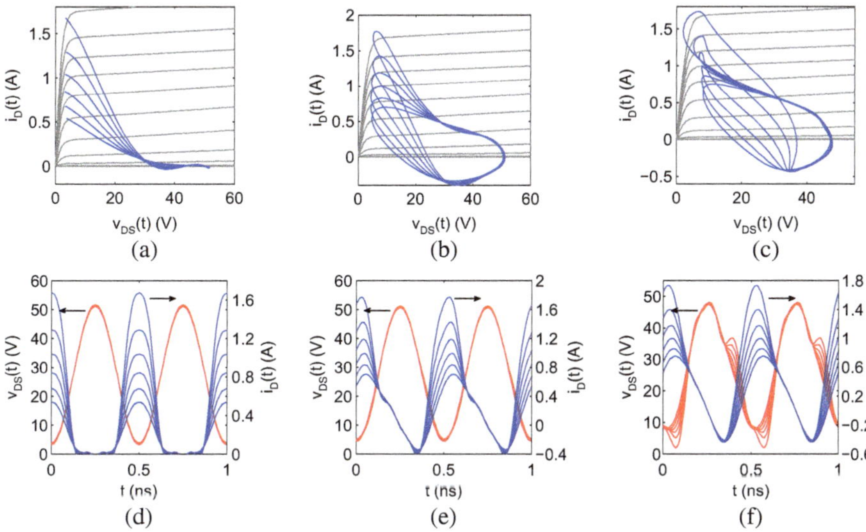

Fig. 3.5 Simulated 2 GHz load lines and waveforms at the current-source reference planes (**a**) and (**d**), the extrinsic reference planes in (**b**) and (**e**), and the package reference planes in (**c**) and (**f**) ([3], reprinted with permission from IEEE)

internal mode of operation can then be verified to be obtained at the current-source reference planes from the original device model when the multi-harmonic gate and drain voltages synthesized by the *embedding device model* at the package reference planes are externally applied to the original device model.

For each load, the load lines are kept above the IV knee at low drain voltages to keep the device operating in the linear region. The intrinsic voltage waveform exhibits a sinusoidal waveform (harmonics shorted) and the current waveforms approach half-rectified sinusoidal waveforms. Directly shaping the intrinsic load lines and waveforms at the current-source reference planes provides for a more effective waveform engineering.

Once the external voltage and currents are obtained, the external source and load impedances can be calculated in the frequency domain for each harmonic (for $n > 1$ on the source side) using the usual formula (assuming the currents are flowing into the device)

$$Z_S(n\omega) = -\frac{V_{GS}(n\omega)}{I_G(n\omega)}, \quad Z_L(n\omega) = -\frac{V_{DS}(n\omega)}{I_D(n\omega)}. \tag{3.1}$$

Instead of applying the synthesized waveforms, the impedance terminations $Z_S(n\omega)$ and $Z_L(n\omega)$ can then be used for the harmonic $n\omega$, when the real part of the impedances is positive. However, after embedding some of the harmonic impedances may exhibit a negative real part and the active injection of these harmonics is required. The *embedding device model* then provides the designer with the precise amplitude and phase of the harmonic signals which need to be applied at the package reference planes to exactly generate the targeted class-B operation at the current-source reference planes. This technique was verified experimentally in Fig. 10 in [3] with the injection of a second harmonic at the input for a class-B PA relying on load modulation.

The source $\Gamma_S(n\omega)$ and load $\Gamma_L(n\omega)$ reflection coefficients generated during the embedding process at each of the reference planes of interest are shown in Fig. 3.6a–f. As can be seen in Fig. 3.6a and d, the second- and third-harmonic source and load reflection coefficients are all shorts at the intrinsic reference planes as expected for class-B operation. The fundamental loads, $\Gamma_L(\omega)$ are resistive (real) at the current-source reference plane as shown in Fig. 3.6a. The six circles (red) correspond to the loads at the six different load and power levels considered. An arrow is used to indicate the direction of increase in input power and associated decrease in load resistance.

As shown in Figs. 3.5b, e and 3.6b, e, the load lines and the reflection coefficients at the extrinsic reference planes are, respectively, distorted and rotated by the extrinsic device parasitics. As shown in Figs. 3.5c, f and 3.6c, f, the load lines and the reflection coefficients at the package reference planes are further distorted and rotated by the package parasitics. Note that the rotation is anti-clockwise indicating that embedding is a non-Foster process.

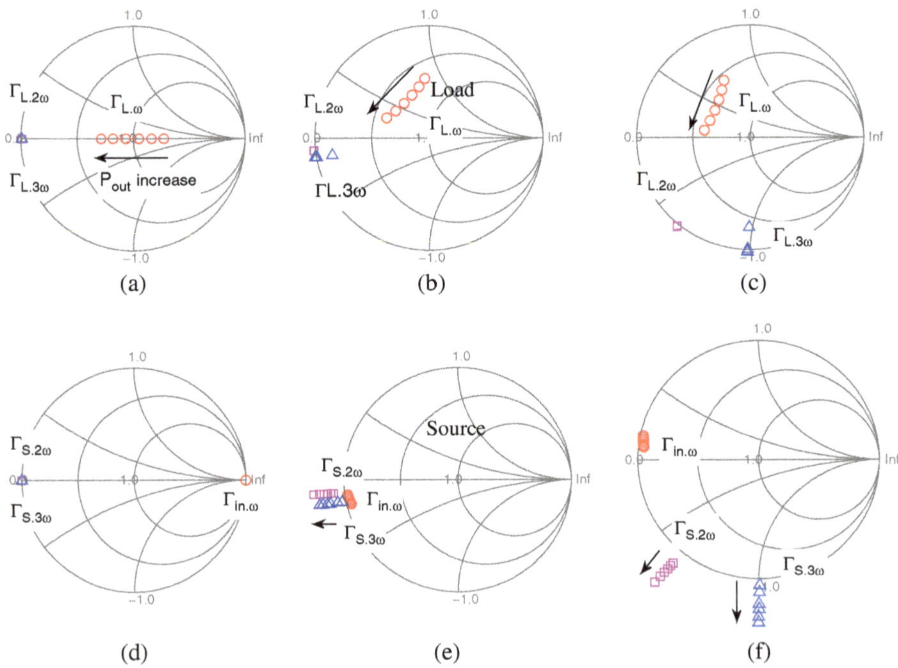

Fig. 3.6 The simulated 2 GHz intrinsic load and source reflection coefficients shown in (**a**) and (**d**), respectively, at the current-source reference planes are successively mapped using the *embedding transfer network* to each of the external reference planes in (**b**) and (**e**) and the package reference planes in (**c**) and (**f**) ([3], reprinted with permission from IEEE)

It is noted that $\Gamma_{in}(\omega)$ in Fig. 3.6f remains inside of the Smith chart due to the lossy parasitic components. However, the second $\Gamma_S(2\omega)$ and third harmonic $\Gamma_S(3\omega)$ reflection coefficients in Fig. 3.6f become larger than one. If the active injection of harmonics outside the Smith chart is not practical, class-B operation can be approached by renormalizing the extrinsic (on-wafer) or package source or load reflection coefficients for $n \geq 1$

$$\Gamma'_{L,ext}(n\omega) = \frac{\Gamma_{L,ext}(n\omega)}{|\Gamma_{L,ext}(n\omega)|},$$

$$\Gamma'_{S,ext}(n\omega) = \frac{\Gamma_{S,ext}(n\omega)}{|\Gamma_{S,ext}(n\omega)|}.$$

To conclude this introduction of the embedding device model we show in Fig. 3.7a and b, respectively, a comparison of the load lines at the current-source and extrinsic reference planes calculated by the device model (thick red lines) and the nonlinear embedding device model (thin black lines) for a SOI MOSFET [5]. In this case a neural network model is used for the SOI device [6]. The device model and the nonlinear embedding device model are found to be in perfect agreement.

3.2 Design of Switch-Mode Power Amplifiers with an Embedding Model

In this section we shall study in more detail the design of class-F PA using nonlinear embedding. The implementation of class F using a nonlinear embedding model is straightforward.

Consider the circuit shown in Fig. 3.8. In the ideal definition of class F, the drain-voltage waveform approaches a square wave. For example, assume that in the interval $[0, T]$ with the T as the RF period, the square drain voltage target is of the form

$$v_{DS} = \begin{cases} V_{on} & \text{for } 0 \geq t > \frac{T}{2} \\ V_{\max} & \text{for } \frac{T}{2} \geq t > T \end{cases}$$

$$= \frac{V_{\max} + V_{on}}{2} + (V_{\max} - V_{on}) \sum_{n=2,\text{even}}^{\infty} \frac{2}{\pi(n-1)} \sin\left((n-1)\omega t + \pi\right),$$

if we account for the knee voltage V_{on}. It results that the DC drain voltage applied should be selected to be

$$V_{DD} = \frac{V_{\max} + V_{on}}{2}.$$

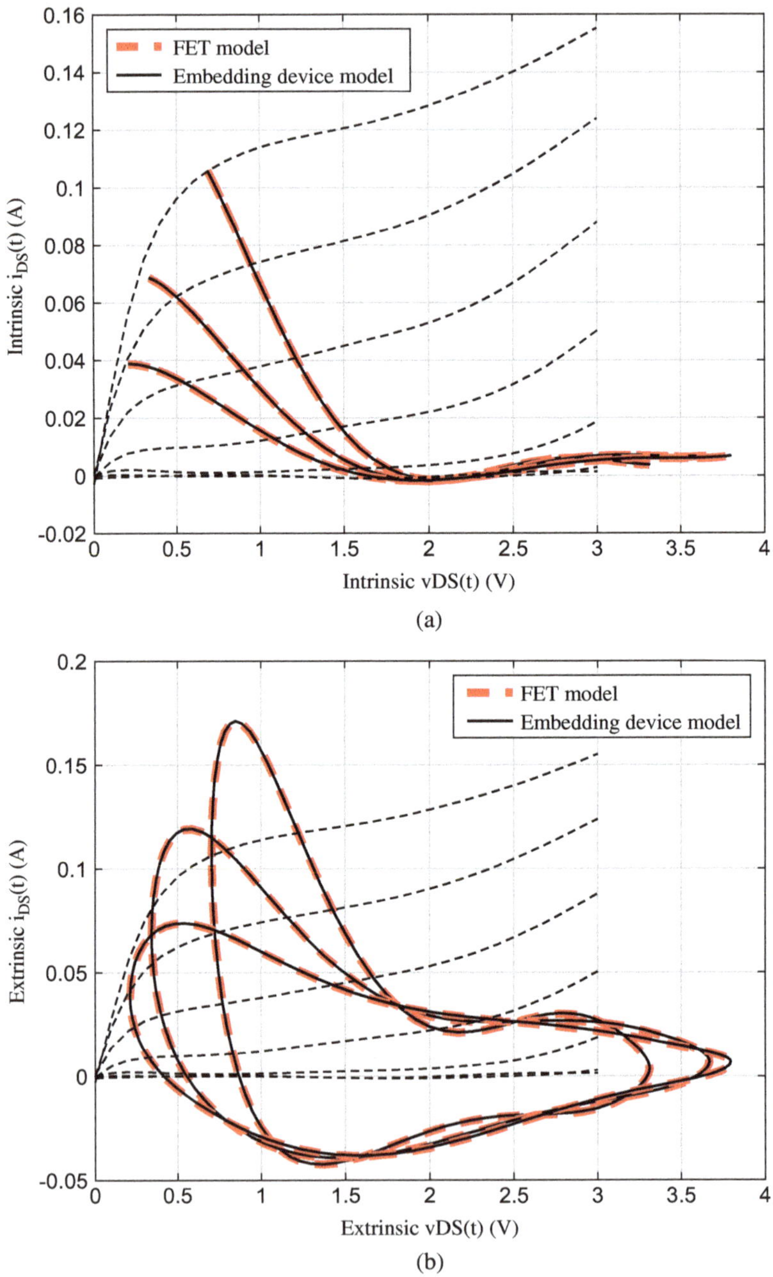

Fig. 3.7 Load lines at (**a**) the current-source reference planes and (**b**) extrinsic reference planes calculated by the device model (thick red lines) and the nonlinear embedding device model (thin black lines) ([5], reprinted with permission from IEEE)

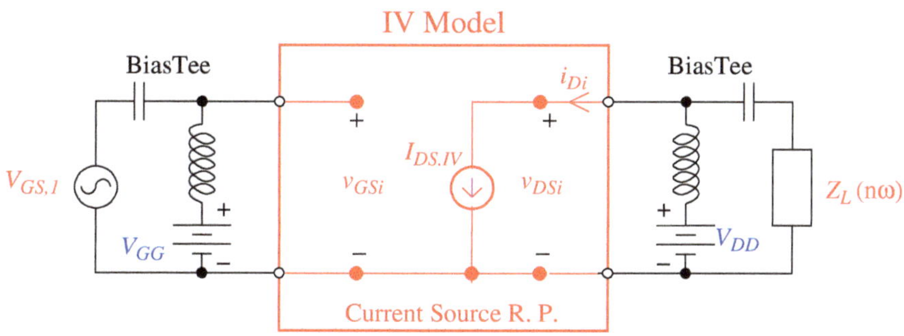

Fig. 3.8 Circuit with frequency-dependent load used to define the class-F operation

Meanwhile, the drain current is targeted to approach a half-rectified sinusoidal waveform

$$i_D = \max\left[0, I_{\max}\sin(\omega t)\right] = \frac{I_{\max}}{\pi} + \frac{I_{\max}}{2}\sin(\omega t) - I_{\max}\sum_{n=2,\text{even}}^{\infty}\frac{2}{\pi(n^2-1)}\cos(n\omega t).$$

Thus the DC gate voltage is at the threshold voltage and the gate voltage waveform is a sine wave of the form

$$v_{GS} = V_{GG} + V_{GS,1}\sin(\omega t) \quad \text{with} \quad V_{GG} = V_T \quad \text{and} \quad V_{GS,1} = \frac{I_{\max}}{g_m}.$$

It is to be noted that the drain current for a perfectly half-rectified sine wave will not support an odd-harmonic current ($I_{D,n\text{ odd}} = 0$ for $n > 1$). It is also noted that the drain voltage does not sustain any even-harmonic voltage ($V_{DS,n\text{ even}} = 0$). Under these conditions the multi-harmonic load $Z_L(n\omega)$ just needs to be selected as

$$Z_L \simeq \begin{cases} R_L = \frac{4}{\pi}\frac{V_{\max}\ V_{on}}{I_{\max}} \text{ for } n = 1 \\ \infty \ \ (\text{open}) \text{ for } n > 1 \text{ odd} \\ 0 \ \ (\text{short}) \text{ for } n \text{ even.} \end{cases}$$

Thus after properly biasing the transistor, it is sufficient to use short terminations for all the even harmonics and open terminations for odd harmonics at the current-source reference planes. The nonlinear embedding model predicts then the multi-harmonic source and load impedance terminations required at the *package* reference planes to sustain the class-F operation at the *current-source* reference planes. This methodology works quite well and will be applied in the next sections to design Doherty and Chireix PAs.

We could thus stop our discussion there. However, there has been a lingering mystery about the origin of the odd harmonic for $n > 1$ in the drain voltage. It has been proposed that in real devices, the odd-harmonic drain currents for $n > 1$ are not exactly vanishing and resistive loads can be used to bring about the required odd-harmonic drain voltages for $n > 1$. The odd-harmonic impedance terminations would then be lossy. However, we shall see that this is not required and the various possible lossless mechanism at the origin of the odd-harmonic voltages in real devices will be identified using the nonlinear embedding model.

For simplicity let us consider the case where only three harmonics (higher harmonics shorted) are considered, such that the class-F waveforms are given by

$$\begin{aligned} v_{GS} &= V_{GG} + V_{GS,1}\sin(\omega t), \\ v_{DS} &= V_{DD} + V_{DS,1}\sin(\omega t) + V_{DS,3}\sin(3\omega t). \end{aligned} \tag{3.2}$$

The gate V_{GS} and drain V_{DD} DC biases and the gate $V_{GS,1}$ and drain $V_{GS,1}$ fundamental voltages define the large signal operating point (LSOP) for the transistor.

It has been determined that for the ideal device (no knee voltage and constant transconductance g_m), the optimal third-harmonic drain voltage giving maximum efficiency [8] and maximum voltage gain [9] requires a third-to-first harmonic ratio of

$$\alpha_3 = \frac{V_{DS,3}}{V_{DS,3}} = \frac{1}{3}.$$

However, as we shall verify later on, that for real devices, different optimal α_3 may be required for different LSOPs.

The output RF power P_{RF} generated by the transistor can be written as

$$P_{\mathrm{RF},1} = P_{\mathrm{DC}} - P_{diss} - \sum_{n=1}^{N} P_{\mathrm{RF},n}, \tag{3.3}$$

with $P_{\mathrm{DC}} = I_D * V_{DD}$ the power provided by the DC supply, $P_{\mathrm{RF},n} = -\mathrm{Re}[V_{DS,n} I^*_{DS,n}]/2$ the output RF power delivered to the load by the transistor at the n-th harmonic, and P_{diss} the average power dissipated by the transistor during a period T

$$P_{diss} = \frac{1}{T}\int_0^T v_{DS}(t) i_D(t) dt. \tag{3.4}$$

Assuming in (3.3) that all the even harmonics are shorted and the odd harmonics of order larger than three have a negligible power contribution. It results that the drain efficiency can be written as

$$\eta_D = \frac{P_{\mathrm{RF},1}}{P_{\mathrm{DC}}} = 1 - \frac{P_{diss}}{P_{\mathrm{DC}}} - \frac{P_{\mathrm{RF},3}}{P_{\mathrm{DC}}}. \tag{3.5}$$

To maximize the drain efficiency η_D, the class-F waveform should be selected to minimize both the power ratios P_{diss}/P_{DC} and $P_{\mathrm{RF},3}/P_{\mathrm{DC}}$ while delivering the targeted output RF power $P_{\mathrm{RF},1}$. The fractional dissipated power P_{diss}/P_{DC} is minimized when the voltage v_{DS} and current i_D waveforms in (3.4) approach orthogonality. The fractional third-harmonic output load power $P_{RF,3}/P_{DC}$ is minimized in class F by setting $I_{D,3} = 0$. This can be achieved by tuning either the bias V_{DD}, the fundamental amplitude $|V_{DS,1}|$, or the third-harmonic amplitude $|V_{DS,3}|$ of the class-F drain-voltage waveform applied to achieve infinite third-harmonic output impedance $Z^{(IV)}_{out,3} = \infty$ as is demonstrated in Figs. 3.9 and 3.10.

Indeed, the third-harmonic component of the drain current $I_{D,3}$ changes sign when tuning either the bias V_{DD} or the fundamental amplitude $V_{DS,1}$ and thus vanishes for the optimal V_{DD} or $V_{DS,1}$ such that no third-harmonic power dissipation is taking place in either the transistor or the external load. In harmonic balance simulations when an open output load is applied at the third-harmonic while using a fixed DC supply V_{DD} and fixed fundamental load, the tuning of the third-harmonic amplitude $V_{DS,3}$ or equivalently α_3 is done automatically by the simulator to enforce $I_{D,3} = 0$.

The question still arises on the origin of the third-harmonic drain voltage since no third-harmonic drain current is generated by the device under these lossless condition of operation at the third-harmonic. In the case of a drain current monotonously increasing with the drain voltage, this is the transistor IV operating as a nonlinear voltage source which provides the required third harmonic voltage. On the other hand, in the ideal case where the transistor drain current is independent of the drain voltage, the third-harmonic voltage must still be provided externally.[1] To

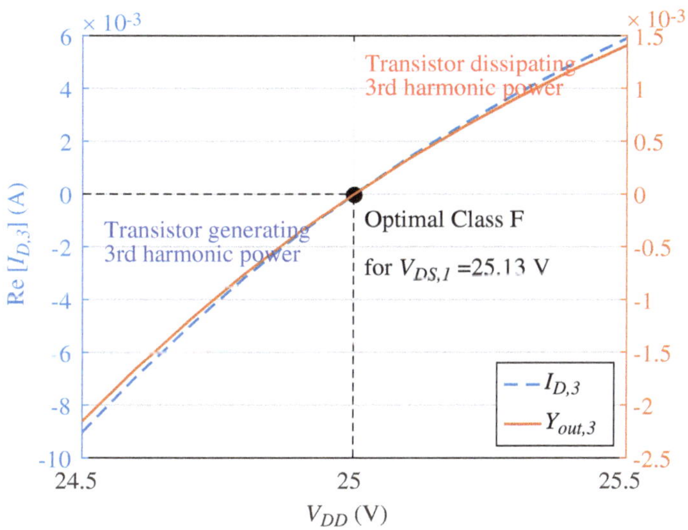

Fig. 3.9 Variation of $I_{D,3}$ and Re[$Y_{out,3}$] versus the drain supply voltage V_{DD} for a GaN HEMT excited by a fundamental drain voltage of $V_{DS,1} = 25.13$ V ([7], reprinted with permission from IEEE)

[1]The author Patrick Roblin would like to acknowledge a discussion on class F with Prof. Paul Tasker of Cardiff University which clarified this issue.

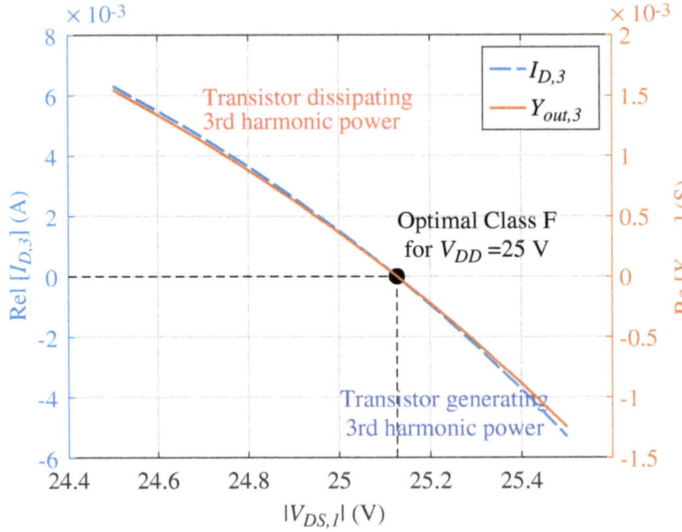

Fig. 3.10 Variation of $I_{D,3}$ and Re[$Y_{out,3}$] versus the fundamental drain voltage $|V_{DS,1}|$ for a GaN HEMT biased with a drain supply voltage of $V_{DD} = 25$ V ([7], reprinted with permission from IEEE)

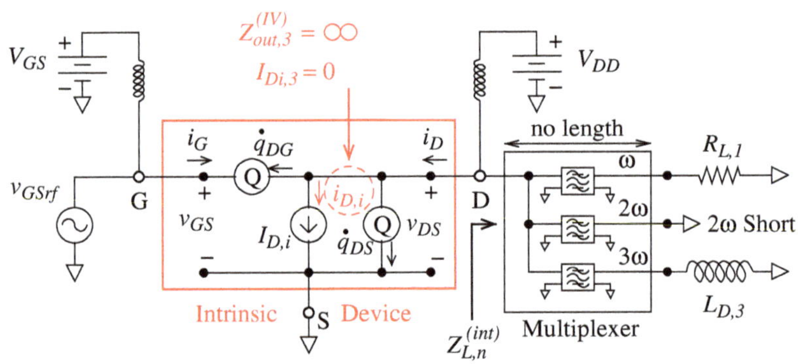

Fig. 3.11 Extrinsic device model including only the nonlinear charges and multi-harmonic loading requirements at the extrinsic reference planes for class-F operation ([7], reprinted with permission from IEEE)

describe in such a case, a potential mechanism where class F is self-sustained with a lossless third-harmonic termination, let us consider the extrinsic device model in Fig. 3.11 where the drain-to-source q_{DS} and drain-to-gate q_{DG} nonlinear charges are included in the model beside the IV current source (I_{Di}). The total displacement drain current sustained by these nonlinear charges is then given by

$$i_{D,displ} = \frac{dq_{DG}(v_{DG})}{dt} + \frac{dq_{DS}(v_{DS})}{dt}$$

$$= C_{DS}\frac{dv_{DS}}{dt} + C_{DG}\frac{dv_{GS}}{dt},$$

with

$$C_{DS} = \frac{\partial q_{DS}(v_{DS})}{\partial v_{DS}} + \frac{\partial q_{DG}(v_{DG})}{\partial v_{DG}},$$

$$C_{DG} = -\frac{\partial q_{DG}(v_{DG})}{\partial v_{DG}}.$$

Expanding the current and charges in a Fourier series we have

$$i_{D,displ} = \mathrm{Re}\left\{\sum_{n=1}^{n_H} I_{D,displ,n}e^{jn\omega t}\right\},$$

$$q_{DG} = \mathrm{Re}\left\{\sum_{n=1}^{n_H} Q_{DG,n}e^{jn\omega t}\right\},$$

$$q_{DS} = \mathrm{Re}\left\{\sum_{n=1}^{n_H} Q_{DS,n}e^{jn\omega t}\right\}.$$

We can then write the identity

$$I_{D,displ,n} = jn\omega\left(Q_{DG,n} + Q_{DS,n}\right).$$

Thus the displacement current components $I_{D,displ,n}$ are all 90° out-of-phase relative to the drain-to-source voltage since the charges $q_{DG}(v_{GS}, v_{DS})$ and $q_{DS}(v_{GS}, v_{DS})$ are both a real function of v_{GS} and v_{DS} and these voltages are mutually 180° out-of-phase. Given the device IV is presenting an open at the third harmonic, the displacement current $I_{D,displ,n}$ must flow outside the device. It results that the device must be terminated at the third harmonic by the following load:

$$Z_{L,3}^{(int)} = \frac{V_{DS,3}}{-I_{D,3}} = \frac{V_{DS,3}}{-I_{D,displ,3}} = \frac{jV_{DS,3}}{3\omega(Q_{DS,3} + Q_{DG,3})}.$$

The optimal load $\Gamma_{L,3}^{(int)}$ predicted by nonlinear embedding is plotted in Fig. 3.12. In this particular case an inductor is required to sustain the third-harmonic voltage needed for class-F operation. The validity of this optimal load $\Gamma_{L,3}^{(int)}$ predicted by nonlinear embedding is then verified using the third-harmonic load-pull contour plot of the efficiency also plotted in Fig. 3.12. Thus in the presence of lossless up-conversion, the device nonlinear charges generate the odd-harmonic displacement currents which must be terminated by the proper lossless harmonic load to sustain the odd-harmonic voltages required for class-F operation.

To experimentally validate the class-F operation predicted by nonlinear embedding, a class-F PA was designed using a realistic transistor model. The drain was

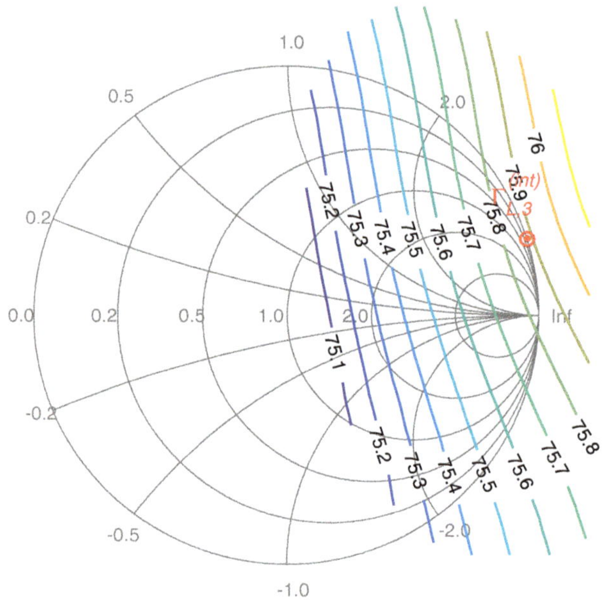

Fig. 3.12 Efficiency (%) contour plot for the third-harmonic load-pull simulation at 2 GHz for the extrinsic device model of Fig. 3.11 confirming the validity of the optimal load $\Gamma_{L,3}^{(int)}$ predicted by nonlinear embedding ([7], reprinted with permission from IEEE)

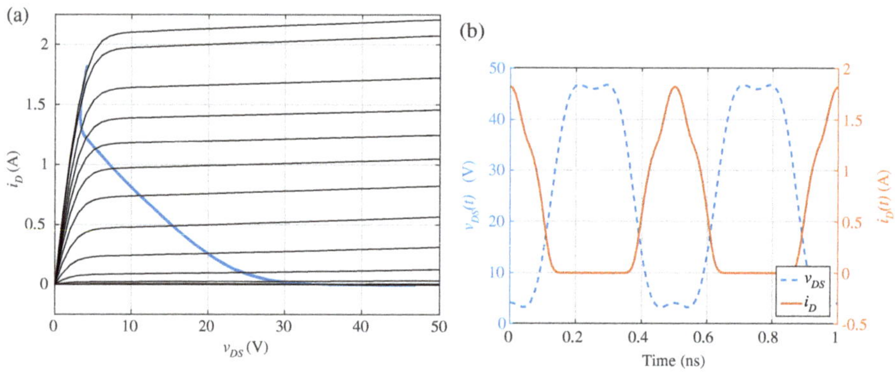

Fig. 3.13 Simulated intrinsic (**a**) load line and (**b**) waveforms ([7], reprinted with permission from IEEE)

biased at $V_{DD} = 25\,\text{V}$ and the gate at $-2.9\,\text{V}$ for a quiescent drain current of 95 mA. The input voltage source was set to a fundamental frequency of 2 GHz and a magnitude of 2.37 V. The magnitude of the drain-to-source voltage was 25.13 V for the fundamental and 4.19 V for the third harmonic.

The resulting intrinsic load line and waveforms calculated from simulation are shown in Fig. 3.13a and b, respectively. The load line is reaching the IV knee at low drain voltages with the drain-to-source voltage approaching a rectangular waveform and the drain current approaching a half-rectified sine wave.

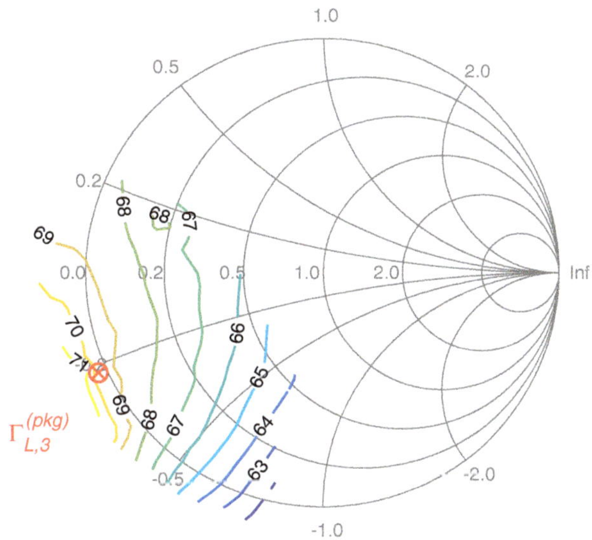

Fig. 3.14 Experimental drain efficiency contour plot obtained from the measured third-harmonic load-pull at the package reference planes. The third-harmonic load reflection coefficient $\Gamma_{L,3}^{(pkg)}$ predicted by the nonlinear embedding device model is verified to be positioned at the location yielding maximum efficiency ([7], reprinted with permission from IEEE)

The calculated load impedances at the extrinsic reference planes in Fig. 3.11 were $Z_{L,1}^{(int)} = 29.1\,\Omega$, $Z_{L,2}^{(int)} = 0\,\Omega$, and $Z_{L,3}^{(int)} = j\,311.9\,\Omega$. For the third harmonic, the load impedance corresponds to an inductive reactance. The embedding model was then used to determine the required input excitation and load for each harmonic at the packaged reference planes. The third-harmonic loads resulting at the package reference planes are plotted (red crosses and circles) in Fig. 3.14. Also shown in Fig. 3.14 is the resulting efficiency contour plot obtained from a third-harmonic load-pull. The third-harmonic load reflection coefficient $\Gamma_{L,3}^{(pkg)}$ predicted by the embedding device model is in good agreement with the load-pull measurements. This experimentally demonstrates the accuracy of the third-harmonic impedance prediction using nonlinear embedding.

In the demonstration example presented above the LSOP (primarily V_{DD} and $V_{DS,1}$) was selected such that the third-harmonic drain voltage verifies $V_{DS,3} = V_{DS,1}/6$. However, the LSOP (V_{DD} and $V_{DS,1}$) can also be tuned by the designer to further reduce P_{diss}/P_{DC} and obtain a higher drain efficiency. The tuning of $V_{DS,1}$ is indirectly achieved by tuning the fundamental resistive load at the current-source reference planes. As described in the introduction, class-F operation is then obtained by simply using an open for the odd harmonics and a short for the even harmonics. When doing so the simulator automatically determines the required third-harmonic to fundamental voltage ratio $\alpha_3 = V_{DS,3}/V_{DS,1}$ providing class-F operation. The objective is to obtain a device dynamic load line which more closely follows the edge of the IV characteristics in the triode region given the reduction of the IV knee voltage at lower instantaneous gate voltages. As shown in Table 3.1, an intrinsic efficiency of 82% and package efficiency of 79% can then be achieved that way at the current-source and package reference planes for the GaN HEMT considered with an optimized LSOP while maintaining about the same

Table 3.1 Simulated efficiencies as a function of α_3 for $V_{DD} = 23.2$ V and $V_{DS,1} = 25.13$ V ([7], reprinted with permission from IEEE)

α_3	P_{DC} (W)	$\frac{P_{RF,3}}{P_{DC}}$ (%)	$\frac{P_{diss}}{P_{DC}}$ (%)	η_D (%)	$\eta_D^{(pkg)}$ (%)	$\lvert\Gamma_{L,3}\rvert$
0.1875	10.79	0.38	17.78	81.84	78.65	0.69
0.1975	10.85	0.00	17.74	82.26	79.04	1.00
0.2075	10.88	−0.40	17.76	82.64	79.39	1.38

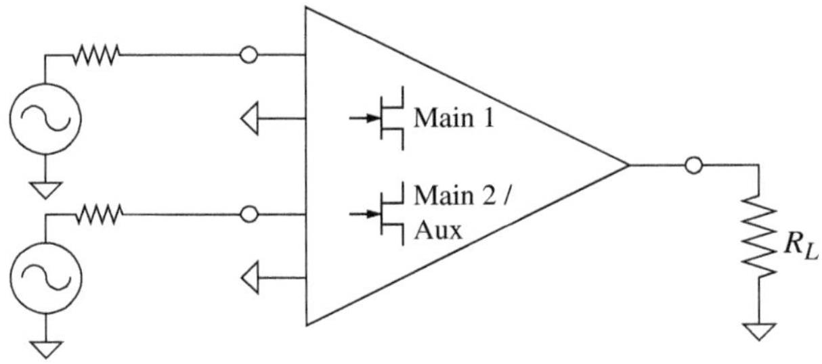

Fig. 3.15 Schematic of a dual-input outphasing power amplifier (OPA) ([10], reprinted with permission from IEEE)

output power. The required third-harmonic to fundamental voltage ratio α_3 is now 0.1975 which departs for the theoretical value of $1/6 \simeq 0.1666$ for the ideal device.

3.3 Design of Load Modulation Power Amplifiers with an Embedding Model

In the previous sections we explored the design of single-transistor PAs. In this section we turn now toward the design of dual-transistor PAs such as Chireix and Doherty PAs. These PAs have the merit of providing a peak of efficiency at both peak power and backoff power to accommodate modern communication signals in which the average operating power is typically 10 dB below the intermittent peak power. To keep the theory the most general, we shall consider the general case where the PA has two inputs so that the phase and amplitude of both input signals can be independently controlled. This kind of PAs belongs to the general class of outphasing power amplifiers (OPA) since not only the input power but also the phase difference between the two inputs controls the PA output power. The general schematic for such a dual-input OPA is presented in Fig. 3.15.

We wish now to apply the nonlinear embedding model to facilitate the design of such PAs. The idea is to first perform the design of the OPA at the current-source reference planes with the memoryless IV characteristics of the FET core before embedding it to the package reference planes. One might wonder on the feasibility of this approach since (1) one can never in practice directly access the FET core and (2) in an OPA the two FET-core devices are load-pulling each other. Consider the general OPA shown in Fig. 3.16a, the FET core marked with the label "IV" in red is buried inside the parasitics networks. However, we can see that the two core FETs are interconnected through the three-port combiner via the parasitic networks

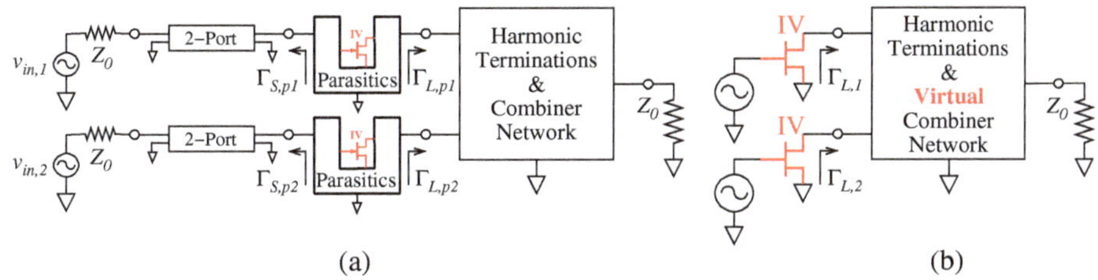

Fig. 3.16 Dual-input outphasing at (**a**) the package and (**b**) the current-source reference planes

of both FETs. Thus, we can first design an optimal virtual combiner network at the current-source reference planes as shown in Fig. 3.16b before embedding the transistor to the package reference network to synthesize the actual physical combiner network as shown in Fig. 3.16a.

Thus, with this embedding approach we can start the design of the dual-input OPA design at the current-source reference planes and more easily control the mode of operation of the core FETs. To rigorously explore the design space available for the design of dual-input OPAs, we shall present here an analytic Doherty–Chireix continuum theory [10] at the current-source reference planes instead of exploring this continuum at the package reference planes [11] and [12].

3.3.1 *Continuum Theory*

To develop a general analytic theory at the current-source reference planes we shall rely on the four current- and voltage-ratio parameters: K_{vm}, K_{im}, K_{ia}, and K_{va}

$$|V_{mp}| = K_{vm}|V_{mb}|, \quad |V_{ap}| = K_{va}|V_{ab}|,$$

$$|I_{mp}| = K_{im}|I_{mb}|, \quad |I_{ap}| = K_{ia}|I_{ab}|.$$

The K-factors rely on the peak and backoff values of the main V_m and auxiliary V_a fundamental voltages and the main I_m and auxiliary I_a fundamental currents defined in Fig. 3.17a. The motivation for selecting K_{vm} and K_{va} is that they directly impact the efficiency of the two devices at peak and backoff as shall be discussed below. The K_{vm}, K_{im}, K_{ia}, and K_{va} parameters are depicted by vertical lines in Fig. 3.17b. The coefficients $K_{va/m}$ and $K_{ia/m}$ are referred to the peak-to-backoff voltage ratio (PBVR) and the peak-to-backoff current ratio (PBCR), respectively. The subscripts m and a refer to the main and auxiliary devices, respectively. The subscripts b and p refer to the backoff and peak power levels, respectively. Note that the $V_{m/a}$ and $I_{m/a}$ refer only to the fundamental voltages and currents of the main and auxiliary devices, respectively. The harmonics of each devices are assumed to be separately defined to establish the desired mode of operation. These terminations are found in

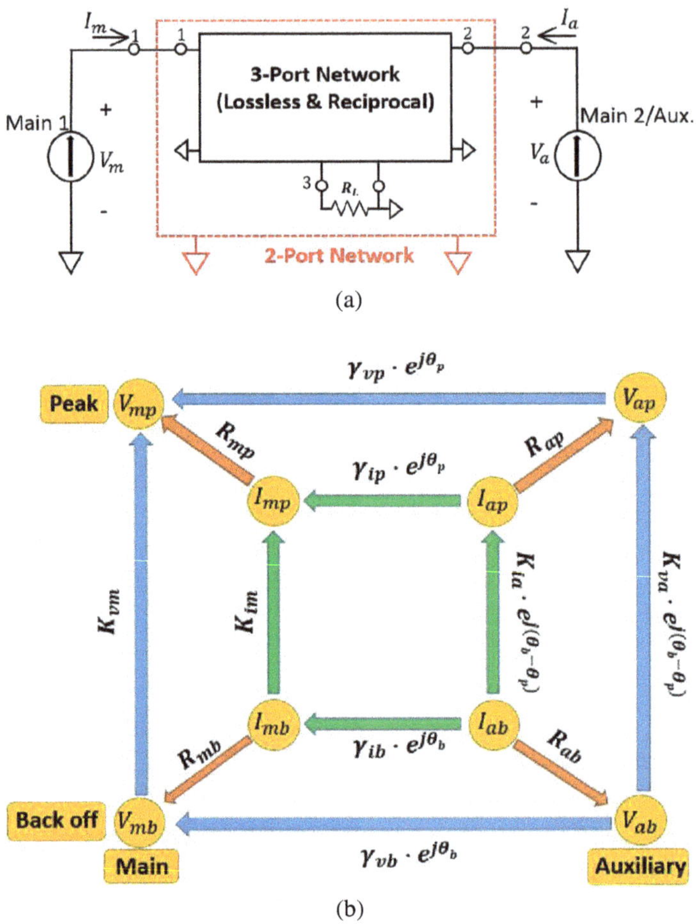

Fig. 3.17 (**a**) Dual-input OPA realized with a lossless and reciprocal three-port network (black solid-line box) terminated with a resistive load R_L at the current-source reference planes. (**b**) Mapping of the connections between fundamental currents and voltages at the two-port output combiner for the main and auxiliary transistors ([10], reprinted with permission from IEEE)

practice to be weakly power dependent and harmonic load-pulling is not needed. For convenience we keep using the Doherty notation of main and auxiliary PAs even though in the pure outphasing mode both transistors will remain on at peak and backoff.

In Fig. 3.17b the horizontal lines are proportional to the γ_{vp}, γ_{vb}, γ_{ip}, and γ_{ip} parameters defined as the amplitude voltage ratios and amplitude current ratios, respectively, between the main and auxiliary devices (Appendix I in [13]) for peak and backoff. The phasors with phase θ_p and θ_b are the outphasing phases applied at peak and backoff. Finally, in Fig. 3.17b the diagonal lines represent the fundamental resistance R_{mp}, R_{mb}, R_{ap}, and R_{ab} defined as the voltage and current ratios of the main and auxiliary devices, at both peak and backoff power. In this initial continuum theory, all the fundamental load impedances are assumed to be real. This is consistent with the fact that usually the main and auxiliary devices are operating in class B, F, or C.

In this work, the main device is assumed to be always on at backoff. Since the main device is always on, maximizing its efficiency at both peak and backoff is of critical importance and the peak and backoff drain voltages of the main device are set to be (a) the same and (b) approximately equal to the supply voltage ($V_{mp} = V_{mb} \simeq V_{DD,m}$). As a result, the main PBVR K_{vm} will be usually kept equal or close to 1.

Since the fundamental load impedances seen by the main and auxiliary devices are assumed here to be real, the fundamental currents and voltages of the main device at peak and backoff power can be expressed in terms of their amplitudes using

$$I_{mp} = |I_{mp}|, V_{mp} = |V_{mp}|, \tag{3.6}$$

$$I_{mb} = |I_{mb}|, V_{mb} = |V_{mb}|, \tag{3.7}$$

$$I_{ap} = |I_{ap}|e^{-j\theta_p}, V_{ap} = |V_{ap}|e^{-j\theta_p}, \tag{3.8}$$

$$I_{ab} = |I_{ab}|e^{-j\theta_b}, V_{ab} = |V_{ab}|e^{-j\theta_b}, \tag{3.9}$$

where θ_b and θ_p are the outphasing angles between the main and auxiliary devices at backoff and peak powers, respectively. Given the voltages and currents, the two-port $\mathbf{Z}$ matrix can be obtained from the peak and backoff boundary conditions

$$\begin{bmatrix} V_{mp} & V_{mb} \\ V_{ap} & V_{ab} \end{bmatrix} = \mathbf{Z} \begin{bmatrix} I_{mp} & I_{mb} \\ I_{ap} & I_{ab} \end{bmatrix},$$

resulting in the following Z-parameters:

$$\mathbf{Z} = \frac{1}{\Delta} \begin{bmatrix} V_{mp}I_{ab} - V_{mb}I_{ap} & V_{mb}I_{mp} - V_{mp}I_{mb} \\ V_{ap}I_{ab} - V_{ab}I_{ap} & V_{ab}I_{mp} - V_{ap}I_{mb} \end{bmatrix}, \tag{3.10}$$

with $\Delta = I_{mp}I_{ab} - I_{mb}I_{ap}$.

Given that the two-port network must be reciprocal ($Z_{12} = Z_{21}$), the following constraint equation is established:

$$V_{mb}I_{mp} - V_{mp}I_{mb} = V_{ap}I_{ab} - V_{ab}I_{ap}. \tag{3.11}$$

Using Eqs. (3.6)–(3.9) in the reciprocity constraint (3.11) gives

$$1 = \frac{V_{mb}I_{mp} - V_{mp}I_{mb}}{V_{ap}I_{ab} - V_{ab}I_{ap}} = e^{j(\theta_p+\theta_b)}\frac{|V_{mb}||I_{mp}| - |V_{mp}||I_{mb}|}{|V_{ap}||I_{ab}| - |V_{ab}||I_{ap}|}$$

$$= e^{j(\theta_p+\theta_b)}\frac{|I_{mp}||I_{mb}|}{|I_{ap}||I_{ab}|}\frac{R_{mb} - R_{mp}}{R_{ap} - R_{ab}},$$

which when using $R_{mp} < R_{mb}$ and $R_{ap} < R_{ab}$ leads to the final identities

$$\theta_p + \theta_b = \pi, \tag{3.12}$$

$$|V_{mb}||I_{mp}| - |V_{mp}||I_{mb}| = |V_{ab}||I_{ap}| - |V_{ap}||I_{ab}|. \tag{3.13}$$

Equation (3.12) implies that once the outphasing angle at peak power θ_p is determined, the outphasing angle at backoff θ_b is obtained and vice versa.

Using identity (3.13), the asymmetry power ratio n between the auxiliary and main devices at peak power is derived as

$$n = \frac{P_{ap}}{P_{mp}} = \frac{|V_{ap}||I_{ap}|}{|V_{mp}||I_{mp}|} = \frac{1}{\gamma_{vp}\gamma_{ip}} = \frac{1/K_{im} - 1/K_{vm}}{1/K_{ia} - 1/K_{va}}. \tag{3.14}$$

For $n = 1$, the two transistors will deliver the same peak power. Similarly, the power ratio m between the auxiliary and main devices at backoff is derived as

$$m = \frac{P_{ab}}{P_{mb}} = \frac{|V_{ab}||I_{ab}|}{|V_{mb}||I_{mb}|} = \frac{1}{\gamma_{vb}\gamma_{ib}} = \frac{K_{im} - K_{vm}}{K_{ia} - K_{va}}. \tag{3.15}$$

The overall peak ($P_{o,peak}$) and backoff ($P_{o,back}$) RF powers delivered to the two-port network equal to the summation of the peak and backoff powers at ports 1 and 2 are thus defined as

$$P_{o,peak} = \begin{cases} P_{ap}\left(1 + \frac{1}{n}\right) \text{ with } P_{ap} = \frac{1}{2}|V_{ap}||I_{ap}| \text{ for } n \geq \\ P_{mp}\,(1 + n) \text{ with } P_{mp} = \frac{1}{2}|V_{mp}||I_{mp}| \text{ for } n < 1, \end{cases} \tag{3.16}$$

$$P_{o,back} = \begin{cases} P_{ab}\left(1 + \frac{1}{m}\right) \text{ with } P_{ab} = \frac{1}{2}|V_{ab}||I_{ab}| \text{ for } n \geq \\ P_{mb}\,(1 + m) \text{ with } P_{mb} = \frac{1}{2}|V_{mb}||I_{mb}| \text{ for } n < 1. \end{cases} \tag{3.17}$$

The peak-to-backoff power ratio (PBPR) also referred as output backoff (OBO) is then given by

$$\text{PBPR} = \frac{P_{o,peak}}{P_{o,back}} = K_{va}K_{ia}\frac{1 + \frac{1}{n}}{1 + \frac{1}{m}}. \tag{3.18}$$

Based on this derivation, it is noted that the PBPR is related to the auxiliary PBVR K_{va}, auxiliary PBCR K_{ia}, n and m factors.

In radio-frequency integrated-circuit (RFIC) processes, the transistors can be sized as desired and the ratio n gives the ratio of the peripheries of the auxiliary to main devices. For the case where the two devices are of the same periphery as is common for packaged devices, it is useful to introduce the following normalized output power for a figure-of-merit:

$$\overline{P}_{o,peak} = \frac{P_{o,peak}}{2P_{\max}(n)} = \begin{cases} \frac{1}{2}\left(1+\frac{1}{n}\right) & n \geq 1 \\ \frac{1}{2}(1+n) & n < 1, \end{cases} \tag{3.19}$$

where $P_{\max}(n)$ is the power of the device (main or auxiliary) providing the most power when the dual-input OPA is operating at peak power. Thus $P_{\max}(n)$ is $|V_{ap}||I_{ap}|/2$ if $n \geq 1$, or $|V_{mp}||I_{mp}|/2$ if $n < 1$. For devices of equal size, $\overline{P}_{o,peak} = 1$ is the optimal value for this figure-of-merit.

The reciprocal lossy two-port combiner seen by the transistors of the OPA at their drain terminals consists of a three-port network terminated by a resistive load as shown in Fig. 3.17a. Applying the condition for a lossless and reciprocal three-port network found in [5, 14] and [15], which is $\mathcal{R}^2\{Z_{12}\}=\mathcal{R}\{Z_{11}\}\mathcal{R}\{Z_{22}\}$, an analytic equation is obtained for the outphasing angle at backoff power

$$\cos^2\theta_b = \frac{(K_{ia}-1)(K_{im}-K_{va})}{(K_{im}+1)(K_{ia}+K_{va})}. \tag{3.20}$$

This equation yields four possible analytic solutions for the outphasing angles at both the backoff and peak power levels

$$\theta_b = \pm\cos^{-1}\left(\pm\sqrt{\frac{(K_{ia}-1)(K_{im}-K_{va})}{(K_{im}+1)(K_{ia}+K_{va})}}\right), \tag{3.21}$$

$$\theta_p = \pi - \theta_b. \tag{3.22}$$

The Z parameters of the two-port circuit in Fig. 3.17a can now be calculated by applying the condition of losslessness and reciprocity to its internal three-port combiner network

$$\begin{aligned} Z_{11} &= \frac{\cos\theta_b R_{mb}(1+K_{ia}) + j\sin\theta_b R_{mb}(K_{ia}-1)}{\cos\theta_b(K_{ia}+K_{im}) + j\sin\theta_b(K_{ia}-K_{im})}, \\ Z_{12} &= \frac{K_{ia}(K_{im}-1)R_{ap}/\gamma_{vp}}{\cos\theta_b(K_{ia}+K_{im}) + j\sin\theta_b(K_{ia}-K_{im})}, \\ Z_{22} &= \frac{\cos\theta_b R_{ab}(K_{im}+K_{va}) + j\sin\theta_b R_{ab}(K_{va}-K_{im})}{\cos\theta_b(K_{ia}+K_{im}) + j\sin\theta_b(K_{ia}-K_{im})}. \end{aligned} \tag{3.23}$$

In the case where the main and auxiliary fundamental peak voltages are selected to be equal ($\gamma_{vp} = 1$), it has been verified that the reciprocal lossless three-port

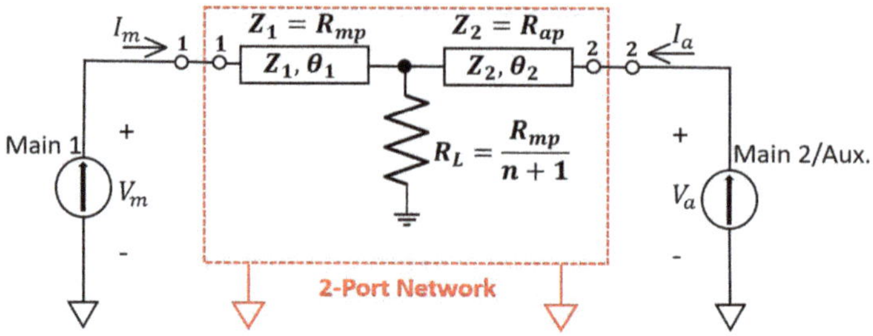

Fig. 3.18 Dual-input OPA prototype realized with two transmission lines connected with a resistive load

combiner for the OPA can be realized with two transmission lines of characteristic impedances $Z_1 = R_{mp}$, $Z_2 = R_{ap}$, connected to a resistive load $R_L = R_{mp}/(n+1)$ as shown in Fig. 3.18 [10]. The line electrical length θ_1, θ_2 are given by

$$\tan\theta_1 = \frac{K_{im}(K_{ia}-1)}{K_{ia}+K_{im}}\tan\theta_b,$$

$$\tan\theta_2 = \frac{K_{ia}(K_{va}-K_{im})}{K_{va}(K_{ia}+K_{im})}\tan\theta_b.$$

In summary after setting $K_{vm} = 1$ for maximum main PA efficiency, we are finding that there exists a three-dimensional continuum in terms of the three variables K_{im}, K_{ia}, and K_{va} that includes all possible solutions for the dual-input OPA combiners. However for a fair comparison, only the OPAs exhibiting the same PBPR (OBO) should be compared, since the motivation for investigating OPAs is to obtain a large efficiency at the targeted backoff power. Under such conditions, the PBPR becomes an input parameter which establishes a relationship between K_{im}, K_{ia}, and K_{va}. For a given PBPR, the main K_{im} obtained from (3.18) is

$$K_{im} = \frac{\mathrm{PBPR}\,(1+K_{va}-K_{ia})-K_{va}K_{ia}}{\mathrm{PBPR}-K_{va}K_{ia}+K_{va}-K_{ia}}. \tag{3.24}$$

By only considering dual-input OPAs with the same PBPR and a main PA operating at high-efficiency ($K_{vm} = 1$), the design space for the dual-input OPA has now reduced to a 2D continuum of possible solutions with K_{ia} and K_{va} the two independent variables. As we shall see this 2D continuum of OPA solutions includes the Doherty and Chireix PAs.

To visualize this 2D continuum, the outphasing angle θ_b and the required main PBCR K_{im} are plotted in Fig. 3.19a and b, respectively, for a PBPR of 9.54 dB and $K_{ia} \in [9, 100]$ and $K_{va} \in [1, 3]$. The corresponding normalized peak power $\overline{P}_{o,peak}(n)$ and asymmetry power ratio n are also plotted in Fig. 3.19c and d, respectively.

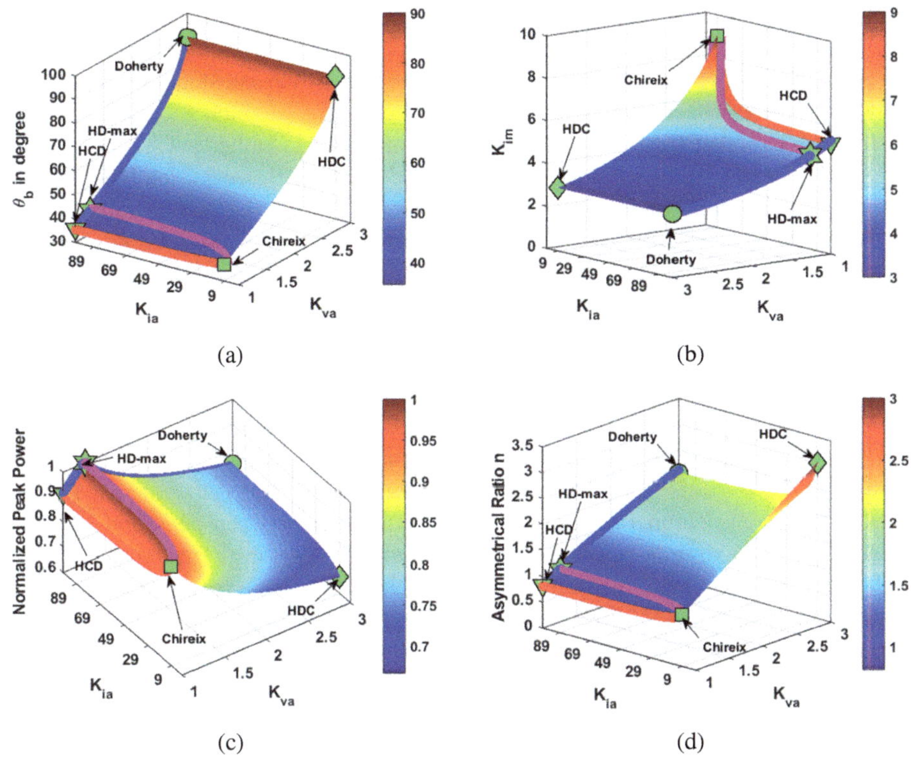

Fig. 3.19 Results of the continuum theory for PBPR = 9.54 dB: (**a**) outphasing angle at backoff power θ_b, (**b**) PBCR of the main device K_{im}, (**c**) normalized peak power $\overline{P}_{o,peak}$, and (**d**) asymmetric power ratio n. The blue line shows the transition from Doherty to hybrid Chireix–Doherty, the red line shows the transition from Chireix to hybrid Chireix–Doherty, and the purple line shows the transition from Chireix to HD-max

As shown in Fig. 3.19a–d a Doherty power amplifier (DPA) with an asymmetry power ratio of $n = 2$ and a PBPR of 9 (9.54 dB) is implemented for $K_{ia} = 100$ and $K_{va} = K_{im} = 3$. The main PBCR K_{im} is equal to $n + 1 = 3$ and the normalized peak power $\overline{P}_{o,peak}$ is 0.75. The outphasing angles at peak and backoff are both 90° as expected. The DPA is then revealed to be a degenerate outphasing PA with the same peak θ_p and backoff θ_b outphasing angles.

When $K_{ia} = K_{im} =$ PBPR and $K_{va} = 1$ a Chireix PA mode is realized. As shown in Fig. 3.19a–d a Chireix PA with a PBPR of 9 (9.54 dB) is implemented for $K_{ia} = K_{im} = 9$. The required outphasing angle at backoff power is 36.9°. This Chireix PA exhibits symmetric power ratios ($n = m = 1$) and its normalized peak power $\overline{P}_{o,peak} = 1$ is the maximum value possible.

A novel hybrid Chireix–Doherty (HCD) PA mode is obtained by setting $K_{va} = 1$ and $K_{ia} = \infty$. As shown in Fig. 3.19a–d the HCD PA with a PBPR of 9 is implemented for $K_{im} = 5$ and $K_{ia} = 100$. The required outphasing angle at backoff power is 35.3°. In an HCD PA, the fundamental drain voltages V_{ap} and V_{ab} presented to the auxiliary transistor at both peak and backoff are kept the same by setting $K_{va} = 1$. The efficiency of the auxiliary PA is then substantially increased between peak and backoff compared to the Doherty PA for which $K_{va} = K_{im} = 3$.

Note that another type of hybrid PA mode noted hybrid Doherty–Chireix (HDC) is realized for $K_{ia} = \text{PBPR}$ and $K_{va} = K_{im} = \sqrt{\text{PBPR}}$. However, the HDC mode is exhibiting a lower efficiency for the secondary main (auxiliary) at backoff.

Also plotted in Fig. 3.19a–d using a purple line is the trajectory of K_{ia} versus K_{va} giving a unity asymmetry power ratio $n = 1$ yielding the maximum normalized peak power $\overline{P}_{o,peak}$ of 1. This trajectory connects the Chireix mode for $K_{ia} = \text{PBPR}$ to the Doherty-HCD (blue line) for $K_{ia} = \infty$ for a PA mode referred as HD-max. This trajectory is defined by the following equation:

$$K_{va} = \frac{\text{PBPR}(K_{ia} - 1)}{\text{PBPR} + K_{ia}(\text{PBPR} - 2)}.$$

The HD-max PA features both an improved efficiency between peak and backoff as well as a maximum output power. HD-max is thus the preferred dual-input OPA choice to maximize the output power when using equal size devices. The key features of the Doherty, Chireix HCD, and HD-max PAs are compared in Table 3.2 together with numerical values for a PBPR of $B = 9 = 9.54\,\text{dB}$.

3.3.2 *Design of the Outphasing Power Amplifier Combiner at the Package Reference Planes*

Once the OPA has been designed at the current-source reference planes, the two-port combiner network at the package reference planes must be designed and the input phase offsets determined. The two-port combiner network generated by nonlinear embedding is usually not directly realizable by a reciprocal and lossless three-port network terminated by a resistor. Various approaches are then possible to obtain a close solution.

In one approach the combiner plus load network can be synthesized using the following approximate two-port $\mathbf{Y_{pkg}}$ matrix:

$$\mathbf{M_1} = \begin{bmatrix} V_{mp,pkg} & V_{ap,pkg} \\ V_{mb,pkg} & V_{ab,pkg} \end{bmatrix},$$

$$\mathbf{Y_{1,pkg}} = \begin{bmatrix} Y_{11,pkg} \\ Y'_{12,pkg} \end{bmatrix} = \mathbf{M_1^{-1}} \begin{bmatrix} I_{mp,pkg} \\ I_{mb,pkg} \end{bmatrix},$$

$$\mathbf{Y_{2,pkg}} = \begin{bmatrix} Y'_{21,pkg} \\ Y_{22,pkg} \end{bmatrix} = \mathbf{M_1^{-1}} \begin{bmatrix} I_{ap,pkg} \\ I_{ab,pkg} \end{bmatrix},$$

Table 3.2 Comparison of canonic dual-input power amplifiers ([10], reprinted with permission from IEEE)

PA	K_{vm}	K_{im}	K_{va}	K_{ia}	$\overline{P}_{o,peak}$	PBPR	n	θ_b	R_{mp}	R_{mb}	R_{ap}	R_{ab}
Doherty	1	$\sqrt{B}$	$\sqrt{B}$	∞	$\frac{0.5\sqrt{B}}{\sqrt{B}-1}$	B	$\sqrt{B}-1$	90°	$\frac{\gamma_{vp}}{\gamma_{ip}} R_{opt}$	$\sqrt{B}\frac{\gamma_{vp}}{\gamma_{ip}} R_{opt}$	R_{opt}	∞
	1	3	3	∞	0.75	9	2	90°				
Chireix	1	B	1	B	1	B	1	$\cos^{-1}\left(\pm\frac{B-1}{B+1}\right)$	R_{opt}	$B\,R_{opt}$	R_{opt}	$B\,R_{opt}$
	1	9	1	9	1	9	1	36.9°				
Hybrid	1	$\frac{B+1}{2}$	1	∞	$\frac{B}{B+1}$	B	$\frac{B-1}{B+2}$	$\cos^{-1}\left(\pm\sqrt{\frac{B-1}{B+3}}\right)$	R_{opt}	$\frac{B+1}{2} R_{opt}$	$\frac{\gamma_{ip}}{\gamma_{vp}} R_{opt}$	∞
(HCD)	1	5	1	∞	9/10	9	8/10	35.3°				
HD-max	1	$\frac{B}{2}$	$\frac{B}{B-2}$	∞	1	B	1	$\cos^{-1}\left(\pm\sqrt{\frac{B^2-4B}{B^2-4}}\right)$	R_{opt}	$\frac{B}{2} R_{opt}$	R_{opt}	∞
	1	4.5	9/7	∞	1	9	1	40.1°				

$$
\begin{aligned}
G_{12,pkg} &= \text{sign}\left[\mathcal{R}\left\{Y_{11,pkg}\right\}\right]\sqrt{\left|\mathcal{R}\{Y_{11,pkg}\}\mathcal{R}\{Y_{22,pkg}\}\right|},\\
B_{12,pkg} &= \text{sign}\left[\mathcal{I}\left\{Y'_{12,pkg}\right\}\right]\sqrt{\left|\mathcal{I}\{Y'_{12,pkg}\}\mathcal{I}\{Y'_{21,pkg}\}\right|},\\
Y_{21,pkg} &= Y_{12,pkg} = G_{12,pkg} + jB_{12,pkg},\\
\mathbf{Y_{pkg}} &= \begin{bmatrix} Y_{11,pkg} & Y_{12,pkg} \\ Y_{21,pkg} & Y_{22,pkg} \end{bmatrix},
\end{aligned}
$$

where the subscript *pkg* refers to the package reference planes.

Alternatively, an OPA with the same power and efficiency at peak and backoff can be obtained by synthesizing a reciprocal and lossless three-port combiner network using a technique similar to that reported in [14, 16] and [15]. The relative amplitude and phase of the voltage and current of each transistor cannot be changed if the same power and efficiency at peak and backoff are to be obtained. However, the outphasing angle ψ_p and ψ_b at peak and backoff can be modified without changing the load impedances and the fundamental voltage and current amplitudes at peak and backoff.

The fundamental drain currents at backoff and peak of the main and auxiliary PAs at the package reference plane are defined as

$$
\begin{aligned}
I_{mb,pkg} &= |I_{mb,pkg}|, \quad I_{ab,pkg} = |I_{ab,pkg}|e^{-j\psi_b},\\
I_{mp,pkg} &= |I_{mp,pkg}|, \quad I_{ap,pkg} = |I_{ap,pkg}|e^{-j\psi_p}.
\end{aligned}
$$

The fundamental drain voltages at backoff and peak of the main and auxiliary PAs at the package reference plane are defined as

$$
\begin{aligned}
V_{mb,pkg} &= |V_{mb,pkg}|e^{j\phi_{mb}}, \quad V_{ab,pkg} = |V_{ab,pkg}|e^{j(\phi_{ab}-\psi_b)},\\
V_{mp,pkg} &= |V_{mp,pkg}|e^{j\phi_{mp}}, \quad V_{ap,pkg} = |V_{ap,pkg}|e^{j(\phi_{ap}-\psi_p)}, \qquad (3.25)
\end{aligned}
$$

where ϕ_{mb}, ϕ_{mp}, ϕ_{ab}, and ϕ_{ap} are the phases of the load impedances at backoff and peak power seen by the main and auxiliary PAs at the package reference planes, respectively,

$$
\begin{aligned}
Z_{mb,pkg} &= \frac{V_{mb,pkg}}{I_{mb,pkg}} = |Z_{mb,pkg}|e^{j\phi_{mb}}, \; Z_{ab,pkg} = \frac{V_{ab,pkg}}{I_{ab,pkg}} = |Z_{ab,pkg}|e^{j\phi_{ab}},\\
Z_{mp,pkg} &= \frac{V_{mp,pkg}}{I_{mp,pkg}} = |Z_{mp,pkg}|e^{j\phi_{mp}}, \; Z_{ap,pkg} = \frac{V_{ap,pkg}}{I_{ap,pkg}} = |Z_{ap,pkg}|e^{j\phi_{ap}}.
\end{aligned}
$$

The Z parameters of the two-port combiner network at the fundamental frequency are then given by

$$Z_{pkg} = \begin{bmatrix} Z_{11,pkg} & Z_{12,pkg} \\ Z_{21,pkg} & Z_{22,pkg} \end{bmatrix}, \tag{3.26}$$

$$Z_{11,pkg} = (V_{mp,pkg} I_{ab,pkg} - V_{mb,pkg} I_{ap,pkg})/\Delta,$$

$$Z_{12,pkg} = (V_{mb,pkg} I_{mp,pkg} - V_{mp,pkg} I_{mb,pkg})/\Delta,$$

$$Z_{21,pkg} = (V_{ap,pkg} I_{ab,pkg} - V_{ab,pkg} I_{ap,pkg})/\Delta,$$

$$Z_{22,pkg} = (V_{ab,pkg} I_{mp,pkg} - V_{ap,pkg} I_{mb,pkg})/\Delta,$$

$$\text{with } \Delta = I_{mp,pkg} I_{ab,pkg} - I_{mb,pkg} I_{ap,pkg}.$$

The two-port network must be reciprocal, which requires $Z_{12,pkg} = Z_{21,pkg}$. Based on (3.26), the following constraint is derived

$$e^{j(\psi_b+\psi_p)} = \gamma,$$

$$\text{with } \gamma = \frac{|V_{ap,pkg}||I_{ab,pkg}|e^{j\phi_{ap}} - |V_{ab,pkg}||I_{ap,pkg}|e^{j\phi_{ab}}}{|V_{mb,pkg}||I_{mp,pkg}|e^{j\phi_{mb}} - |V_{mp,pkg}||I_{mb,pkg}|e^{j\phi_{mp}}}. \tag{3.27}$$

Equation (3.27) requires among other thing that the relationship between the input phase offset ψ_b at backoff power and ψ_p at peak power be given by

$$\psi_b + \psi_p = \angle\gamma\,. \tag{3.28}$$

The reciprocal two-port network shown in Fig. 3.17a is realized with a lossless three-port network terminated with a 50 Ω load. To enforce the losslessness of the three-port network, the two-port Z parameters must verify $\mathcal{R}^2\{Z_{12}\} = \mathcal{R}\{Z_{11}\}\mathcal{R}\{Z_{22}\}$ [16] (see [5] for a derivation). The input phase offsets ψ_b and ψ_p which satisfy this criterion can then be numerically computed [14, 15]. Note that requirement $|\gamma| = 1$ needs also to be enforced. An approximate solution is then obtained by using the following geometrical average:

$$Z'_{21,pkg} = Z'_{12,pkg} = \sqrt{|Z_{21,pkg} Z_{12,pkg}|}\frac{Z_{12,pk}}{|Z_{12,pkg}|}.$$

3.3.3 *Dual-Input Doherty Power Amplifier*

Given its present dominance in basestation applications, let us focus now on the DPA. A prototype for realizing an ideal DPA at the current-source reference plane is presented in Fig. 3.20. The fundamental drain currents of the main and auxiliary transistors are represented by two current sources. They are connected to the output load R_L using a $\lambda/4$ transformer with a characteristic impedance of Z_T. Analyzing the Doherty circuit, we readily derive the amplitude of the main voltage to be given by

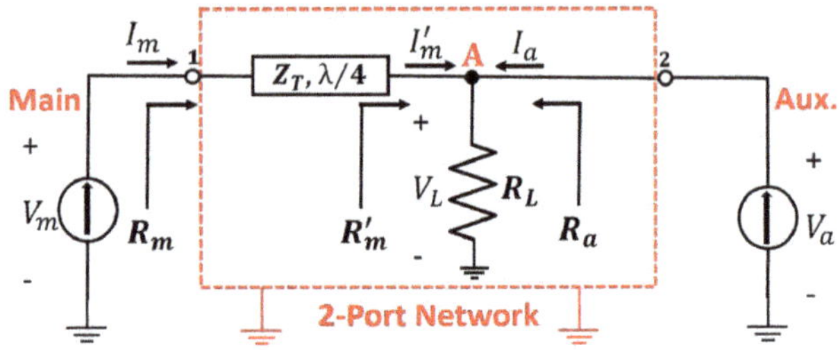

Fig. 3.20 Ideal circuit for a Doherty power amplifier (DPA) at the current-source reference plane ([17], reprinted with permission from IEEE)

$$|V_m| = -jZ_T|I_a| + |I_m|\frac{Z_T^2}{R_L}. \tag{3.29}$$

In the traditional DPA, the auxiliary transistor is assumed to be off at backoff $K_{ia} = \infty$. In practice, it is not desirable to fully turn off the auxiliary transistor (non-zero I_{ab}) and I_{ab} provides an additional degree of freedom to explore since we are going to embed the PA from the current-source to the package reference planes. The amplitude of the fundamental voltages $|V_{mp}|$ and $|V_{mb}|$ are then obtained from (3.29) to be given by

$$|V_{mb}| = -jZ_T|I_{ab}| + |I_{mb}|\frac{Z_T^2}{R_L},$$

$$|V_{mp}| = -jZ_T|I_{ap}| + |I_{mp}|\frac{Z_T^2}{R_L}. \tag{3.30}$$

To further increase the degrees of freedom available to the designer we shall also relax the $K_{vm} = 1$ assumption in the continuum theory and thus use slightly different amplitudes for the backoff $|V_{mb}|$ and peak $|V_{mp}|$ fundamental drain voltages: $|V_{mp}| = K_{vm}|V_{mb}|$. Solving the system of Eq. (3.30), generalized expressions for the characteristic impedance Z_T of the $\lambda/4$ transformer and the common load R_L are easily derived

$$Z_T = \frac{|V_{mb}||I_{mp}| - |V_{mp}||I_{mb}|}{|I_{mb}||I_{ap}| - |I_{mp}||I_{ab}|}, \tag{3.31}$$

$$R_L = \frac{(|V_{mb}||I_{mp}| - |V_{mp}||I_{mb}|)^2}{(|I_{mb}||I_{ap}| - |I_{mp}||I_{ab}|)(|V_{mb}||I_{ap}| - |V_{mp}||I_{ab}|)}. \tag{3.32}$$

The asymmetry peak power ratio n reduces now to

$$n = \frac{K_{im} - K_{vm}}{K_{vm}\left(1 - \frac{K_{im}}{K_{ia}}\right)}. \tag{3.33}$$

In the conventional DPA theory $K_{vm} = 1$ and $K_{ia} = \infty$ and the asymmetry peak power ratio n simplifies to $K_{im} - 1$.

The peak-to-backoff power ratio (PBPR) of the DPA reduces to

$$\mathrm{PBPR} = \frac{P_{o,peak}}{P_{p,back}} = K_{im}^2 = \left(\frac{K_{vm}(n+1)}{1+n\frac{K_{vm}}{K_{ia}}}\right)^2. \tag{3.34}$$

In the conventional DPA theory, $K_{vm} = 1$ and $K_{ia} = \infty$ and the PBPR simplifies to $(n+1)^2$.

The general DPA design theory derived establishes the relationships between the eight voltage and current parameters ($|V_{mp}|$, $|V_{mb}|$, $|I_{mp}|$, $|I_{mb}|$, $|V_{ap}|$, $|V_{ab}|$, $|I_{ap}|$, and $|I_{ab}|$) and the six design parameters ($V_{\max}$, $I_{\max}$, n, γ_{vp}, K_{ia}, and K_{vm}).

The design of the DPA at the current-source reference planes proceeds now as follows. First, the amplitude of fundamental drain voltage at peak power for the auxiliary PA $|V_{ap}|$ is taken to be approximately half the maximum drain voltage $V_{\max}$: $|V_{ap}| = V_{\max}/2$. Similarly, the amplitude of fundamental drain current at peak power for the auxiliary PA $|I_{ap}|$ is taken to be approximately half the maximum drain current $I_{\max}$: $|I_{ap}| = I_{\max}/2$. The amplitude $|V_{mp}|$ of the main drain voltage at peak power is then given by: $|V_{mp}| = \gamma_{vp}|V_{ap}|$. The amplitude $|I_{mp}|$ of the main drain current at peak power is then given by

$$|I_{mp}| = \gamma_{ip}|I_{ap}| = \frac{1}{n\gamma_{vp}}|I_{ap}|.$$

Note that γ_{vp} is usually selected to be one. However, using $\gamma_{vp} = 1/n$ reduces the main voltage V_{mp} and thus reduces the charging of traps degrading the IV characteristics. The PBCR K_{im} is then found from (3.33) to be given by

$$K_{im} = K_{va} = \frac{K_{vm}(n+1)}{1+n\frac{K_{vm}}{K_{ia}}}.$$

The main and auxiliary voltages and currents at backoff are then given by

$$|I_{mb}| = \frac{|I_{mp}|}{K_{im}},\quad |I_{ab}| = \frac{|I_{ap}|}{K_{ia}},\quad |V_{ab}| = \frac{|V_{ap}|}{K_{va}},\quad \text{and } |V_{mb}| = \frac{|V_{mp}|}{K_{vm}}.$$

The load impedance (R_{mp}, R_{mb}, R_{ap}, and R_{ab}) seen by the main and auxiliary PAs at peak and backoff can now also be determined.

Having acquired all the needed fundamental voltages, currents for the main and auxiliary transistors, it is necessary next to establish the gate and drain biases and RF gate drives (LSOP) for the main and auxiliary transistors. This is readily done using single-transistor simulations using the realistic IV characteristics of the FET core as shown in Fig. 3.8. In the case of DPA, class-F operation is targeted for the main PA and class C for the auxiliary PA. The required harmonic load impedance terminations are just applied accordingly to the RF core. The drain supply for the main transistor is set to $V_{DD,m} = |Vmp| + V_{on}$ and the gate supply to the threshold voltage $V_{GG,m} = V_T$. The peak $V_{GS,mp}$ and backoff $V_{GS,mb}$ RF gate voltages required to obtain the desired fundamental peak and backoff drain voltages V_{mp} and V_{mb} are then obtained by sweeping the RF gate drives $V_{GS,m}$ in harmonic balance simulations using the FET core. The drain supply for the auxiliary transistor is set to $V_{DD,a} = |V_{ap}| + V_{on}$ and the gate supply to the threshold voltage $V_{GG,a}$ is set below the threshold voltage V_T. The required peak $V_{GS,ap}$ and backoff $V_{GS,ab}$ RF gate voltages required to obtain the desired fundamental peak and backoff drain voltages V_{ap} and V_{ab} are then obtained by sweeping the RF gate drives $V_{GS,a}$ in harmonic balance simulation of the FET core. Thus, four rapid harmonic balance power sweeps are required in total to define the LSOP (gate drives) for the main and auxiliary transistors.

The characteristic impedance of the $\lambda/4$ transformer Z_T and the load R_L is obtained from (3.31) and (3.32). The DPA prototype at the current-source plane is then established and can be simulated using the nonlinear embedding model. This entire design algorithm can be implemented in a MATLAB code for the automatic design of DPAs [10]. The MATLAB code calls the above five harmonic balance simulations which are performed by running ADS netlists under script control.

A MATLAB graphical user interface (GUI) as shown in Fig. 3.21a–k was created to facilitate the automated design of DPA prototypes at the package reference planes based on the input parameters selected by the PA designer. The design takes typically 24 s to complete the design (first four harmonic balance simulations) on a laptop computer and 31 s to complete the performance verification (fifth harmonic balance simulation) for the DPA prototype by sweeping the input RF voltages. The input design parameters in the top left red box in Fig. 3.21a–k include the operating frequency f_0, the maximum drain voltage $V_{\max}$, the maximum drain current $I_{\max}$, the desired asymmetry peak power ratio n, the main-to-auxiliary PA peak voltage ratio γ_{vp}, the auxiliary peak-to-back current ratio $K_{ia} = 1/\delta$, and the main peak-to-backoff voltage ratio $K_{vm} = \nu$. The output parameters for the DPA prototype outlined by the blue box are the calculated $\lambda/4$ characteristic impedance Z_T, the common load R_L, and the RF gate voltages. The results shown in the GUI are listed in the left side. On the right side of the GUI, the load lines at the current-source reference planes seen by the main and auxiliary PAs are presented in Fig. 3.21a and b, respectively. The Doherty fundamental impedances seen by the main and auxiliary transistors are shown in Fig. 3.21c and d, respectively. The drain voltage and current waveforms are plotted in Fig. 3.21e and f, respectively. The drain efficiency and gain predicted by the DPA prototype at the package reference planes are shown in Fig. 3.21g. The S-parameters of the output combiner circuits calculated at the package reference planes after embedding are shown in Fig 3.21h. The input fundamental and second-harmonic impedance are shown in Fig. 3.21h. The total output power and the output power for main and auxiliary PAs are plotted in Fig. 3.21i versus input power.

This software program enables to quickly generate multiple DPA prototypes with different performances using parametric sweep simulations of the design parameters. The reader is referred to Ref. [10] for a more detailed analysis. A demonstrator PA was reported in Ref. [10] using for design parameters $n = 1.78$,

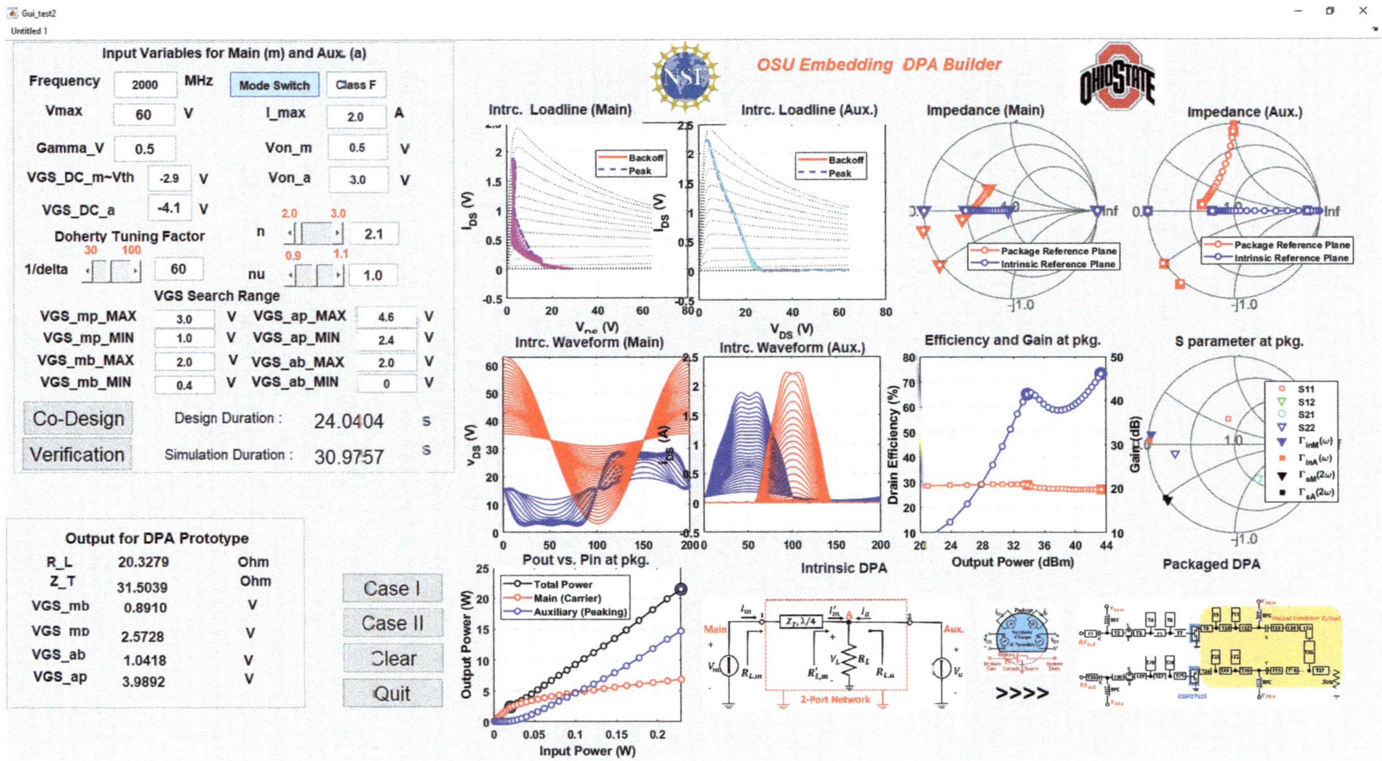

Fig. 3.21 MATLAB GUI used to design the DPA prototype (24 s) and verify the DPA (31 s). Note that in the GUI $1/\delta = K_{ia}$ and $nu = K_{vm}$. (**a**) and (**b**) show the load lines at the current-source reference planes seen by the main and auxiliary PAs, respectively. (**c**) and (**d**) show the Doherty fundamental impedances seen by the main and auxiliary transistors, respectively. (**e**) and (**f**) show the drain voltage and current waveforms, respectively. (**g**) shows the drain efficiency and gain predicted by the DPA prototype at the package reference planes. (**h**) shows the S-parameters of the output combiner circuits calculated at the package reference planes after embedding. (**h**) shows the input fundamental and second-harmonic impedance. (**i**) shows the total output power and the output power for main and auxiliary PAs versus input power. (**j**) shows the DPA at the current-source reference planes. (**k**) shows the DPA at the package reference planes ([17], reprinted with permission from IEEE)

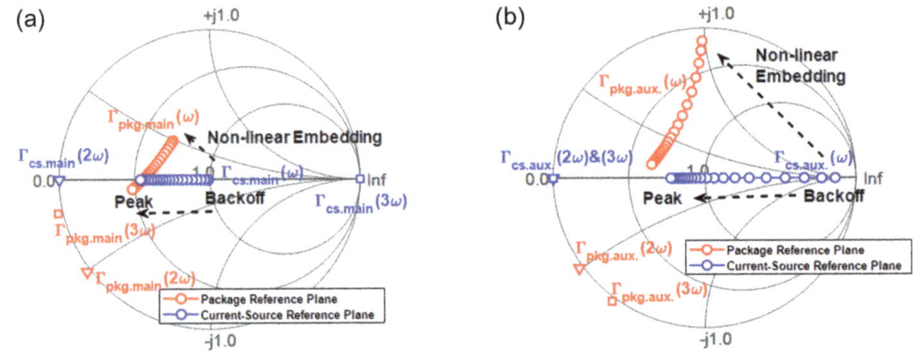

Fig. 3.22 The simulated load modulation trajectories including the impedance $\Gamma_{cs}(n\omega)$ (blue) at the device current-source reference planes and the corresponding impedance $\Gamma_{pkg}(n\omega)$ (red) at the device package reference planes presented to the (**a**) main and (**b**) auxiliary PAs ([17], reprinted with permission from IEEE)

$\nu = K_{vm} = 1.1$, $\delta = 1/K_{ia} = 0.015$, $|\gamma_{vp}| = 0.56$, $I_{\max} = 2.1$ A, and $V_{\max} = 60.0$ V for the prototype. This design yields an OBO of PBPR $= 9.4$ dB. The simulated fundamental load modulation behaviors for the main and auxiliary PAs based on the DPA prototype at the current-source reference planes are presented in Fig. 3.22a and b, respectively. The optimal second- and third-harmonic impedances at the output package reference planes seen by the main and auxiliary transistors are also presented in Fig. 3.22a and b, respectively. The output harmonic filter circuits are realized using offset transmission lines and open stubs to match the harmonic impedance goals established by the nonlinear embedding model shown in Fig. 3.22a and b. The input matching networks are also designed. The conjugate matching is applied for the fundamental input impedances and the input second-harmonic impedances are set as predicted by the nonlinear embedding device model. Note that the source and load harmonic terminations established by the nonlinear embedding model are found in practice not to vary much from backoff to peak.

Figure 3.23 shows the optimal measured power added efficiency (PAE) envelope versus output power. The measured associated input phase offsets, drain efficiency, and gain are also plotted in Fig. 3.23. A 65% measured drain efficiency and 62% measured PAE are achieved at 8-dB backoff with a peak power of about 42.9 dBm and a measured saturation gain of about 15 dB. It is noted that the dual-input gain (in dB) is defined here as the difference between the output power (in dBm) and the combined input power (in dBm) from the two RF input ports.

3.3.4 Chireix Power Amplifier

In this subsection we turn now toward the design of Chireix PAs using nonlinear embedding. Figure 3.24 shows the circuit topology of the prototype outphasing

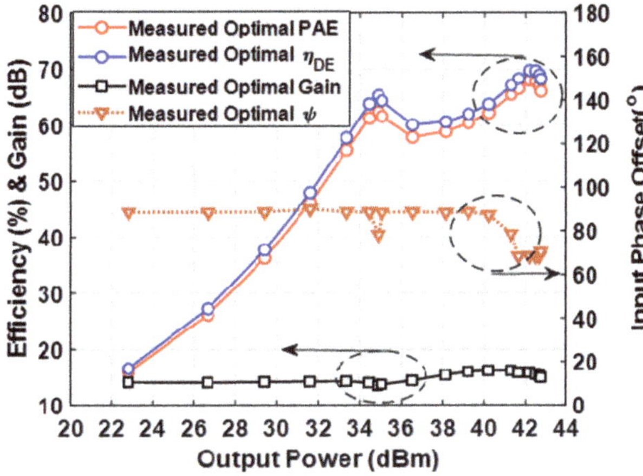

Fig. 3.23 Measured drain efficiency and power added efficiency (PAE), gain and input phase offset angles versus output power at 2 GHz for the DPA fabricated ([17], reprinted with permission from IEEE)

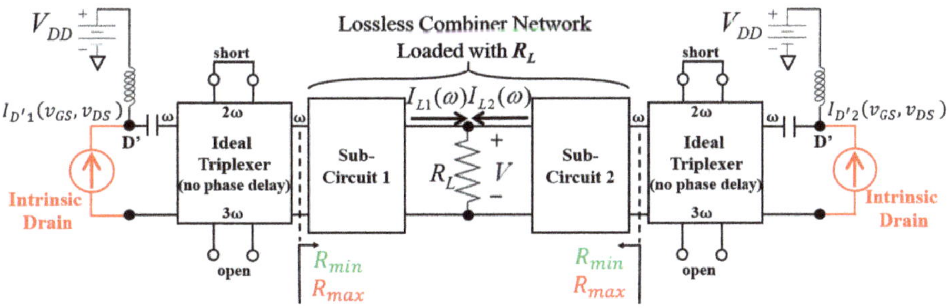

Fig. 3.24 Prototype of a Chireix outphasing amplifier at the current-source reference planes. The ideal triplexers are used for setting the class-F multi-harmonic impedance terminations without introducing any delay or phase shift ([18], reprinted with permission from IEEE)

combining network to be used with the FET cores implemented with the embedding FET device model. As shown in Fig. 3.24, the left and right current sources represent the IV characteristics $I_{D'}(v_{GS}, v_{DS})$ generating the intrinsic drain current at the current-source reference planes of each transistor. The outphasing combiner is implemented with two lossless two-port sub-circuits connected to a common load R_L. Two ideal triplexers (multi-harmonic terminations with zero phase delay) are implemented in order to define the PA's class of operation and establish the transistor dynamic load lines providing high efficiency. In this work, a class-F implementation is used at both peak and backoff in order to provide a higher efficiency. Figure 3.25 shows the resulting class-F load lines for the intrinsic fundamental loads R_{min} and R_{max} at peak and backoff power, respectively.

Different gate drives are selected for the two transistors at peak and backoff to achieve a higher efficiency. This is referred as *mixed-mode outphasing* since the control of the dual-input Chireix PA involves not only the outphasing phase between the two inputs but also a variation of the input power between backoff and peak.

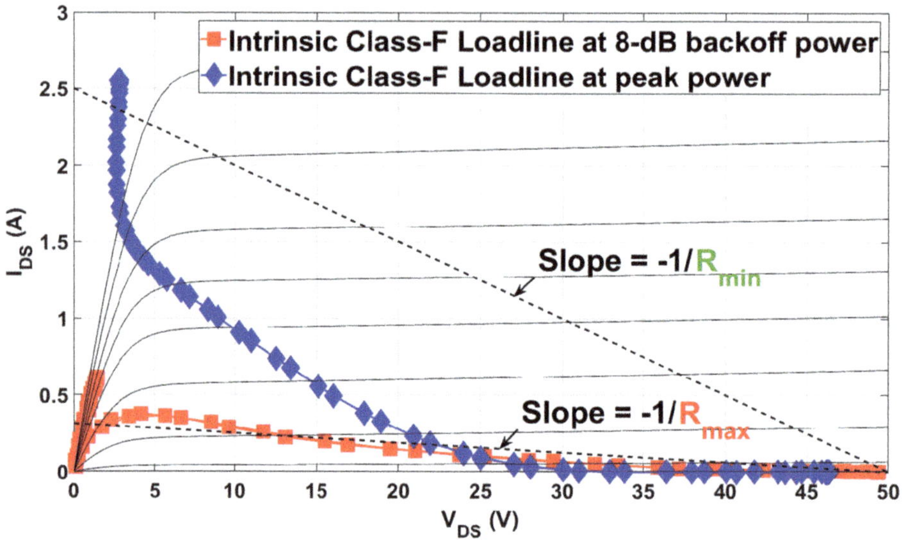

Fig. 3.25 Class-F PA load lines at 8 dB backoff power ($R_{\max}$) and peak power ($R_{\min}$) ([18], reprinted with permission from IEEE)

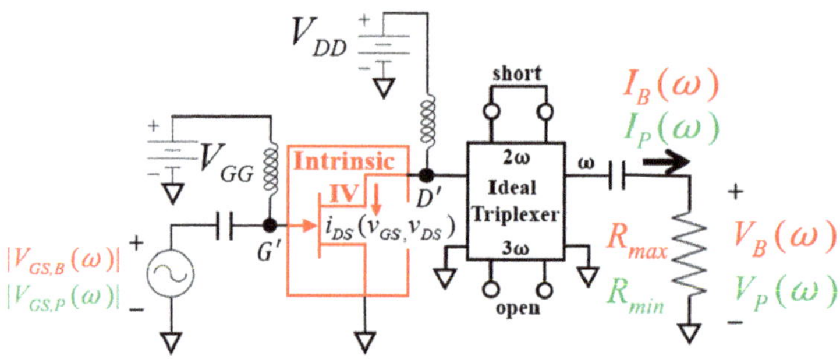

Fig. 3.26 Single-transistor circuit used for optimizing the peak and back operations ([18], reprinted with permission from IEEE)

The two transistors will operate with the same resistive drain fundamental loads $R_{\min}$ and $R_{\max}$, and the same amplitude for the input fundamental RF gate drives $|V_{GS,P}(\omega)|$ and $|V_{GS,B}(\omega)|$ at peak (P) and backoff (B), respectively. Indeed in the continuum theory, the Chireix model is associated with the same fundamental drain voltages ($K_{vm} = K_{va}$) and the same fundamental drain currents ($K_{im} = K_{ia}$) at peak and backoff. It results that before doing any two-transistor simulations of the Chireix PA, the designer can rely on single-transistor simulations with a fixed real fundamental load at the current-source reference planes (see Fig. 3.26) to determine the fundamental operating point (LSOP) at peak and backoff providing the optimal Chireix output power P_o and efficiency, for the selected peak-to-backoff power ratio (PBPR).

To implement the class-F operation, ideal open and short harmonic terminations are used at the current-source reference planes for the even and odd harmonics, respectively, as shown in Fig. 3.26. The design of a class-F PA with optimal performance at peak and backoff can be pursued using the selection of the real load R_{min} and R_{max} and the optimal fundamental gate voltages $V_{GS,P}(\omega)$ and $V_{GS,B}(\omega)$ at the current-source reference planes providing the optimal output power P_o and drain efficiency for the desired PBPR at the package reference planes. To do so the following four performance parameters are monitored: (1) the peak output power $P_{o,peak}$, (2) the backoff output power $P_{o,back}$, (3) the drain efficiency $\eta_{D,peak}$ at peak, and (4) the drain efficiency $\eta_{D,back}$ at backoff, all at the package reference planes.

Figure 3.27a and b show the simulation results obtained with a single transistor under (a) backoff and (b) peak power operations, respectively, as a function of the drain loads R_{min} or R_{max} and the fundamental amplitude of the RF gate voltages $|V_{GS,P}(\omega)|$ or $|V_{GS,B}(\omega)|$. The gate voltages $|V_{GS}(\omega)|$ are swept for various R_{max} (backoff operation), and (b) R_{min} (peak operation). The optimal peak $P_{o,peak}$ and backoff $P_{o,back}$ power at the package reference planes can be then selected from the resulting design space in Fig. 3.27a and b to achieve a high peak power while maintaining optimal efficiencies at both peak and backoff for the selected peak-to-backoff power ratio (PBPR). As shown in Fig. 3.27, using yellow triangles, the loads $R_{\text{min}} = 20\,\Omega$ and $R_{\text{max}} = 160\,\Omega$ can be selected to deliver 42.1 dBm peak power and 34.2 dBm backoff power while maintaining about 80% drain efficiency and 8 dB backoff (PBPR) at the package reference planes.

Outphasing must be established at the common load R_L at the current-source reference planes. In mixed-mode operation the load current is given by

$$\begin{aligned} I_{L1}\left(|V_{GS}(\omega)|, \theta_L\right) &= |I_{L1}\left(|V_{GS}(\omega)|\right)| \exp\left(+j\theta_L/2\right), \\ I_{L2}\left(|V_{GS}(\omega)|, \theta_L\right) &= |I_{L2}\left(|V_{GS}(\omega)|\right)| \exp\left(-j\theta_L/2\right), \\ I_L\left(|V_{GS}(\omega)|, \theta_L\right) &= I_{L1} + I_{L2}, \end{aligned} \tag{3.35}$$

where θ_L is the outphasing angle at the common load R_L. Note that I_{L1} and I_{L2} are both functions of $|V_{GS}(\omega)|$ in addition to the outphasing angle θ_L to allow for the mixed-mode operation.

At peak power ($\theta_L = \theta_{L,P} = 0°$) and at backoff power ($\theta_L = \theta_{L,B}$) the two transistors operate identically in terms of loads and power and the PBPR for mixed-mode operation at the common load R_L is given by

$$\text{PBPR} = \frac{\left|I_L\left(|V_{GS,P}(\omega)|, \theta_{L,P}\right)\right|^2}{\left|I_L\left(|V_{GS,B}(\omega)|, \theta_{L,B}\right)\right|^2} = \frac{|I_{L,P}|^2}{|I_{L,B}|^2 \cos^2(\theta_{L,B})}, \tag{3.36}$$

where we define $|I_{L1,P}| = |I_{L2,P}| = |I_{L,P}|/2$ and $|I_{L1,B}| = |I_{L2,B}| = |I_{L,B}|/2$.

The ideal lossless and reciprocal outphasing networks connected between the planes D′ and the common load R_L in Fig. 3.28 need now to be designed so that

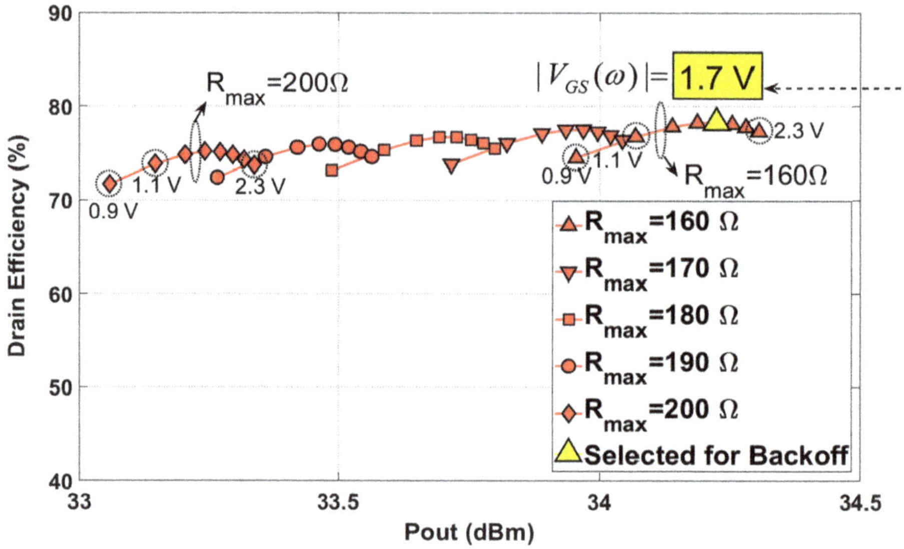

(a) Backoff design space

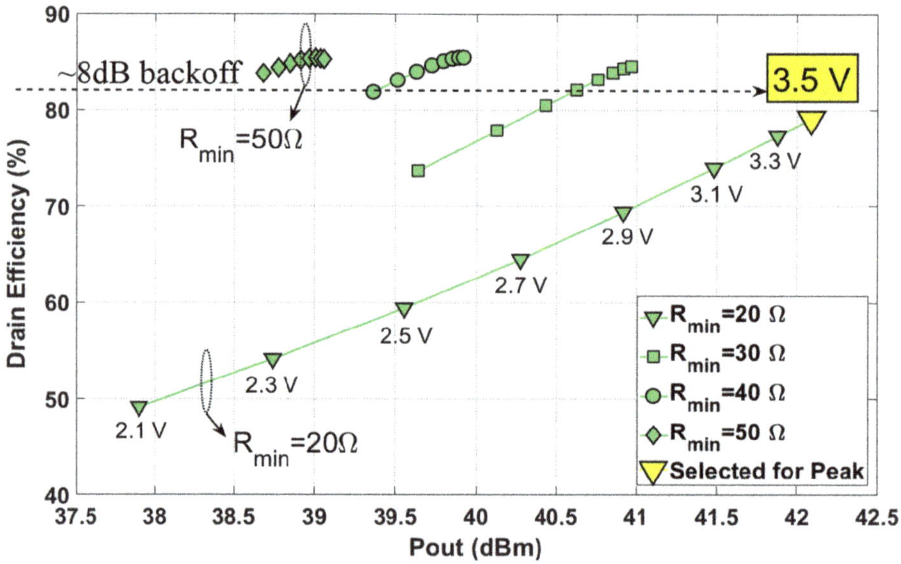

(b) Peak design space

Fig. 3.27 Efficiency versus output power at the package reference planes at (**a**) backoff power $(P_{o,back})$ and (**b**) peak power $(P_{o,peak})$. The efficiencies at the package reference planes are calculated using the embedding model by applying excitations at the current-source reference planes (FET core) in the single-transistor circuit shown in Fig. 3.26 ([18], reprinted with permission from IEEE)

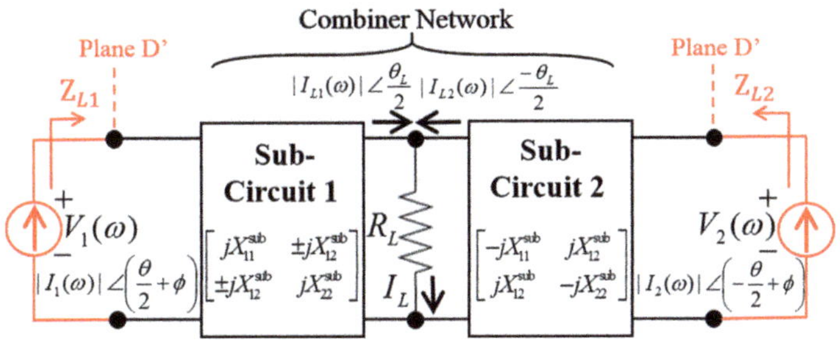

Fig. 3.28 Circuit implementation of the Chireix outphasing combiner. $|I_1(\omega)|$ and $|I_2(\omega)|$ are the amplitudes of the fundamental *drain* currents ([18], reprinted with permission from IEEE)

the combiner presents the impedance $R_{\max}$ at backoff and $R_{\min}$ at peak operation to both transistors at planes D′. The equations in (3.23) derived in terms of the K parameters can be used to implement the combiners. Alternately, as derived in Ref. [18], the pure-imaginary Z-parameters ($Z_{ij} = jX_{ij}$) of the lossless and reciprocal outphasing combiners can be expressed in terms of the load R_L and the intrinsic peak $R_{\min}$ and backoff $R_{\max}$ loads as

$$X_{11}^{\text{sub}} = -\sqrt{\frac{R_{\min}^3}{R_{\max}}}, \tag{3.37}$$

$$X_{12}^{\text{sub}} = \pm\sqrt{2R_L R_{\min}\left[\frac{R_{\min}}{R_{\max}} + 1\right]}, \tag{3.38}$$

$$X_{22}^{\text{sub}} = -2R_L\sqrt{\frac{R_{\min}}{R_{\max}}}. \tag{3.39}$$

Note that the selection of the load R_L is arbitrary since this PA prototype is only used in simulation to facilitate the design of the PA at the package reference planes.

Next, the outphasing angles θ_P and θ_B at peak and backoff between the two branches need to be calculated. These outphasing angles can be calculated in terms of the K factors using (3.22). Alternatively, the following analytic equations expressed in terms of $R_{\min}$ and $R_{\max}$ derived in [18] can be used:

$$\theta_B = \pm\cos^{-1}\left(\pm\frac{R_{\max} - R_{\min}}{R_{\max} + R_{\min}}\right), \tag{3.40}$$

$$\theta_P = \pi - \theta_B. \tag{3.41}$$

In the example selected with $R_{\min} = 20\,\Omega$ and $R_{\max} = 160\,\Omega$, this yields four possible outphasing angles: ±38.94° and ±141.06° for backoff power. The solution −141.06° is associated with $X_{12,}^{\text{sub1}} = X_{12}^{\text{sub2}}$ for the combiner. The solution −38.94° is associated with $X_{12,}^{\text{sub1}} = -X_{12}^{\text{sub2}}$ for the combiner. The two other solutions 141.06° and 38.94° result simply from the exchange of the sub-circuits

1 and 2 of the combiner. The two choices for the left (top) and right (bottom) two-port impedance matrices of the combiners used in the Chireix PA prototype are summarized in Fig. 3.28.

The intrinsic drain loads Z_{L1} (top) and Z_{L2} (bottom) labeled as $Z_{L,intr}(\omega)$ at the current-source reference planes are plotted (black dots) in the Smith chart in Fig. 3.29 for the outphasing angle varying from backoff to peak. The fundamental drain loads $Z_{L,pkg}(\omega)$ at the package reference planes calculated by the embedding device model are also plotted (green dots) in the Smith chart in Fig. 3.29.

Figure 3.30 shows a possible implementation of the outphasing Chireix PA circuit after nonlinear embedding from the current-source to package reference planes. In this implementation a two-port harmonic termination network is designed to provide the harmonic impedances $Z_{L,pkg}(2\omega)$ and $Z_{L,pkg}(3\omega)$ at the package references shown in Fig. 3.29. The three-port combiner used for the fundamental waveforms is then designed using (1) the transistor drain voltages V_D and currents I_D at the package drain reference planes D calculated by the embedding model and (2) the procedure described in Sect. 3.3.2.

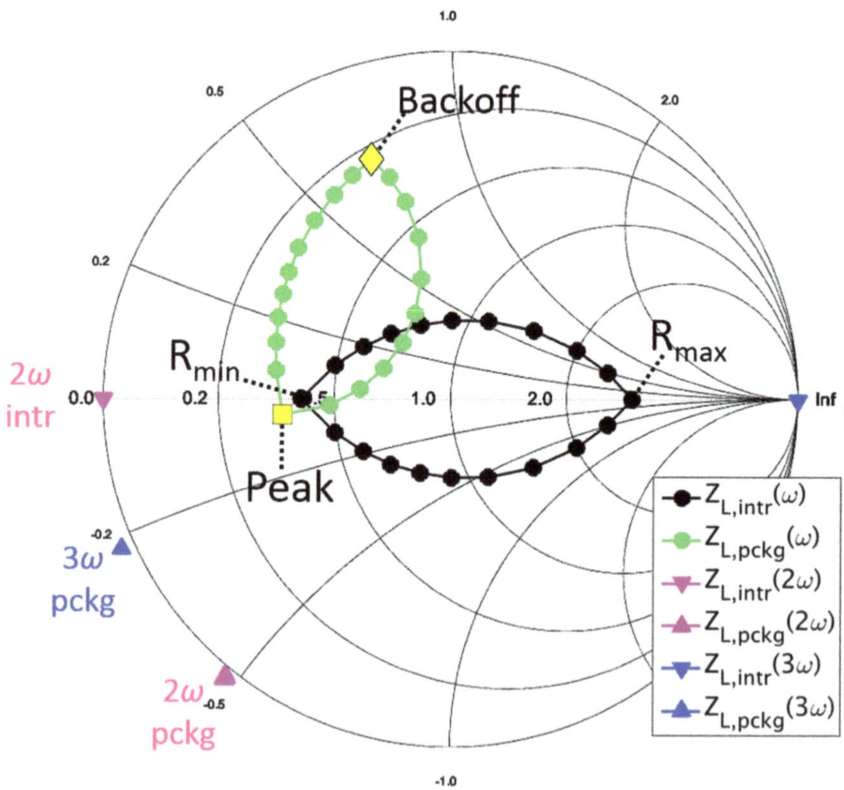

Fig. 3.29 Intrinsic $Z_{L,intr}(\omega, 2\omega, 3\omega)$ and package $Z_{L,pkg}(\omega, 2\omega, 3\omega)$ load obtained in simulation for a mixed-mode outphasing PA using the embedding device model ([18], reprinted with permission from IEEE)

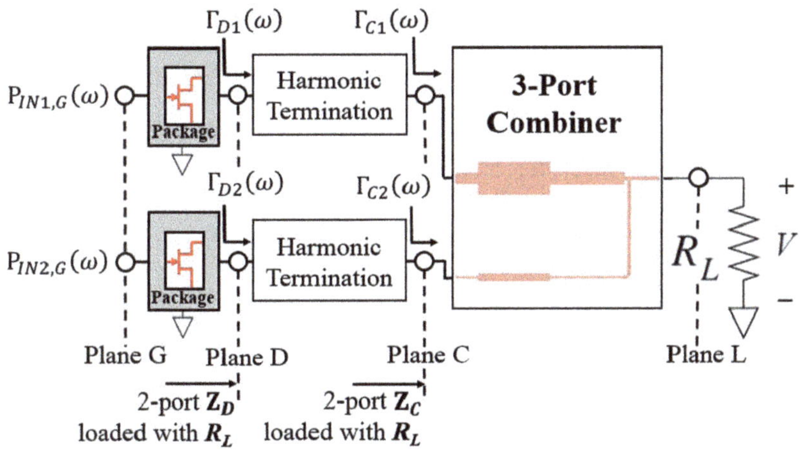

Fig. 3.30 Dual-input OPA circuit realized at the package reference planes after the nonlinear embedding process. Plane G: package gate reference planes. Plane D: package drain reference planes. Plane C: physical combiner reference planes ([18], reprinted with permission from IEEE)

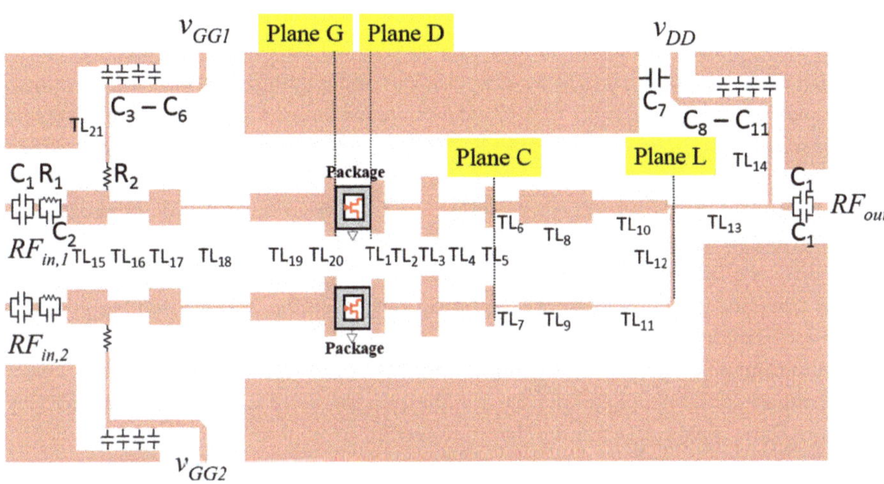

Fig. 3.31 Layout of the demo Chireix outphasing amplifier ([18], reprinted with permission from IEEE)

The layout of the demo outphasing PA reported in [18] is shown in Fig. 3.31. The continuous wave (CW) drain efficiency and gain obtained from CW measurements at 1900 MHz are shown in Fig. 3.32. Each red triangle curve represents a plot of the efficiency versus output power obtained while sweeping the outphasing angle from $-140°$ to $-40°$ in $10°$ steps and keeping the input power levels P_{in} constant. The same incident power levels are used for the two PA inputs with the input power P_{in} ranging from 10 to 28 dBm. The P_{in} step is set to be 2 dB below 18.5 dBm and 1 dB above 18.5 dBm. The bottom black triangle lines give the corresponding gain curves.

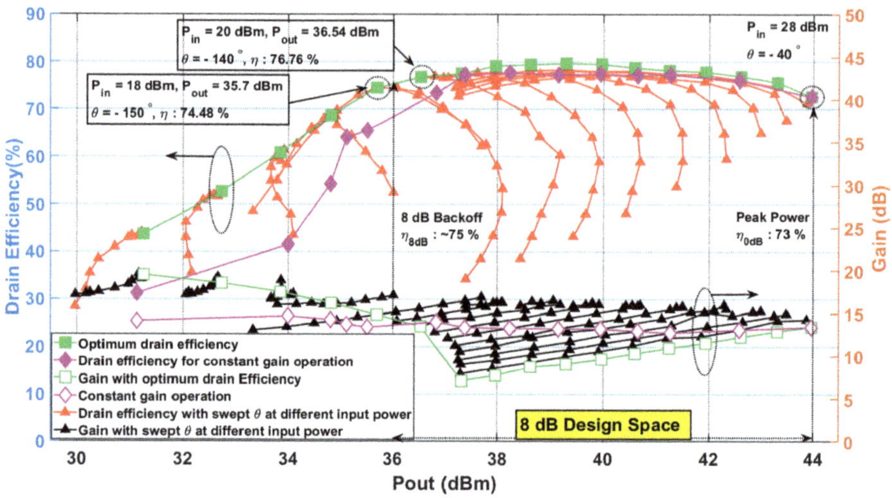

Fig. 3.32 Continuous wave (CW) measurements of the drain efficiency and gain versus the outphasing angle θ for 15 different input power levels at 1900 MHz ([18], reprinted with permission from IEEE)

Two different types of mixed-mode operations are highlighted: (1) the optimum-efficiency (green square line) and (2) the constant-gain modes (magenta diamond line). The bottom green and magenta hollow circles show the corresponding power gains associated with optimum drain efficiency. A look-up table (LUT) for the incident power levels, outphasing angles, and output power levels is created to sustain the constant-gain mode of operation. This constant-gain based LUT will be used for the linearization discussed next.

The constant-gain mixed-mode operation shown in Fig. 3.32 is used for generating an LUT including the incident power $P_{in,1/2}$, outphasing angles θ at plane G, and the output power P_o. With the modulated RF input signal expressed in discrete-time form as $x(n) = |x(n)| \exp[j\phi_x(n)]$, the two inputs of the PA are controlled by the following signals $x_1(n)$ and $x_2(n)$, defined as

$$x_1(n) = \sqrt{P_{in,1}\left[P_o(n)\right]} \exp j \left\{\phi_x(n) + \frac{1}{2}\theta\left[P_o(n)\right]\right\},$$

$$x_2(n) = \sqrt{P_{in,2}\left[P_o(n)\right]} \exp j \left\{\phi_x(n) - \frac{1}{2}\theta\left[P_o(n)\right]\right\},$$

where $P_{in,1/2}[P_o(n)]$ are the incident powers at ports 1 and 2 and $\theta[P_o(n)]$ is the outphasing angle, both of them a function of the output power $P_o(n) = P_{o,peak} \times |x(n)|^2$. Note that $x(n)$ is normalized so that its peak amplitude is one.

The Chireix PA was linearized by predistorting the input signal $x(n)$ using the technique of direct learning. Figure 3.33 shows the output spectra of the PA AM/AM, AM/PM performance with and without DPD when using 10 MHz LTE signal as a benchmark. The adjacent channel leakage power ratio (ACLR) after

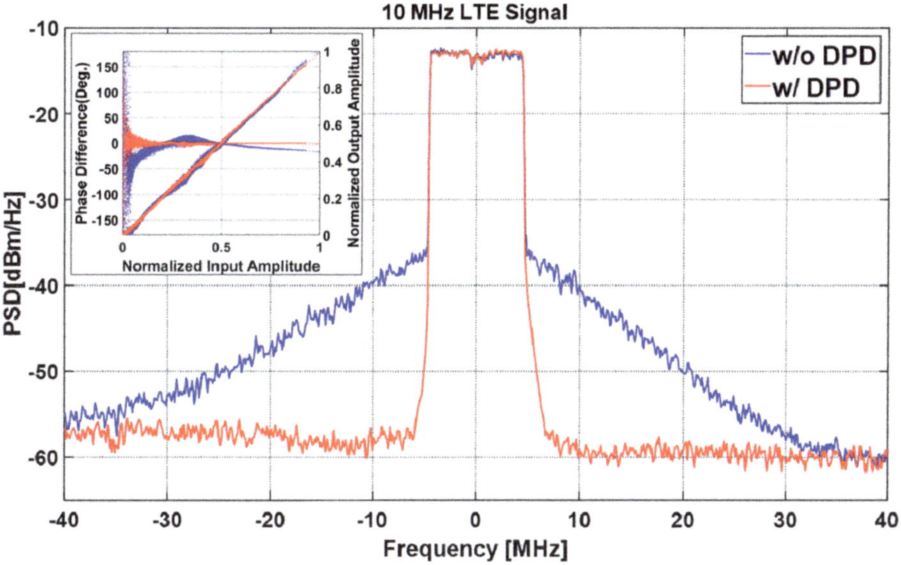

Fig. 3.33 Power spectral density, AM/AM, and AM/PM results for the Chireix OPA designed before digital predistortion (DPD) (blue) and after DPD (red) at 1.9 GHz implemented with a 10 MHz LTE signal ([18], reprinted with permission from IEEE)

linearization is −45.33 dBc for the LTE case, which indicates that this PA satisfies the linearity requirements of 3GPP TS 36.141 for LTE when DPD is applied. The average efficiency obtained with this PA remains above 59% for the 10 MHz LTE considered after DPD is applied.

3.3.5 Hybrid Chireix–Doherty Power Amplifier

In Sect. 3.3.1, we have introduced a 2D design space for the realization of dual-input two-transistor outphasing PAs with a fixed backoff (PBPR). This 2D continuum included the Chireix and Doherty modes of operation. The 2D design space presented in Fig. 3.19 revealed also the existence of two new types of hybrid dual-input PA, labeled HCD and HDmax with high performance. In this section, we study the characteristics of the HCD mode with the design of a 2 GHz demo HCD PA.

The HCD PA mode is obtained in the continuum theory presented in Sect. 3.3.1 by (1) setting $K_{va} = 1$ for V_{ap} and V_{ab} to be equal like in a Chireix PA and (2) selecting a very large PBCR K_{ia} to turn off the auxiliary PA at backoff like in a DPA. In the demo design K_{ia} is set to 50 yielding a PBPR of 10.37 dB and an outphasing angle at backoff power of 32.9° from the continuum theory. Several single-transistor sweep simulations are then performed with the embedded model as outlined in [10] to determine the LSOP providing an optimal tradeoff for the peak power and efficiency at the package reference planes for both the main and

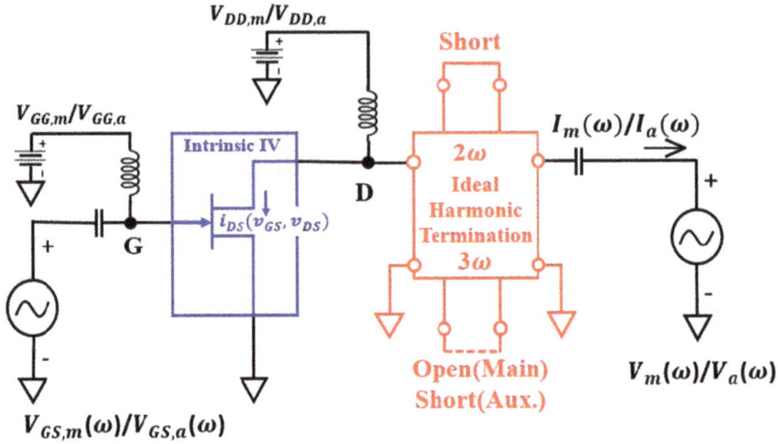

Fig. 3.34 Schematic of a single amplifier terminated with RF drain and gate voltage sources. The intrinsic embedding device model is used in the blue box labeled "Intrinsic IV" ([10], reprinted with permission from IEEE)

auxiliary transistors. In these single-transistor simulations, the drain voltages V_m or V_a and gate voltages $V_{GS,m}$ or $V_{GS,a}$ are applied as boundary conditions for the main or auxiliary transistors as shown in Fig. 3.34.

The main transistor is operated in class F with its DC gate voltage at the threshold voltage. The parameters sweeps for the main transistor lead to the selection of the drain fundamental voltage $V_{mp} = V_{mb} = 21$ V and drain bias $V_{DD,m} = 18$ V. This corresponds to the gate fundamental voltages $V_{GS,mp} = 4.08$ V and $V_{GS,mb} = 0.94$ V for peak and backoff, respectively. The auxiliary transistor is operating in class C with its drain fundamental voltage set at $V_{ap} = V_{ab} = V_m \gamma_v = 23.5$ V corresponding to γ_{vp} of 0.894. The drain bias voltage is selected to be $V_{DD,a} = 25$ V allowing for a 1.5 V knee voltage. The parameter sweeps for the main transistor lead to the selection of a DC gate bias of $V_{GG,a} = -4.1$ V for high efficiency. The gate fundamental voltages of $V_{GS,ap} = 4.00$ V and $V_{GS,mb} = 1.16$ V for peak and backoff, respectively, are then needed to obtain the peak I_{ap} and backoff I_{ab} fundamental drain currents required by the embedding theory.

The load impedances presented to the transistors under load modulation as obtained in simulation are presented in Fig. 3.35. As designed, the fundamental load impedances seen by the main and auxiliary transistors at the current-source reference plane are purely real at both backoff and peak powers. However, the modulated fundamental load impedances in between are complex, due to the variation of the outphasing angle between backoff and peak. The fundamental and harmonic loads seen by the main and auxiliary transistors at the package reference planes marked in red are presented in Fig. 3.35. The loads at the package reference planes are then used to design the physical output combiner and harmonic termination circuits using the procedure already outlined in Sect. 3.3.2.

Figure 3.36 compares the measured and simulated efficiency versus output power. A 61% measured drain efficiency is achieved at 9-dB backoff with a peak

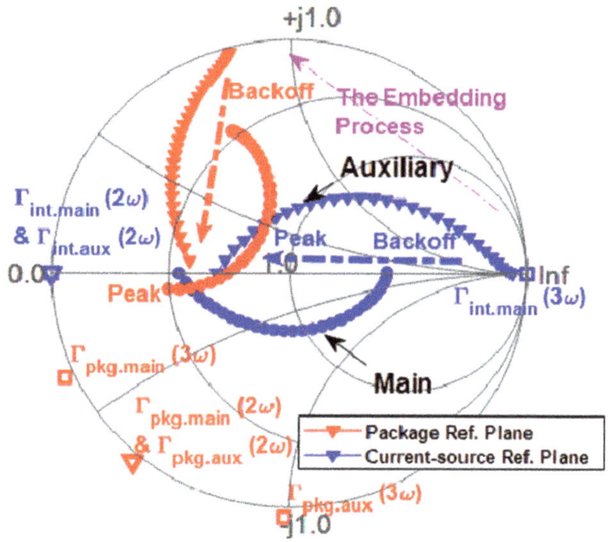

Fig. 3.35 Simulated load modulation trajectories for the HCD OPA including the intrinsic impedance $\Gamma_{int.main/aux}(n\omega)$ (blue) presented to the main and auxiliary devices and the corresponding impedance at the device package reference plane $\Gamma_{pkg.main/aux}(n\omega)$ (red) ([10], reprinted with permission from IEEE)

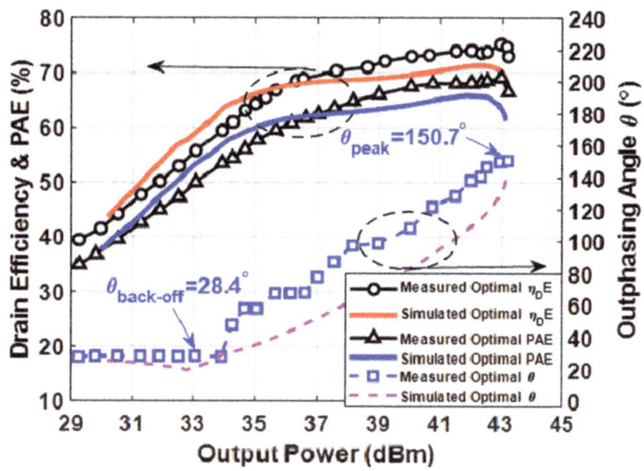

Fig. 3.36 Drain efficiency and the associated gain for the HCD OPA ([10], reprinted with permission from IEEE)

power of about 43 dBm and a measured saturation power gain of about 10 dB. Note that the power gain (in dB) is defined as the difference between the output power (in dBm) and the sum of the input powers (in dBm) at the two RF input ports. The efficiency increases monotonously with output power unlike that of the DPA in Fig. 3.23. This is due to the fact that both the main and auxiliary drain voltages exhibit little variation between peak and backoff as shown in Fig. 3.37. This originates from the selection of $K_{va} = K_{vm} = 1$.

The dynamic response of the HCD PA is verified with both 10 MHz and 20 MHz input LTE signals exhibiting 9.6 dB and 9.5 dB peak-to-average power ratio (PAPR), respectively. The modulated signal measurements are summarized in Table 3.3. When excited with a 20 MHz LTE signal with 9.5 dB PAPR, the dual-input PA yields a 60.0% average drain efficiency and −48.1 dBc ACLR after linearization.

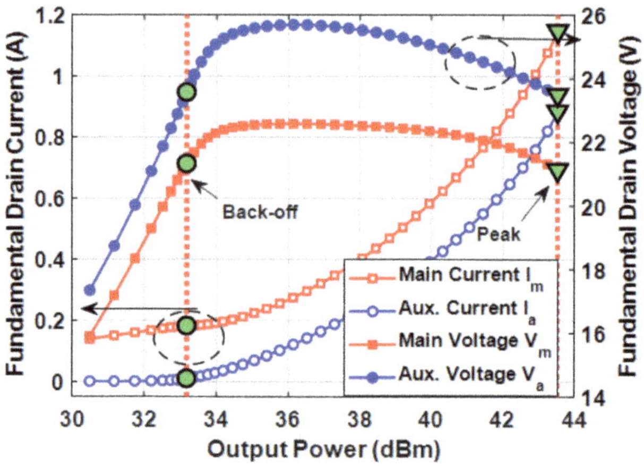

Fig. 3.37 Simulated amplitudes of the intrinsic fundamental currents and voltages versus output power for the HCD OPA ([10], reprinted with permission from IEEE)

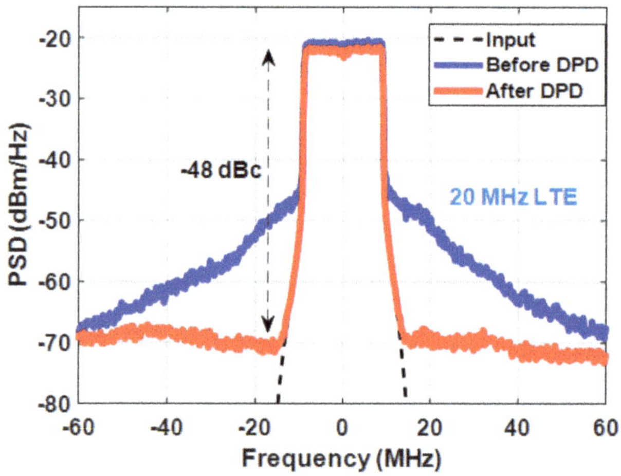

Fig. 3.38 Spectral density before (blue) and after DPD (red) 20 MHz LTE signals centered at 2 GHz for the HCD OPA ([10], reprinted with permission from IEEE)

The spectral density before (blue) and after DPD (red) 20 MHz LTE signals are shown in Fig. 3.38. In summary, the HCD PA yields an improved output power and efficiency compared to the Doherty PA but this comes at the price of using two inputs relying on outphasing.

3.3.6 *Conclusion*

The various examples presented in this chapter demonstrate how using an embedding device model can streamline and accelerate the design of high-performance PAs. For example, a prototype DPA was found to be automatically designed at the package reference planes with the help of a GUI (see Fig. 3.21) in less than 1 min when using a laptop.

Table 3.3 Modulated signal measurement at 2 GHz for an HCD OPA ([10], reprinted with permission from IEEE)

	Signal	PAPR (dB)	$P_{inc,avg.}$ (dBm)	$P_{out,avg.}$ (dBm)	$P_{out,peak}$ (dBm)	$\eta_{D,avg.}$/PAE$_{avg.}$ (%)	ACLR$_{L,H}$ (dBc)	NMSE (dB)
Before DPD	10 MHz LTE	8.4	22.7, 21.7	34.4	42.8	63.0/55.4	−30.7, −30.0	−15.2
After DPD	10 MHz LTE	9.6	21.2, 20.3	33.3	42.9	60.3/53.4	−53.2, −51.2	−35.4
Before DPD	20 MHz LTE	8.5	22.7, 21.7	34.4	42.9	62.3/54.9	−31.7, −30.5	−15.8
After DPD	20 MHz LTE	9.4	21.5, 20.5	33.4	42.8	60.0/52.9	−48.1, −47.7	−32.5

Preliminary investigations in simulation on the application of the nonlinear embedding model to the design of four-way Chireix PAs and class-E PAs have been reported in Refs. [19] and [20], respectively. The nonlinear embedding model offers also some promising applications in the design of novel types of broadband amplifiers. Design of broadband continuous class-J and class-F amplifiers has already been reported in [21] and [22] and will be explored in the next chapter. Both PAs were implemented with a 15 W GaN HEMT transistor. The class-J PA provided a measured drain efficiency between 63 and 72% in the frequency range of 1.3–2.4 GHz. The continuous class-F PA provided a measured drain efficiency between 60 and 75% in the frequency range of 1.5–2.5 GHz. Preliminary simulation work has also been reported for a generalized class-J PA which features Foster fundamental and harmonic loads [23]. The reader is invited to follow the literature for novel applications of the embedding model.

References

1. Raffo, A., Scappaviva, F., & Vannini, G. (2009). A new approach to microwave power amplifier design based on the experimental characterization of the intrinsic electron-device load line. *IEEE Transactions on Microwave Theory and Techniques, 57*(7), 1743–1752.
2. Vadalà, V., Avolio, G., Raffo, A., Schreurs, D. M. M.-P., & Vannini, G. (2012). Nonlinear embedding and de-embedding techniques for large-signal FET measurements. *Microwave and Optical Technology Letters, 54*(12), 2835–2838.
3. Jang, H., Roblin, P., & Xie, Z. (2014). Model-based nonlinear embedding for power amplifier design. *IEEE Transactions on Microwave Theory and Techniques, 62*(9), 1986–2002.
4. Angelov, I., Thorsell, M., Andersson, K., Rorsman, N., & Zirath, H. (2012). Recent results on using LSVNA for compact modeling of GaN FET devices. In *IEEE International Microwave Symposium Workshop (WMB): On "Device Model Extraction form Large-Signal Measurements,"* Montreal, June 2012.
5. Roblin, P., Martinez-Rodriguez, F., Chang, H. C., Xie, C. G., & Martinez-Lopez, J. I. (2015). Transistor characterization and modeling and the use of embedding device models for the design of microwave power amplifiers. In *IEEE Proceedings of Integrated Nonlinear Microwave Millimetre-Wave Circuits Workshop*, October, pp. 1–4.
6. Ko, Y., Roblin, P., Zarate-de Landa, A., Reynoso-Hernandez, A., Nobbe, D., Olson, C., et al. (2014). Artificial neural network model of SOS-MOSFETs based on dynamic large-signal measurements. *IEEE Transactions on Microwave Theory and Techniques, 62*(3), 491–501.
7. Martinez-Rodriquez, F. J., Roblin, P., Popovic, Z., & Martinez-Lopez, J. I. (2017). Optimal definition of class F for realistic transistor models. *IEEE Transactions on Microwave Theory and Techniques, 65*(10), 3585–3595.
8. Cripps, S. (2006). *RF amplifiers for wireless communication* (2nd ed., Table 6.2, p. 154). Norwood: Artech House.
9. Colantonio, P., Giannini, F., Leuzzi, G., & Limiti, E. (1999). On the class-F power-amplifier design. *RF Microwave Computer-Aided Engineering, 32*(2), 129–149.
10. Liang, C., Roblin, P., Hahn, Y., Popovic, Z., & Chang, H.-C. (2019). Novel outphasing power amplifiers designed with an analytic generalized Doherty-Chireix continuum theory. *IEEE Transactions on Circuits and Systems I, 66*(8), 2935–2948.
11. Andersson, C. M., Gustafsson, D., Cahuana, J. C., Hellberg, R., & Fager, C. (2013). A 1-3-GHz digitally controlled dual-RF input power-amplifier design based on a Doherty-outphasing continuum analysis. *IEEE Transactions on Microwave Theory and Techniques, 61*(10), 3743–3752.

12. Hellberg, R. (2017). *Composite power amplifier*. U.S. Patent 7 145 387, December 5, 2017.
13. Jang, H., Roblin, P., Quindroit, C., Lin, Y., & Pond, R. D. (2014). Asymmetrical Doherty power amplifier designed using model-based nonlinear embedding. *IEEE Transactions on Microwave Theory and Techniques, 62*(12), 3436–3451.
14. Özen, M., Andersson, K., & Fager, C. (2016). Symmetrical Doherty power amplifier with extended efficiency range. *IEEE Transactions on Microwave Theory and Techniques, 64*(4), 1273–1284.
15. Özen, M., van der Heijden, M., Acar, M., Jos, R., & Fager, C. (2017). A generalized combiner synthesis technique for class-E outphasing transmitters. *IEEE Transactions on Circuits and Systems—Part I: Regular Papers, 64*(5), 1126–1139.
16. Özen, M., & Fager, C. (2014). Symmetrical Doherty amplifier with high efficiency over large output power dynamic range. In *IEEE MTT-S International Microwave Symposium Digest*, Tampa, June 2014, pp. 1–4.
17. Liang, C., Roblin, P., & Hahn, Y. (2019). Accelerated design methodology for dual-input Doherty power amplifiers. *IEEE Transactions on Microwave Theory and Techniques, 67*(10), 3983–3995.
18. Chang, H. C., Hahn, Y., Roblin P., & Barton, T. W. (2018). New mixed-mode design methodology for high-efficiency outphasing Chireix amplifiers. *IEEE Transactions on Circuits and Systems—Part I: Regular Paper, 66*(4), 2935–2948.
19. Roblin, P., Liang, C., Chang, H.-C., & Rawat, K. (2019). Class-E PA prototype using an embedding model. In *IEEE 20th Wireless and Microwave Technology Conference (WAMI-CON)*, Cocoa Beach, 9 May 2019, pp. 1–6.
20. Roblin, P., Barton, T. W., Chang, H.-C., Liang, C., & Sear, W. (2017). Design of a 4-way Chireix amplifier using a nonlinear embedding device model. In *2017 IEEE 18th Annual Wireless and Microwave Technology Conference (WAMICON)*, pp. 1–5.
21. Saxena, S., Rawat, K., & Roblin, P. (2016). Continuous class-B/J power amplifier using nonlinear embedding technique. *IEEE Transactions on Circuits and Systems II: Express Briefs, 64*(7), 837–841.
22. Aggrawal, E., Rawat, K., & Roblin, P. (2017). Investigating continuous class-F power amplifier using nonlinear embedding model. *IEEE Microwave and Wireless Components Letters, 27*(6), 593–595.
23. Roblin, P., Chang, H.-C., Gomez-Perez, J.-M., & Martinez-Lopez, J. I. (2019). Generalized class-J theory. In *Latin America Microwave Conference*, Arequipa, December 2019, pp. 1–3.

Chapter 4
Broadband Techniques in Power Amplifiers

4.1 Introduction

RF PA with broadband characteristics are always in demand. However, designing a broadband PA with good gain, efficiency, linearity, etc., is indeed a challenging task. For instance, when the bandwidth of PA is multifold, its gain drops automatically by virtue of constant gain bandwidth product. Consequently, the efficiency also drops [1]. A number of bandwidth enhancement designs/architectures can be found in the literature for broadband PAs [2–101]. At RF frequencies, the device parasitic plays an important role and their effect must be compensated in order to achieve broadband response. One of the well-established techniques in literature is distributed power amplifiers [2]. In this case, the device parasitic capacitances are absorbed by the shunt capacitances of artificial transmission lines. Two artificial transmission lines are used for the gate and drain of the FET device. Between these two lines a number of amplifiers are connected at regular intervals, such that the output current of each amplifier gets added in phase to deliver an amplified output current to the load. The term "distributed" is used since the overall arrangement of amplifiers resembles a distributed model of transmission line. The broadband nature of distributed amplifier is possible because the effective circuit of such PA is equivalent to that of a transmission line. In general, if device parasitics are known, the matching network can compensate for the frequency-dependent effect of device parasitics to obtain an optimum load at the current generator plane (CGP) of the device [3, 4]. In order to obtain wideband frequency response, this optimum load realization requires a load trajectory at the package plane to be matched over a wide bandwidth. This load trajectory is based on the device parasitic and it may be sometimes difficult to match over a wide bandwidth. For example, anticlockwise movement of such load trajectory with increasing frequency points in Smith chart requires non-foster circuits, which cannot be realized using passive

K. Rawat et al., *Bandwidth and Efficiency Enhancement in Radio Frequency Power Amplifiers for Wireless Transmitters*, Analog Circuits and Signal Processing,
https://doi.org/10.1007/978-3-030-38866-9_4

matching network. The situation becomes more critical when harmonic matching over wide bandwidth is required especially in the case of waveform-engineered classes of operation. In such cases, continuous class PAs are proposed which explore a design space in terms of fundamental and harmonic loads at the CGP of the devices. Each point in such design space gives nonoverlapping voltage–current waveform and hence results in a highly efficient operation. Due to the availability of the design space with several choices for optimum loads at the CGP, one can choose the solutions which can result in feasible load trajectory even after the effect of package parasitic. The load trajectory at the package plane must follow foster behavior such that it can be easily realized with passive matching network [3]. This technique may relax constraints over broadband passive matching network synthesis technique and enables simplified real frequency techniques (SRFT) [5] to be easily employed. One can also think of obtaining load trajectory over frequency range at package plane using load-pull-based simulation/measurement. However, this CAD-based approach may be very complex and iterative for waveform engineering-based PA design where harmonic manipulation is inevitable. In such a case, it is also difficult to ascertain that the designed PA is working in the particular class.

The shunt feedback technique is also employed to enhance the bandwidth of PA [6]. These techniques are used for multi-octave PAs; however, due to possibility of oscillation, such techniques are sometimes not among the popular choices for broadband PA design at high frequencies. In the case of on-chip MMIC PA design, it is easy to obtain broadband matching due to low-parasitic effect. In such instance, broadband spatial power-combining technique maybe sometimes useful for matching as well as combining the output power of several PAs using reactively matched power combining technique [7–12]. In case of harmonic rejection architecture such as push-pull amplifiers, a wideband passive balun is required in addition to matching network in order to combine the two PAs operating in push-pull mode [13].

In the case of load modulation-based PA, the bandwidth is extended by employing broadband load combiners along with broadband PAs. Several attempts have been done in extending the bandwidth of DPA as well as Chireix PA [14]. Recently, continuous mode PAs are also used in realizing broadband load modulation-based PA. In addition, reconfigurable multiband Chireix PA design is also investigated as one of the methods for expanding the operating frequency range of the outphasing operation [15].

This chapter describes the design challenge and methodology to achieve the best bandwidth out of continuous mode PAs with feasible matching networks. The bandwidth expansion scheme for load modulation-based PA such as DPA and Chireix outphasing PA are also described in this chapter. In addition, for multi-octave PA design, the reactively matched PAs and distributed PAs are discussed in detail.

4.2 Broadband Continuous Class of Operation

Continuous class of operation can provide broadband solution by selecting different modes of solution for different frequencies over wide bandwidths. As discussed in Chap. 1, each of these modes can provide a constant level of output power and efficiency; therefore, one can get a wide design space to choose different modes presenting a practical technique for broadband operation. In such a case, when a new parameter is added to the expression of voltage of any particular mode X, the resultant continuous modes will be termed as continuous class XV and vice versa [16]. Generally, the added function is in the form of $(1\text{–}\delta\ \sin\theta)$ where δ is the parameter to vary a waveform in order to maintain constant output as discussed in Chap. 1. Although continuous class of operation presents several modes to the designer, a careful selection of these modes over the frequency band can guarantee practical implementation of broadband PA. For example, depending on the device parasitic, several modes chosen at particular frequencies may result in conditions which are either difficult or unrealizable with passive matching networks [3]. Moreover, in order to get unclipped operation, several solutions among design space are not viable [17]. This section discusses the methods for appropriate selection of different modes at different frequencies to get feasible matching and best response in terms of power, efficiency, and linearity. A recently introduced class E continuum is also discussed in this section where nonlinear model-based embedding can also be employed to obtain feasible design space in terms of practical matching [18, 19].

4.2.1 Design Strategy for Continuous Class J Power Amplifiers

The class J PA continuum was first introduced by Cripps in [20]. However, this concept of high-efficiency continuous modes were utilized in [17] for the first time to realize a broadband class J PA. Furthermore, the clipping behavior when voltage waveform reaches knee voltage is also mapped in impedance space. This is possible since the coefficients of voltage parameters are directly proportional to impedances corresponding to fundamental and second harmonic frequencies for class AB/J type. Figure 4.1 illustrates the impedance space on Smith chart presenting all possible impedance values resulting in a linear operation. While analyzing this, second harmonic impedance value is considered fixed for ease of analysis.

The shaded region differentiates between the load conditions which gives a voltage grazing zero condition and voltage higher than zero condition. Thus, it provides the designer a choice to find load conditions in the design space for linear application. However, the achieved power and efficiency are also determined by the fundamental impedance. Therefore, one should also consider these performance metrics while selecting the appropriate impedance at fundamental frequency. For the case when harmonic is also varied, the clipping region also changes accordingly. Figure 4.2 shows this variation of clipping region as harmonic impedance Z_{2f0} varies

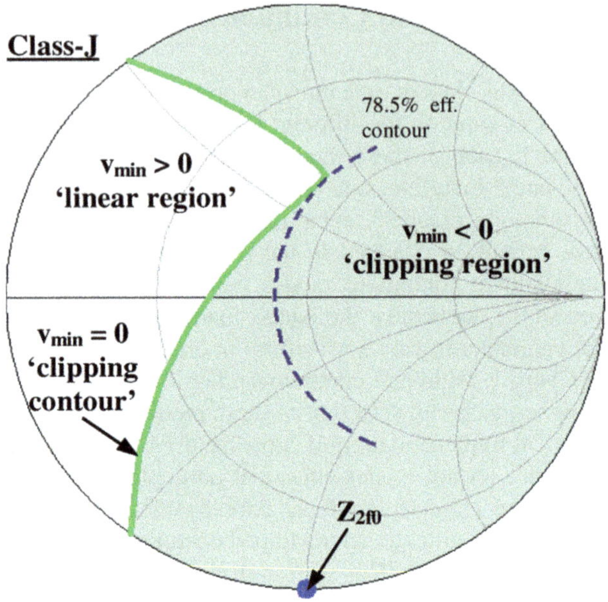

Fig. 4.1 Class J clipping contour separating clipping and non-clipping regions indicating loads corresponding to fundamental frequency for zero-grazing condition and fixed harmonic load ([17], reprinted with permission from IEEE)

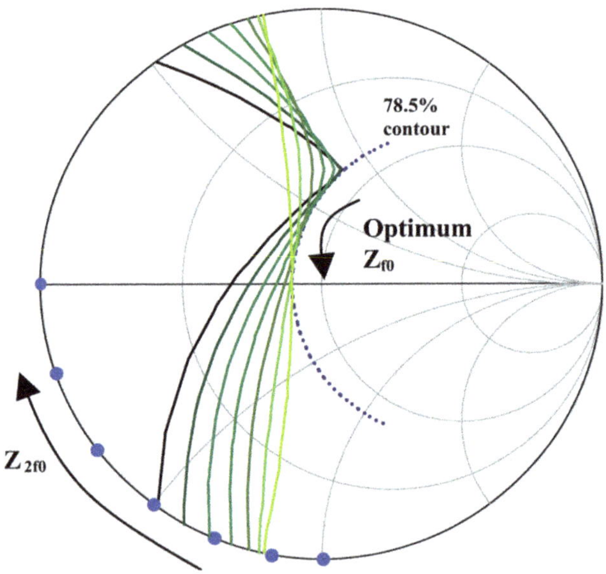

Fig. 4.2 Class J clipping contour specifying optimum fundamental loads for efficiency with variation of harmonic impedance between conditions of class B and class J ([17], reprinted with permission from IEEE)

clockwise. These plots define an impedance solution for continuous class B/J design methodology for obtaining the best efficiency with linear behavior.

Based on this analysis along with harmonic load-pull result, a broadband PA prototype is developed. This prototype has improved input matching and therefore its gain was 10.2–12.2 dB across the bandwidth of 1.4–2.7 GHz [17] as shown in Fig. 4.3a. One can see that the power-added efficiency (PAE) is better than

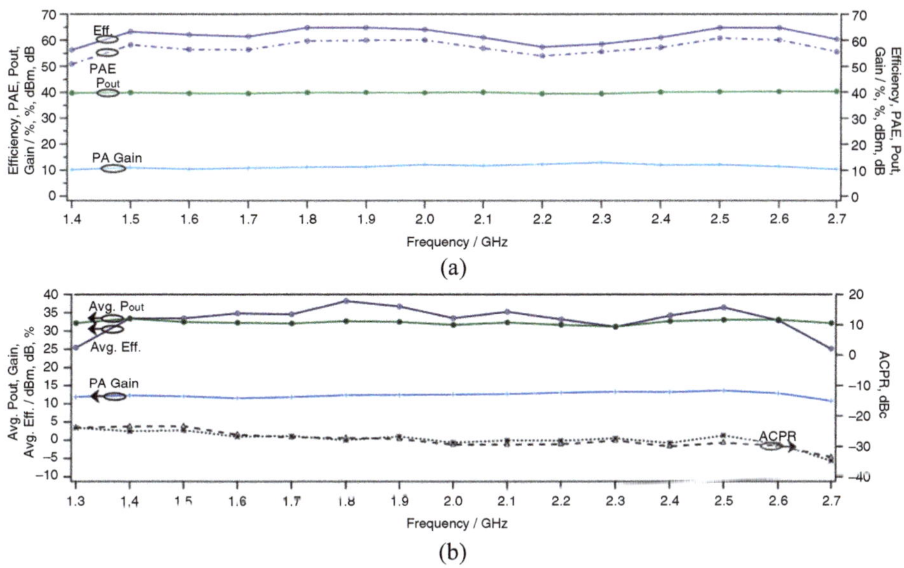

Fig. 4.3 Measured performance of power amplifier (PA) prototype with modified output matching and wideband input matching: (**a**) efficiency, output power, and gain with continuous wave (CW) stimulus, (**b**) average efficiency, average output power, and ACPR with modulated signal ([17], reprinted with permission from IEEE)

50% over this bandwidth. Figure 4.3b shows ACPR for lower and upper channel over the frequency range of 1.3–2.7 GHz. These ACPR values are calculated using 5 MHz bandwidth WCDMA signal with PAPR of 8.51 dB and without applying any predistortion [17].

Figure 4.3b also shows average power, efficiency, and gain. All these measurements are performed by driving PA at 2 dB peak compression. One can see that ACPR is better than −30 dBc over the frequency range of 1.4–2.6 GHz. The average output power is better than 32 dBm and average efficiency is better than 32% over this frequency range.

The analysis in Chap. 1 does not include the considerable effect of nonlinear parasitic. The effect of nonlinear capacitance on the theory of class B/J was considered in [21]. In such a case, the load impedance of fundamental frequency is same as that derived in Chap. 1 but the higher harmonic impedances are considered to be governed by the output capacitance. Moreover, the prediction of loads at package reference plane (PRP) using nonlinear-embedding technique in Chap. 3 takes this parasitic into account.

Although continuous class J provides a wide impedance space giving the designer a good flexibility for implementing wideband PA, all these solutions may not yield an appropriate matching profile especially at device package plane. It is observed that several load trajectories for the class B/J mode at package plane show anticlockwise rotation. But all passive components used in matching network can lead to only Foster reactance exhibiting clockwise rotation on the Smith chart with increasing frequency. Along with this, the general tendencies of harmonic impedances are to disperse more on Smith chart. To overcome this challenge, the "design space" have been explored in order to select the appropriate variation of the design parameter δ with frequency. This selection ensures that the second harmonic has clockwise rotation at extrinsic package plane. In such cases, the nonlinear-embedding model is very useful where the designer can start the design at the current generator reference plane (CRP) while maintaining the required class of operation. As discussed in Chap. 3, the nonlinear-embedding model can expedite such design where appropriate loads among design space must be set at intrinsic CRP while keeping the matching feasibility in check at the external package reference plane (PRP). On the contrary, some of the conventional approaches utilize load-pull measurements for obtaining optimal external load locations [22]. Later, these loads are de-embedded using a suitable device model available to reach the CRP. A design example is presented in [3] where, a nonlinear-embedding device model of a 15 W CGH27015F GaN HEMT CREE device is used to project loads of continuous class J from CRP to External package reference plane (ERP). The projection of theoretically calculated loads Z_1 and Z_2 at fundamental and second harmonic frequencies are shown in Fig. 4.4. The load trajectory in red describes the load calculated for class B/J mode at CRP for $0.5 < \delta < 1$. The load trajectory in blue represents projected loads at PRP for 1.3–2.4 GHz using the model-based nonlinear embedding as discussed in Chap. 3 [3].

One can see that for this range of δ and its allocation to various equally spaced frequency points between 1.3 and 2.4 GHz results in both fundamental and harmonic loads rotating anticlockwise. It can be observed that, the second harmonic impedance expands more than the fundamental one. This varies from $-j37\ \Omega$ at 1.3 GHz to $j102\ \Omega$ at 2.4 GHz. Since this is counterclockwise rotation, it is not possible to match using passive matching network. Therefore, the design space is explored by varying the design parameter δ with frequency to obtain clockwise rotating second harmonic loads on the Smith chart. This design space analysis is performed in terms of two parameters: first, the gradient of δ with frequencies which

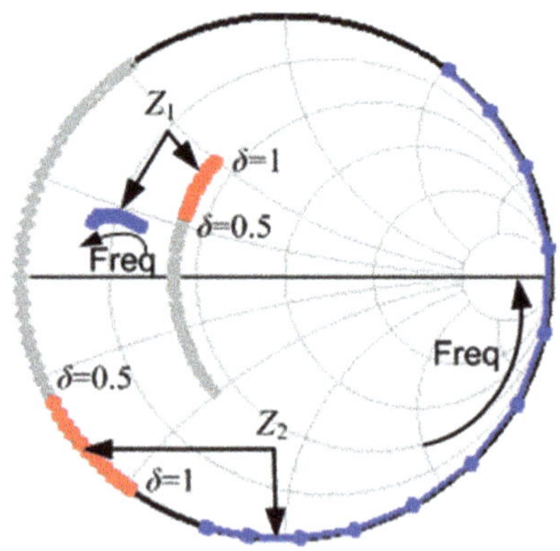

Fig. 4.4 Family of fundamental and second harmonic loads for class B/J operation at current generator reference plane (CRP, in gray) and extrinsic load at package reference plane (PRP, in blue) corresponding to CRP load (in red) for $0.5 < \delta < 1$ across the 1.3–2.4 GHz [3]

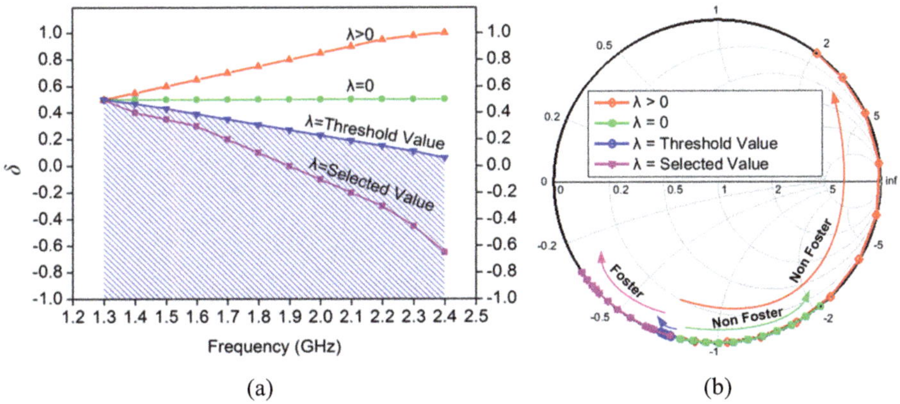

Fig. 4.5 (**a**) Variation of δ with different gradient λ for the same starting value of $\delta = 0.5$, (**b**) corresponding second harmonic loads Z_2 at PRP projected by embedded model [3]

is represented as λ; and second, the range of values of δ that can be chosen over the band. Figure 4.5a shows variation of δ starting from 0.5 for different values of gradient λ. Figure 4.5b shows corresponding loads at second harmonic frequencies for various λ projected at PRP using the nonlinear-embedding model [3].

Figure 4.5b shows a case where λ is positive which results in the impedance Z_2 following a highly non-foster rotation which extends from $-j37\ \Omega$ to $j102\ \Omega$ corresponding to 1.3–2.4 GHz. With $\lambda = 0$, the extent of this counterclockwise rotation is reduced over the frequency range as shown in Fig. 4.5b. Furthermore, λ is reduced implying δ is reduced from 0.5 to lower values with frequency varying from 1.3 to 2.4 GHz. Figure 4.5b also shows that $Z_{2\text{ext}}$ will stop moving anticlockwise after λ is sufficiently decreased below zero to its threshold value. However, at this threshold value of λ, the Z_2 is extremely concentrated around a single point, which is also not favorable for the circuit designer. In fact, it is more beneficial to have sufficient clockwise spread in the Z_2 with frequency such that it can be matched along with the fundamental load by a passive matching network. This is possible by further decreasing λ beyond the threshold value. Therefore, the shaded region in Fig. 4.5a corresponds to a family of foster solutions where λ is below the threshold. Furthermore, one can keep λ decreasing throughout the frequency range. However, this family of foster solutions for δ is not unique and will vary with the device, the initial value δ, and the operating frequency. Therefore, depending on the desired frequency of operation and the selected device, any designer should investigate this design space with foster solutions. As discussed earlier, the impact of selecting appropriate range of the values of δ is also studied [3]. Figure 4.6a shows different range of δ chosen with frequency. One can see that these three sets of ranges of δ are $1 \leq \delta \leq -0.2$, $0.5 \leq \delta \leq -0.65$, and $0.2 \leq \delta \leq -1$. The gradient λ is also varied with frequency in this analysis where the range of its variation is shown in Fig. 4.6a. The corresponding loads trajectory obtained at PRP of the device predicted using nonlinear-embedding model of 15 W GaN HEMT devices is shown in Fig. 4.6b

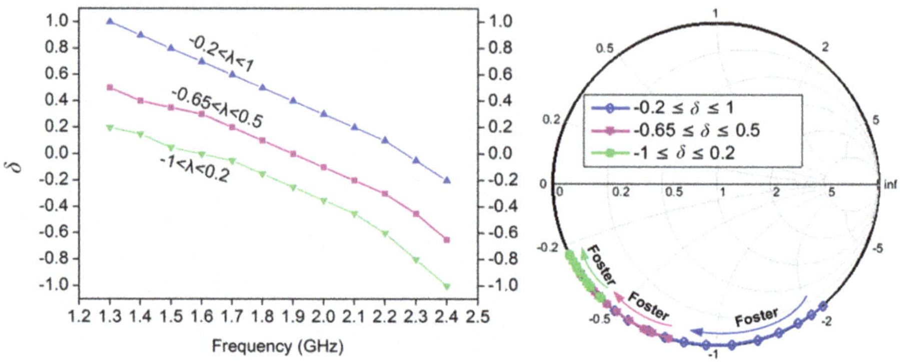

Fig. 4.6 (**a**) Different ranges of δ with variable gradient over frequency and their corresponding (**b**) second harmonic loads projected by nonlinear embedded model [3]

[3]. All the trajectories correspond to the clockwise rotating extrinsic PRP loads which can be realized with Foster matching networks. Therefore, the designer gets an extra degree of freedom to choose a solution which is easy for him to realize with the matching network. For example, the solution where variation of δ is from 0.5 to -0.65 is chosen for designing continuous class J PA presented in [3].

Since, δ can only vary between -1 and 1, step size should be considered small with frequency. If δ is reduced in larger steps, the lowest value cannot be assigned to farther frequency point. For instance, if $\delta = 1$ at 1.3 GHz and step size of 1 is considered for each 100 MHz increment in frequency, only three values of δ, that is, 1, 0, and –1, can be possibly assigned to the frequencies that are 100 MHz apart. Therefore, the PA will operate from 1.3 to 1.5 GHz with only 300 MHz bandwidth corresponding to 11% fractional bandwidth. This is not a good choice for obtaining best bandwidth out of class B/J continuum. However, as δ decreases adequately with frequency, Z_2 rotates sufficiently in a clockwise manner as shown in Fig. 4.6b which will ease the designing of matching network. Since decreasing δ sufficiently with increase in frequency will ease the matching design whereas achieving broadband demands δ to decrease with a small step size, a trade-off exists between feasible matching network and achieving broadband response. Thus, a designer should carefully understand this trade-off in order to design a broadband continuous class B/J PA. This analysis has been explored further to design a 15 W broadband continuous class B/J PA using 15 W GaN HEMT device. This PA operates from 1.3 GHz to 2.4 GHz and utilizes broadband input matching network as presented in [23]. Figure 4.7a shows the photograph of the prototype developed using this design scheme. The circuit is fabricated on Rogers substrate RO4350B. The substrate has dielectric constant, thickness, and loss tangent as 3.66, 20 mil, and 0.0037, respectively. Figure 4.7b shows simulation and measurement results of the prototype using continuous wave (CW) measurements.

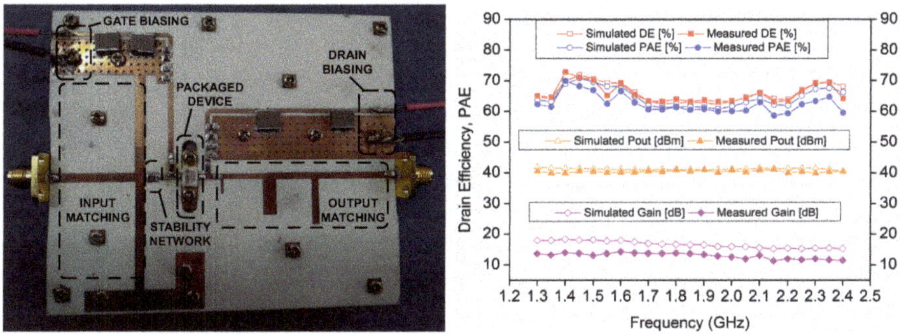

(a) (b)

Fig. 4.7 15 W GaN HEMT PA prototype developed based on nonlinear-embedding model. (**a**) Photograph, (**b**) measured and simulated results ([3], reprinted with permission from IEEE)

One can see that DE varies between 63 and 72% corresponding to a fractional bandwidth of 59% over the frequency band of 1.3–2.4 GHz. The output power and measured gain varies from 40.1 to 41.1 dBm and 11.4 to 14.3 dB, respectively. The measurement results are comparable with the simulation results. A similar design space analysis should be performed in continuum of other classes to obtain best bandwidth solutions in practice.

4.2.2 Design Strategy for Continuous Class F Power Amplifier

Continuous class F mode of operation maintains constant high efficiency and output power over a continuous range of impedances. The theoretical formulation for voltage waveform of continuous class F mode has been described in Chap. 1, where the design space was explored for the identified range of parameter δ. This chapter will discuss the extended continuous class F mode with the analysis of design space in depth. This will provide wider design space with feasible load solution which further increases the flexibility for PA designers for maintaining the optimum device performance over wide band of operation. The general formulation of the extended continuous class F mode is presented in Eqs. (1.116)–(1.128) in Chap. 1. The continuous class F mode discussed in Chap. 1 delivers a maximum efficiency of 90.7% because α and β were kept constant and only δ was allowed to vary. This leads to second harmonic impedance to be varied on the edge of the Smith chart while fundamental impedance is swept to the circle at constant resistance maintaining a constant and maximum output power and DE. In order to expand the design space, Carubba et al. has demonstrated the extended continuous class F mode by varying the parameters α and β along with the parameter δ of the voltage [24]. This leads to significantly wider design space with a DE maintained to be >75%. Therefore, continuous class F mode of operation can provide a wide range of solutions to attain high efficiency over a wider bandwidth. One of the limitations in continuous mode of PA design is counterclockwise variation of impedances at package plane when represented in Smith chart. Such kind of trajectory is difficult to realize and infeasible with passive matching network. This restricts the practical implementation of continuous class F mode over wider frequency range. To overcome this limitation, investigation of design space for feasible matching solution has been discussed recently in [4]. Therefore, section "Extended Continuous Class F Mode" first describes the extended continuous class F mode and section "Investigating Feasible Design Space" illustrates the investigation of design space for feasible impedance solution.

Extended Continuous Class F Mode

Two different approaches are followed in [24] to extend the continuous class F mode for wider design space. In the first approach, keeping β at a constant value of $\beta = \alpha/2$, the other two parameters α and δ are varied. This condition will lead to the variation of second harmonic impedance along the edge of the Smith chart and fundamental impedance varies in both magnitude and phase. An efficiency greater than 75% is maintained while varying α and δ between $0.75 \leq \alpha \leq 1$ and $-1 \leq \delta \leq 1$ respectively. Figure 4.8 shows the theoretical extended continuous class F impedance range and efficiency contour plot for fundamental and second harmonic component. The step of α and δ are taken to be 0.5 and 0.25, respectively.

Figure 4.9 shows the variation of efficiency and output power with parameter α for a constant value of $\delta = 0$. One can observe from Fig. 4.9 that the efficiency varies from 75 to 90.7% with the highest efficiency at $\alpha = 2/\sqrt{3}$. This corresponds to the class F condition while considering three harmonic components. As the value of α increases for $\delta = 0$, the efficiency reduces due to drop in fundamental voltage. This restricts the range of α to be varied between $0.75 \leq \alpha \leq 1.5$ for maintaining efficiencies greater than 75%. This analysis extends the choice of fundamental

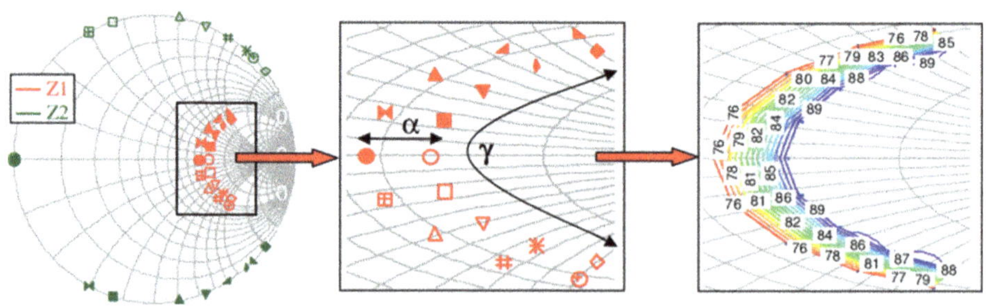

Fig. 4.8 Impedance range and efficiency contour corresponding to the first two harmonic impedances (the third is kept open) for continuous class F mode when α and δ vary between $0.75 \leq \alpha \leq 1.25$ and $-1 \leq \delta \leq 1$ with α at a step of 0.5 and δ at a step of 0.25 ([24], reprinted with permission from IEEE)

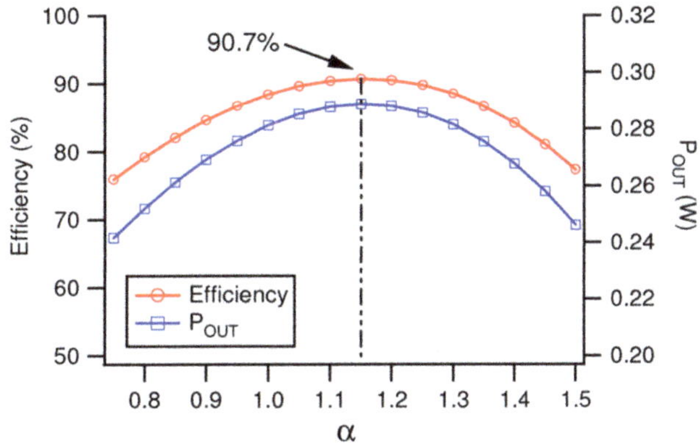

Fig. 4.9 DE and output power when β and δ are kept at constant $\beta = \alpha/2$ and $\delta = 0$ for a function of α, where α is varied between $0.75 \leq \alpha \leq 1.25$ with steps of 0.05 ([24], reprinted with permission from IEEE)

Table 4.1 Design space for $\beta \geq \alpha/2$

$\beta = \alpha/2$	$\beta = \alpha/1.9$	$\beta = \alpha/1.8$	$\beta = \alpha/1.7$	$\beta = \alpha/1.6$	$\beta = \alpha/1.5$	$\beta = \alpha/1.4$
$0.75 \leq \alpha \leq 1.5$	$0.75 \leq \alpha \leq 1.45$	$0.8 \leq \alpha \leq 1.45$	$0.8 \leq \alpha \leq 1.35$	$0.85 \leq \alpha \leq 1.3$	$0.9 \leq \alpha \leq 1.2$	$\alpha = 1.05$
$-1 \leq \delta \leq 1$	$-1 \leq \delta \leq 1$	$-1 \leq \delta \leq 1$	$-0.9 \leq \delta \leq 0.9$	$-0.5 \leq \delta \leq 0.5$	$-0.2 \leq \delta \leq 0.2$	$\delta = 0$

impedance to a wider range. This range of efficiency and output power shown in Fig. 4.9 is maintained at a constant value over the range of $-1 \leq \delta \leq 1$.

In the second approach, all the three parameters α, β, and δ are varied. This is a more general case, where second harmonic impedances can be selected inside the Smith chart instead of varying only at the edge of Smith chart as done in the first approach. This approach provides a wider design space and assures a minimum efficiency of 75%. To vary second harmonic impedance inside the Smith chart, the parameter a_2 in voltage Eq. (1.116) in Chap. 1 is kept greater than zero. Using Eq. (1.118) leads to the condition: $\beta > \alpha/2$. When the voltage waveform is maintained to be positive and DE greater than 75%, the range of all three parameters is given in Table 4.1. One can observe from Table 4.1 that the range of α and δ decreases with the increase in the value of parameter β. This range is restricted in order to maintain the voltage waveform positive and efficiency greater than 75%. Figure 4.10 shows the variation in efficiency for the parameter α for different ranges of parameter β, when the value of $\delta = 0$. For $\beta = \alpha/2$, the range of α varies from 0.75 to 1.5 in order to achieve efficiency greater than 75%. However, for $\beta = \alpha/1.4$, this range of α reduced to a single point where $\alpha = 1.05$ and $\delta = 0$.

By selecting the parameter β inside the Smith chart, a wide design space is explored. The design space for one of the cases with $\beta = \alpha/1.9$, when varying α and δ between $0.75 \leq \alpha \leq 1.45$ and $-0.5 \leq \delta \leq 0.5$ with step of 0.1 for both α and δ is shown in Fig. 4.11 [24]. This whole design space provides an efficiency greater than 75% with a maximum efficiency of 89.5%.

The extended continuous class F mode demonstrated above provides a very wide design space while achieving a targeted predetermined value of DE. This allows an

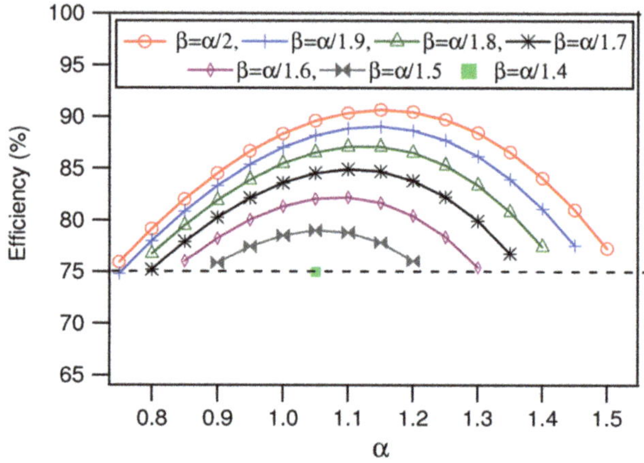

Fig. 4.10 DE as a function of α and β, for $\delta = 0$ ([24], reprinted with permission from IEEE)

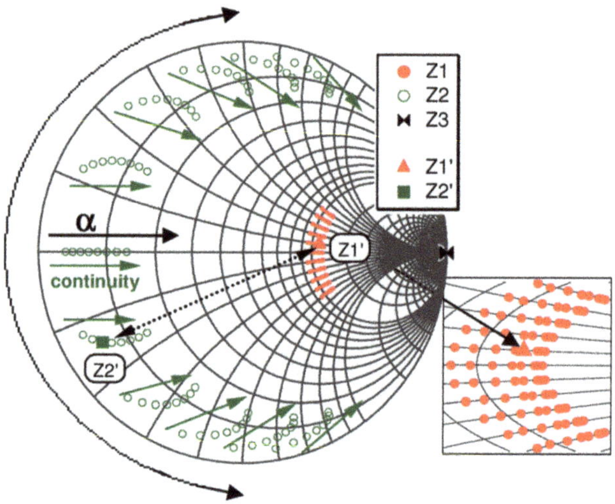

Fig. 4.11 Continuous class F impedance range for the first two harmonic impedances (the third is kept open) with $\beta = \alpha/1.9$ when varying α and δ between $0.75 \le \alpha \le 1.45$ and $-0.5 \le \delta \le 0.5$ with step size of 0.1 for both α and δ ([24], reprinted with permission from IEEE)

increased flexibility for the PA designer to realize a matching network. The design and realization of broadband PA becomes less complex with this type of analysis. Hence, a step from standard continuous class F to an extended continuous class F is beneficial for the broadband PA designers.

Investigating Feasible Design Space

The analysis of extended continuous class F mode described in section "Extended Continuous Class F Mode" significantly expands the design space and increases the choice of impedances for matching network design. Still, the realization of matching network for broadband PA is challenging due to the infeasible behavior of load impedances at the PRP of the device. The theoretical analysis demonstrated for extended continuous class F mode is valid at CRP of the device and its behavior at package plane cannot be predicted until the designer has the knowledge of parasitic between the CRP and package plane of the device. If the load impedances projected to package plane of the device are non-foster, then the matching circuit realization with passive network becomes infeasible. Hence, this section focuses on the investigation of feasible design space while using nonlinear-embedding device model [25] which facilitates the projection of load from CRP to the PRP of the device. Instead of extended continuous class F mode, the standard continuous class F described in Chap. 1 is used in this analysis.

The load impedances for continuous class F mode in Eqs. (1.126)–(1.128) in Chap. 1 represent the load at CRP. For the design of continuous class F mode broadband PA, these loads must be projected to the package plane of the device, while maintaining the correct load impedances at the CRP [26]. The suitable device model with all linear and nonlinear parasitic elements is required to compute the load impedances at the package plane of the device. The load impedances projection at package plane of the device over a broad range of frequency becomes unfeasible many times [3, 4, 26]. The theoretical loads at current generator plane are widespread over the frequency and based on the nonlinear parasitic elements may rotate in an anticlockwise rotation on the Smith chart over the frequency band. Unfortunately, all passive components present Foster reactance, which leads to clockwise rotation on the Smith chart with increasing frequency. Therefore, it is necessary to obtain a strategy to select appropriate set of loads at current generator plane that translates into fundamental and second harmonic impedance pairs with clockwise trajectory on the Smith chart across the frequency band. Among the various solutions, these loads lead to the feasible matching network using passive components.

In order to project loads at PRP of device from CRP, the nonlinear-embedding device model can be used [25, 27]. Its embedding transfer network (ETN) facilitates the projection of theoretical loads from CRP to the PRP of device as shown in Fig. 4.12. ETN synthesizes the required multi-harmonic source and load terminations at the external reference planes considering all device parasitic and capacitances [25]. Z_{int1} and Z_{int2} are fundamental and second harmonic loads at the current generator plane which are obtained for a range of $-1 < \delta < 1$ using Eqs. (1.126)–(1.128) in Chap. 1. These loads are then projected to package plane of the device using ETN. The projected fundamental and second harmonic loads at package plane of the device are denoted by Z_{ext1} and Z_{ext2} as shown in Fig. 4.12. These impedances at device package plane correspond to ideal waveforms of continuous class F mode at CRP. The design of continuous class F PA using model-based nonlinear-embedding technique is presented in the section "Broadband Continuous Class F PA Design".

The family of solutions obtained for continuous class F PA design depends upon the parameter δ. Figure 4.13a shows the design space where multiple impedance solutions are possible for each frequency point. These solutions are based on the step of δ, range of δ, and frequency of operation. To overcome the problem of non-foster behavior of load at the package plane, design space must be investigated

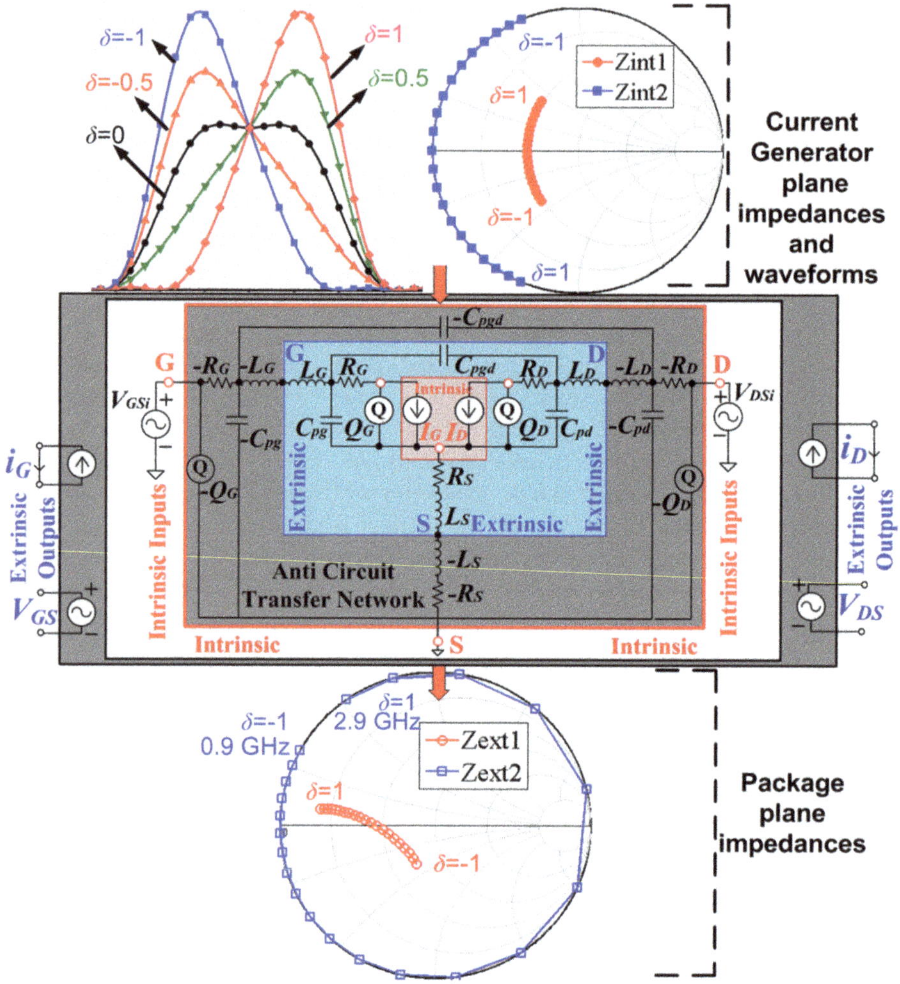

Fig. 4.12 Nonlinear-embedding model and continuous class F loads and waveforms [4]

and values of parameter δ with frequency should be selected to provide realizable matching network as well as higher operating bandwidth. Since the projection of second harmonic loads to the package plane of the device is significantly expanded on the Smith chart compared to fundamental loads for a specified frequency range, its foster behavior for matching requirement becomes more critical. The optimum load can be decided by selection of range of δ with frequency and step size of δ with every change in frequency [4]. Step size of δ is defined as parameter k. Figure 4.13a shows the different cases with different values of k for the selection of δ with frequency for different k. In this design space analysis, the frequency points are considered to be 100 MHz apart. The selection of k has a direct impact on the

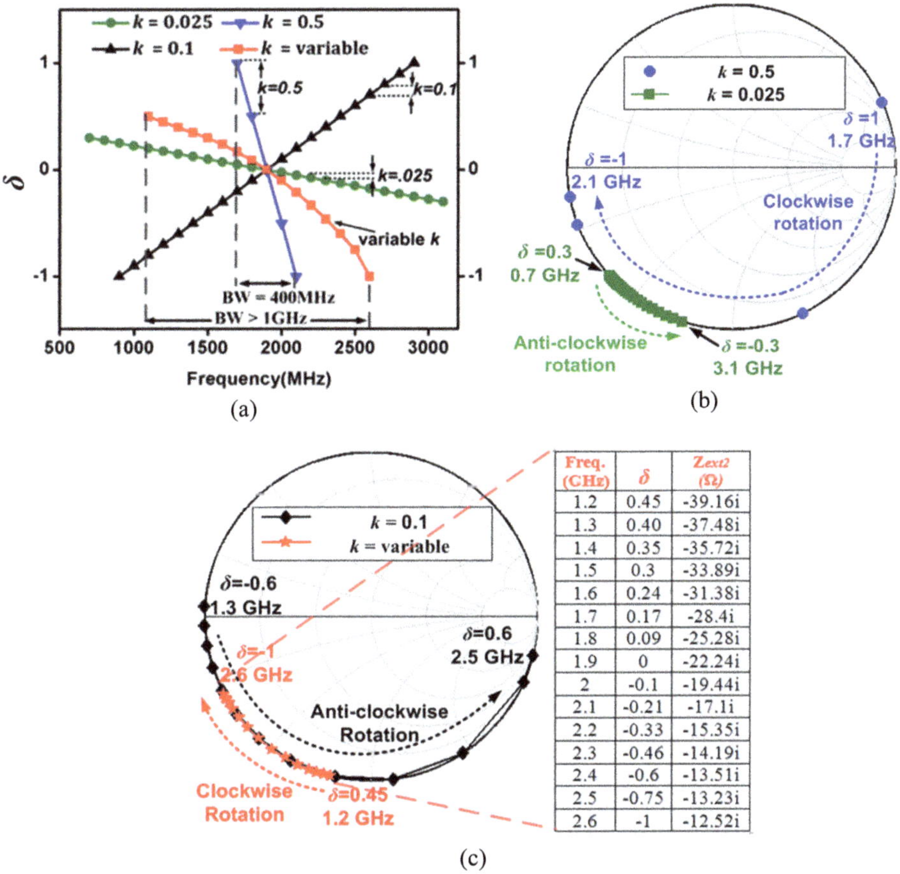

Freq. (GHz)	δ	Zext2 (Ω)
1.2	0.45	-39.16i
1.3	0.40	-37.48i
1.4	0.35	-35.72i
1.5	0.3	-33.89i
1.6	0.24	-31.38i
1.7	0.17	-28.4i
1.8	0.09	-25.28i
1.9	0	-22.24i
2	-0.1	-19.44i
2.1	-0.21	-17.1i
2.2	-0.33	-15.35i
2.3	-0.46	-14.19i
2.4	-0.6	-13.51i
2.5	-0.75	-13.23i
2.6	-1	-12.52i

Fig. 4.13 (**a**)Variation of parameter δ with frequency for different k; (**b**) extrinsic impedances projected by embedding transfer network (ETN) for $k = 0.5$ and $k = 0.025$; (**c**) extrinsic impedances projected by ETN for $k = 0.1$ and for variable k [4]

bandwidth. One can see in Fig. 4.13a that the slope of δ can be taken positive or negative with increasing frequency. Also, step size of δ, that is, k, can be taken as a constant fixed value or variable. Based on the step size (k), the design space analysis is divided into two categories: (a) fixed value of k and (b) variable k.

Three cases of fixed step size k are shown in Fig. 4.13a and corresponding second harmonic loads at package plane of device are plotted in Fig. 4.13b, c. In case of very small value of k, a large operating bandwidth can be achieved, but as the step size becomes smaller, the extent of an anticlockwise rotation of Z_{ext2} with frequency increases. For example, consider the case with $k = 0.025$, where δ is decreasing from 0.3 to -0.3 with negative slope and covers a very large bandwidth as shown in Fig. 4.13a. However, this case does not provide feasible loads for matching network design due to anticlockwise movement of second harmonic loads at the package plane of device as shown in Fig. 4.13b. For a case where k is 0.5 with δ decreasing from 1 to -1, Z_{ext2} shows the foster behavior with frequency on Smith chart but the bandwidth coverage is only 400 MHz. In both the above discussed cases, slope of δ is negative with increasing frequency. Third case with positive slope of δ with increasing frequency is also shown in Fig. 4.13a. In this case k is taken as 0.1. The corresponding trajectory of Z_{ext2} on Smith chart is shown in Fig. 4.13c. One can see that this case resulted into anticlockwise rotation of Z_{ext2} with frequency. Thus, the positive gradient of δ with increasing frequency does not provide a feasible solution.

It is concluded that fixed step size leads to the compromise between bandwidth and realizable impedance solution. Hence, a variable step size case is examined where step size k is varied with increasing frequency. One can see from Fig. 4.13a that k is kept smaller at lower frequency points and it increases slightly at each frequency point till it reaches $\delta = -1$. This set of δ covers the bandwidth more than 1 GHz and Z_{ext2} corresponding to this case shows foster movement on smith chart as shown in Fig. 4.13c. Thus, the variable step size is the optimum solution which can provide feasible matching solution along with large band of operation.

Many more feasible solutions can be achieved by following the similar strategy of design space analysis. Depending upon the designer's matching requirement, one can choose the optimum load which can result in the easy implementation of matching network.

Broadband Continuous Class F PA Design

Broadband continuous class F PA is designed using 15 W GaN HEMT transistor CGH27015F from CREE [4]. The transistor is biased in deep class AB region with a drain current of 84 mA and a drain voltage of 28 V. The proposed broadband PA is fabricated using Rogers RO4350 substrate with dielectric constant 3.66 and height of 20 mil. The optimum load impedances selected in the above analysis are projected to the package plane of the device using ETN used in nonlinear embedded model. The second harmonic impedance which is more scattered on the Smith chart compared to fundamental impedance follows the clockwise trajectory with increasing frequency as shown in Fig. 4.14a. These loads are realized with broadband matching network using SRFT [28]. Figure 4.14b shows the schematic of continuous class F PA with input and output matching network. The output matching network comprises three segments of microstrip lines as shown in Fig. 4.14b. The broadband input matching network is realized using an unequally terminated band-pass filter [23]. Figure 4.14c shows the photograph of the fabricated circuit.

PA is tested with single-tone CW signal from 1.5 to 2.5 GHz. Simulated and measured results are shown in Fig. 4.15. Simulated DE varies between 60% and 75% over the entire operating frequency range. The measured DE lies between 61% and 73.5%. These results are obtained at 3–3.5 dB gain compression point. The fractional bandwidth of 50% is achieved with 1 GHz bandwidth. The simulated and measured output power at saturation varies between 9 and 14 W all over

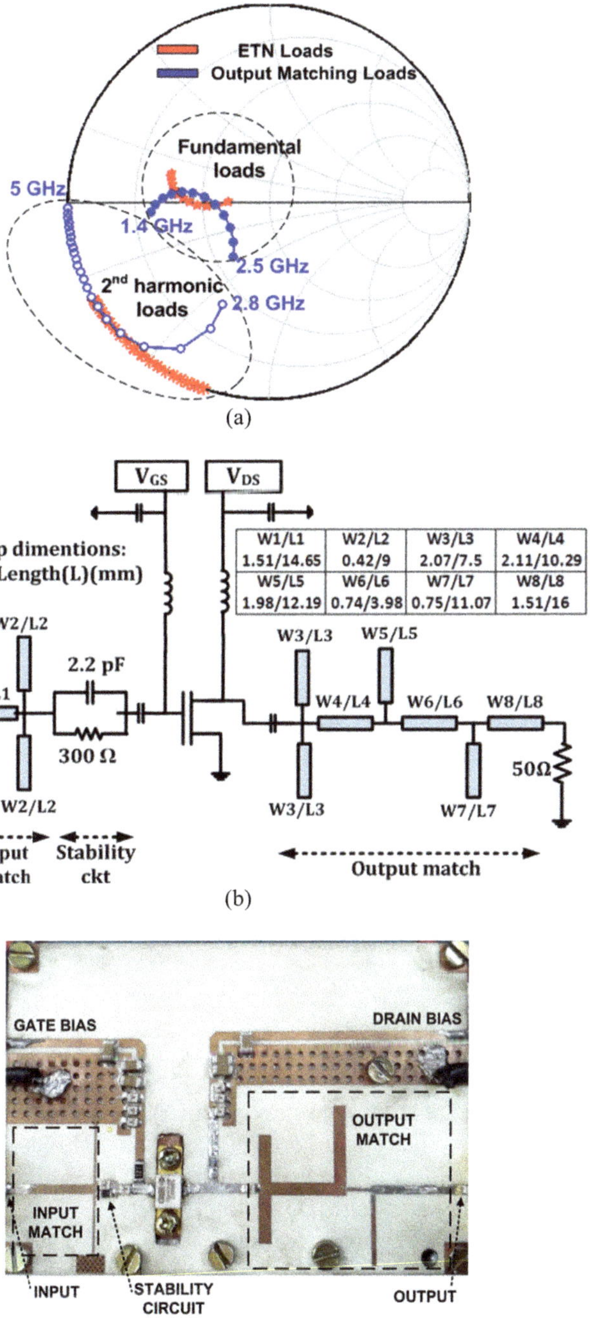

Fig. 4.14 (**a**) Impedances projected on the extrinsic plane by ETN (red) and impedances synthesized using matching network (blue), (**b**) schematic of continuous class F PA, and (**c**) photograph of the fabricated circuit ([4], reprinted with permission from IEEE)

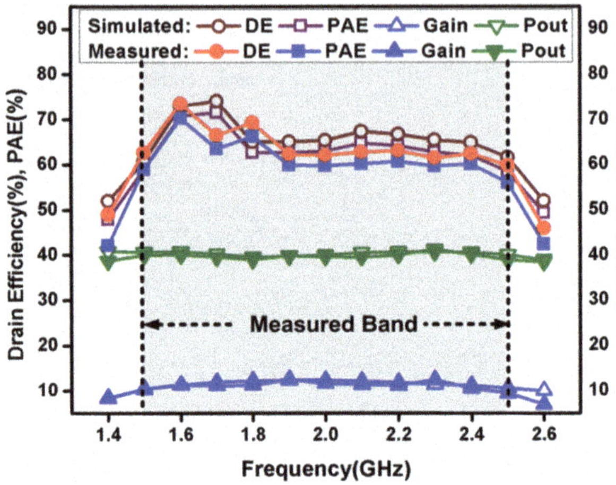

Fig. 4.15 Measured and simulated drain efficiency (DE), power-added efficiency (PAE), output power, and gain ([4], reprinted with permission from IEEE)

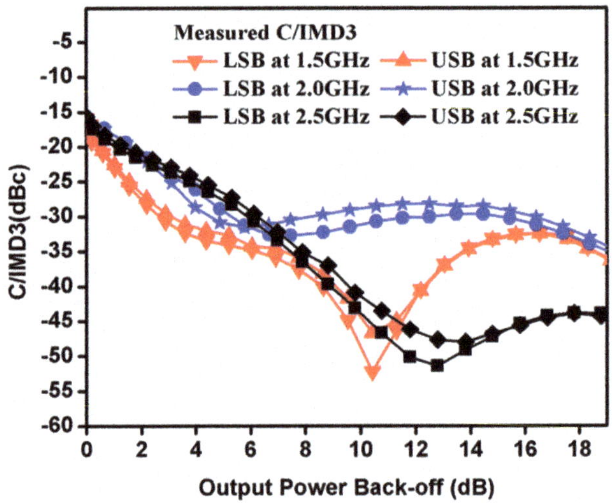

Fig. 4.16 Measured C/IMD3 for center and corner frequencies ([4], reprinted with permission from IEEE)

the frequency range. Simulated gain lies between 10.6 and 12.6 dB whereas the measured gain varies from 9.5 to 12.7 dB. Figure 4.16 shows the measured carrier to third-order intermodulation suppression ratio (C/IMD3) with output power back-off from saturation at center and corner frequencies. This is measured using two tone with 5 MHz spacing and result shows that C/IMD3 is better that −15 dBc for both the upper- and lower-side IMD3 product.

Thus, it is possible that entire design space of continuous class F does not provide feasible solution in terms of possible matching with passive network. A careful design space analysis as discussed above is important in understanding the feasible loads. A nonlinear-embedding model is found to be very useful in such analysis and expedites the CAD design of these broadband PAs.

4.2.3 *Reactive Compensation Scheme in Class E Power Amplifier for Broadband Application*

There are several methodologies proposed for enhancing the bandwidth of class E PA [18, 19, 29–33]. When broadly classified, there are two approaches, one based on reactance compensation and the second based on describing the continuum of class E operation. This section primarily describes both the approaches with few schemes for enhancing class E PA based on filter design.

Reactance Compensation Technique

The analysis of conventional class E topologies, namely, shunt capacitor and parallel circuit, is described in Chap. 1 in section "Class E Power Amplifier". While analyzing these topologies, it has been assumed that series resonant circuit L_S and C_S is tuned to fundamental frequency ω_0. Therefore, the primary role seems to be matching at fundamental frequency. Hence, while analyzing the equivalent circuit at fundamental frequency, this L_S–C_S resonator is considered short. In order to analyze the bandwidth response of class E PA, one should consider the behavior of this series L_S–C_S resonator at frequencies around its resonant frequency which is the fundamental frequency of operation. For example, considering the class E parallel topology as shown in Fig. 1.18 of Chap. 1, the reactance of shunt L–C circuit can be compensated by the series resonant circuit L_S–C_S. This is possible since the reactance of series resonant circuit increases with frequency and for shunt circuit this reactance decreases. Figure 4.17 shows this trend, where the slope of reactance variation for series and parallel resonant circuit are positive and negative, respectively. Therefore, the overall reactance slope of the class E load network can be reduced by choosing appropriate values of L_S and C_S [18]. The value of the other elements L and C will be chosen according to the class E mode as discussed in Chap. 1.

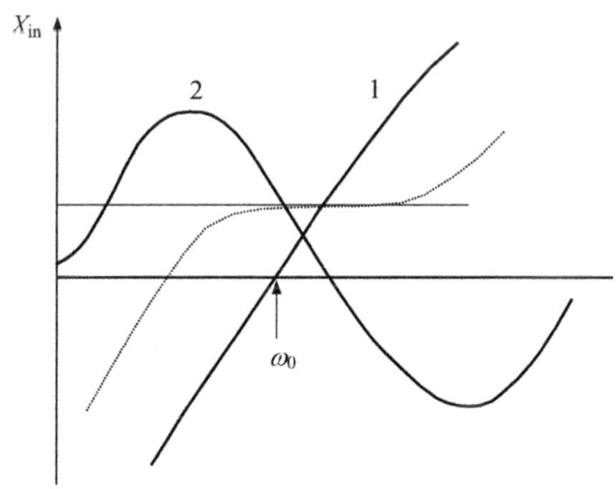

Fig. 4.17 Illustration of reactance compensation principle ([18], reprinted with permission from IEEE)

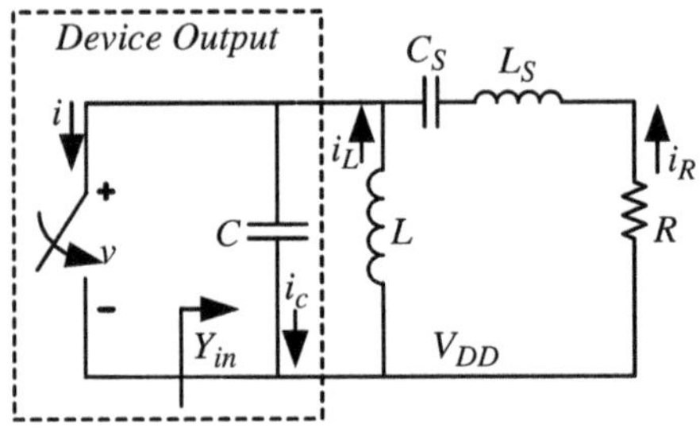

Fig. 4.18 Reactive compensation in class E parallel circuit topology

Considering the series L_S–C_S circuit in Fig. 1.18, one can redraw this circuit including this series resonator as shown in Fig. 4.18 [18].

At any other frequency ω, offset from the fundamental frequency ω_0 at which L_S–C_S resonates, the overall reactance of series L_S–C_S resonator is represented as $j\omega/L_S$. The angular frequency $\omega' = \omega - \frac{\omega_0^2}{\omega}$, where $\omega_0 = 1/\sqrt{L_S C_S}$ is the resonant frequency. The input admittance Y_{in} is given by,

$$Y_{in} = j\omega C - j\frac{1}{\omega L} + \frac{1}{R + j\omega' L_S} \tag{4.1}$$

Separating the imaginary part to obtain susceptance B_{in} and applying following condition on B_{in} to maximize bandwidth around resonant frequency ω_0,

$$\left.\frac{dB_{in}(\omega)}{d\omega}\right|_{\omega=\omega_0} = 0 \tag{4.2}$$

This results in the following equation,

$$C + \frac{1}{\omega^2 L} - \frac{2L_S}{R^2} = 0 \tag{4.3}$$

Using L and C from the analysis of class E parallel topology as given in Eqs. (1.71) and (1.72) of Chap. 1, one can obtain the series resonator elements as,

$$L_S = 1.026\frac{R}{\omega} \tag{4.4}$$

$$C_S = \frac{1}{\omega_0^2 L_S} \tag{4.5}$$

By employing the reactance compensation technique, a class E PA is designed to operate over a band of 136–174 MHz corresponding to 24.5% fraction bandwidth [18]. Figure 4.19 shows the photograph along with simulated and measurement results. A printed circuit board (PCB) is developed on FR4 board which has a dielectric constant of 4.5 and a thickness of 14 mil. The circuit uses LDMOS

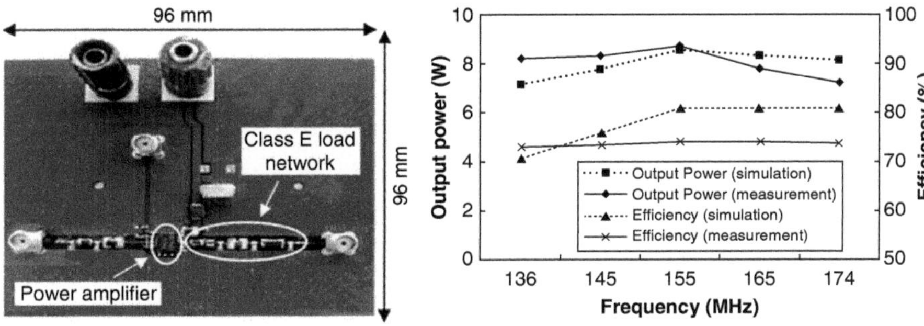

(a) (b)

Fig. 4.19 Prototype developed based on reactance compensation of class E parallel circuit topology: (**a**) photograph and (**b**) measured results ([18], reprinted with permission from IEEE)

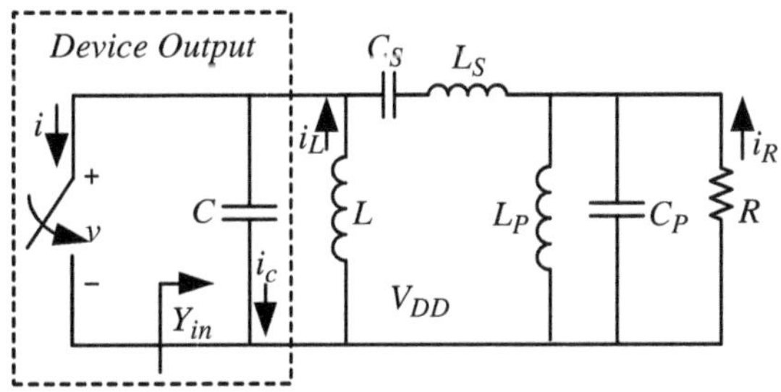

Fig. 4.20 Double reactance compensation in class E parallel circuit topology

transistor and the board size is 96 × 96 mm as shown in Fig. 4.19a. The simulated and measured results are shown in Fig. 4.19b. One can see that the maximum drain efficiency of 74.5% with an output power of 8.77 W was achieved at 155 MHz. The DE varies within 1.3% over the entire frequency band of 136–174 MHz. Similarly, power varies within 0.7 dB across the entire frequency range while the gain drops by 0.7 dB at 136 MHz and 1.6 dB at 174 MHz from its maximum value of 12.2 dB at 155 MHz.

In order to obtain wider bandwidth, an additional resonant circuit L_P and C_P was added in shunt in Fig. 4.18 in order to obtain the double resonant reactance compensation circuit as shown in Fig. 4.19 [18].

Similar to single-section reactance compensation circuit, the admittance of double-section reactance circuit Y_{in} as shown in Fig. 4.20 can be written as,

$$Y_{\text{in}} = j\omega C - j\frac{1}{\omega L} + \frac{1}{j\omega' L_S + \frac{R}{j\omega' C_P R + 1}} \tag{4.6}$$

where the shunt L_P–C_P also resonates at $\omega_0 = 1/\sqrt{L_P C_P}$ and its equivalent susceptance is $j\omega' C_P$. The angular frequency ω' is same as the previous case, that is, $\omega' = \omega - \frac{\omega_0^2}{\omega}$. The susceptance part of Eq. (4.6) is given by,

$$B_{\text{in}}(\omega) = \omega C - \frac{1}{\omega L} + \frac{\omega' C_1 R^2 \left[1 - (\omega')^2 L_S C_1\right] - L_S}{R^2 \left[1 - (\omega')^2 L_S C_1\right]^2 + (\omega' L_S)^2} \tag{4.7}$$

where applying the following two conditions on B_{in} to maximize bandwidth around resonant frequency ω_0,

$$\left.\frac{\mathrm{d}B_{\text{in}}(\omega)}{\mathrm{d}\omega}\right|_{\omega=\omega_0} = \left.\frac{\mathrm{d}^3 B_{\text{in}}(\omega)}{\mathrm{d}\omega^3}\right|_{\omega=\omega_0} = 0 \tag{4.8}$$

the following two equations are obtained as,

$$C + \frac{1}{\omega^2 L} - 2\frac{C_P R^2 - L_S}{R^2} = 0 \tag{4.9}$$

$$\frac{1}{\omega^2 L} + \frac{C_P R^2 - L_S}{R^2} - 8\omega^2 L_S \left[C_P^2 + \frac{\left(C_P R^2 - L_S\right)\left(L_S - 2C_P R^2\right)}{R^4}\right] = 0 \tag{4.10}$$

Using L and C from the analysis of class E parallel topology as given in Eqs. (1.71) and (1.72) of Chap. 1 respectively, one can obtain the series and shunt resonator elements as,

$$L_S = \frac{R}{\omega} 1.618; \quad C_S = \frac{1}{\omega_0^2 L_S} \tag{4.11}$$

$$C_P = \frac{L_S}{R^2} 0.382; \quad L_P = \frac{1}{\omega_0^2 C_P} \tag{4.12}$$

This proposed technique was used in [19] to design class E PA with efficiency more than 80% from 900 to 1500 MHz with 50% fractional bandwidth. In order to improve harmonic suppression in reactance-compensated parallel class E topology of Fig. 4.19, a harmonic trap is introduced to the reactance-compensated parallel class E topology as described in [29].

So far, a reactance compensation technique is applied in parallel class E topology. In order to apply this reactance compensation in class E with shunt capacitance topology, a shunt resonance circuit L_P–C_P tuned to fundamental frequency is introduced in place of series L_S–C_S circuit. This topology is termed as class E PA with shunt capacitor and shunt filter and shown in Fig. 4.21. This shunt resonator performs matching/filtering of fundamental frequency as well as present short at second and higher harmonics. The analysis of this topology is described in [30]. Figure 4.22 shows reactance compensation scheme in class E with shunt capacitor and shunt filter. The component values of shunt L_P–C_P filter have been chosen such that they resonate at fundamental frequency ω_0 and also compensate the reactance to achieve broadband response. At resonance, the equivalent circuit of Fig. 4.22 is

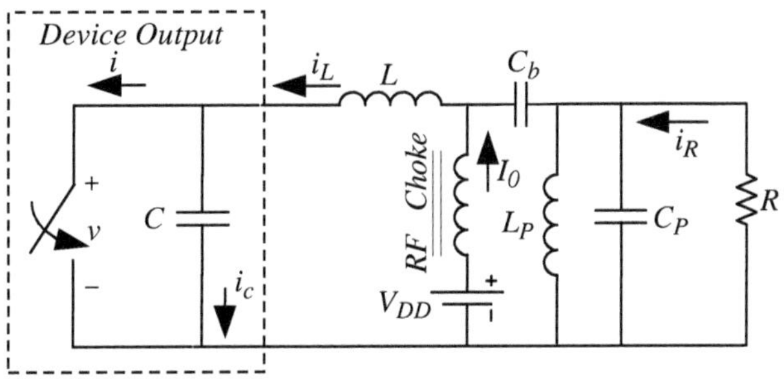

Fig. 4.21 Circuit topology of class E PA with shunt capacitor and shunt filter

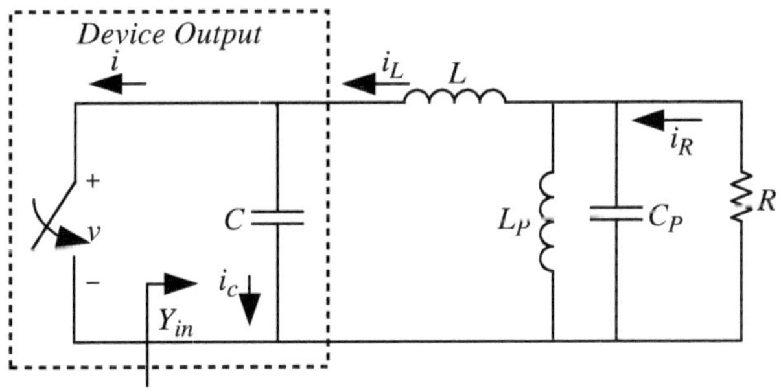

Fig. 4.22 Reactance compensation scheme applied in equivalent circuit of class E with shunt capacitor and shunt filter

Table 4.2 Load network parameters for different class E topology

Normalized load network parameters	Conventional Class E topology with shunt capacitance	Class E topology with shunt capacitance and shunt filter
$\frac{\omega L}{R}$	1.1525	1.4836
ωCR	0.1836	0.261
$\frac{P_{\text{out}}R}{V_{\text{DD}}^2}$	0.5768	0.4281

similar to the equivalent circuit of tuned class E PA with shunt capacitance as shown in Fig. 1.13. However, an additional blocking capacitance C_b which is shown in Fig. 4.21 is considered during analysis as described in [30].

The analysis however assumes C_b as short in steady-state condition at fundamental frequency, but may contribute toward the initial condition used in circuit analysis. Therefore, the analysis in this case at fundamental frequency is different from the one performed in Chap. 1 in the instance of tuned class-E PA with shunt capacitance topology. The summary of load network parameters as obtained from this analysis [30] is listed in Table 4.2. This table also shows the load network component values for conventional tuned class E topology shunt capacitance as discussed in Chap. 1 for comparison purpose [30].

As discussed earlier, the shunt L_P–C_P also resonates at $\omega_0 = 1/\sqrt{L_P C_P}$ and its equivalent susceptance is $j\omega' C_P$, where $\omega' = \omega - \frac{\omega_0^2}{\omega}$. In such case the susceptance B_{in} is calculated as imaginary part of admittance Y_{in} as,

$$B_{\text{in}}(\omega) = \omega C + \frac{\omega' C_{\text{P}} R^2 \left[1 - j\omega\omega' L C_{\text{P}}\right] - \omega L}{R^2 [1 - j\omega\omega' L C_{\text{P}}]^2 + (\omega' L)^2} \tag{4.13}$$

Applying the following condition on B_{in} to maximize bandwidth around resonant frequency ω_0,

$$\left.\frac{\mathrm{d}B_{\text{in}}(\omega)}{\mathrm{d}\omega}\right|_{\omega=\omega_0} = 0 \tag{4.14}$$

This results in an additional equation

$$2\omega_0 C_{\text{P}} R - \frac{\omega_0 L}{R} - \omega_0 C R \frac{\left(\frac{\omega_0 L}{R}\right)^2 + 1}{\left(\frac{\omega_0 L}{R}\right)^2 - 1} = 0 \tag{4.15}$$

Using load network parameters *L*, *C*, and *R* from Table 4.2, one can obtain L_{P} and C_{P} as,

$$C_{\text{P}} = \frac{1.0896}{\omega_0 R} \tag{4.16}$$

$$L_{\text{P}} = \frac{1}{\omega_0^2 C_0} \tag{4.17}$$

Figure 4.23a shows the photograph of broadband class-E GaN HEMT PA based on this topology implemented using a CREE CGH40010P device [30]. The circuit is fabricated on a 20 mil RO4360 substrate. Figure 4.23b shows the measured results of a broadband class-E GaN HEMT PA with shunt filter [30].

Among all the topologies discussed so far, an additional resistance transformation circuit should be added to transform the 50 Ω load to the required optimum load for class E mode. In general, the value of this optimum load is small for high-power devices compared to 50 Ω resulting in two three-stage impedance transformer for wideband transformation. For example, [18] uses a three-stage low-pass filter as resistance transformation network. This will increase circuit complexity in the output matching network. If a reactance compensation network can inherently provide the impedance transformation, then there is no requirement for additional impedance transformation which simplifies the output matching network for a wideband design. An improved reactance compensation technique was proposed in [31] where the reactance compensation network can translate to the network with larger optimum load resistance and hence can lead to simple resistance transformation network. In fact, one can avoid the additional resistance transformation network if the magnitude of the desired drain impedance of the device lies between 12.5 and 50 Ω [31]. PAE more than 70% with fractional bandwidth of 51% is achieved using this methodology, whereas with conventional

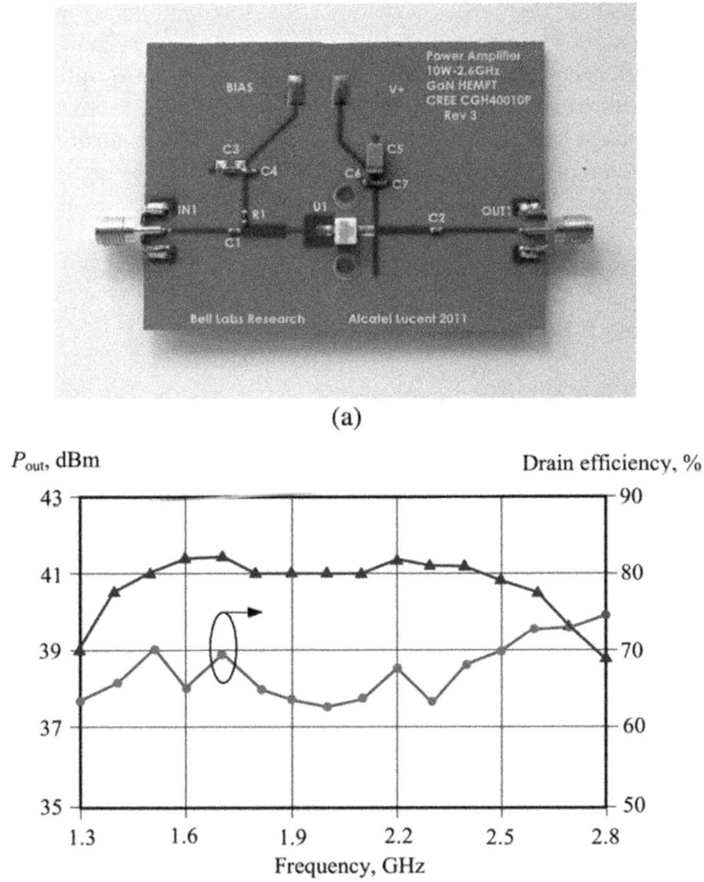

Fig. 4.23 Prototype developed based on class E topology with shunt capacitor and shunt filter: (**a**) photograph and (**b**) measured results ([30], reprinted with permission from IEEE)

reactance compensation technique the fractional bandwidth achieved was only 42% [31].

Another attempt in realizing broadband Class-E PA with shunt capacitor topology and higher order low-pass filter is presented in [32]. The broadband characteristics are achieved by solving two problems of conventional class E mode with a single network. The load network acts both as the impedance transformer across the continuous spectrum of frequencies and as a broadband filter for passband frequencies, thus reducing the circuit complexity. This network is also designed to provide appropriate harmonic impedances in order to give high efficiency. The filter should be designed such that its impedance is different for fundamental and higher harmonics and with minimum in-band ripple. A low-pass matching network can fulfill these requirements with minimum in-band ripple and broadband characteristics with higher order topology. The low-pass matching network is designed in two steps. First, design of a three-stage real-to-real Chebyshev low-pass matching network is completed, which is then followed by optimization of the real-to-real transformer to a real-to-complex transformer. Hence, a broadband class E PA with a three-stage low-pass matching network other than conventional reactance compensation technique is presented in [32]. The class E PA is implemented using GaN HEMT with an efficiency of more than 80% over the fractional bandwidth of 50%.

Continuous Class E Power Amplifier

The continuous class E mode presented in Chap. 1 has the capability to achieve broadband class E performance. The broadband class E can be achieved by assuming different modes of continuum for different frequencies. The switched impedances $Z_1{}^S$ and $Z_2{}^S$ at fundamental and second harmonic frequencies can be achieved by considering the parallel combination of the respective impedances presented by load matching circuit, that is, Z_1 and Z_2 with C as obtained in section "Class E Continuum" of Chap. 1. Therefore, one can write [33],

$$Z_1^S(\phi_1) = Z_1 \parallel j\omega C = R\frac{(1 + jx_1)\pi}{\pi + j2(1 + jx_1)\cos^2\phi_1} \tag{4.18}$$

$$Z_2^S(\phi_1) = Z_2 \parallel j\omega C = R\frac{jx_2\pi}{\pi - 4x_2\cos^2\phi_1} \tag{4.19}$$

where

$$R = \frac{V_{DD}^2}{P_{out}}\frac{8\sin^2\phi_1}{\pi^2} \tag{4.20}$$

One can calculate $Z_1{}^S$ and $Z_2{}^S$ by varying ϕ_1 from 43.3° to 78° which gives the fundamental and second harmonic switching impedances for continuous class E mode as shown in Fig. 4.24 [33].

By assuming the loads presented in Fig. 4.24, output power comparable to conventional class *E* can be obtained. One can observe that the second harmonic loads vary from open to short with increase in ϕ_1. When $\phi_1 = 43.3°$, the second harmonic switched impedance is open similar to class E/F$_2$ mode and with $\phi_1 = 78°$, the second harmonic switched impedance is short similar to class EF$_2$ mode [33, 34]. For the ϕ_1 in between 43.3°and 78°, the second harmonic impedance has negative values in between short and open. Considering the set of fundamental and second harmonic loads for a particular mode at each frequency, the broadband PA can be designed. Hence the continuous class E mode has the capability to realize broadband high-efficiency switched-mode PA. The continuous class E PA can also be designed using the nonlinear-embedding model similar to continuous

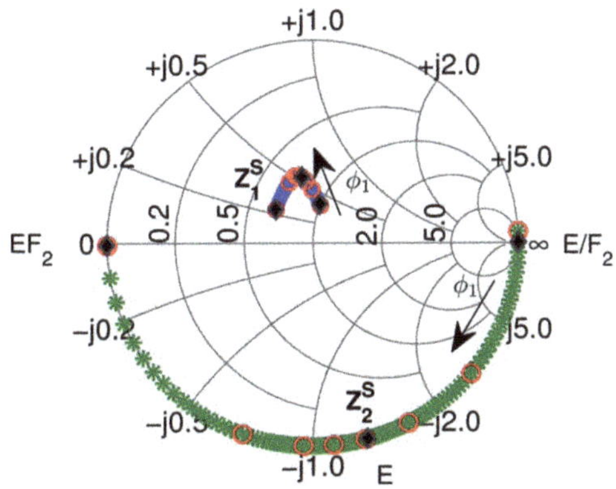

Fig. 4.24 Fundamental and second harmonic impedance seen by switch for $\phi_1 \epsilon [43.3°, 78°]$ ([33], reprinted with permission from IEEE)

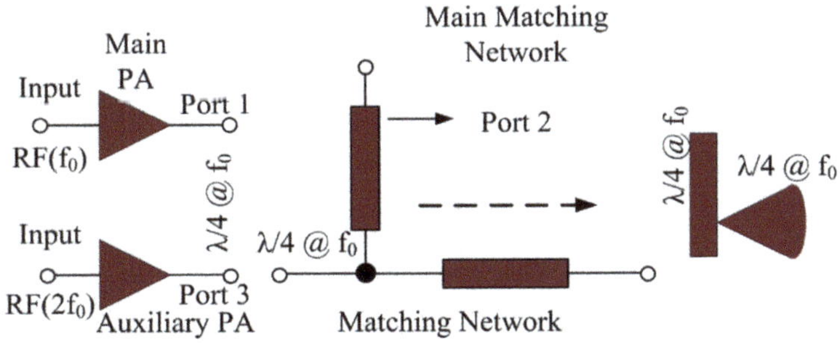

Fig. 4.25 Diplexer topology for broadband design

class B/J and continuous class F modes. The appropriate selection of frequency for a particular mode in continuum can be obtained using nonlinear-embedding model. This technique will give a good prediction of the feasible load at the package plane of the device, hence helpful in realizing practical matching network. Such design methodology has been presented in [35], where the mapping of ϕ_1 with frequency has been introduced in order to obtain feasible loads over a wide frequency range.

4.3 Bandwidth Enhancement in Harmonic Injection Power Amplifiers

Increasing the bandwidth of harmonic injection PAs (HI-PA) presented in Sect. 1.3.4 can result in broadband high-efficiency PAs. The diplexer topology shown in Fig. 4.25 exhibits narrow band characteristics as it uses multiple quarter-wave transformers. In order to enhance the bandwidth of the HI-PA, the quarter-wave open stub can be replaced by a radial stub as shown in Fig. 4.25 [36]. The bandwidth can also be improved by using a hybrid HI-PA concept where both passive and active HI can be done [36]. Such design integrates the dual-band PA with active HI to achieve high efficiency across wide bandwidth. A resistively loaded class B mode

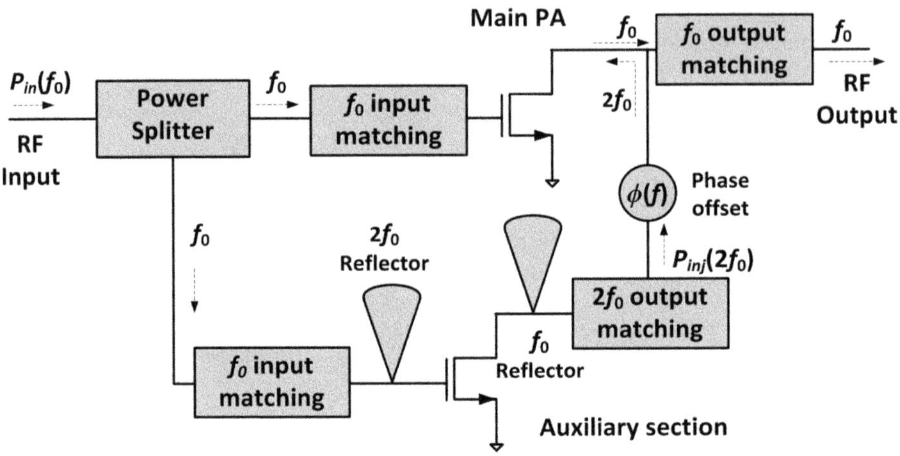

Fig. 4.26 Block diagram of harmonic injection-power amplifier (HI-PA) topology with active harmonic injection

can be used at the lower frequencies and a class J mode at higher frequencies while the transition between these two modes can be compensated by active HI at the mid-band [36].

Instead of using a broadband diplexer, the broadband HI-PA can also be achieved by designing an auxiliary doubler along with the main PA. This section presents the design example of broadband continuous Class B/J mode HI-PA, where the second harmonic signal is generated on the same PCB using an on-board frequency doubler [37]. The block diagram of the HI-PA with active HI is shown in Fig. 4.26.

The injected second harmonic signal should be of a particular phase and amplitude for proper voltage–current wave shaping, or else the performance of HI-PA would be worse than the conventional class AB mode [37]. Hence, the design space of amplitude and phase of injected second harmonic is analyzed to design HI-PA. Diplexer is generally used to avoid the signal at f_0 to leak toward the auxiliary path. Therefore, in this design the doubler is designed with two objectives. First, it should be able to generate sufficient amount of harmonic signal ($P_{\text{inj}}(2f_0)$) to change the harmonic impedance seen by main PA. Second, it should be able to inject a weak fundamental signal ($P_{\text{inj}}(f_0)$) toward the main path to minimize or cancel out leakage of f_0 signal from main to auxiliary path.

The HI-PA is simulated in Keysight's Advanced Design System (ADS). The desired harmonic terminations at the device CRP are set by appropriately choosing amplitude and phase of injected harmonics at $2f_0$. The continuous mode of operation is obtained by analyzing these harmonic impedances. A 15 W Wolfspeed CGH27015F GaN HEMT is used for the validation of this analysis. For the design space analysis, the input power $P_{\text{in}}(f_0)$ is kept constant and main PA is operated in saturation. Whereas the phase ϕ of injected signal and its power at $2f_0$, i.e., $P_{\text{inj}}(2f_0)$, are varied from $-180°$ to $180°$ and 15–35 dBm, respectively. The phase ϕ is defined in reference to the output signal of the main PA. Figure 4.27a, b shows the respective

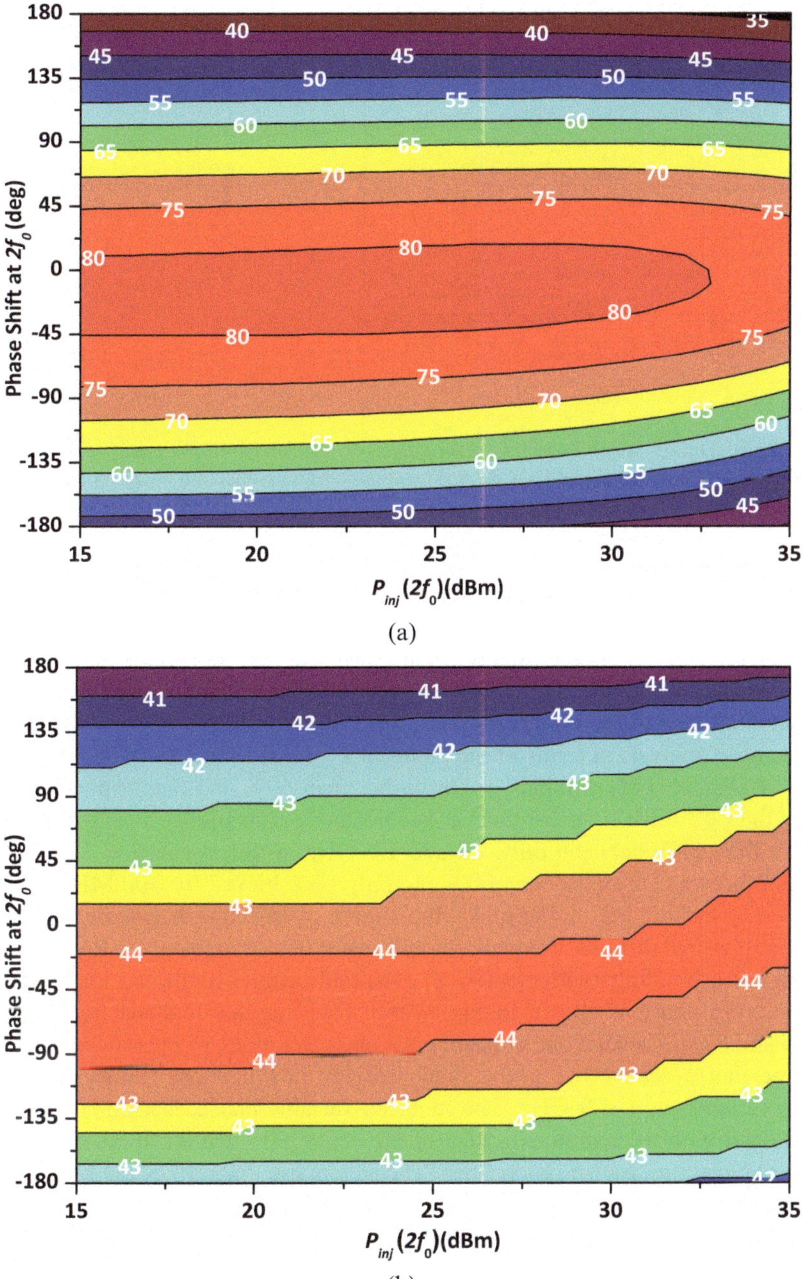

Fig. 4.27 Contour plots of (**a**) simulated DE and (**b**) fundamental output power $P_{\text{out}}(f_0)$ at 1.5 GHz

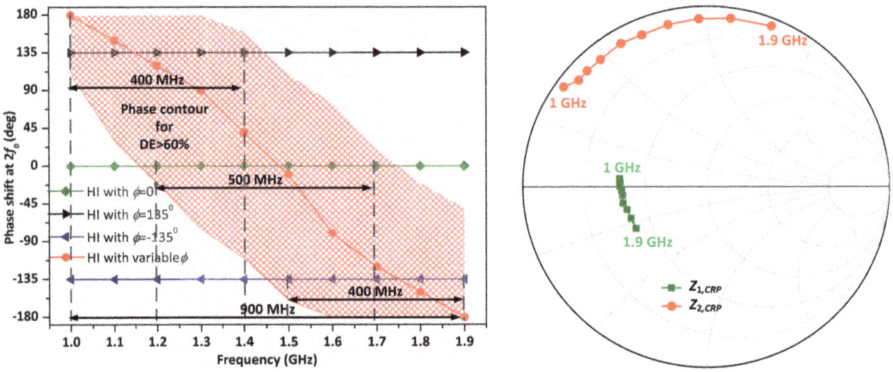

(a) (b)

Fig. 4.28 (**a**) Design space in terms of the phase of second harmonic injection for optimum operation and (**b**) loads presented at CRP with second harmonic injection in appropriate phase

contours of DE and $P_{out}(f_0)$ with simultaneous variation in injected harmonic power $P_{inj}(2f_0)$ and phase of the injected harmonic at 1.5 GHz.

A similar optimization of output power $P_{out}(f_0)$ and DE is carried out from 1 to 1.9 GHz. This determines the appropriate phase and amplitude of the injected second harmonic signal for efficiency enhancement. One can observe from Fig. 4.27 that the amplitude of the HI-PA is less dependent on the amplitude of $P_{inj}(2f_0)$. The design space analysis in terms of the phase of $P_{inj}(2f_0)$ from 1 to 1.9 GHz is shown in Fig. 4.28a. This corresponds to DE higher than 60% and output power $P_{out}(f_0)$ higher than 42 dBm. If a constant ϕ is selected, for example, $\phi = 0°$(blue), DE is more than 60% between only 1.2 and 1.7 GHz. Similarly, if $\phi = 135°$(purple) or −135°(brown) DE better than 60% can only be achieved for 400 MHz band. As one closely inspect Fig. 4.28a(pink), the choice of injecting harmonic signal with different phase at different frequencies will result in a broadband HI-PA.

The corresponding fundamental ($Z_{1,CRP}$) and second harmonic loads ($Z_{2,CRP}$) with varying phase is shown in Fig. 4.28b. One can see that the trajectories of $Z_{1,CRP}$ and $Z_{2,CRP}$ is set close to continuous class B/J mode of operation. Therefore, ensuring that ϕ always lies in the shaded region in Fig. 4.28a results in the class B/J operation over 1–1.9 GHz. The required variable-phased harmonic signal can be obtained with the combined effect of the active HI from the frequency doubler and the connecting transmission lines. The schematic diagram of the active HI-PA is shown in Fig. 4.29.

A broadband frequency doubler uses 10 W Wolfspeed CGH40010F GaN HEMT device in the design as shown in Fig. 4.29. In order to generate significant amount of second harmonic component, the doubler is biased with a drain voltage and current of 8 V and 48 mA, respectively. Due to the nonlinear operation of the doubler, it may generate fundamental and higher harmonic components along with the required second harmonic component. Hence, the doubler output section is designed with f_0

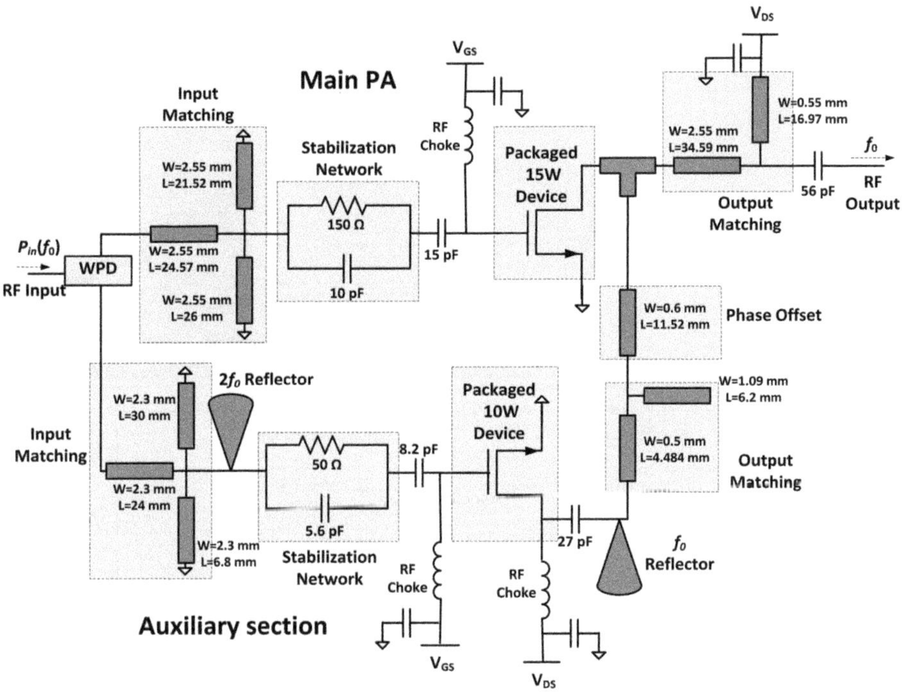

Fig. 4.29 Schematic diagram of proposed active HI-PA (all dimensions in mm)

reflector followed by the output matching for $2f_0$ as shown in Fig. 4.29. Similarly, the input section of the doubler consists of the input matching for f_0 followed by a $2f_0$ reflector to avoid the generated second harmonic leaking toward the input side. For broadband performance, radial stubs are used for the design of reflectors instead of quarter-wave transformers. As mentioned earlier, a small portion of fundamental signal is allowed at the output of the doubler for f_0 isolation in the auxiliary path. Since the phase of the transmission line varies linearly with frequency, it is used to achieve variable phase harmonic signal. The length of transmission line along with the output matching of the doubler provides the required phase shift to fit in design space.

The main PA is designed with 15 W Wolfspeed CGH27015F GaN HEMT. The device is set at drain bias voltage and current of 27.5 V and 90 mA, respectively. The device is unconditionally stable and a broadband matching network is used in the input and output matching. As shown in Fig. 4.29, the generated harmonic signal from the doubler is injected at the drain terminal of main PA. The input matching network is realized using unequally terminated band-pass filter topology [23]. A conventional equal-split Wilkinson power divider (WPD) splits the input signal to the main and auxiliary sections. RO4350B substrate with dielectric constant 3.66 and height 20 mil is used to fabricate the printed circuit board (PCB) as shown in Fig. 4.30.

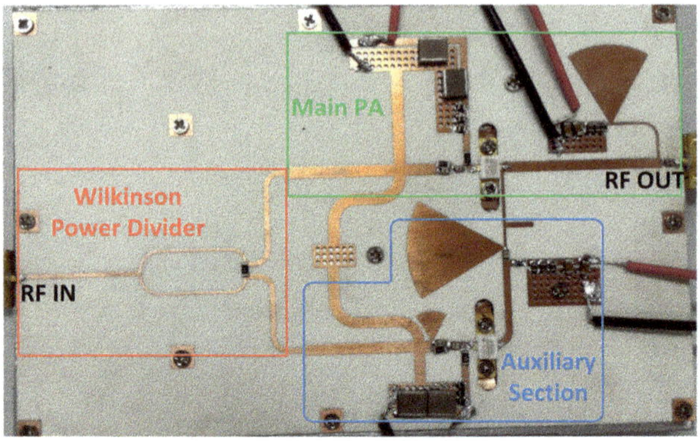

Fig. 4.30 Photograph of the fabricated HI-PA

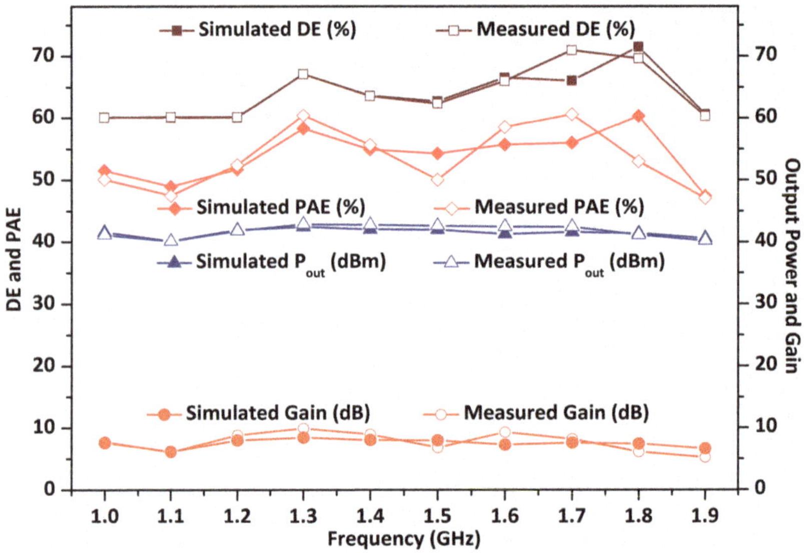

Fig. 4.31 Simulated and measured DE, PAE, output power, and gain of HI-PA

The simulation and measurement results of HI-PA with CW RF signals is shown in Fig. 4.31. The DE of the HI-PA is calculated considering the DC consumption of main PA and the auxiliary section. The simulated and measured DE varies between 60% and 71.49% over the designed band of 900 MHz as shown in Fig. 4.31. The output power is in the range of 40.17–42.6 dBm in simulation and measurement in the entire band. The average value of gain is around 7.57 dB in simulation. The corresponding value is 7.78 dB in measurement. Figure 4.31 shows simulated and measured PAE accounting the loss of WPD.

Along with the fundamental load termination using passive output matching network, the efficiency of the active HI of the second harmonic can be further enhanced if the optimized third harmonic load is provided to the HI-PA [38]. Such topology will work similar to the inverse class F mode which has theoretically 100%

efficiency. This topology is presented in [38], where harmonic load-pull simulation is used to find the optimum third harmonic load. Class J_2 mode can also be used for efficiency enhancement where the reduced conduction angle, even-order harmonics are injected at the drain node of the main PA. This mode can result in efficiency and output power greater than the conventional class J mode [39].

4.4 Enhancing the Operating Range and Bandwidth of Load Modulation Schemes

Being the most popular architecture for load modulation, several attempts have been carried out in the past for enhancing the bandwidth of DPA. Broadband DPAs have been reported in the recent past in literature. The basic DPA architecture is already discussed in Chap. 1. Broadband DPA design faces limitation in bandwidth extension due to the quarter-wave impedance transformer and the output capacitance of the transistors [40]. Extensive research is done to solve this narrowband issue in DPA architecture [40–43, 48–54, 56–62, 64–66 and 69]. Many of these attempts, though conceptually similar, differ in design approach. This section summarizes some important attempts which broadly classify different approaches adopted by designers to enhance the DPA scheme.

4.4.1 Architectures for Broadband Doherty Power Amplifier

In the past, there have been several efforts carried out to enhance the bandwidth of DPA. These bandwidth enhancement schemes can be broadly categorized in three ways: first is incorporating modified load combiner topology, second is by using continuous mode topologies with load combiners providing appropriate matching and load modulation, and the third is by using digital techniques to compensate imbalances along with broadband PA design scheme. One of the key challenges is to present an open circuit condition to the main PA current such that it should not leak to the auxiliary PA path when the auxiliary PA is *off*. A simplest technique is to present shunt inductor which can resonate with the output capacitances of the transistor to present an open circuit condition. Sometimes bias inductors can also be used for this purpose. However, this solely will not solve the purpose of achieving wide bandwidth perhaps due to limited bandwidth of resonator circuit. Thus, the design of load combiner must be modified to achieve load modulation as well as this open circuit condition over the wide bandwidth. One such effort involves design of quasi-lumped impedance inverter while absorbing the effect of output capacitances and bond-wires in case of bare die transistors [41]. A 20 W Doherty PA was designed using this technique which operates over 1.7–2.3 GHz with drain efficiency of 40–50% at saturation as well as back-off [41]. Although the performance is good in this case, yet it finds limited application due to the use of unpackaged device and mixed-signal setup adding significant complexity to the design. Unlike this work, in [42] the authors have discussed a broadband DPA with complete analog architecture. In general, the conventional DPA design involves calculation of load combiner design parameters Z_0 and R_L based on specified current profiles of the main and auxiliary PA in DPA configuration. However, one can also start from specifying Z_0 and R_L and deriving the corresponding current profiles. The later approach gives a freedom to optimize the bandwidth response of the load combiner by appropriately choosing the design parameters Z_0 and R_L. For example, the conventional architecture of DPA as discussed in Chap. 1 requires the quarter-wave transformer characteristic impedance Z_T as R_{opt} and the common load R_L as βR_{opt}. The load modulation factor β is 0.5 in case efficiency enhancement at 6 dB back-off is targeted [42]. R_{opt} is the optimum load that should be presented to the main PA in order to operate with optimum power and efficiency. In such a case, the load seen by the main PA at back-off is frequency dependent and the frequency characteristic of quarter-wave transformer decides the load modulation over the frequency range in conventional load combiner. This results in a narrow band response in terms of efficiency enhancement at back-off. In order to further analyze this, Eq. (1.170) in Chap. 1 is rewritten for back-off operation (where auxiliary PA is off, i.e., $I_A = 0$) for any frequency f around the center frequency f_c.

$$Z_{M,BO} = \frac{V_m}{I_m} = R_L \left[\frac{\cos\left(\frac{\pi}{2}\frac{f}{f_c}\right) + j\frac{Z_T}{R_L}\sin\left(\frac{\pi}{2}\frac{f}{f_c}\right)}{\cos\left(\frac{\pi}{2}\frac{f}{f_c}\right) + j\frac{R_L}{Z_T}\sin\left(\frac{\pi}{2}\frac{f}{f_c}\right)} \right] \tag{4.21}$$

One can see that if Z_T and R_L are chosen as same, the load seen by main PA at back-off as shown in Eq. (4.21) becomes independent of frequency. This is also evident from the fact that for any transmission line terminated with a constant load equal to its characteristic impedance will have an input impedance equal to the load impedance regardless of the frequency. According to ideal class B theory, if the main PA requires R_{opt} load at saturation, the load impedance required at back-off is βR_{opt} in order to achieve optimum efficiency same as saturation. Therefore, R_L in Eq. (4.21) must be set as βR_{opt}. Similarly, the corresponding Z_T is chosen as βR_{opt}. Therefore, in such a case the main PA will see optimum load at back-off, independent of the frequency. This is validated for efficiency enhancement at 6 dB back-off, where a constant load of $2R_{opt}$ at the main PA output is presented in the complete frequency range [42]. The corresponding current of main and auxiliary PA and load seen by them with the input voltage drive is shown in Fig. 4.32. One can see from Fig. 4.32a that the load seen by the auxiliary PA is $4R_{opt}$ at saturation. This will reduce the current at the output of the auxiliary PA at saturation well below the main PA as shown in Fig. 4.32b. Moreover, the load seen by the main PA is $2R_{opt}$ at back-off over a fractional bandwidth of 40% which is quite high when compared to the conventional two-stage DPA. This however slightly changes the loads from their required values at saturation without affecting the performance significantly.

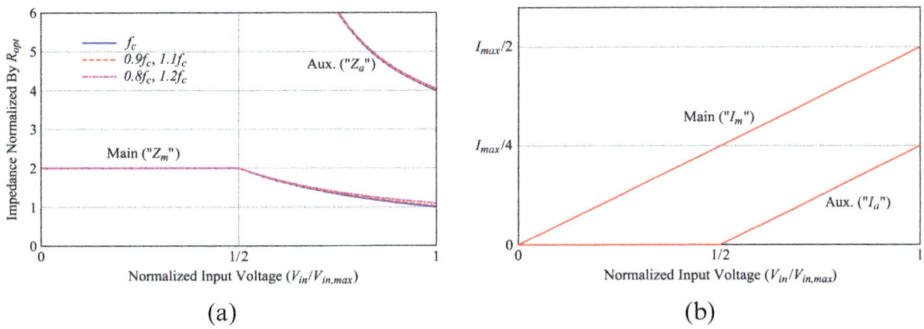

Fig. 4.32 Broadband Doherty Power Amplifier (DPA) with modified load combiner: (**a**) load seen by main and auxiliary PA and (**b**) the current profile

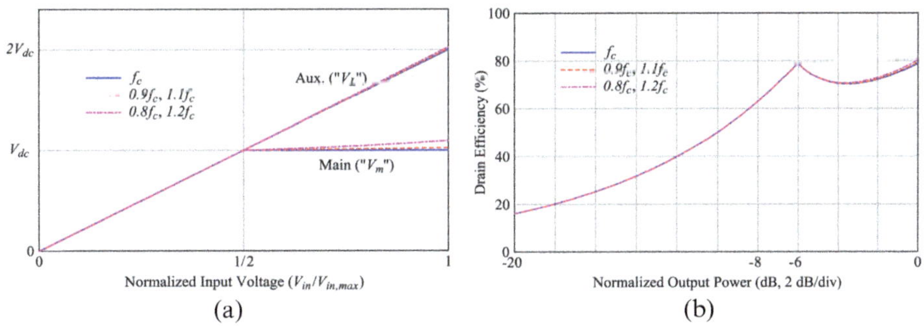

Fig. 4.33 Voltage and efficiency of broadband DPA with modified load combiner: (**a**) voltage profile and (**b**) efficiency characteristic ([42], reprinted with permission from IEEE)

One can use Z_T and R_L as $2R_{opt}$ in Eq. (1.170) to obtain currents at saturation. Therefore, the current profile as shown in Fig. 4.32b can be obtained by deriving mathematical expression for the main and auxiliary currents similar to Eqs. (1.172) and (1.173). Moreover, high loads presented to the auxiliary PA at saturation results in high-voltage swing across the device used. Therefore, the bias voltage is scaled to two times more than the main PA which can be possible with GaN devices capable of handling high voltage. Figure 4.33a shows the voltages at the output of both main and auxiliary PA with the input drive. Figure 4.33b shows the expected efficiency performance with this scheme considering the main PA as ideal class B PA. One can see that 40% fractional bandwidth is expected from this scheme with slight deterioration at saturation pertaining to nonoptimal load seen by auxiliary PA at saturation. Along with the broadband load combiner design, [42] utilizes Klopfenstein taper matching network at the output side for broadband matching. With this scheme, a 90 W GaN broadband DPA prototype was designed and fabricated to operate from 700 to 1000 MHz (35.3% bandwidth) [42].

The scheme discussed above is good for 6 dB back-off. In case of higher back-off, the load modulation factor β increases, and therefore the value of characteristic impedance Z_T should be set very high. The maximum value of this characteristic impedance depends on the minimum width of the microstrip line that can be realized with the given technology. Such thin line may have low power handling and maybe difficult to realize with conventional PCB fabrication process. In such case, if R_L is also allowed to be chosen, it will add an extra degree of freedom which relaxes the constraint over the choice of the characteristic impedance Z_T to be very high [43]. Such analysis provides a design space where the next best solution which is optimum, in terms of bandwidth as well as feasibility for fabrication in microstrip, can be selected. Considering an ideal class B operation, the best value of load at back-off for enhancing efficiency is βR_{opt}. This can be used in (Eq. 1.170) to obtain,

$$Z_M = \beta R_{opt} + \frac{R_L \frac{I_A}{I_M}}{\cos\left(\frac{\pi}{2}\frac{f}{f_c}\right) + j\frac{R_L}{Z_T}\sin\left(\frac{\pi}{2}\frac{f}{f_c}\right)} \tag{4.22}$$

At saturation, the load seen will become R_{opt} which can be used in Eq. (4.22) to obtain,

$$\frac{I_{A,sat}}{I_{M,sat}} = \frac{p(1-\beta)}{R_L}\left[Z_T \cos\left(\frac{\pi}{2}\frac{f}{f_c}\right) + jR_L \sin\left(\frac{\pi}{2}\frac{f}{f_c}\right)\right] \tag{4.23}$$

where $p = R_{opt}/Z_T$ is an intermediate variable. At center frequency of the quarter-wave transformer, if the angle between main and auxiliary PA currents are set such that $I_{A,sat}$ lags $I_{M,sat}$ by 90°, Eq. (4.23) reduces to,

$$\alpha = (\beta - 1)\,p \tag{4.24}$$

where $\alpha = |I_{A,\,sat}|/|I_{M,sat}|$ represents the ratio of magnitudes of currents supplied by the main PA and auxiliary PA at saturation. Equation (4.24) explains the relationship between current ratio α and characteristic impedance of load combiner Z_T in terms of R_{opt} and β. For a conventional DPA, it is generally observed that for a specified back-off, choice of current profiles is to be fixed. If one can choose any value of α for this specified back-off, there are number of possible solutions for design parameters which can give bandwidth-enhanced architecture. The designer can choose an optimum solution among them depending on different parameters such as bandwidth and physical realizability. Fig. 4.34a shows variation of back-off efficiency with frequency for different values of p/α. These graphs are plotted for targeting efficiency enhancement at 9.54 dB back-off which corresponds to $\beta = 3$. For this analysis, R_{opt} is considered as 50 Ω. Figure 4.34a shows $p = 0.335$ corresponding to $\alpha = 0.667$ presents the best bandwidth solution. This results into $Z_T = R_L = 3R_{opt}$, where the characteristic impedance $Z_T = 150$ Ω is very high value if $R_{opt} = 50$ Ω.

Therefore, the analysis presented in [43] provides a design space where the next best solution can be opted. This solution can provide reasonable value of characteristic impedance of a quarter-wave transformer that can be easily realizable

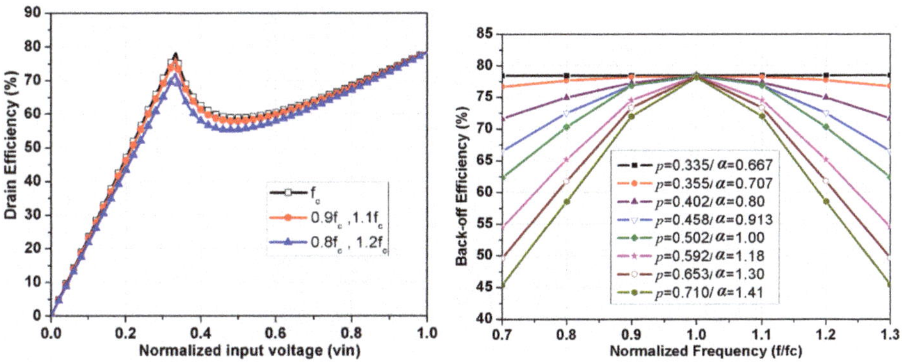

(a) (b)

Fig. 4.34 (**a**) Design space analysis for load combiner parameter in terms of back-off efficiency over frequency range, (**b**) DE profile of DPA with optimized broadband combiner ([43], reprinted with permission from WILEY)

with a microstrip line without sacrificing much in terms of bandwidth. One can see from Fig. 4.34a, that $p = 0.458$ and $\alpha = 0.913$ gives $Z_0 = 2.18R_{\text{opt}}$ which is quite less when compared to the case of best bandwidth. Therefore, it is comparatively easy to realize such load combiner with the fabrication technology available [43]. Figure 4.35a shows the circuit photograph of broadband DPA as designed in [43]. The circuit is fabricated in R04350 substrate which has dielectric constant and height as 3.66 and 20 mil, respectively. Figure 4.35b shows the measured gain and DE with respect to output power at different frequencies. One can see that for all frequencies, DE measured is 66–73.8% at saturation. Figure 4.35c shows the variation of gain at saturation over the frequency band 1.7–2.025 GHz between 6.8 and 11.0 dB.

The performance in terms of linearity is also evaluated for this circuit. With single-channel WCDMA signals of 9.54 dB PAPR, the measured ACPR curves are shown in Fig. 4.36. Figure 4.36a, b shows the ACPR at 5 MHz offset around 1.9 GHz frequency. One can see that at an average power of 37 dBm, the measured ACPR is less than −24 dBc over the entire frequency range as shown in Fig. 4.36a, b.

Limitation in terms of bandwidth enhancement in DPA is also discussed in [40]. This work also discusses the opportunity of choosing common load R_L optimally, such that the impedance transformation ratio of quarter-wave transformer is minimized. The bandwidth of impedance transformer depends on the transformation ratio [44]. Lower the transformation ratio, higher the bandwidth of the transformer [44]. However, this will result in the impedance of quarter-wave transformer to be chosen accordingly. This choice depends on the realization of minimum or maximum width of the microstrip line based on the quarter-wave transformer. The bandwidth of DPA in such a case has shown 11% improvement than that of the conventional DPA, resulting in 44% fractional bandwidth [40]. Further, [5] proposed a SRFT for broadband matching with the desired frequency-dependent optimum

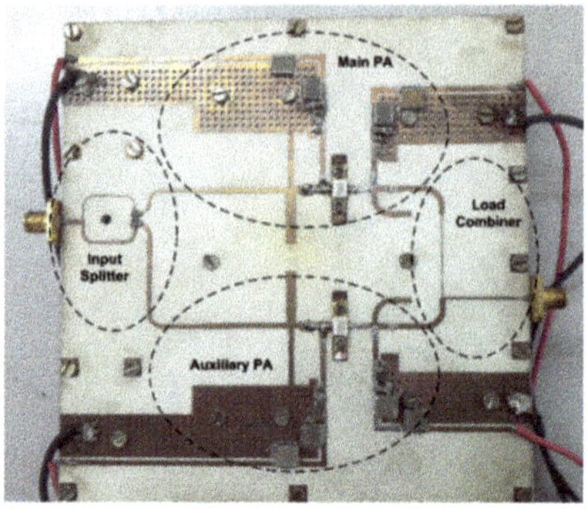

(a)

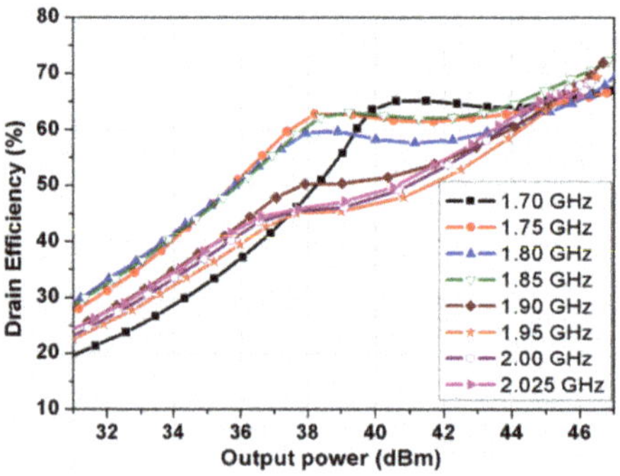

(b)

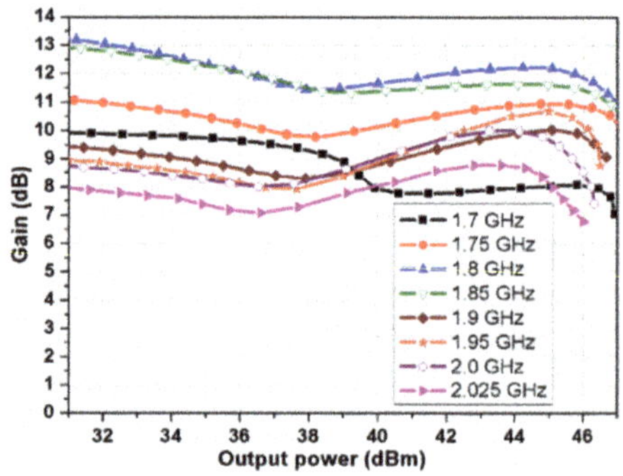

(c)

Fig. 4.35 (**a**) Photograph of the fabricated circuit, (**b**) measured efficiency curves, and (**c**) measured gain curves ([43], reprinted with permission from Wiley)

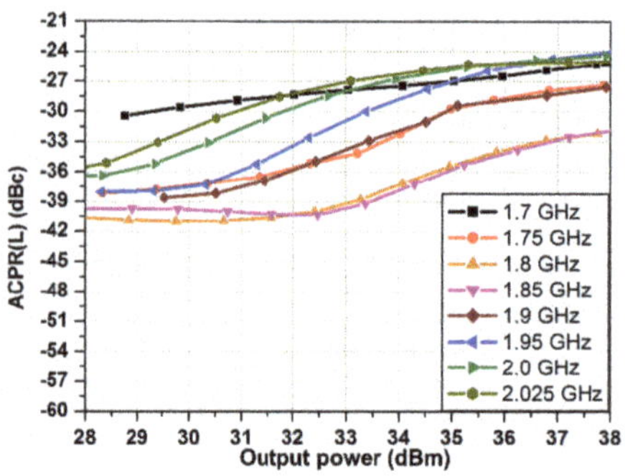

(a)

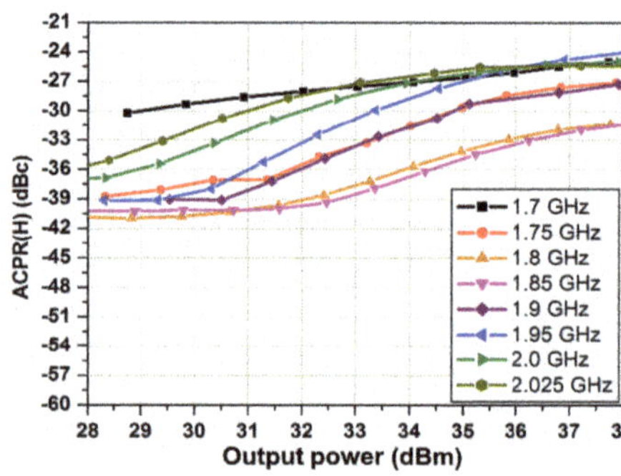

(b)

Fig. 4.36 Measured ACPR characteristics at the (**a**) lower frequency side and (**b**) higher frequency side ([43], reprinted with permission from Wiley) impedances. The frequency-dependent property of an ideal lossless reciprocal two-port network is represented by scattering matrix in the Belevitch canonic form [45]. The transducer power gain is defined for this scattering parameters. An LC low-pass filter is used as the optimization prototype. Different parameters of the prototype are optimized using Levenberg-Marquardt algorithms for the transducer power gain. The optimized values are obtained to keep this gain as high and as flat as possible over the given frequency interval. There are two options explored based on whether the performance optimization was carried out for the back-off or for the saturation termed as option I and option II in [5]. Option I (II) suggests that the optimum modulated impedance of the main PA can be only attained at the back-off (saturation) power level. On the other side, the modulated impedance at the saturation (back-off) power level takes only a suboptimum value due to the nonideal load modulation. The required load modulation criteria are specified for both the power levels and the combiner S-matrix is optimized to achieve them. In such case bandwidth is also limited in terms of broadband matching which are described by Bode [46] and Fano [47]. In addition, use of particular optimization methods may also result into narrow band solutions [5]. A broadband DPA is designed using CREE GaN CGH4000P transistors and SRFT technique for both PAs [5]. This DPA operates at the center frequency of 2.6 GHz. A good performance in terms of efficiency at 5–6 dB back-off over 2.32–2.825 GHz and 2.2–2.96 GHz has been achieved in [5].

Another novel method for broadband DPA design is presented in [48], by taking care of electrical memory effect of the broadband DPA employing LC resonant circuits in the drain bias networks of DPA. The effect of the integrated Doherty combiner is compensated by the broadband output matching networks, which are designed by applying the SRFT. A resonator-type bias network has been introduced and it has been shown that such biasing network exhibits reduced baseband impedance and also a lower impedance variation. To verify the proposed theory, the authors have designed the circuit with minimum baseband impedance, that is, about 3.5–7 Ω. One of the key points to remember for this technique is that the output matching network should be judiciously designed, as the low RF impedance of the resonant-type bias network will cause power leakage into the bias tee and this will vary the impedance presented to the transistor. The design procedure is elaborately explained in [48]. This scheme provides 100 MHz instantaneous bandwidth; however, the overall operating frequency range varies from 698 to 862 MHz.

In continuation to various efforts for introducing broadband matching and impedance inversion using broadband filter-topologies, a post matching-based broadband DPA design is introduced in [49]. Starting from the idea of [40] which proposes to use impedance inverter with low impedance transformation ratio, a low-order impedance inverter is used in [49] for bandwidth enhancement. In order to use low-order impedance inverter, the common load (load presented at point where both main and auxiliary branch combine) must be modified for correct load modulation. A post-matching network will later convert the standard 50 Ω load to the common load. The idea is similar to [40]; however, a filter-based matching topology is utilized

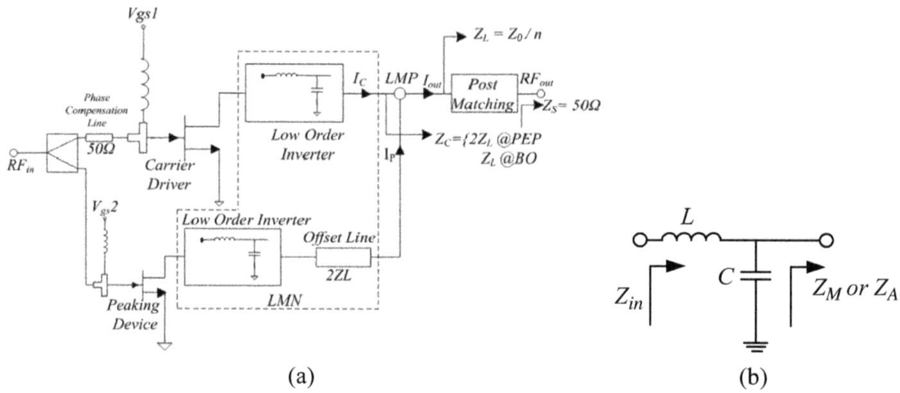

Fig. 4.37 (**a**) Architecture of broadband DPA using low-order impedance inverter and post-matching network, (**b**) low-pass prototype of impedance inverter/matching network ([49], reprinted with permission from IEEE)

for the design of impedance inversion, matching, and post-matching. Figure 4.37a shows the architecture, where the common load at load modulation point (LMP) is Z_L realized by post-matching network. Due to load modulation, Z_M and Z_A which are the respective loads seen by the main and auxiliary PA are same as $2Z_L$ at saturation. At back-off, the Z_M changes to Z_L, whereas Z_A has a very high value close to infinity as auxiliary PA is off. However, this is opposite to the requirement where main PA should see higher load at back-off to drive the main device at voltage saturation during lower input power drive. In conventional DPA, as discussed in Chap. 1, a quarter-wave transformer is later used as an impedance inverter to provide correct load variation from $2Z_L$ to Z_L between back-off and saturation at main PA output. Later, this load variation is converted to $2R_{opt}$ and R_{opt} at the device plane. In general, the R_{opt} is load required at intrinsic CRP of the device as described earlier. For a package transistor these R_{opt} and $2R_{opt}$ may correspond to two different complex impedances Z_{opt} and Z^*_{opt} at the PRP of the transistor. One way to find these impedances is to use embedding model as discussed in Chap. 3.

The other common method is to use load-pull at peak power to obtain Z_{opt} and at 3 dB back-off power to obtain Z^*_{opt}. Such simulation has been carried out for 10 W and 25 W GaN HEMT transistors from CREE in [49]. A low-pass filter prototype can be used to match Z_L to Z_{opt} for the main PA. Similarly, another prototype can be used to match $2Z_L$ to $Z_{opt,Aux}$ for the auxiliary PA. In both the cases, a single low-pass filter prototype can provide matching as well as impedance inversion in place of quarter-wave transformer as used in conventional DPA [49]. Figure 4.37b shows this low-order low-pass impedance transformer. The $Z_{opt,Aux}$ is the optimum impedance required at the output of auxiliary PA at saturation. This impedance may be slightly different from the one required for the main PA at saturation, as it is obtained considering class C operation in auxiliary PA. Several load-pull simulations may have to be carried out for various gate voltages below pinch-off to design this optimum load. In case of 10 W and 25 W GaN HEMT transistors from CREE, it is found that the real value of Z_{opt} and Z^*_{opt} does not change much, whereas the imaginary part varies between 1.7 and 2.6 GHz frequency range. Thus, a low-order filter (single-section *L–C* low-pass prototype) as presented in Fig. 4.37b is analyzed, which when terminated by a purely real impedance (i.e., resistive in nature) realizes an input impedance where the real part is independent of variation in the terminating load. This purely real terminating load Z_{M} and Z_{A} in the case of main and auxiliary PA varies while operating at back-off and at saturation due to load modulation. Their values depend on the common load Z_{L} terminating at node LMP which is also purely real. It is worth mentioning that the low-order filter prototype used here can present high bandwidth if its impedance transformation ratio is kept smaller. This is only possible if the designer is free to choose Z_{L}. This flexibility is possible due to the presence of a post-matching network which is later used to match 50 Ω to the common load Z_{L}. Taking a design example from [49], the optimum impedance for the main PA changes from the area $Z_{\text{opt}} = 10.2 + j14.6\ \Omega$ to $Z^*_{\text{opt}} = 9.6 + j6.9\ \Omega$. This means that the impedance transformer should keep the real part nearly invariant while decreasing the imaginary part with change in load Z_{M}. In case Z_{L} is purely real, that is, *R*, then, Fig. 4.38a, b shows low-pass filter prototype of order 1. As discussed earlier, *R* is a free parameter and can be realized by post-matching network at node LMP in Fig. 4.37a, b. The input impedance provided by this filter is given by,

$$Z_{\text{in}} = \frac{R}{1 + R^2\omega^2C^2} + j\left(\omega L - \frac{\omega C}{\omega^2C^2 + \frac{1}{R^2}}\right) \tag{4.25}$$

Considering the variation of *R* to 2*R* from back-off to saturation, the real part of Eq. (4.25) should not vary at certain frequency ω_0 and this results in following condition,

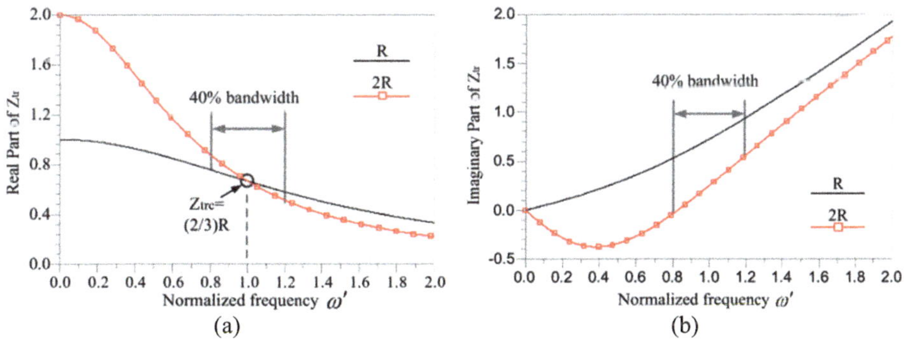

Fig. 4.38 Normalized impedance variation of low-order low-pass impedance transformer with normalized frequency and different load conditions: (**a**) normalized real part of impedances and (**b**) normalized imaginary part of impedances ([49], reprinted with permission from IEEE)

$$\frac{R}{1+R^2\omega^2C^2}=\frac{2R}{1+4R^2\omega^2C^2} \tag{4.26}$$

Solving Eq. (4.26), one can obtain ω_0 as,

$$\omega_0=\sqrt{\frac{1}{2R^2C^2}} \tag{4.27}$$

Therefore, using Eq. (4.27) in Eq. (4.25), the real part of the Z_{in}, that is, $\mathrm{Re}\{Z_{in}\}$, is realized as $2R/3$ at angular frequency ω_0 which is independent of load variation from R to $2R$ from back-off to saturation. Figure 4.38a shows this variation of $\mathrm{Re}\{Z_{in}\}$ with normalized frequency $\omega^{'}=\omega/\omega_0$. Similarly, $\mathrm{Re}\{Z_{in}\}$ is also normalized with respect to R. One can see at $\omega^{'}=1$, that is, at $\omega=\omega_0$, $\mathrm{Re}\{Z_{in}\}=2R/3$ and does not change with R changing to $2R$. Moreover, one can see that the variation in $\mathrm{Re}\{Z_{in}\}$ is less than 16% between $0.8\le\omega^{'}\le1.2$ corresponding to 40% fractional bandwidth. Figure 4.38b shows the variation of imaginary part of the Z_{in}, that is, $\mathrm{Im}\{Z_{in}\}$, with normalized frequency $\omega^{'}$. One can see that imaginary part is becoming smaller over the $0.8\le\omega^{'}\le1.2$ with load variation from R to $2R$. This is same as the requirement for main PA matching.

For example, as given in [49], the real parts of optimum impedance Z_{opt} and Z^*_{opt} for the GaN CGH40025F CREE device operating in class AB bias varies from 7 to 11.4 Ω. Taking the average value, the required $\mathrm{Re}\{Z_{in}\}=9.2\ \Omega$ which is equivalent to $2R/3$ at ω_0 as computed from Eq. (4.25). This gives R as 13.8 Ω which is nearly approximated as 14 Ω [49]. Considering ω_0 is the center frequency of operation, the element C of low-pass prototype can be calculated using Eq. (4.27). Therefore, $C=3.74$ pF is calculated for the value of $R=14\ \Omega$ at center frequency of 2.15 GHz using Eq. (4.27). The inductor element L is later tuned to best fit the Z_{in} to the load-pull data over the frequency band. In the case of auxiliary PA, the load-pull is performed with class C operation. Here, an average value of 11 Ω is obtained as the matching requirement of $\mathrm{Re}\{Z_{in}\}$ at ω_0. Since R is already finalized as 14 Ω during main PA design, a low-pass filter prototype for auxiliary PA chain requires to realize real part matching from $2R=28$ to 11 Ω. The element value C in this case is obtained by equating $\mathrm{Re}\{Z_{in}\}$ in Eq. (4.25) to 11 Ω which results in the value of 3.2 pF. The inductor L is tuned to 1 nH to match the load-pull data [49]. An offset-line in auxiliary path is added to avoid any leakage from main to auxiliary PA, when main PA is *off*. This line should have characteristic impedance same as $2R$ such that it will be transparent to the load $2R$ applied at the output of the impedance inverter in the auxiliary PA at saturation. This offset line may not be able to present high impedance over the entire band to the main PA current when auxiliary is *off*. However, due to the presence of low-value common load $Z_L=R=14\ \Omega$ appearing in shunt, the main current may not flow to the auxiliary PA path [49]. The elements of these impedance inverters are then converted into equivalent transmission line elements by applying appropriate transformation [44]. Figure 4.39a, b shows the measured performance of the prototype developed in [49] with CW signal.

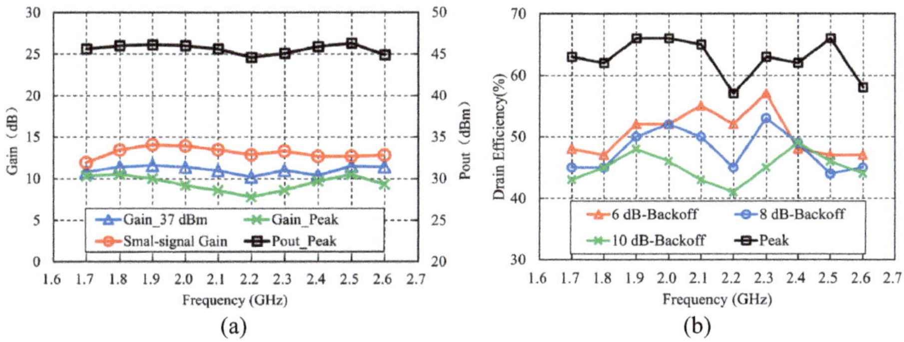

Fig. 4.39 Measured performance of prototype developed based on low-order impedance inverter and post-matching technique: (**a**) gain and output power with frequency, (**b**) DE at saturation and back-off with frequency ([49], reprinted with permission from IEEE)

Figure 4.39a shows that the output power varies between 44.9 and 46.3 dBm over a frequency range of 1.7–2.6 GHz. The gain is within the range of 10.2–11.6 dB at 8–9 dB back-off and 8.6–10.5 dB at saturation in this frequency range. Figure 4.39b shows DE varying between 47% and 57% at 6 dB back-off over the frequency range of 1.7–2.6 GHz. Even at 10 dB back-off the DE is better than 40%. At saturation, the DE is above 57% across the band.

The above schemes depend on optimizing the conventional load combiner parameters to obtain best bandwidth out of it. However, some of the works present a new load combiner architecture comprising several quarter-wave transformers [50–54]. Apparently, these combiners look like the passive-branch line hybrid couplers and therefore named here as hybrid-type load combiner. However, their network parameters (e.g., ABCD) are quite different from the conventional branch line hybrid coupler. This is perhaps due to one port being left open in these load combiners as shown in Fig. 4.40a–c. Therefore, these load combiners are three-port networks, where the two ports connected to the main and auxiliary PA are non-isolated unlike branch line hybrid coupler. This non-isolation will make load modulation possible in these load combiners.

The load combiner in Fig. 4.40a–c is asymmetric structures; therefore, they can be analyzed using the scheme presented in [51, 55]. This scheme computes the impedances presented by the combiner at the two ports connected to the main and auxiliary PAs. These impedances are calculated from effective node voltages and currents at these two ports. This is calculated by relating effective node voltages and currents using ABCD parameters of various quarter-wave transformers used in these load combiner topologies [51]. For both the topologies as described in Fig. 4.40a–c, the impedances presented to the main and auxiliary PA is given as [51, 52],

$$Z_{\mathrm{M}} = A + \frac{I_{\mathrm{A}}}{I_{\mathrm{M}}} B \tag{4.28}$$

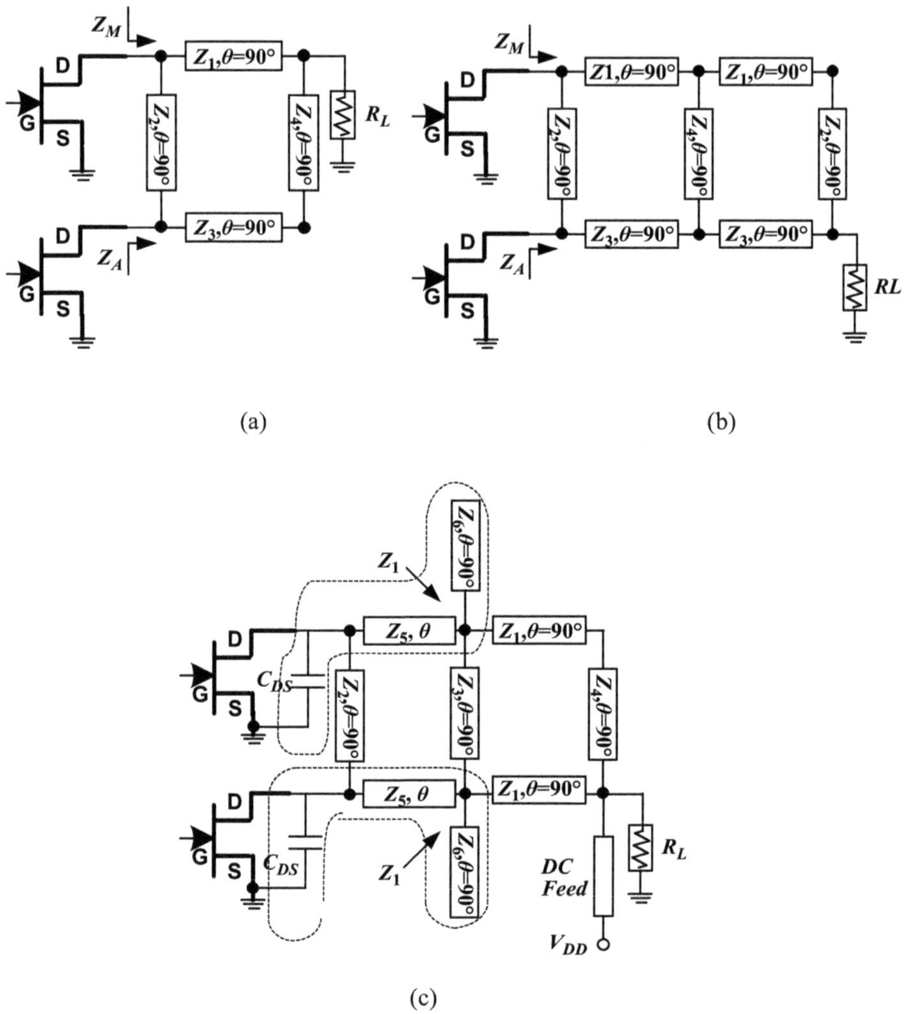

Fig. 4.40 Architecture of different load combiners: (**a**) single-section hybrid-type combiner, (**b**) two-section hybrid-type combiner, and (**c**) two-section optimized hybrid combiner with drain to source capacitance compensation

where A and B depend on the characteristic impedances of the quarter-wave transformers used in the combiner. The Eq. (4.28) also assumes that a quadrature phase relation between the main and auxiliary PA currents, that is, $I_{\mathrm{A}} = -jI_{\mathrm{M}}$, has been already set over the frequency band. Considering $Z_1 = Z_3$ and $Z_2 = Z_4$, for the topology described in Fig. 4.40a, the A and B in Eq. (4.28) are given by [52],

$$A = \frac{Z_1^2}{R_{\mathrm{L}}}\left(\frac{Z_2^2}{Z_1^2 - Z_2^2}\right)^2 \tag{4.29}$$

$$B = \frac{Z_1^2}{Z_2}\left(\frac{Z_2^2}{Z_1^2 - Z_2^2}\right) \tag{4.30}$$

Similarly, considering $Z_1 = Z_3$, the A and B in Eq. (4.28) for the topology described in Fig. 4.40b are given by [51],

$$A = \frac{Z_2^4 Z_4^2}{R_L\left(Z_1^2 - 2Z_2Z_4\right)^2} \tag{4.31}$$

$$B = Z_2\left(\frac{Z_1^2 - Z_2Z_4}{Z_1^2 - 2Z_2Z_4}\right) \tag{4.32}$$

A DPA prototype based on the topology of Fig. 4.40a is designed to operate over the frequency range from 1.95 to 2.25 GHz using 8 W CGH60008D GaN die device from CREE, corresponding to 14% of fractional bandwidth [52]. Similarly, the topology in Fig. 4.40b is used to design DPA operating over frequency range with 1.7–2.15 GHz using 10 W GaN package device from CREE. This corresponds to fractional bandwidth of 23.4%.

Figure 4.41a shows the photograph of the DPA circuit developed using the topology of Fig. 4.40b.

The circuit is fabricated over Rogers 4350B substrate with dielectric constant of 3.66 and height 20 mil. Figure 4.41b shows the measured performance of this DPA. One can see that over the frequency range from 1.7 to 2.15 GHz, the DE is between 40% and 52% at 6 dB back-off with power around 37.5–38.2 dBm. The

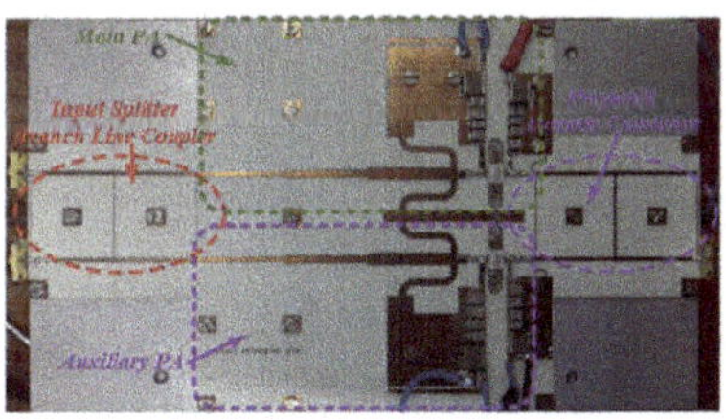

(a)

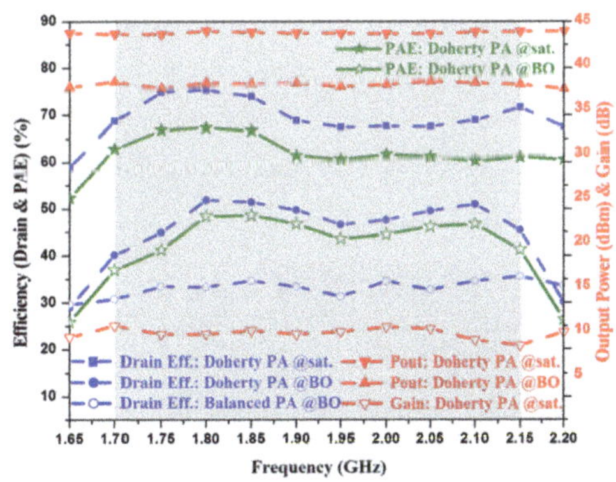

(b)

Fig. 4.41 (**a**) Photograph of designed prototype and (**b**) measured performance over the frequency range ([51], reprinted with permission from IEEE)

DE at saturation over this band of operation is 67–75% delivering output power around 43.6–43.9 dBm. Later, this topology of Fig. 4.40b is modified to obtain a better bandwidth [53]. In this case, the quarter-wave transformer appearing in series and shunt have different characteristic impedances as shown in Fig. 4.40c. Apparently, the first series quarter-wave transformer of characteristic impedances Z_1 in Fig. 4.40b is realized using pi-type CLC structure in Fig. 4.40c to compensate for the drain to source capacitance C_{ds} of the device. Moreover, the impedances of the two corner shunt branches are different in Fig. 4.40c unlike the topology described in Fig. 4.40b. The DC feed shown in Fig. 4.40c is high impedance quarter-wave which imitates the inductor (RF choke) and theoretically does not have impact on the topology. This topology is used in DPA design as shown in Fig. 4.42a.

The circuit is fabricated on RT-duroid (RT-5880 from Rogers) substrate with dielectric constant of 2.2 and thickness of 15 mil. Figure 4.42b shows the measured results in comparison to [52]. One can see that the output power varies from 39.5 to 41.5 dBm over frequency range of 1.7–2.4 GHz at 1.5 dB gain compression with input drive of 33 dBm. Figure 4.42b shows that the DE varies between 55% and 75% over the frequency range of 1.7–2.4 GHz at this gain compression. At 6-dB back-off, that is, 27 dBm input power drive, the output power varies from 34 dBm to around 36.3 dBm over the frequency range of 1.7–2.4 GHz. The efficiency at this input drive varies between 40% and 55% over the frequency range of 1.7–2.4 GHz.

Later, the topologies in Fig. 4.40a are modified where the characteristic impedances of the output combiner have been optimized to maintain the output loads seen by the active devices within the 1 dB constant power contour in a frequency range as large as possible around 2 GHz [54]. In such case, the impedances of each branch are related using the following expressions:

$$Z_1 = \frac{Z_2 k R_L}{\frac{1}{\beta} + \left(k^2 Z_2 R\right)} \tag{4.33}$$

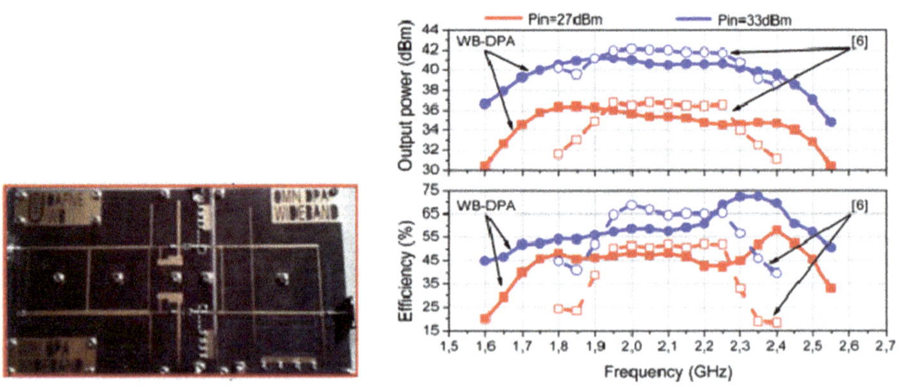

(a) (b)

Fig. 4.42 (**a**) Photograph of designed prototype and (**b**) measured performance over the frequency range ([53], reprinted with permission from IEEE)

$$Z_3 = \frac{Z_4}{\beta k R_L} \tag{4.34}$$

where R_L is the load to the DPA load combiner, β is the load modulation factor relating the back-off and k is given by,

$$k = \frac{1}{\beta\,(V_{DD} - V_K)}\sqrt{\frac{2P_{pk}}{R_L}} \tag{4.35}$$

where P_{pk} is the DPA output power; V_{DD} and V_K are the respective drain bias voltage and the knee voltage of the device used in designing the main PA. The CGH60008D GaN HEMT die device by CREE Inc. is selected to design both main and peaking amplifier [54]. The optimum resistance R_{opt} of this transistor is assumed as 45.6 Ω. In order to achieve efficiency enhancement at 6 dB back-off, β is set to 2. The drain bias is set as $V_{DD} = 28$ V and the knee voltage is 5 V. If the peak power is assumed to be 11.48 W and $R_L = 50\ \Omega$ one can find $k = 0.0147$ from Eq. (4.35). The combiner should present $R_{opt} = 45.6\ \Omega$ and $2R_{opt} = 91.2\ \Omega$ to the main PA at saturation and back-off, respectively. While performing optimization [54], the higher value of Z_2 results in larger bandwidth; therefore, this line was removed considering Z_2 tends to infinity for best bandwidth solution [54]. In such a case Eq. (4.33) results in $Z_1 = 1/k$. This corresponds to $Z_1 = 67.7\ \Omega$. If Z_4 is selected as 121 Ω, Eq. (4.34) can be used to obtain $Z_3 = 82\ \Omega$. Figure 4.43a shows the frequency behavior of the impedance presented to the main PA at the break point with the above calculated values of Z_1, Z_2, and Z_3 (red line with square symbols), with the 1 dB constant power contour. One can see that the impedance seen by the main PA, that is, Z_M, is purely resistive and equal to R_{opt} at f_0, whereas the Z_M crosses the 1 dB constant power contour at ±500 MHz at around $f_0 = 1.9$ GHz. This corresponds to 52.6% fractional bandwidth. However, one can achieve wider bandwidth if a smaller impedance value is allowed to be selected (touching the left edge of the 1 dB constant power contour as depicted by blue line with circular symbol in Fig. 4.43a). In such case Z_1 is calculated as,

$$Z_1 = \sqrt{R_L \frac{2R_{opt}}{10^{0.1}}} \tag{4.36}$$

where $2R_{opt}/10^{0.1}$ is the resistance value obtained by Z_M (which is lying at the left edge of the 1 dB constant power contour) at back-off. The expected bandwidth is ±700 MHz as shown in Fig. 4.43a. From Eq. (4.36), Z_1 is computed as 60.18 Ω. This corresponds to $k = 1/Z_1 = 0.0167$ (if Z_2 tends to infinity). If Z_4 is selected as 121 Ω, one can obtain $Z_3 = 73\ \Omega$ from Eq. (4.34). Figure 4.43b shows the actual topology used in designing broadband load combiner in [54].

The quarter-wave transformer of characteristic impedance Z_1 and Z_3 is realized using C–L–C pi-type network. This will also integrate the devices' drain-to-source parasitic capacitance ($C_{ds} = 0.75$ pF) as described in [53]. Figure 4.44a shows the photograph of the circuit. The auxiliary PA is referred here as peaking PA. The

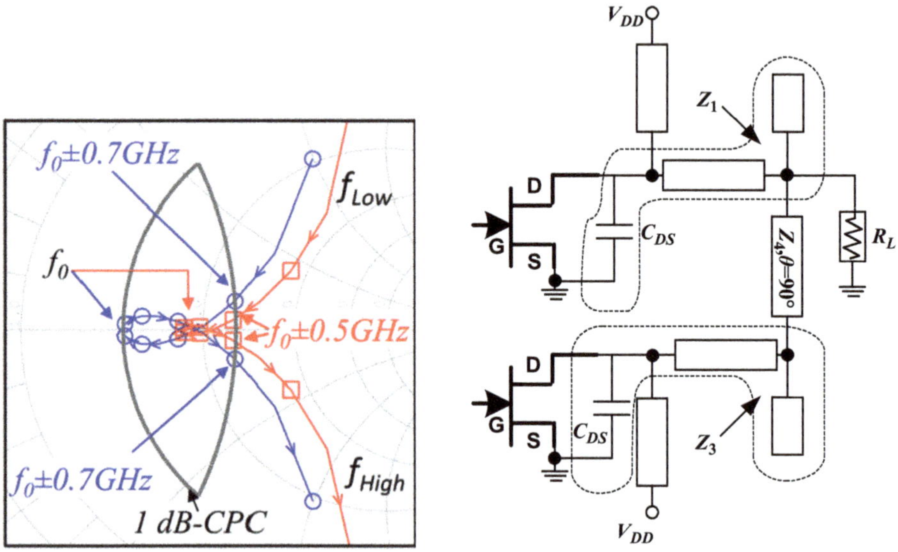

(a) (b)

Fig. 4.43 (**a**) Circuit topology of broadband load combiner, (**b**) load behavior presented to the main PA with 1 dB constant power contour ([54], reprinted with permission from IEEE)

circuit is fabricated on substrate with dielectric constant of 2.17 and height of 30 mil. Figure 4.44b shows the corresponding measured performance over the frequency range obtained using CW signal as the input drive. One important point in the topology described in Fig. 4.43b is that the main PA needs 90° lag with respect to the auxiliary PA for correct load modulation unlike conventional DPA as discussed in Sect. 1.4.1. Therefore in Fig. 4.44a, RF input has been fed through the Lange coupler branch which is in front of the auxiliary PA, in order to obtain the desired 90° delay for the main amplifier.

4.4.2 Broadband Multistage Doherty Power Amplifier with High Dynamic Range

The bandwidth enhancement scheme in multistage DPA is similar to the approach described in [42, 43]. In such an approach, the DPA design starts from specifying optimum load combiner parameters for best bandwidth, and later the corresponding current profiles are derived and realized by choosing appropriate devices' size and bias. This approach has been discussed in the beginning of Sect. 1.4.1, where it is employed in two-stage DPA to extend its bandwidth. Later, a similar approach is used in extending bandwidth of two common topologies of multistage DPA as describe in Sect. 1.4.2 of Chap. 1.

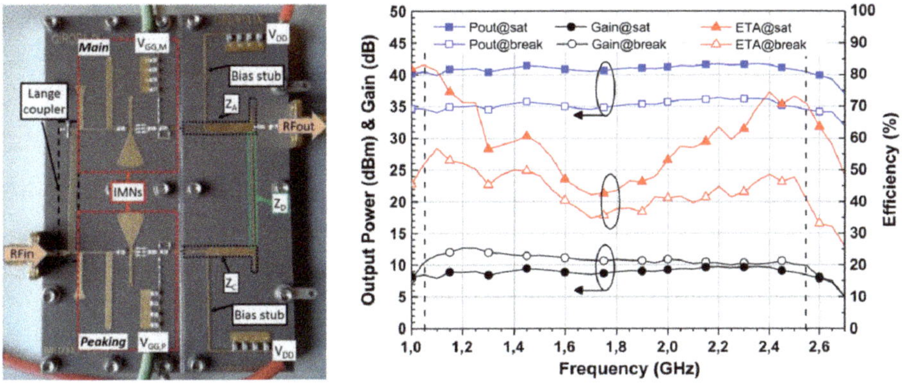

(a) (b)

Fig. 4.44 Prototype designed based on: (**a**) photograph and (**b**) measured performance over the frequency range ([54], reprinted with permission from IEEE)

Bandwidth Enhancement in Multistage DPA Topology-1

The first step for bandwidth enhancement in topology-1 of multistage DPA has already been performed while establishing the relation expressed in Eq. (1.196) in Chap. 1. When both the auxiliary PAs are off, the overall architecture of multistage DPA topology-1 is shown in Fig. 1.38a. The expression of Eq. (1.196) is obtained from a notion similar to Sect. 4.4.1 and [42], where the input impedance of the transmission line is same as its characteristic impedance irrespective of the operating frequency provided that the line is also terminated by its characteristic impedance. Therefore, exercising this idea in Fig. 1.38, one must follow Eq. (1.196) for initial step of relating Z_{01}, Z_{02}, and R_L. In a conventional three-stage DPA of topology-1 as shown in Fig. 1.38a, it is assumed that all the three PAs shall see equal voltage at saturation as shown in Fig. 1.39b in order to obtain high efficiency at two-back-offs and saturation. However, one can relax this condition and obtain optimum load combiner parameters along with appropriate sizes of the devices which can provide broadband operation in the three-stage DPA.

Using Eq. (1.196) in Eqs. (1.197) and (1.198), one can obtain

$$\alpha_1 = \frac{\beta_2 (\beta_1 - 1)}{\beta_1} \tag{4.37}$$

$$\alpha_2 = (\beta_2 - 1) \frac{Z_{01}}{R_L} \tag{4.38}$$

The impedance $Z_{M,sat}$ seen by main PA at saturation (i.e., $v_{in} = 1$) is determined using Eq. (1.193), where the currents of auxiliary PAs are substituted as $I_{A1} = \alpha_1 I_M$, $I_{A2} = \alpha_2 I_M$. Equation (1.193) is further solved using Eq. (1.196–1.198) to obtain following relation:

$$Z_{\mathrm{M,sat}} = (\beta_2 - \alpha_1)\, Z_{02} \tag{4.39}$$

Using above equations, the load combiner parameters given in Eqs. (1.205)–(1.207) can be rewritten as:

$$Z_{01} = \left(\frac{\beta_2 - 1}{\alpha_2}\right)\left(\frac{Z_{\mathrm{M,sat}}}{\beta_2 - \alpha_1}\right) \tag{4.40}$$

$$Z_{02} = \frac{Z_{\mathrm{M,sat}}}{\beta_2 - \alpha_1} \tag{4.41}$$

$$R_{\mathrm{L}} = \left(\frac{\beta_2 - 1}{\alpha_2}\right)^2\left(\frac{Z_{\mathrm{M,sat}}}{\beta_2 - \alpha_1}\right) \tag{4.42}$$

For efficiency enhancement at 6 and 12 dB back-off, the values of β_1 and β_2 are obtained as 4 and 2, respectively. Using these values in Eqs. (4.37) and (4.38), one can obtain

$$\alpha_1 = 1.5 \tag{4.43}$$

$$\alpha_2 = \frac{Z_{01}}{R_{\mathrm{L}}} = \frac{Z_{02}}{Z_{01}} \tag{4.44}$$

Equations (4.43) and (4.44) show that the value of α_1 is fixed but the value of α_2 depends on the load combiner values. Therefore, this current ratio α_2 can be optimized around the center frequency to obtain the best bandwidth solution while keeping the value of α_1 fixed. If the optimum output impedance of the main PA at saturation, that is, $Z_{\mathrm{M,sat}}$, is set as $R_{\mathrm{opt,M}}$ as obtained from the load-pull, then Z_{02} can be obtained from Eq. (4.41), when the value of $\alpha_1 = 1.5$. Therefore, the value of Z_{01} and R_{L} as expressed in Eqs. (4.40) and (4.42) can be obtained for wider bandwidth operation by choosing optimum value of α_2. The optimum value of parameter α_2 selected for best bandwidth will be further used to calculate optimum load combiner parameters Z_{01}, Z_{02}, and R_{L} for broadband three-stage DPA implementation. This bandwidth analysis is done in terms of calculating drain efficiencies over the frequency at saturation as well as at back-offs for different values of parameter α_2 which correspond to different sets of load combiner parameters. An ADS simulation test bench is used for this analysis, where the three voltage-dependent current sources are used to depict the behavior of three PAs with current profile as described in Fig. 1.39a. The main current source is initially ON, following which first auxiliary current source turns ON at 12 dB back-off ($v_{\mathrm{in}} = 0.25$) and later the second auxiliary current source turns ON at 6 dB back-off ($v_{\mathrm{in}} = 0.5$). The load combiner values were obtained using Eqs. (4.40)–(4.42) keeping α_1, β_1, and β_2 fixed as 1.5, 4, and 2, respectively. The optimum impedance of the main PA, $R_{\mathrm{opt,M}}$ is obtained as 20 Ω from load-pull simulation. By setting the value of $Z_{\mathrm{M,sat}} = R_{\mathrm{opt,M}} = 20$ Ω, Z_{02} is

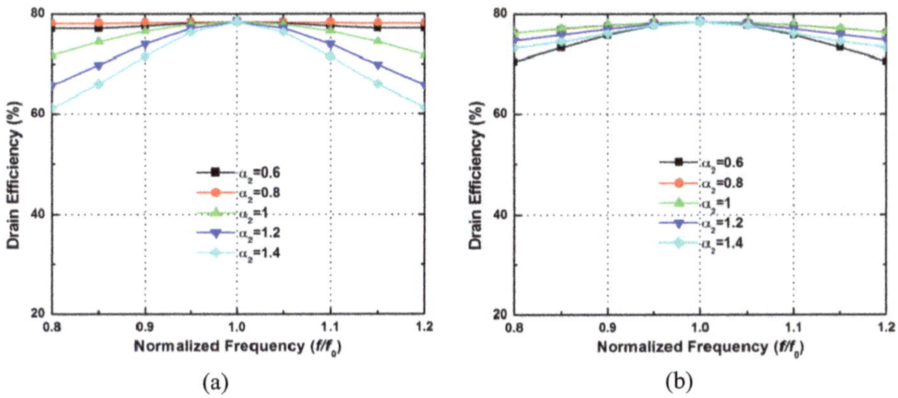

Fig. 4.45 Drain efficiency curves with respect to frequency for different values of α_2: (**a**) at 12 dB back-off and (**b**) at 6 dB back-off ([56], Reprinted with permission from IEEE)

calculated as 40 Ω from Eq. (4.41). Therefore, the other load combiner parameter Z_{01} and R_L are expressed in terms of α_2 as Z_{02}/α_2 and $Z_{02}/\alpha_2{}^2$ using Eqs. (4.40 and 4.42), respectively. These values are used in the simulation bench of ADS and the DE variation is observed about the center frequency of 750 MHz. Figure 4.45a, b shows the DE over the frequency at 12 and 6 dB back-offs, respectively. Considering the bandwidth as the frequency range in which DE falls within 10% of its maximum value, one can observe that $\alpha_2 = 1$ provides the considerable choice for wideband efficiency enhancement at 12 and 6 dB back-off. Figure 4.45 shows DE variation over the frequency at saturation and $\alpha_2 = 1$ also provides the best solution in terms of high efficiency over the band of frequency at saturation. From Figure 4.45a, b and 4.46 one can observe that the choice of $\alpha_2 = 1$ can provide 40% fractional bandwidth over which DE is better than 70%. Therefore, the optimum bandwidth performance at all the back-offs and saturation is obtained for $\alpha_1 = 1.5$ and $\alpha_2 = 1$.

Figure 4.47a shows the current profile of the three-stage DPA with these selected values of current ratio. Figure 4.47b shows the corresponding load variation with respect to normalized input voltage. One can see from this figure that at saturation, the main PA and the first and second auxiliary PA see a load of 20, 26.67, and 80 Ω, respectively. The corresponding optimum values of load combiner parameters are $Z_{01} = Z_{02} = R_L = 40$ Ω.

The bandwidth performance of the selected load-combiner is also analyzed in terms of variation of the load seen by each PA with frequency. Figure 4.48 shows variation of load seen by the main PA at back-offs as well as at saturation. This figure also shows the 1 dB power contour plotted for load variation at saturation as well as at back-offs [56]. One can see from Fig. 4.48 that the load of the main PA lies within 1 dB power contour at 12 dB back-off for all frequencies. However, at saturation, the load seen by main PA remains within the 1 dB power contour for the frequency range of 525–975 MHz corresponding to fractional bandwidth of 60%. Similarly,

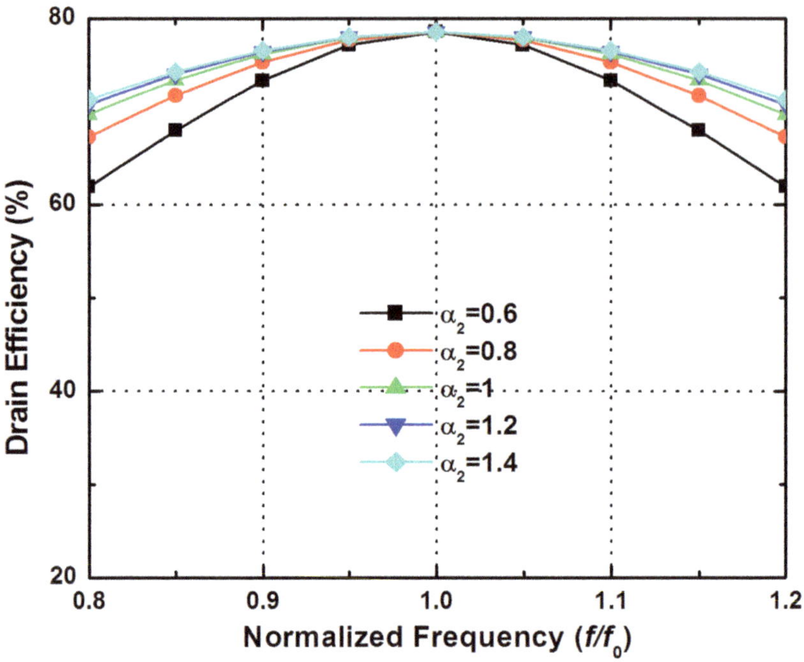

Fig. 4.46 Drain efficiency curves with respect to frequency at saturation for different values of α_2 ([56], reprinted with permission from IEEE)

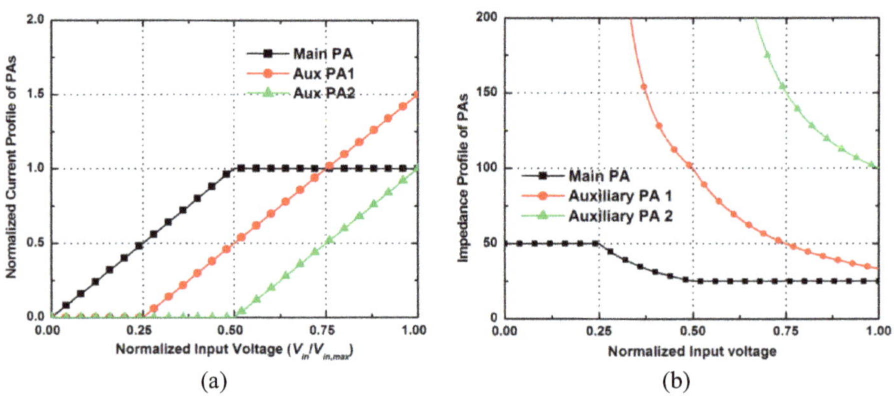

Fig. 4.47 Broadband three-stage DPA (topology-1) with respect to normalized input voltage for $\alpha_1 = 1.5$ and $\alpha_2 = 1$: (**a**) normalized current profile and (**b**) load seen by each PA ([56], reprinted with permission from IEEE)

at 6 dB back-off, the load variation is within 1 dB power contour for the frequency range of 625–875 MHz which corresponds to 33.3% fractional bandwidth.

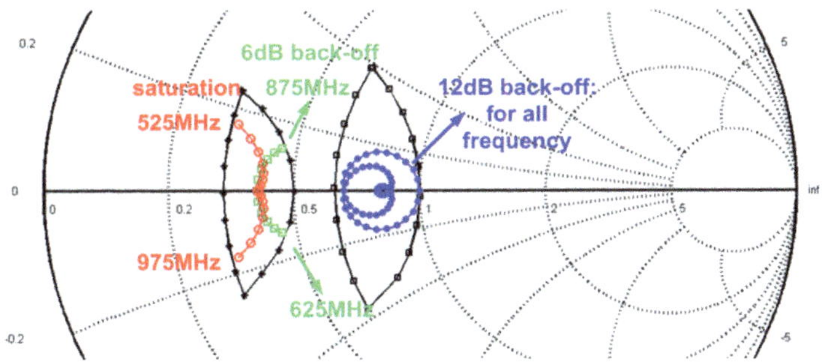

Fig. 4.48 Variation of impedance seen by main PA at 12 dB back-off, 6 dB back-off, and at saturation with 1 dB power contours ([56], reprinted with permission from IEEE)

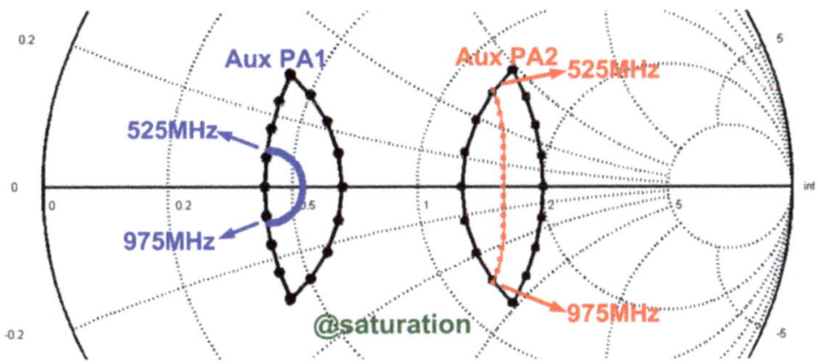

Fig. 4.49 Variation of impedance seen by first and second auxiliary PAs at saturation with 1 dB power contours ([56], reprinted with permission from IEEE)

Figure 4.49 shows load variation seen by the auxiliary PAs at saturation over the frequency range. This figure also shows the 1 dB power contours for load variation seen by each auxiliary PA at saturation. One can see from Fig. 4.49 that the variation of load seen by first and second auxiliary PA is within the 1 dB power contour for frequency range of 525–975 MHz which corresponds to 60% fractional bandwidth.

The device size ratio can be estimated by knowing output current and impedance seen by each PA with respect to the main PA at saturation as shown in Fig. 4.47. The first auxiliary PA should have a device periphery three times as that of the main PA, whereas the second auxiliary PA should have device periphery equal to four times of the main PA.

This bandwidth enhancement scheme is validated in experiment by developing a prototype of broadband three-stage DPA as shown in Fig. 4.50. This prototype is tested with both CW sinusoidal and modulated signals. Figure 4.51a, b shows the measured output power, gain, and DE for different sets of operating frequencies with CW signals as input. At 12 dB back-off, the DE is between 50% and 61.8%. The gain at saturation is more than 7 dB over the band.

Figure 4.52 shows the output power, DE, PAE at back-off as well as saturation over the frequency range of 600–900 MHz. The figure also shows gain at saturation.

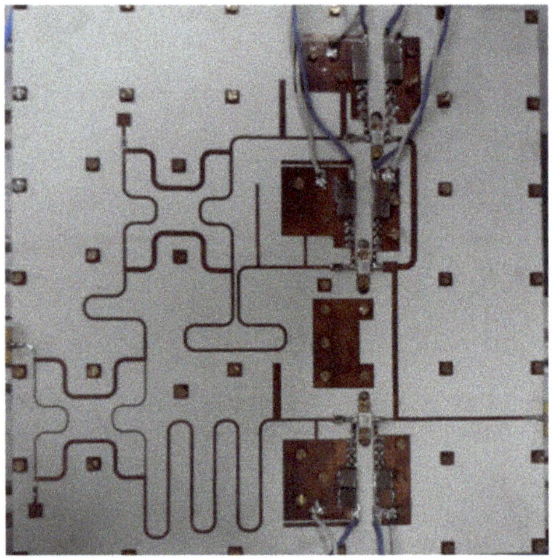

Fig. 4.50 Photograph of the fabricated DPA ([56], reprinted with permission from IEEE)

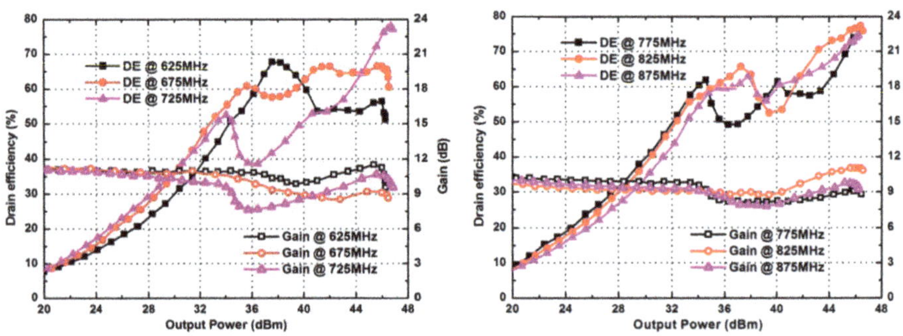

(a) (b)

Fig. 4.51 Measured DE and gain curves at (**a**) lower range of frequencies and at (**b**) upper range of frequencies with respect to output power ([56], reprinted with permission from IEEE)

One can see that at 6-dB back-off, the DE is between 51.9% and 66.2% over the entire range of frequency. Whereas at 12 dB back-off, the DE is better than 50% over the entire frequency range. The corresponding DE at saturation is between 51.1% and 78% over the band.

At saturation, the corresponding gain is between 7 and 11 dB and PAE is between 45.2% and 70.8%, over the frequency band. Similarly, over the frequency range of 600–900 MHz, the output power variation at saturation is within ±0.5 dB corresponding to fractional bandwidth of 40%.

The prototype is also tested with long-term evolution (LTE) signal of 10 MHz bandwidth having a PAPR of 11.75 dB at 700, 750, and 800 MHz. The ACPR is measured at an offset of 10 MHz. Figure 4.53 shows that the average efficiency of LTE at 700 MHz is 57.93% at an average power of 34.42 dBm with ACPR below −23.96 dBc.

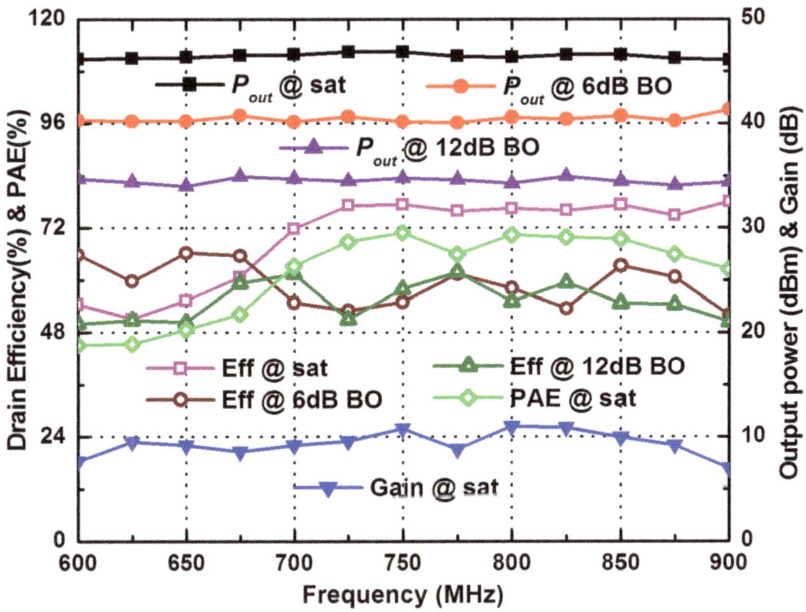

Fig. 4.52 Measured performance parameters across the frequency band ([56], Reprinted with permission from IEEE)

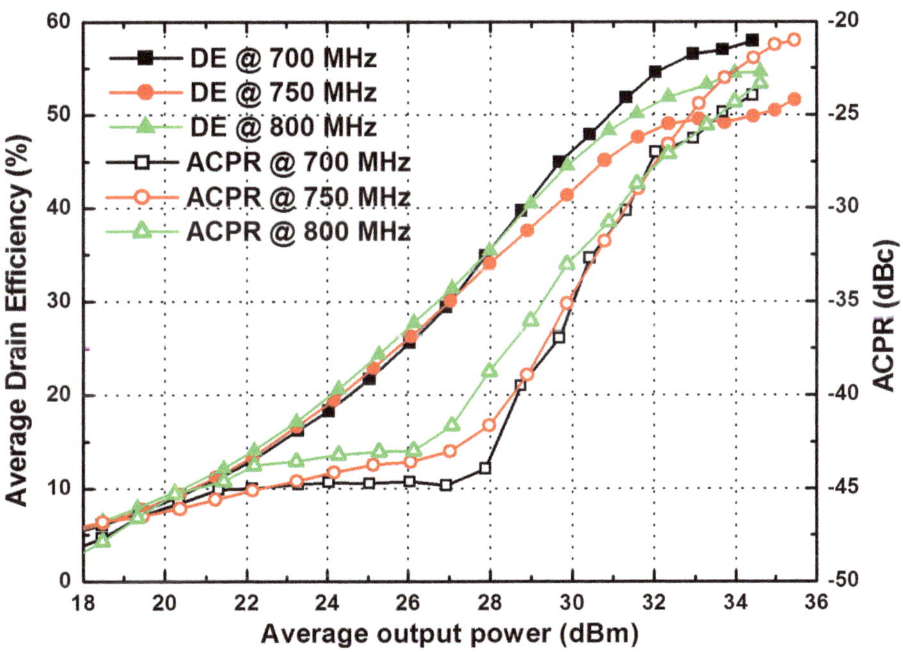

Fig. 4.53 Measured DE and ACPR for a 10 MHz long-term evolution (LTE) signal with a PAPR of 11.75 dB ([56], reprinted with permission from IEEE)

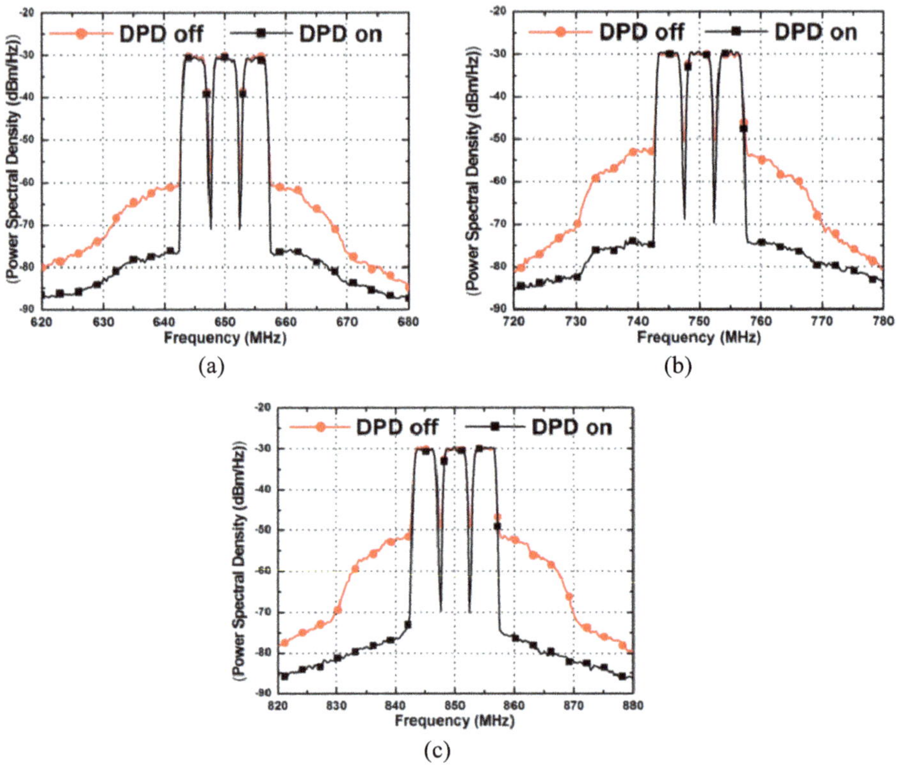

Fig. 4.54 Output signal spectrum of three carrier WCDMA signals before and after digital predistortion (DPD) at (**a**) 650 MHz, (**b**) 750 MHz, and (**c**) 850 MHz ([56], reprinted with permission from IEEE)

Table 4.3 Summary of DPD linearization

Frequency (MHz)	Average power (dBm)	Average DE (%)	ACLR at 15 MHz	
			DPD OFF	DPD ON
650	34.23	52.3	−34.5/−35.8	−47.7/−48.3
750	35.52	51.7	−27.4/−28.6	−46.3/−46.8
850	35.7	52.2	−26.9/−27.1	−49/−50.2

At a frequency of 750 MHz and at an average power of 35.45 dBm,the average efficiency is reported as 51.6% and theACPR is less than −21 dBc. An average efficiency of 54.5% is observed at an average power of 34.61 dBm where the ACPR is around −23.3 dBc at 800 MHz. Figure 4.54a–c shows the output spectra for three carrier 15 MHz WCDMA signals with a PAPR of 10.6 dB at 650, 750, and 850 MHz before and after applying digital predistortion (DPD).

Table 4.3 shows the summary of DPD linearization with average DE and ACPR after linearization. The average output power and average DE are almost same for DPD ON and DPD OFF case. It is worth mentioning that the DPA topology suffers from linearity issue due to the constant current profile of the main PA after second back-off. Therefore, to obtain best linearization, the active drive profiling may be used with this topology [57, 58].

Nevertheless, the DPD-based linearization as shown in Fig. 4.54a–c and summarized in Table 4.3, shows that the output signal qualifies the spectral mask. The corresponding ACPR values also qualify spectral mask specifications. Therefore, the presented broadband three-stage DPA is linearizable.

Bandwidth Enhancement in Multistage DPA Topology-2

In the first chapter, the load combiner impedances of conventional three-stage DPA are derived considering the condition of current ratios as $\alpha_1 = \alpha_2 = 1$ along with the condition that the voltage of all PAs in DPA at saturation is equal. The characteristic impedances of the quarter-wave transformers used in the load combiner are obtained in Eqs. (1.219–1.221). These values of characteristic impedances of load combiner provide a unique solution which improves efficiency at 6 dB and 9.54 dB back-offs, respectively. These load combiner impedances do not provide a broadband solution and the efficiency is enhanced for limited bandwidth around center frequency only. The bandwidth of DPA not only depends on load combiner impedance values but also on current ratios [42, 59]. Therefore, a relationship must be developed between load combiner impedances (Z_{01}, Z_{02}, and Z_{03}) and current ratios (α_1 and α_2). The relationship obtained can be examined for different sets of load combiner impedances which may provide an optimum solution in terms of bandwidth. For such analysis, it is important to relate the load combiner characteristic impedances and current ratios. The load impedance of main PA, Z_M given by Eq. (1.211) varies from $\beta_1 R_{opt}$ at first back-off to R_{opt} at saturation in order to obtain high efficiency at back-off levels. Therefore, at first back-off, Z_M can be expressed as,

$$Z_{M,BO} = \frac{Z_{01}^2}{R_L} = \beta_1 R_{opt} \tag{4.45}$$

Using Eq. (4.45) and the corresponding value of Z_M at saturation as R_{opt}, one can write [59],

$$\alpha_1 = (\beta_1 - 1)\left(\frac{Z_{03}}{Z_{01}}\right)\left(\frac{R_{opt}}{Z_{02}}\right) \tag{4.46}$$

where α_1 is the current ratio of first auxiliary PA with respect to main PA as defined in Eq. (1.192).

Similarly, α_2 can be obtained by equating the voltages seen by the first auxiliary PA at saturation as well as second back-off.

$$\alpha_2 = \left(\frac{\beta_2 - 1}{\beta_2}\right)\left(\frac{Z_{01}}{Z_{03}}\right) \tag{4.47}$$

Equations (4.46) and (4.47) establish the relation between the characteristic impedances of the quarter-wave transformers of the load combiner and the current ratio α_1 and α_2 of topology-2 as described in Fig. 1.38b. Moreover, according to Eqs. (4.46 and 4.47), the current ratio α_1 and α_2 also depend on the back-off level β_1 and β_2. Therefore, unlike the conventional three-stage DPA topology-2 as described in Chap. 1, the current ratios α_1 and α_2 for the desired back-off levels can be determined along with the choice of load combiner parameters Z_{01}, Z_{02}, and Z_{03}. This may provide the designer an opportunity to optimize them for enhanced bandwidth.

In the case where the efficiency enhancement is required at 9.54 and 6 dB back-off, the respective back-off parameters are set as $\beta_1 = 3$ and $\beta_2 = 2$ and the following ratios k and m can be defined as

$$k = \frac{Z_{01}}{Z_{03}} \tag{4.48}$$

$$m = \frac{Z_{02}}{R_{\text{opt}}} \tag{4.49}$$

Using these intermediate variables k and m in Eqs. (4.46 and 4.47), the respective current ratios α_1 and α_2 can further be written as,

$$\alpha_1 = \left(\frac{2}{k}\right)\left(\frac{1}{m}\right) \tag{4.50}$$

$$\alpha_2 = \frac{k}{2} \tag{4.51}$$

The current ratios as stated in Eqs. (4.50 and 4.51) show its dependency on m and k. Further, the m and k depends on load combiner characteristic impedances such that the three-stage DPA enhances the average efficiency at the back-off levels of 9.54 and 6 dB, respectively. Figure 4.55 shows the dependence of α_1 on α_2, with different values of m.

The region where α_1 and α_2 are greater than 1, asymmetric devices will be used [60]. However, the region where α_1 and α_2 are less than or equal to 1 (shaded region in Fig. 4.55), the symmetric devices can be used. It is worth mentioning that choosing the value of α_1 and α_2 less than 1 corresponds to the auxiliary devices of same sizes as that of the main device, except the currents of the auxiliary PAs are less than the current of the main PA at saturation. This corresponds to the higher impedance seen by the auxiliary PA at saturation, and therefore a high voltage will be developed across the device terminal. This indeed requires high drain bias for the auxiliary PA which has a current that is less than the maximum value of main PA current at saturation [42, 43, 56, 59]. Moreover, the quadrature phase relationships as shown in Fig. 1.38b are kept same by choosing an analog input splitter topology with sufficient bandwidth. One can also see from Fig. 4.55 that, in order to choose the value of α_1 and α_2 less than or equal to 1, a range of k is available such that,

$$\frac{2}{m} \leq k \leq 2 \tag{4.52}$$

This range of m and k in Fig. 4.55 provides the designer a choice of different sets for the values of Z_{01}, Z_{02}, and Z_{03}, which can be analyzed for obtaining best bandwidth solution. Therefore, in order to obtain the optimized values of these characteristic impedances for enhancing bandwidth, one must select the values of m and k judicially. Once the best choice is made in terms of m and k which corresponds to the best choice in terms of characteristic impedances Z_{01}, Z_{02}, and Z_{03}, the corresponding current ratios can be obtained from Fig. 4.55.

In order to analyze the bandwidth of the load combiner, its design parameters, Z_{01}, Z_{02}, Z_{03}, and R_L should be selected in relation to m and k. In order to analyze the various values of m and k for bandwidth enhancement, a simulation bench is developed in Keysight's ADS [59]. The ideal transmission lines of $\theta - 90^\circ$ are used for realizing branches of the load combiner of Fig. 1.38b at 825 MHz, which is the center frequency of operation. The main and auxiliary PAs are modeled as ideal equation-based current sources with current profile as defined mathematically in Eqs. (1.214)–(1.216) of Chap. 1. The same simulation setup is also used to plot

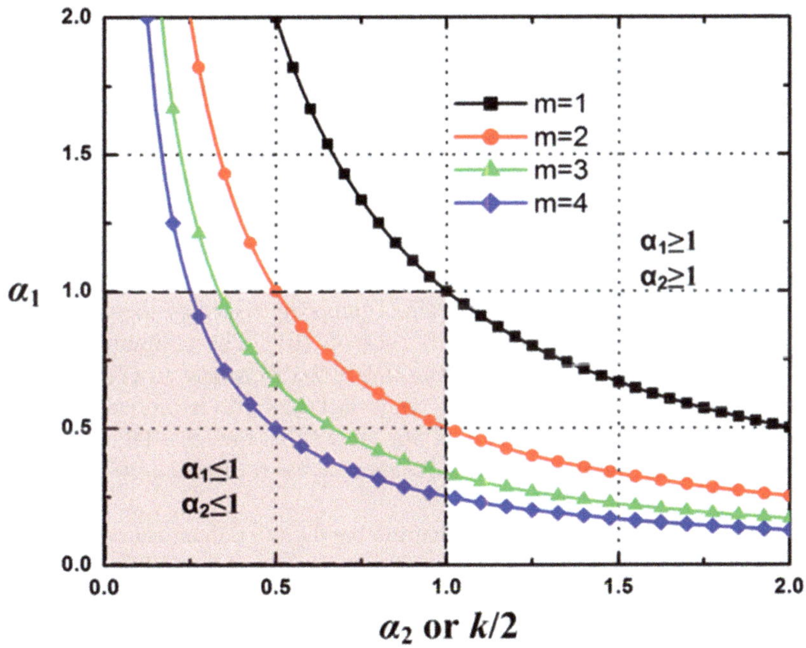

Fig. 4.55 Current ratios and their dependence on load combiner parameters m and k ([59], reprinted with permission from IEEE)

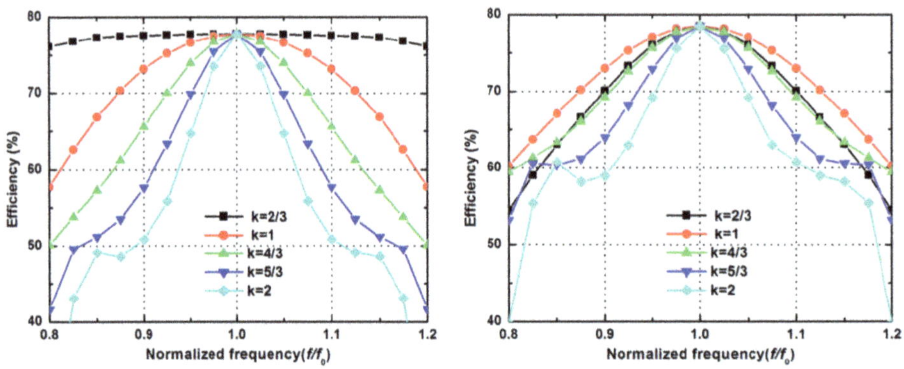

(a) (b)

Fig. 4.56 Efficiency variation with frequency (for $m = 3$ and for different value of k ranging from 2/3 to 2 for $R_L = 47\ \Omega$): (**a**) at 9.54 dB back-off and (**b**) at 6 dB back-off ([59], reprinted with permission from IEEE)

variation of ideal DE characteristics with frequency for various choices of m and k. These efficiency characteristics are plotted using ideal voltage and current waveform relations as described in [59]. In order to start the search for the optimum values of m and k, m is first set to a fixed value in order to see the impact of different values of k across the bandwidth. Figure 4.56a shows the ideal DE variation over the frequency band for $m = 3$ and a range of k at 9.54 dB back-off power. According to Eq. (4.49), the value of Z_{02} depends on m and R_{opt}, where R_{opt} is considered as fixed, based on the device. Therefore, for a given value of m, Z_{02} can be considered fixed for plotting Fig. 4.56a. From Eq. (4.45), one can observe that Z_{01} is dependent on R_L only, since β_1 and R_{opt} are fixed as 3 and 50 Ω which is based on first back-off level and device selection, respectively. Therefore, for a given value of m, different values of k are obtained which corresponds to different values of Z_{03}, since the value of Z_{01} is decided by the value opted for R_L. Moreover, an optimum value of R_L is also obtained such that the variation of impedance seen by the main amplifier over frequency is minimum at 9.54 dB back-off [61]. The optimum value of R_L is 47 Ω for $m = 3$ and k ranging from 2/3 to 2 (according to (Eq. 4.52)). The ideal DE variation over the frequency range in Fig. 4.56a is plotted only for this optimum value of R_L. Similarly, Fig. 4.56b shows that the ideal DE variation over the frequency band for $m = 3$ and a range of k for 6 dB back-off power for an optimum value of $R_L = 47\ \Omega$.

From Fig. 4.56a, b, one can infer that the bandwidth enhancement is maximum for the case of minimum possible value of k. According to Eq. (4.52), the minimum value of k depends on the value of m chosen for the design. Thus, depending on the value of m, the corresponding choice of minimum possible value of k provides the best bandwidth. Knowing this behavior, the next step is to decide the best value of m. Fig. 4.57 analyzes frequency variation of DE at 9.54 and 6 dB back-offs, respectively, for different values of m. Accordingly, the corresponding values of k

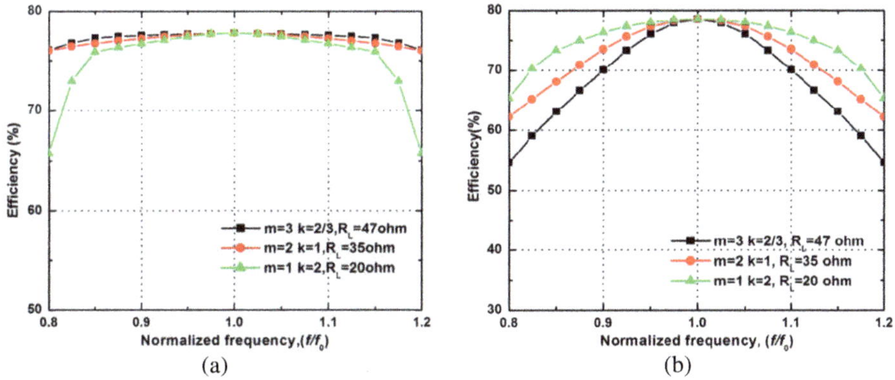

Fig. 4.57 Efficiency variation with frequency. Efficiency at 9.54 dB back-off (for different values of m for minimum value of k obtained within the range), (**a**) at 9.54 dB back-off and (**b**) at 6 dB back-off ([59], reprinted with permission from IEEE)

are chosen to be minimum as calculated from Eq. (4.52) for each value of m. These figures also show the effect of choosing different values of load impedance R_L for a given choice of m and k. It can be inferred from Fig. 4.57a that at 9.54 dB back-off, bandwidth is almost same for $m = 2$ and $m = 3$ for respective values of R_L at 35 and 47 Ω. However, from Fig. 4.57b one can observe that at 6 dB back-off, DE variation with frequency is less for $m = 2$ than when $m = 3$. The choice $m = 1$ corresponds to the maximum bandwidth for 6 dB back-off as shown in Fig. 4.57b.

However, this choice results in comparatively lesser bandwidth for 9.54 dB back-off as shown in Fig. 4.57a. Therefore, the final design of load combiner in [59] selects the parameter $m = 2$, $k = 1$, and the corresponding value of $R_L = 35$ Ω. Considering the bandwidth as the frequency range over which efficiency drops by 10%, one can observe that this choice results in a fractional bandwidth of more than 40% at 9.54 dB input back-off, as shown in Fig. 4.57a. However, at 6 dB back-off, this choice results in the fractional bandwidth of 30% as shown in Fig. 4.57b. One should note that the values of $\alpha_1 = \alpha_2 = 1$ correspond to a unique value of $m = 1$ and $k = 2$, which eventually does not correspond to the best bandwidth solution as verified from Fig. 4.57a, b. Thus, the resulting choice of the characteristic impedances given by Eqs. (1.219–1.221) is not a bandwidth optimized solution, and hence the above analysis is important to obtain the best set of load combiner impedances of the three-stage DPA (topology-2) in terms of bandwidth.

The load combiner parameters $m = 2$, $k = 1$ as finalized above correspond to $\alpha_1 = 1$ and $\alpha_2 = 0.5$ in Fig. 4.57a, b. Therefore, the current profile for the main and auxiliary PAs corresponding to this choice is shown in Fig. 4.58a.

Accordingly, the impedance seen by second auxiliary PA at saturation can be calculated using (Eq. 1.213) which is equal to $4R_{opt}$. This develops high voltage at saturation across the second auxiliary device. Figure 4.58b shows that the voltage developed across the second auxiliary PA at saturation is twice the voltage

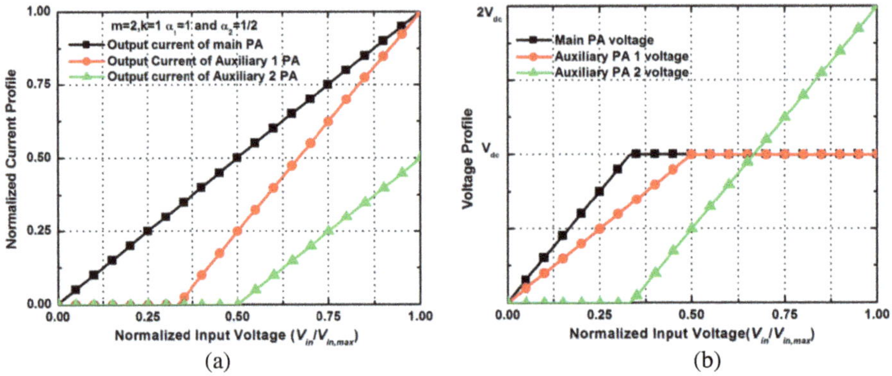

Fig. 4.58 Current and voltage of main and auxiliary PAs with input voltage drive when choice of the design parameters are $m = 2$, $k = 1$; (**a**) current profile and (**b**) voltage profile ([59], reprinted with permission from IEEE)

experienced by the main PA at saturation. Thus, the drain bias voltage of the second auxiliary PA needs to be twice than that of the main PA. Moreover, the second auxiliary PA is set to deep class C bias point which limits its output current due to the low gain characteristic. This limitation is indeed useful in the proposed scheme, since auxiliary PA is required to present less output current at saturation as compared to the main PA for achieving a current ratio of $\alpha_2 = 0.5$.

The effect of the proposed bandwidth enhancement on the output power is also analyzed in terms of 1 dB power contours as shown in Fig. 4.59 [59]. The Smith chart of Fig. 4.59 shows the variation of load impedance experienced by the main amplifier at 9.54 back-off, 6 dB back-off and at saturation over the frequency band.

These impedance variations are obtained with the load combiner designed for $m = 2$, $k = 1$, $R_L = 35\ \Omega$, and $R_{opt} = 40\ \Omega$. One can see from Fig. 4.59 that the impedance seen by the main PA at saturation (represented by R_{opt} in Fig. 4.59) is within the 1 dB power contour for the frequency range of 675–975 MHz. This corresponds to the fractional bandwidth of 36.36%. Similarly, the impedance seen by the main PA at 6 dB back-off, represented by $2R_{opt}$ in Fig. 4.59, falls within the 1 dB power contour across the frequency range of 650–1000 MHz corresponding to the fractional bandwidth of 42.42%. The impedance seen by the main PA at 9.54 dB back-off represented by $3R_{opt}$ in Fig. 4.59 lies within the 1 dB power contour over the frequency range of 625–1025 MHz corresponding to the fractional bandwidth of 48.48%. Thus, in terms of power variation, the overall bandwidth is restricted to 36.36% which is the fractional bandwidth of the load seen by the main PA at saturation, such that it remains within the 1 dB power contour. It is worth mentioning that theoretically there is no restriction over the proposed bandwidth enhancement scheme to be applied for low R_{opt} values in the case of high-power devices.

Figure 4.60 shows the photograph of the prototype of broadband three-stage DPA developed for the proof of concept. The circuit is implemented on Rogers RO4350B substrate. This substrate has thickness of 20 mil with dielectric constant of 3.66

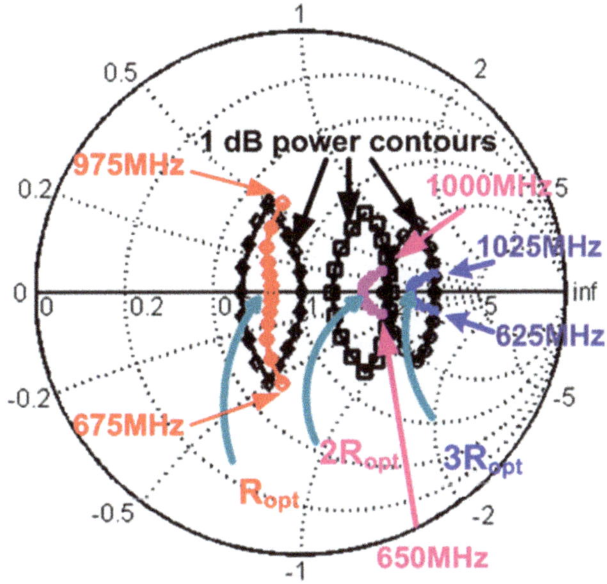

Fig. 4.59 Variation of impedance seen by the main amplifier at saturation, first back-off, and second back-off with 1 dB power contours ([59], reprinted with permission from IEEE)

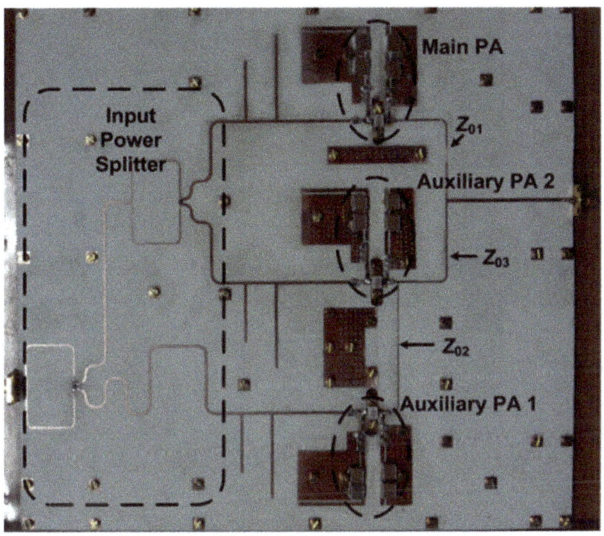

Fig. 4.60 Photograph of fabricated broadband three-stage DPA ([59], reprinted with permission from IEEE)

and a loss tangent of 0.0037. The circuit is tested with both modulated and CW sinusoidal excitation.

Figure 4.61a, b shows the CW measurements where DE is plotted with the output power drive at different frequencies of operation. From Fig. 4.61a, b, one can see that DE is more than 50.16% at saturation and up to 9.54 dB output power back-off (OPBO) for all frequencies of operation.

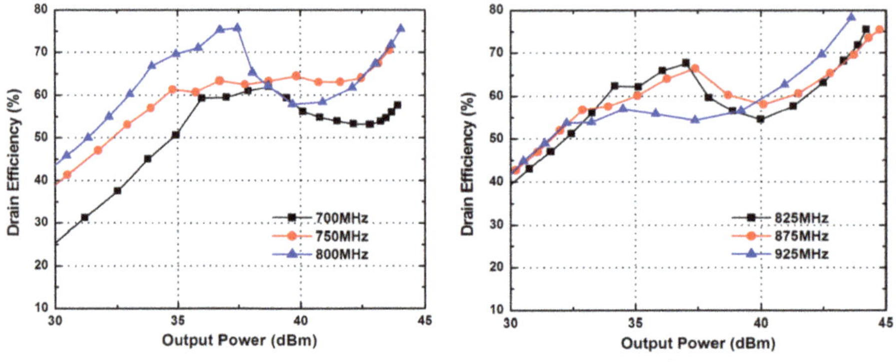

Fig. 4.61 Measured DE curves for (**a**) lower frequency range and (**b**) upper frequency range ([59], reprinted with permission from IEEE)

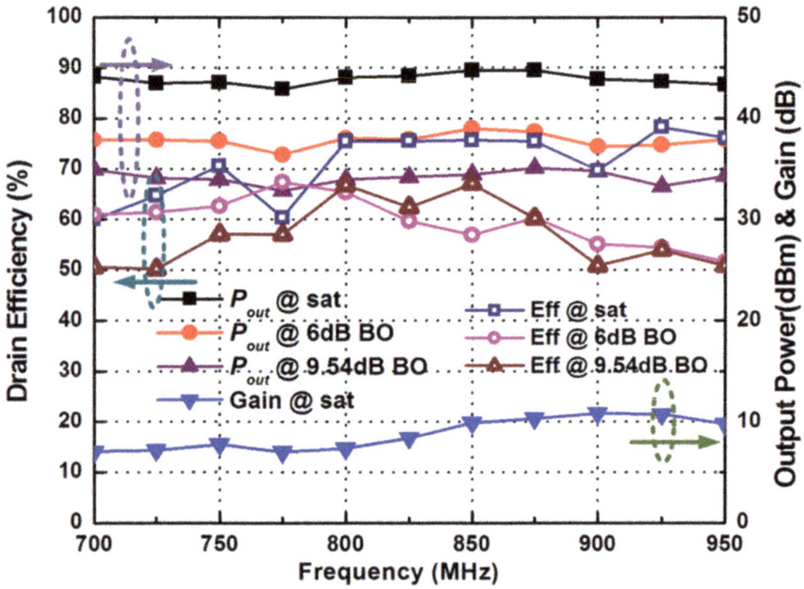

Fig. 4.62 Measured response of broadband three-stage DPA across the frequency range ([59], reprinted with permission from IEEE)

Figure 4.62 shows the measured output power, gain, and DE (at saturation, 6 dB and 9.54 OPBO) of the broadband three-stage DPA over the band of frequencies.

At 6 dB OPBO, the DE is more than 51.6% over the band of 250 MHz. This corresponds to fractional bandwidth of 30.3%. The corresponding DE is more than 50.16% over this band of operation. The output power variation is within ±1 dB limit at saturation as well as at both the OPBOs. At saturation, the minimum value of gain is around 7.1 dB over the frequency range of 700–950 MHz.

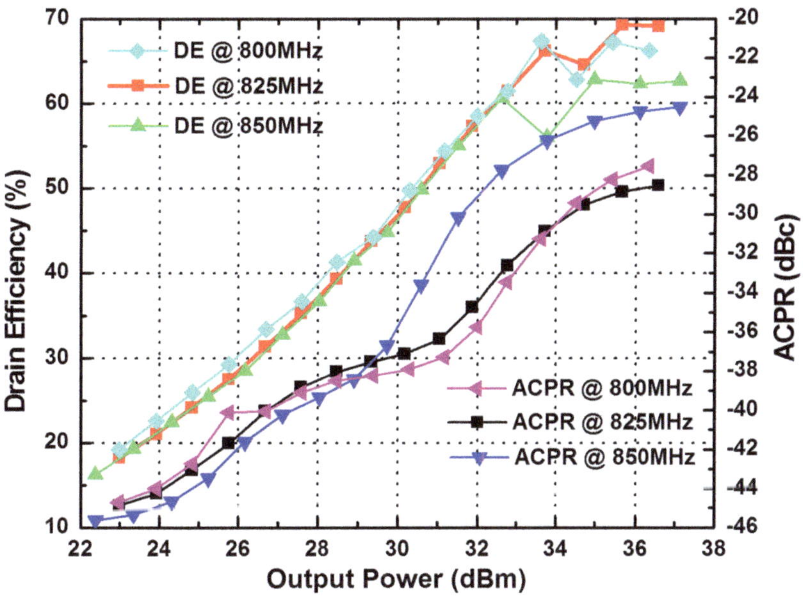

Fig. 4.63 Measured average DE and ACPR variation with average output power for a 5 MHz WCDMA signal as input, at different frequencies ([59], reprinted with permission from IEEE)

The prototype is also tested using modulated signals at three frequencies, 800, 825, and 850 MHz, respectively. A single carrier WCDMA signal of bandwidth 5 MHz and PAPR of 7.32 dB is used as modulated signal for these measurements. Figure 4.63 shows the ACPR measured at 5 MHz offset and the average DE. From Fig. 4.25, one can see that the average efficiency and ACPR are 61.59% and −28.53 dBc, respectively, which are measured at 825 MHz with an average output power of 36.08 dBm. Similarly, at 800 MHz, the average efficiency is measured as 59.93% at an average power of 34.92 dBm. The corresponding ACPR at 5 MHz offset measured at the same average output power is below −27.54 dBc. Figure 4.63 also shows the average efficiency and ACPR as 55.84% and −24.53 dBc, respectively, measured at an average output power of 36.61 dBm while operating the PA at 850 MHz.

Figure 4.64a, b shows the output spectrum of the broadband three-stage DPA at two different frequencies of operation with and without linearization. The PA is characterized using 5 MHz WCDMA signal with PAPR of 9.65 dB and linearized using DPD at 825 and 950 MHz.

Table 4.4 reports summary of the DPD performance. A memory polynomial with 12th-order nonlinearity and three memory taps have been used for modeling nonlinearity in this PA.

One can see the output of PA qualifies the spectrum mask only when DPD is applied. Table 4.4 also shows average efficiencies measured at the average power back-off when DPD is ON and OFF.

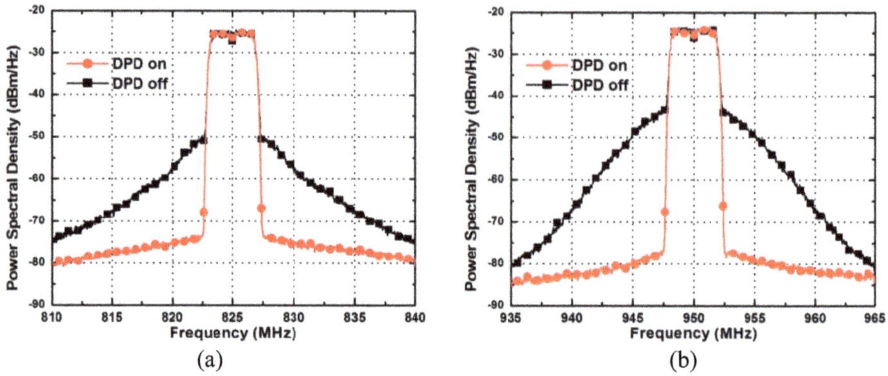

Fig. 4.64 Spectrum of 5 MHz WCDMA signal amplified using broadband three-stage DPA with and without linearization at (**a**) 825 MHz and (**b**) 950 MHz ([59], reprinted with permission from IEEE)

Table 4.4 Summary of DPD linearization of broadband DPA

Signal type	Bandwidth (MHz)	PAPR (dB)	Frequency (MHz)	Average power (dBm)	Average DE (%)	ACPR at 5 MHz offset (dBc)	
						DPD OFF	DPD ON
WCDMA	5	9.65	825	33.97	57.57%	−27.72	−45.66
WCDMA	5	9.65	950	32.57	46.1%	−22.17	−49.51

4.4.3 *Broadband Generalized Load Combiner for Doherty Power Amplifier*

The design of DPA is not just limited to the design of load combiner, but matching has an equally important role to achieve best performance in terms of load modulation. Moreover, there are several other practical issues such as current leakage toward auxiliary PA when it is OFF due to non-open circuit output impedance of auxiliary PA, slow turn-on effect, low auxiliary gain, etc., which reduces the performance of DPA when implemented [62]. This requires several optimizations to achieve desired performance out of DPA design. Matching is especially very critical in such cases. Moreover, extending the bandwidth or operating frequency range in case of DPA makes the matching design more critical. In general, the load-combiner in DPA is a three-port network, and therefore its design can be generalized to take care of all the requirements in DPA design. Such a generalized load-combiner design is first proposed in [62]. Figure 4.65 shows this three-port output matching/combining network (OMCN). This is a nonsymmetrical, non-isolated power combiner.

Here port 1 is connected to main transistor, port 2 is connected to the auxiliary transistor, and port 3 is connected to the common load. Figure 4.65 also presents the normalized incident and reflected power waves represented, respectively, as

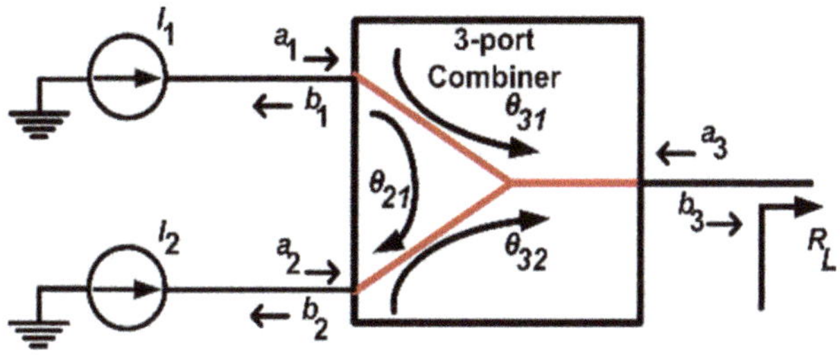

Fig. 4.65 Schematic of generalized three-port combiner network with output matching ([65], reprinted with permission from IEEE)

a_1 and b_1 which are used to derive the generalized scattering parameters of the 3-port network [63]. For DPA operation, it is assumed that the main transistor is biased at class AB or class B and the auxiliary transistor is biased at class C. This combiner also realizes the impedance matching from any arbitrary real- or complex-valued load impedance to any complex-valued optimum impedances at the output of transistors at saturation. Optimum performance of DPA is also ensured at output power back-off by this combiner. The assumption of arbitrary complex load impedance (Z_L at port 3) also presents an opportunity where PA can be directly connected with complex-valued input impedance of a circuit such as an antenna. This scheme is not proposed for broadband design, yet it is the first scheme which presented load combiner design in a generalized manner with relaxation of criticality in matching network design [62]. Similarly, the three-port input matching/divider network (IMDN) discussed in this chapter is a nonsymmetrical, non-isolated power divider. This divider also matches the input impedances of the two transistors to the source impedance. In addition, the IMDN provides an arbitrary phase difference between the two transistor paths, and the output combiner combines the output powers from the two transistors in any arbitrary ratio. Therefore, unlike conventional DPA design, the offset lines at the input of the amplifiers can be avoided in such case [64]. Moreover, an appropriate selection of the parameters of the IMDN may help in optimizing the performance of the amplifier by utilizing the nonlinear behavior of the transistor input impedances. The matching to any arbitrary complex load helps in minimizing the overall size of the circuit and the number of elements used in the circuit, by removing the need to match to common 50 Ω value. The S-parameters of the proposed OMCN are calculated using some assumptions as follows [65],

1. The proposed OMCN is a passive, lossless, and reciprocal network.
2. Proposed OMCN presents optimum load impedance to both transistors at saturation; thus, the power from the two transistors at any arbitrary ratio can be added. The two optimum load impedances can be any complex-valued impedances inside the Smith chart.
3. The amplifier's optimum load can be of any arbitrary value, real or complex, and lies within the Smith chart.
4. Proposed OMCN guarantees maximum possible efficiency at the desired output power back-off.

In order to start S-parameter calculation, a set of reference impedances have been selected. For port 1, optimum impedance is the complex conjugate of the optimum impedance of the main transistor as obtained from load pull at saturation. Port 2 is terminated with a complex conjugate of the optimum load of peaking transistor. Thus, if one desires to present the optimum impedance of the main and auxiliary transistors at port 1 and 2, there will be no reflection from port 1 and 2 (since the reference impedances are assumed to be conjugate of the impedances presented at port 1 and 2). Therefore, when both the transistors are ON, one can use $b_1 = b_2 = 0$. Moreover, assuming port 3 of the combiner is matched, with zero power reflection, that is, $a_3 = 0$, one can write,

$$S_{\mathrm{p}11}a_1 + S_{\mathrm{p}12}a_2 = 0 \tag{4.53}$$

$$S_{\mathrm{p}21}a_1 + S_{\mathrm{p}22}a_2 = 0 \tag{4.54}$$

$$S_{\mathrm{p}31}a_1 + S_{\mathrm{p}32}a_2 = b_3 \tag{4.55}$$

where $S_{\mathrm{p}ij}$ are the generalized S-parameters defined in terms of power waves at ith and jth ports. At back-off, if the two transistors in Fig. 4.65 provide powers with k:1 ratio at any arbitrary phases, one can write,

$$a_2 = \frac{S_{\mathrm{p}31}}{S_{\mathrm{p}32}}\frac{1}{k}a_1 \tag{4.56}$$

The above equation is valid for instances when auxiliary PA has just started conducting current. At saturation, the two transistors will provide same power at the same phase. Similarly, Eqs. (4.54 and 4.55) can further be rearranged as

$$S_{\mathrm{p}11} = \frac{S_{\mathrm{p}31}}{S_{\mathrm{p}32}}\frac{1}{k}S_{\mathrm{p}12} \tag{4.57}$$

$$S_{p22} = \frac{S_{p32}}{S_{p31}}kS_{p21} \tag{4.58}$$

Considering the 3-port load combiner network in Fig. 4.65 as lossless and reciprocal, one can write,

$$S_{\mathrm{p}11}S_{\mathrm{p}12}^{*} + S_{\mathrm{p}21}S_{\mathrm{p}22}^{*} + S_{\mathrm{p}31}S_{\mathrm{p}32}^{*} = 0 \tag{4.59}$$

$$\left|S_{\mathrm{p}11}\right|^2 + \left|S_{\mathrm{p}21}\right|^2 + \left|S_{\mathrm{p}31}\right|^2 = 1 \tag{4.60}$$

$$\left|S_{\mathrm{p}21}\right|^2 + \left|S_{\mathrm{p}22}\right|^2 + \left|S_{\mathrm{p}32}\right|^2 = 1 \tag{4.61}$$

$$|S_{p31}|^2 + |S_{p32}|^2 + |S_{p33}|^2 = 1 \tag{4.62}$$

Using Eqs. (4.56–4.58) in Eqs. (4.59–4.62), one can write,

$$|S_{p31}| = \sqrt{\frac{k}{1+k}} \tag{4.63}$$

$$|S_{p32}| = \frac{1}{\sqrt{1+k}} \tag{4.64}$$

$$|S_{p11}| = \frac{k}{1+k} \tag{4.65}$$

$$|S_{p21}| = \frac{\sqrt{k}}{1+k} \tag{4.66}$$

$$|S_{p22}| = \frac{1}{1+k} \tag{4.67}$$

Using magnitude information of S-parameters from Eqs. (4.63 to 4.67) and phase information from Eqs. (4.56 to 4.58), the S-parameter matrix for 3-port network can be expressed as,

$$\begin{bmatrix} -\frac{k}{1+k}e^{j(\theta_{21}+\theta_{31}-\theta_{32})} & \frac{\sqrt{k}}{1+k}e^{j\theta_{21}} & \sqrt{\frac{k}{1+k}}e^{j\theta_{31}} \\ \frac{\sqrt{k}}{1+k}e^{j\theta_{21}} & -\frac{1}{1+k}e^{j(\theta_{21}+\theta_{32}-\theta_{31})} & \frac{1}{\sqrt{1+k}}e^{j\theta_{32}} \\ \sqrt{\frac{k}{1+k}}e^{j\theta_{31}} & \frac{1}{\sqrt{1+k}}e^{j\theta_{32}} & 0 \end{bmatrix} \tag{4.68}$$

One can note that the phase of each S-parameters is expressed in terms of free variables θ_{21}, θ_{31}, and θ_{32}, which are the angles of S-parameters S_{p21}, S_{p31}, and S_{p32}. Moreover, negative sign in S_{p11} and S_{p22} ensures the lossless conditions such as in Eq. (4.59).

For any arbitrary back off power, if Γ_{A22} is the reflection coefficient at the output of auxiliary PA in OFF condition, using Eqs. (4.57 and 4.58), the load combiner produces a load to the main PA with reflection coefficient Γ_M which is given by [63],

$$\Gamma_M = S_{p11} + \frac{S_{p21}S_{p12}\Gamma_{A22}}{1 - S_{p22}\Gamma_{A22}} \tag{4.69}$$

Using Eqs. (4.57 and 4.58) in Eq. (4.69), one can get,

$$\Gamma_M = \frac{S_{p11}}{1 - S_{p22}\Gamma_{A22}} \tag{4.70}$$

Using the values of S-parameters from Eq. (4.68) in Eq. (4.69), one can compute the reflection coefficient presented at port 1 of the 3-port network in Fig. 4.65. This is the same as the load reflection coefficient presented to the main PA at back-off which is given by,

$$\Gamma_{\mathrm{M}} = \frac{-k\mathrm{e}^{j(\theta_{21}+\theta_{31}-\theta_{32})}}{(k+1) + k\left|\Gamma_{\mathrm{A22}}\right|\mathrm{e}^{j(\theta_{\mathrm{A22}}+\theta_{21}-(\theta_{31}-\theta_{32}))}} \tag{4.71}$$

Γ_{A22} and power ratio k both are constant for a particular back-off and with predefined bias conditions. Therefore, Γ_{M} at the output of main PA can be maintained at any arbitrary load using free parameters ψ and θ_{31}, where ψ is the relative phase difference θ_{31}–θ_{32}.

The magnitude and phase of Γ_{M} can be obtained from Eq. (4.71) as,

$$\left|\Gamma_{\mathrm{M}}\right|^2 = \frac{k^2}{(k+1)^2 + \left|\Gamma_{\mathrm{A22}}\right|^2 + 2(k+1)\left|\Gamma_{\mathrm{A22}}\right|\cos(\xi)} \tag{4.72}$$

$$\angle\Gamma_{\mathrm{M}} = \pm\pi + (\theta_{21} + \theta_{32} - \theta_{31}) - \phi \tag{4.73}$$

where angle ϕ is given by,

$$\phi = \tan^{-1}\left(\frac{\left|\Gamma_{\mathrm{A22}}\right|\sin(\xi)}{(k+1) + \left|\Gamma_{\mathrm{A22}}\right|\cos(\xi)}\right) \tag{4.74}$$

and angle ξ is defined as,

$$\xi = \theta_{\mathrm{A22}} + \theta_{21} + \theta_{32} - \theta_{31} \tag{4.75}$$

The range of $\mid \Gamma_{\mathrm{M}} \mid$ is based on maximum and minimum value of cosine of angle ξ in Eq. (4.72) which is given by,

$$\frac{k}{(k+1) + \left|\Gamma_{\mathrm{A22}}\right|} \leq \left|\Gamma_{\mathrm{M}}\right| = \frac{k}{(k+1) - \left|\Gamma_{\mathrm{A22}}\right|} \tag{4.76}$$

From Eqs. (4.72 and 4.75) one can calculate value of angle ξ in terms of known design parameters as,

$$\xi = \pm\cos^{-1}\left(\frac{\frac{k^2}{\left|\Gamma_{\mathrm{M}}\right|^2} - (k+1)^2 - \left|\Gamma_{\mathrm{A22}}\right|^2}{2(k+1)\left|\Gamma_{\mathrm{A22}}\right|}\right) \tag{4.77}$$

From Eqs. (4.73 and 4.77), one can deduce,

$$\theta_{21} = \frac{\angle\Gamma_{\mathrm{M}} + \phi + \xi - \theta_{\mathrm{A22}} \pm \pi}{2} \tag{4.78}$$

$$\psi = \theta_{31} - \theta_{32} = \frac{\angle\Gamma_{\mathrm{M}} + \phi - \xi + \theta_{\mathrm{A22}} \pm \pi}{2} \tag{4.79}$$

In order to obtain desired Γ_{M} as given in Eq. (4.71), the phase of θ_{21} should be fixed to a set of specific values based on the signs selected in Eqs. (4.77 and 4.79). Similarly, Eq. (4.79) shows that the phase difference $\psi = \theta_{31}–\theta_{32}$ should have a set of specific values based on the signs selected in Eqs. (4.77 and 4.79). The selection of the signs in Eqs. (4.77 and 4.78) can be selected independent to each other; however, one should choose similar signs in Eqs. (4.78 and 4.79). Therefore, for certain values of Γ_{M}, k, Γ_{A22}, etc., there are four possible set of solutions for solving Eqs. (4.77–4.79).

The desired value of Γ_{M} should be obtained either from load-pull analysis or analytically for specific requirements of the matching conditions for the main PA along with power leakage into the auxiliary transistor at power back-off.

One can define power leakage factor γ as the ratio of the power leakage into the auxiliary transistor to the power delivered to the load at back-off, when auxiliary PA is OFF. This can be written as,

$$\gamma = \left.\frac{P_{\mathrm{leak}}}{P_{\mathrm{load}}}\right|_{\mathrm{Back\text{-}off}} = \frac{|b_2|^2 - |a_2|^2}{|b_3|^2} \tag{4.80}$$

As discussed earlier, port 3 is assumed to be matched to its reference impedance; therefore, $a_3 = 0$ is considered while expressing Eq. (4.80).

The reflected wave b_2 and b_3 in Fig. 4.65 can be expressed as,

$$b_2 = \frac{S_{\mathrm{p21}}}{1 - S_{\mathrm{p22}}\Gamma_{\mathrm{A22}}} a_1 \tag{4.81}$$

$$b_3 = S_{\mathrm{p11}} + \frac{\left|S_{\mathrm{p21}}\right|\left(1 - |\Gamma_{\mathrm{A22}}|^2\right)}{\left|S_{\mathrm{p31}} + \Gamma_{\mathrm{A22}}\left(S_{\mathrm{p32}}S_{\mathrm{p21}} - S_{\mathrm{p31}}S_{\mathrm{p22}}\right)\right|^2} \tag{4.82}$$

Using Eqs. (4.81 and 4.82), one can express γ in terms of circuit parameters as,

$$\gamma = \frac{k\left(1 - |\Gamma_{\mathrm{A22}}|^2\right)}{(k+1)\left(1 + |\Gamma_{\mathrm{A22}}|^2 + 2\,|\Gamma_{\mathrm{A22}}|\cos\xi\right)} \tag{4.83}$$

The range of leakage factor γ is based on maximum and minimum value of cosine of angle ξ in Eq. (4.83) which is given by,

$$\frac{k\,(1 - |\Gamma_{\mathrm{A22}}|)}{(k+1)\,(1 + |\Gamma_{\mathrm{A22}}|)} \le \gamma \le \frac{k\,(1 + |\Gamma_{\mathrm{A22}}|)}{(k+1)\,(1 - |\Gamma_{\mathrm{A22}}|)} \tag{4.84}$$

This concept allows one to design DPA for optimum performance using any real- or complex-valued load. Now, the incorporation of continuous modes in DPA topology requires to present certain loads at the output of main PA. This concept

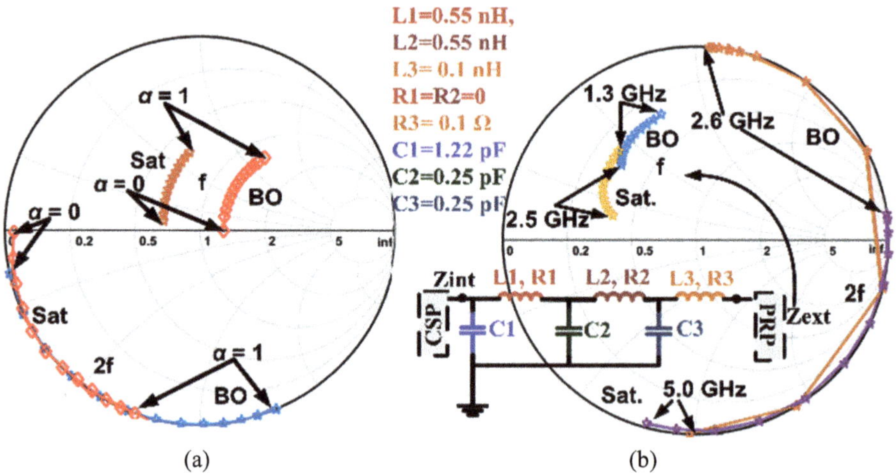

Fig. 4.66 Load trajectories at (**a**) intrinsic current source plane (CSP) and (**b**) extrinsic PRP ([65], reprinted with permission from IEEE)

is used with the consideration of non-open output impedance of auxiliary PA in designing broadband load combiner in [65]. In such a case where the main PA operates in continuous class B/J mode, the load-impedance trajectories for fundamental, as well as harmonic, are calculated and demonstrated at intrinsic current source plane as shown in Fig. 4.66a, b.

For continuous class B/J mode load conditions, the voltage and current equations are already defined in section "Design Strategy for Continuous Class J Power Amplifiers". The fundamental and second harmonic impedances at intrinsic CRP are already expressed in Eqs. (1.114 and 1.115). Figure 4.66a shows that these required impedances for design parameter α ranging between 0 and 1. R_{opt} is the optimum load resistance and can be obtained from load line. For DPA operation, the impedances need to be modulated in a certain manner with input power. Therefore, in order to calculate the appropriate load values at back-off, the voltage conditions are considered to remain at a constant level, whereas the current values are considered to be changed according to the input power. This gives the load values at back-off as,

$$Z_{f,\text{BO}} = \frac{V_{\text{DD}}\,(1 + j\alpha)}{I_{\max}/2\beta} = \beta R_{\text{opt}}\,(1 + j\alpha) \tag{4.85}$$

$$Z_{2f,\text{BO}} = \frac{V_{\text{DD}}\,(-j\alpha/2)}{I_{\max}/\beta} = j\,\frac{3\pi\alpha\beta R_{\text{opt}}}{8} \tag{4.86}$$

where β is termed as the back-off factor and defined as the ratio of input voltage drive at saturation and back-off ($V_{\text{in,sat}}/V_{\text{in,BO}}$). Considering R_{opt} as 32 Ω at saturation for a commercially available 10 W GaN HEMT device [49, 65], the appropriate loads are calculated for Class B/J mode continuum at saturation and back-off using (Eqs. 1.114, 1.115 and 4.85, 4.86), respectively. The calculated values of loads for $\alpha = 1$ to 0 are shown in Fig. 4.66a. It is observed from Fig. 4.66a that the load values vary from $32.6 + j31.7\ \Omega$ to $33.9 + j3.66\ \Omega$ for saturation and from $64.0 + j64.0\ \Omega$ to $64.0 + j0\ \Omega$ for operation at 6 dB back-off. The corresponding load trajectories for the second harmonic is also shown in the same figure.

One can see from Eq. (4.71) that the load impedances at the output of main PA can be maintained at the impedance values of class B/J continuum given by Eqs. (4.85 and 4.86) using free parameters θ_{21} and θ_{32}–θ_{31}. Table 4.5 shows the selected values of parameters θ_{21} and θ_{32}–θ_{31}. Here, for each value of θ_{21} such as 0°, 45°, 90°, 135°, and 180°, the difference θ_{32}–θ_{31} is varied from 0° to ±360° in order to achieve high efficiency over the entire frequency band. The resultant load trajectories are shown in Fig. 4.67a.

One can see from Fig. 4.67a that with different combinations of design parameters θ_{21} and θ_{32}–θ_{31}, the extrinsic loads at PRP can take any value on the Smith chart. Therefore, the designer can choose the appropriate values for broadband matching. For example, the case, $\theta_{21} = 135°$ and θ_{32}–θ_{31} varying between 50° and −320°, offers a good fitting to the load trajectory desired for class B/J mode of operation as shown in Fig. 4.66b.

Table 4.5 Variation of parameters θ_{21} and θ_{32}–θ_{31} for obtaining desired load trajectory

Angle	Values in degree (°)				
θ_{21}	0	45	90	135	180
θ_{32}–θ_{31}	−340 to 150	−340 to 125	−340 to 62	−340 to 50	−50 to 310

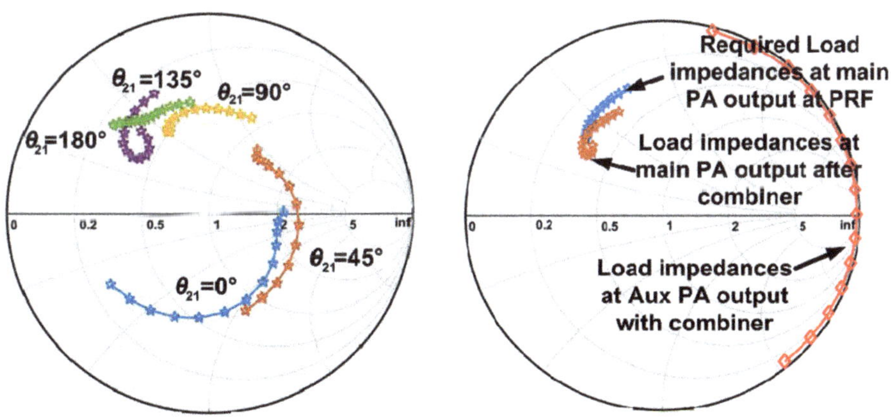

(a) (b)

Fig. 4.67 Load trajectory setting with different combiner parameters. (**a**) An example of a variety of load values possible and the (**b**) load trajectories at intrinsic CSP and extrinsic PRP for 10 W GaN HEMT device with R_{opt} as 32 Ω at saturation ([65], reprinted with permission from IEEE)

Figure 4.67b shows the optimum choice of combiner parameters in order to present class B/J load conditions to the main PA. In addition this figure also shows the load trajectory desired for class B/J mode of operation as obtained in Fig. 4.66b. A slight optimization of θ_{21} to 107° while varying θ_{32}–θ_{31} between −320° and 50°, one can obtain a reasonable match to the desired load trajectory of 10 W GaN HEMT device, as depicted in Fig. 4.66b. Similar analysis can be performed analytically where one will get specific solutions for θ_{21} and θ_{32}–θ_{31} for a given Γ_{A22} and k using Eq. (4.71) [62]. However, among several sets of analytical solutions, the designer should choose the solutions viable in terms of layout design.

The previous section describes the design methodology of a generalized broadband load combiner. In case of a fully analog DPA architecture, the phase offset added by the phase difference between the two branches of the load combiner is compensated by offset lines at the input of the main and auxiliary PAs [43, 49].

The two amplifiers used as main and auxiliary have different classes of operation which results in inappropriate power combining at the output of the proposed load combiner. This will reduce the output power, and therefore the efficiency of the overall DPA. In order to characterize such imperfections, an additional input phase is introduced to the signals fed to the main and auxiliary PA. This is possible with hybrid digital/RF continuous mode DPA (CM-DPA) to further enhance its performance [65]. These independent additional phases at the inputs are provided in digital domain in the hybrid digital/RF CM-DPA system [65]. In such a case, one can define the main and auxiliary PA current as,

$$I_1 = \sigma_1 I_{\text{max}} e^{j\phi_1} \tag{4.87}$$

$$I_2 = \sigma_2 I_{\text{max}} e^{j\phi_2} \tag{4.88}$$

where ϕ_1 and ϕ_2 are the phases which are added to the input currents fed to the two branches of the load combiner in Fig. 4.65. σ_1 and σ_1 are the magnitude control parameters. From Eq. (4.57), one can write,

$$\frac{|a_1|}{|a_2|} = \frac{|S_{p32}|}{|S_{p31}|} \tag{4.89}$$

where $\delta = \phi_2$–$\phi_1$is the phase offset presented digitally. From Eq. (4.89), one can see that this input phase offset δ should compensate for the phase difference between the two branches of load combiner, that is, θ_{32}–θ_{31}. In addition, it will also compensate for any additional imperfections due to dispersive characteristics of the transmission line as well as PA and phase discrepancies between main and auxiliary PAs due to their class of operation. Therefore, by setting phase offset δ digitally, one can get optimum performance in terms of power and efficiency over a wide bandwidth. Due to nonlinear phase compression in both main and auxiliary PA, the phases of the main and auxiliary paths should be characterized for all power levels after the auxiliary PA is turned on [66]. This power-related phase offset can be adjusted using

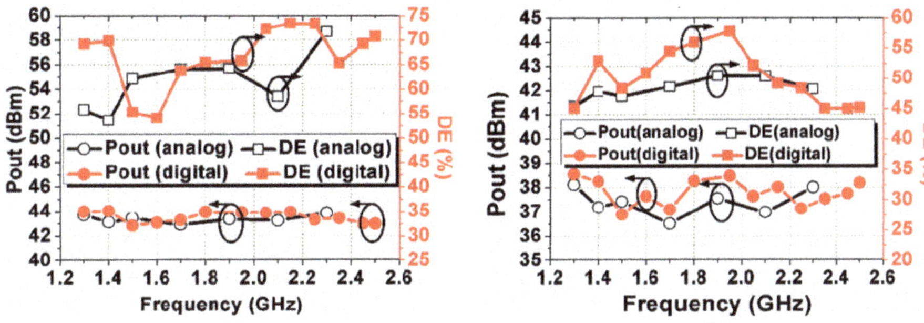

Fig. 4.68 Efficiency improvement across the band by varying δ: (**a**) at saturation and (**b**) at back-off ([65], reprinted with permission from IEEE)

Table 4.6 Setting of input phase offset in digital domain with frequency over the proposed frequency range

Frequency (GHz)	Phase offset (°) (ϕ_2–ϕ_1)
1.3	40
1.35	35
1.55	10
1.75	0
1.95	0
2.15	0
2.35	0
2.5	−10

a power-indexed lookup table (LUT). This LUT helps to correct the phase difference at all power levels where the total output power of the PA comes from the main and auxiliary PAs [66].

Moreover, the amplitude in terms of σ_1 and σ_2 of the two input currents can also be optimized. This may not affect the load-combiner performance but can compensate for the limitations such as the soft turn-on effect and low gain of the auxiliary PA operating in class C bias. Figure 4.68a, b shows the effect of applying the phase offset at the inputs of the main and auxiliary PAs, in terms of output power and efficiency.

One can see from these figures that both the available output power and efficiency are improved with digital compensation. Table 4.6 summarizes the phase offset applied at the inputs for compensating the analog imperfections to obtain performance in Fig. 4.68a, b. A slight amplitude offset is introduced in favor of auxiliary PA (with $\sigma_2 > \sigma_1$) at corner frequencies to improve the performance.

One can see from Fig. 4.68a, b that independent input control enhances the bandwidth of the proposed DPA by 200 MHz in this case. With the appropriate phase compensation, the CM-DPA is able to achieve efficiency enhancement in the frequency band from 1.3 to 2.5 GHz in simulation [65]. One can also see from these figures that the output power is enhanced by 0.5–1 dB in the band at both power

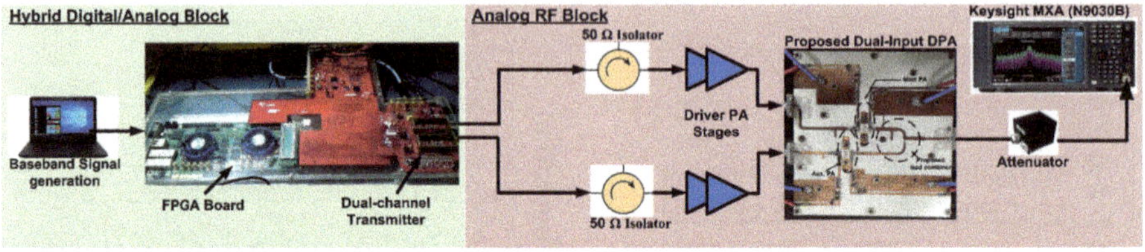

Fig. 4.69 Characterization setup for the proposed hybrid analog/digital continuous mode DPA (CM-DPA) with the photograph of the hardware prototype developed ([65], reprinted with permission from IEEE)

levels: back-off and saturation. The corresponding DEs are improved up to 6% at saturation and 2% at back-off.

A prototype based on this scheme has been developed using the nonlinear model of Wolfspeed 10 W transistor in Keysight's ADS software [65]. Both of the transistors are biased with 28 V drain bias. The main PA is biased with drain current of 86 mA and the gate bias of auxiliary PA is set at −5.5 V. Figure 4.69 shows the measurement setup along with the photograph of hardware prototype.

The hardware prototype is fabricated on a Rogers R4350B substrate with dielectric constant, height, and loss tangent of 3.66, 20 mil, and 0.0037, respectively. One can see that there is a provision of feeding the main and auxiliary PA independently using two separate inputs. Figure 4.69 also shows the setup comprises Altera Arria V-GT field programmable gate array (FPGA) platform, isolators, driver PAs, and the fabricated DPA. The FPGA platform will generate two signals with phase-offset for driving the main and auxiliary PAs independently. One can generate two digital modulated signals with desired phase offset which will be fed to the digital to analog converters (DACs) of two-channel transmitter (TSW30SH84) from Texas instruments (TI). These TI transmitters comprise DACs and quadrature modulators. The two baseband analog signals after DACs are upconverted to RF signals using the two quadrature modulators. The two channels are synchronized at baseband level using the same clock. Therefore, an accurate phase difference (δ) can be set digitally between the two signals upconverted by the two-channel transmitters. Similarly, the amplitude of these RF signals can also be adjusted independently in digital domain. These two RF signals are then fed to the main and auxiliary path of DPA unit as shown in Fig. 4.69. This setup can generate modulated signals using the IQ data generated in MATLAB. This IQ data is later stored in the random access memories (RAM) of FPGA using Quartus software and continuously sent to DAC. It is worth mentioning that the setup can also provide CW signals if I and Q data are chosen as sinusoidal signals. In order to set required phase difference between the two channels, the system should be calibrated first. This can be done by using a power combiner which combines the output from the two channels of the transmitter. A phase offset between the two channels will be set such that the signal at combiner output is canceled by 50–60 dBc. This phase value corresponds to the case when the two channels are out-of-phase, that is, 180° phase difference. This offset will

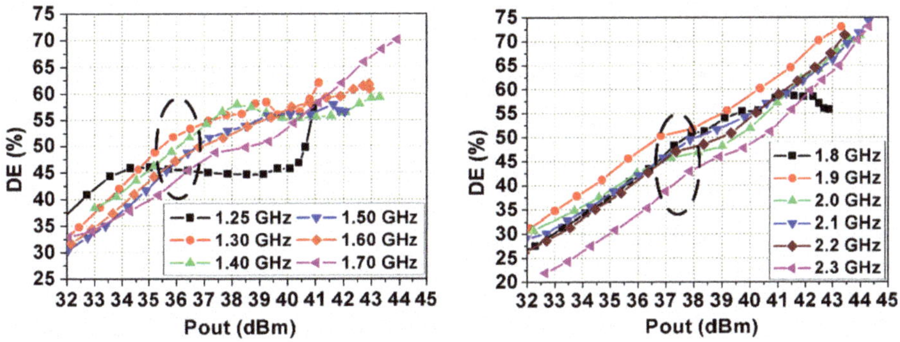

(a) (b)

Fig. 4.70 Continuous wave measurements results of the proposed two-input CM-DPA at (**a**) 1.25–1.70 GHz and at (**b**) 1.8–2.3 GHz ([65], reprinted with permission from IEEE)

Fig. 4.71 Summarized results for CW measurements with frequency for the hardware prototype of hybrid analog/digital CM-DPA ([65], reprinted with permission from IEEE)

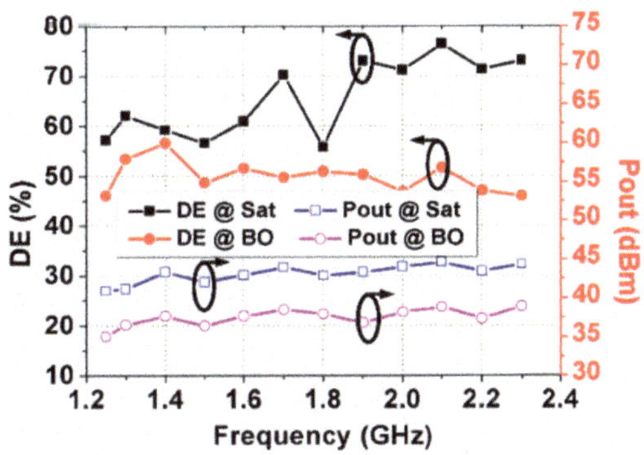

be added to one of the channels to set equal phase between the two channels at the desired power level. Once the calibration is completed, the two channels will be in equal phase. After this, the desired phase offset can be added in any of the channel. The detailed procedure is given in [67]. After calibration, a phase difference δ can be applied between the two channels feeding the main and auxiliary PA of the CM-DPA. Figure 4.70a, bshows the measured DE under CW test. Figure 4.70a, b shows that the measured DEis between 56% and 75.4% and the corresponding power is 41.0–44.6 dBm, at saturation. At 6 dB back-off, the output power varies between 35.7 and 38.5 dBm and the DE varies from 45% to 56.57% in the frequency range of 1.25–2.3 GHz.

Figure 4.71 illustrates the measured performance of hybrid analog/digital CM-DPA over the frequency band from 1.25 to 2.3 GHz. From Fig. 4.71 one can see that the DE and the output power are reduced at corner frequency at lower side. The DE is higher than 45% at specified back-off over the entire band. The minimum value of DE occurs at 2.3 GHz and maximum value reaches up to 56.57% at 1.4 GHz. However, at saturation the minimum value of DE is 56% at 1.8 GHz. The maximum value of DE at saturation is 75.4% at 2.1 GHz. Therefore, the implemented hybrid analog/digital CM-DPA offers highly efficient broadband performance corresponding to fractional bandwidth of 59.15%.

The setup is also characterized using a 5 MHz LTE signal with PAPR of 9.65 dB. The modulated signal is split into two equal phased time-aligned signals and transmitted through drivers. The output signal from the DPA is received using a receiver (TSW1266) from Texas instruments with a baseband sample rate of 307.2 M samples/s. The receiver is synchronized with the transmitter using the same clock. Therefore, the input and output signals are time aligned with each other. The proposed DPA is also linearized using DPD method which utilizes neural network-based behavioral modeling [67, 68] at three frequencies within the band of operation, that is, 1.9, 2.1, and 2.2 GHz. The linearized performance is evaluated using the same 5 MHz LTE signal. Figure 4.72a–c shows the measured output spectrums at these frequencies, with and without DPD. One can see from Fig. 4.72a–c that at these three frequencies, the proposed DPA provides average DE of 36.60, 32.31, and 35.67% at an average output power of 33.35, 33.63, and 32.4 dBm, respectively.

From the above figures one can also see that the proposed DPA offers ACPR level lower than −30.7, −31.4, and −31.9 dBc at the abovementioned power

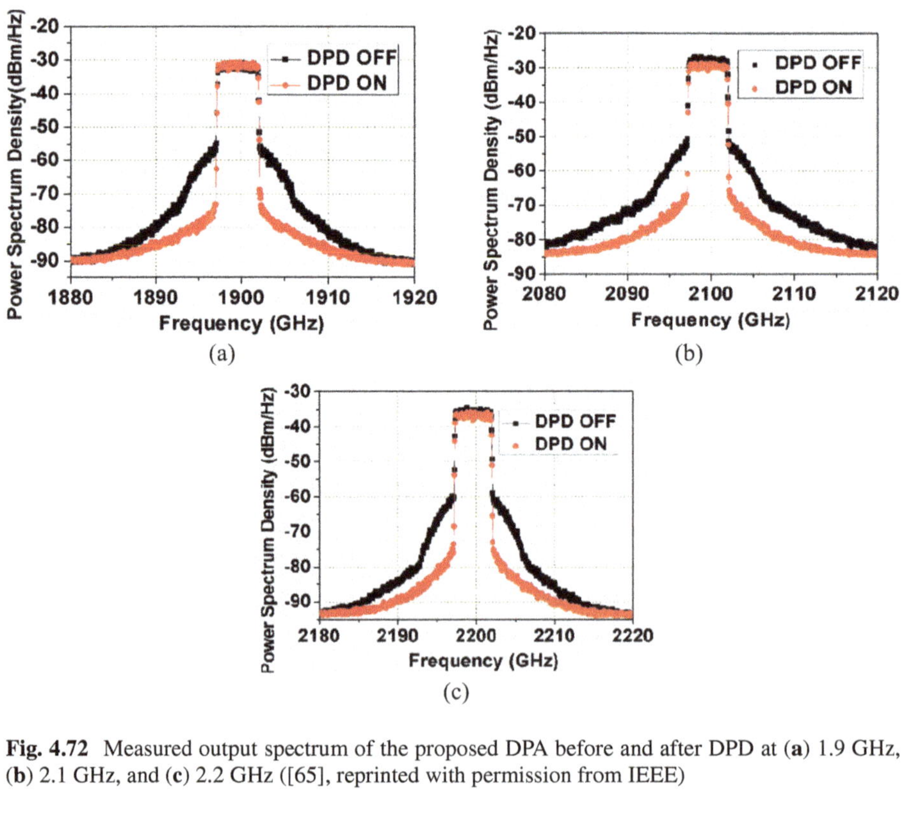

Fig. 4.72 Measured output spectrum of the proposed DPA before and after DPD at (**a**) 1.9 GHz, (**b**) 2.1 GHz, and (**c**) 2.2 GHz ([65], reprinted with permission from IEEE)

Table 4.7 Summary of modulated measurement and linearization

Bandwidth of LTE signal (MHz)	PAPR (dB)	Frequency (GHz)	Average power (dBm)	Average DE (%)	ACPR at 5 MHz offset (dBc)	
					DPD OFF	DPD ON
5	9.65	1.9	33.35	36.60	−30.7	−45.9
5	9.65	2.1	33.63	32.31	−31.4	−45.3
5	9.65	2.2	32.4	35.67	−31.9	−45.2

levels when DPD is OFF. Whereas these values reach below −45.9, −45.3, and −45.2 dBc corresponding to −15.2, −13.9, and −13.3 dB improvement at 1.9, 2.1, and 2.2 GHz, respectively, after DPD is put ON. Table 4.7 represents performance of the prototypes while employing DPD linearization technique. The measurements are performed at three frequencies over the band using 5 MHz LTE signals where the ACPR is measured at 5 MHz offset. The table also lists the average efficiencies measured at average power back-off level with and without putting DPD ON.

Another class of broadband DPA design has been introduced recently in literature [69] by taking advantage of harmonic components. In a conventional DPA, it is observed that the two transistors are prevented from modulating each other at harmonic frequencies by employing harmonic isolation. On the contrary, this type of DPA produces a number of high-efficiency DPA modes over a continuous band of frequency by using the two transistors to modulate each other at harmonic frequencies by using a properly designed post-harmonic tuning network. This leads to a broadband continuous mode DPA. Since load modulation is occurring at fundamental as well as harmonic frequency, it is difficult to analyze the proposed architecture analytically. Therefore, a Monte Carlo-type technique is proposed to do numerical analysis on such harmonically linked DPAs. The theory was validated with a prototype developed which works over the frequency range of 1.65–2.75 GHz.

4.4.4 *Broadband and Multiband Chireix Outphasing Power Amplifier*

In order to co-op with the requirement of upcoming wireless communication standards, the RF designers have started investigating outphasing transmitters for multiband or broadband operation [15, 70–74]. The major limiting factor in expanding the bandwidth of outphasing transmitter is the design of Chireix combiner based on quarter wave transmission lines and reactance compensation circuit which restricts the bandwidth. Moreover, the conventional Chireix outphasing PA is analyzed by considering the devices as constant voltage sources which is not correct in the practical scenario. Since, the instantaneous output voltage across the active device changes with the variation in the outphasing angle, the assumption that the individual device efficiency is constant and the overall efficiency of the outphasing PA is merely a product of this efficiency and combiner efficiency is not correct. This limitation is solved by analyzing Chireix PA with mixed mode of operation, where each branch PA is driven with phase and amplitude modulated drives as discussed in Chap. 3 [73]. In addition, a generalized combiner is used which can obtain desired load conditions required for any mode of operation of PAs in each branch of Chireix PA [71]. The following sections describe generalized outphasing combiner design using continuous class E mode of operation. Later, a reconfigurable architecture offering Chireix outphasing PA operation in multiple bands is also discussed.

The conventional class E theory gives a single set of designer parameters for generalized Chireix combiner as shown in Fig. 1.50 of Chap. 1. However, in case of continuous class E, the switching condition for class E can be satisfied for a continuum of fundamental and second harmonic loads [33]. This also gives wide design space for selecting network parameters of the combiner for broadband realization of outphasing PA. From the discussion in section "Continuous Class E Power Amplifier", by varying the design parameter ϕ_1 ranging between 43° and 78°, different sets of fundamental and harmonic loads can be obtained. These loads provide same output power levels comparable to the conventional class E output power level. Figure 4.73 shows the fundamental load trajectory with output power for different design parameters ϕ_1 [71].

One can see that these trajectories are constant Q lines, where the ratio of real and imaginary part of the impedance is constant along the line [71]. In order to achieve continuous class E loads from the generalized combiner discussed in section "Class E Continuum" and shown in Fig. 1.50 of Chap. 1, the network parameters of this combiner are computed by setting $Z_1^L(\psi_P) = Z_1^R(\psi_P) = Z_1^S(\phi_1^P)$. The $Z_1^S(\phi_1^P)$ is the load calculated for continuous class E operation from Eq. (4.18) for phase angle of fundamental component ${\phi_1}^P$ calculated at peak power. In such a case, resistance R in continuous class E load network as given by Eq. (4.20) (of section "Continuous

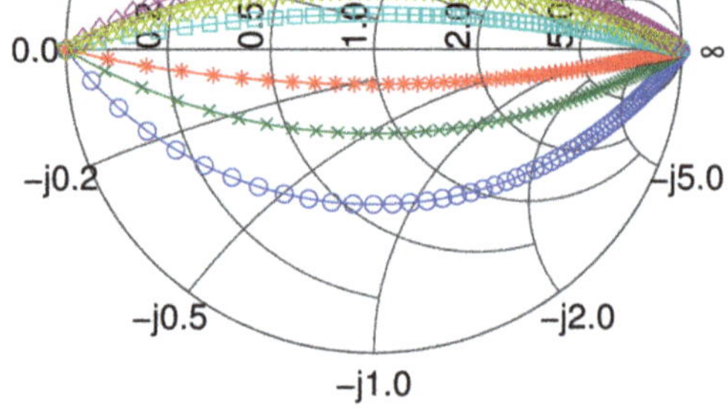

Fig. 4.73 Fundamental load trajectory for continuous class E with output power for various values of design parameter ϕ_1 ([71], reprinted with permission from IEEE)

Class E Power Amplifier") is calculated for ${\phi_1}^P$. Similarly, when output power is backed-off by a factor γ at $\psi = \psi_{\mathrm{B}}$, the load combiner parameters are calculated by setting $Z_1^L(\psi_{\mathrm{B}}) = Z_1^R(\psi_{\mathrm{B}}) = Z_1^S(\phi_1^B)$, where $Z_1^S(\phi_1^B)$ is the load calculated for continuous class E operation from Eq. (4.18) for phase angle of fundamental component ${\phi_1}^B$ calculated at back-off power. In such a case, R in continuous class E load network is given by [71],

$$R\left(\psi_{\mathrm{P}}, \phi_1^B\right) = \gamma \frac{8\sin^2\left(\phi_1^B\right)}{\pi^2} \frac{V_{\mathrm{DD}}^2}{P_{\mathrm{out}}} \tag{4.90}$$

where P_{out} is peak envelope power. One can see that different values for the design parameter ϕ_1 have been chosen at peak power level $\phi_1 = {\phi_1}^P$ and back-off $\phi_1 = {\phi_1}^B$. This implies that the impedance seen by the devices are on different trajectory in Fig. 4.73. This introduces two new degrees of freedom in calculation of network parameter of the combiner. Each combination of ${\phi_1}^P$ and ${\phi_1}^B$ will produce different set of network parameter for the combiner providing large design space for the combiner. This large design space enhances the possibility of wideband realization of this combiner. Based on this theory a prototype has been developed and tested for validating wideband performance. Figure 4.74 shows the photograph of broadband outphasing PA prototype.

The circuit is implemented on a Duroid 5870 substrate with dielectric constant of 2.3 and thickness of 15 mil. The input drives for the two branches are generated using a VSG (Keysight 83650A) and an RF synthesizer (Keysight E4438C) which are synchronized with the same 10 MHz reference signal. Figure 4.75 shows the measured DE over a frequency range of 750–1050 MHz.

One can see that the DE at 7.5 dB back-off is better than 52% over the band. Figure 4.76a shows the DE for various back-off levels over the frequency range of 0.7–1.05 GHz. One can see that at lower frequencies, the efficiency drops faster with the increase in back-off level. Figure 4.76b shows the peak output power level with frequency.

Fig. 4.74 Photograph of the broadband outphasing PA prototype ([71], reprinted with permission from IEEE)

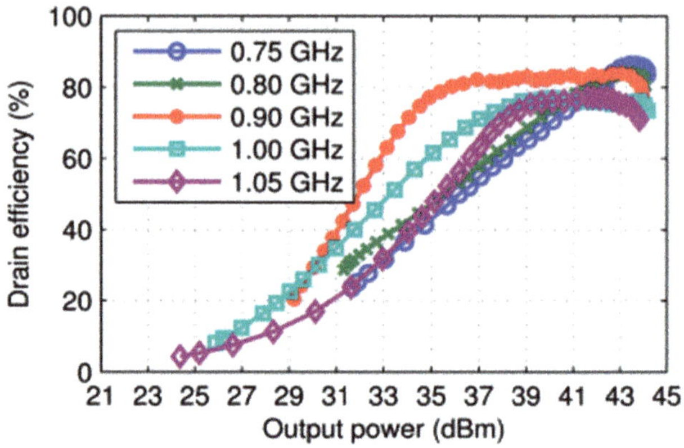

Fig. 4.75 Measured DE with output power at different operating frequencies ([71], reprinted with permission from IEEE)

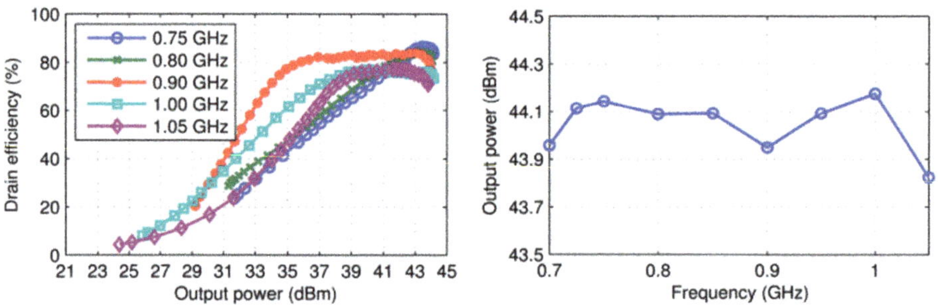

Fig. 4.76 Measurement of broadband Chireix outphasing PA with frequency: (**a**) DE at various back-off levels and (**b**) peak output power ([71], reprinted with permission from IEEE)

One can see from this figure that peak output power is more than 49.3 dBm over the entire frequency range of 0.7 GHz–1.05 GHz. Therefore, the Chireix PA based on generalized load combiner as presented in [71] offers a broadband solution for Chireix outphasing PA with continuous class E mode of operation.

Apart from expanding the bandwidth of Chireix PA, an alternate approach by introducing reconfigurability is also proposed recently in Chireix outphasing PA [15]. In this case, a reconfigurable outphasing transmitter is designed, where Chireix outphasing PA can be reconfigured to three operating frequencies of operation at 1.65, 2.1, and 2.35 GHz with ±25 MHz band around each operating frequency. Figure 4.77 shows the block diagram of reconfigurable outphasing transmitter. The decomposition of digital baseband signal is implemented using FPGA platform. This provides two synchronized outphased signals for driving the Chireix outphasing amplifier. One can see from Fig. 4.77 that the reconfigurable Chireix combiner consists of reconfigurable reactance compensation network and power combining. A reconfigurable reactance compensation network is realized to cancel the imaginary part of load seen by each branch PA due to load modulation at different frequency points.

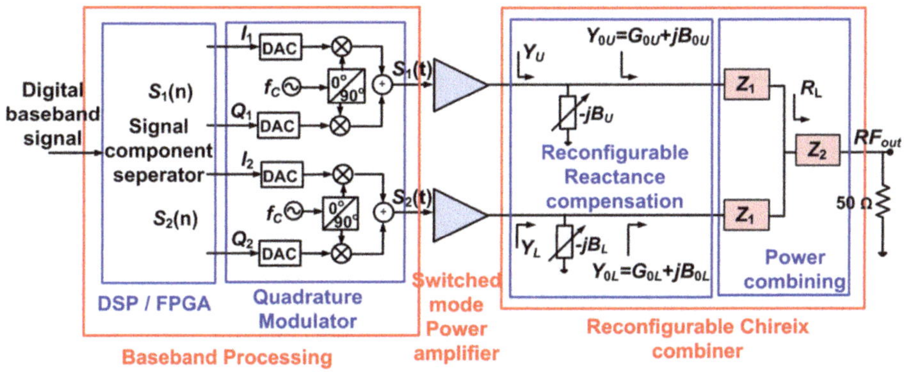

Fig. 4.77 Outphasing transmitter utilizing reconfigurable Chireix combiner ([15], reprinted with permission from IEEE)

In Fig. 4.77, $Y_{0\mathrm{U}}$ and $Y_{0\mathrm{L}}$ are the load impedances presented to upper and lower PAs, respectively, without applying reactance compensation. These loads can be represented as,

$$Y_{0\mathrm{U}} = G_{0\mathrm{U}}(\psi) - jB_{0\mathrm{U}}(\psi) \tag{4.91}$$

$$Y_{0\mathrm{L}} = G_{0\mathrm{L}}(\psi) + jB_{0\mathrm{L}}(\psi) \tag{4.92}$$

where $G_{0\mathrm{U}}$ and $G_{0\mathrm{L}}$ represents the real part of the load impedance, and $B_{0\mathrm{U}}$ and $B_{0\mathrm{L}}$ represents the imaginary part of the load impedance presented to upper and lower branches, respectively. Using ABCD parameters of the quarter-wave transformer, one can write G_0 and B_0 as,

$$G_0 = \frac{2R_\mathrm{L}\left[R_\mathrm{L}\sin 2(\pm\psi)\cot\left(\frac{\pi}{2}\frac{f}{f_0}\right) + Z_1\cos^2(\pm\psi)\right]}{Z_1\left[4{R_\mathrm{L}}^2\cos^2\left(\frac{\pi}{2}\frac{f}{f_0}\right) + Z_1^2\sin^2\left(\frac{\pi}{2}\frac{f}{f_0}\right)\right]} \tag{4.93}$$

$$B_0 = \frac{\cot\left(\frac{\pi}{2}\frac{f}{f_0}\right)}{Z_1}\left[\frac{4{R_\mathrm{L}}^2\cos^2(\pm\psi)}{\left(4{R_\mathrm{L}}^2\cos^2\left(\frac{\pi}{2}\frac{f}{f_0}\right)+Z_1^2\sin^2\left(\frac{\pi}{2}\frac{f}{f_0}\right)\right)} - 1\right] - \frac{R_\mathrm{L}\sin 2(\pm\psi)}{\left(4{R_\mathrm{L}}^2\cos^2\left(\frac{\pi}{2}\frac{f}{f_0}\right)+Z_1^2\sin^2\left(\frac{\pi}{2}\frac{f}{f_0}\right)\right)} \tag{4.94}$$

where f_0 is the center frequency of operation, at which the electrical length of each transformer is 90°. The power combining part as shown in Fig. 4.77 which transforms R_L to G_0 is realized in a way to get a wideband response at saturated power. At saturation, when $\psi = 0$ the condition $(Z_1/R_\mathrm{L}) = 2$ is used to design the combiner for wideband response. This condition will make the conductive part of the load, independent of the operating frequency at saturation. The imaginary part of the load impedances represented by Eq. (4.94) is responsible for the efficiency degradation at back-off. Therefore, the reactance compensation block in Fig. 4.77 which compensates the imaginary part jB_0 can provide the efficiency peak at OPBO. For efficiency enhancement at the desired OPBO, reactance must be compensated at a fixed outphasing angle called compensating angle (ψ_c) and its relation with OPBO is given as:

$$\mathrm{OPBO} = -20\log(\cos\psi_c) \tag{4.95}$$

Figure 4.78 shows the combiner response at three operating frequencies with ±25 MHz band around each frequency. Load presented by this Chireix combiner is shown in Fig. 4.78a–c and combiner efficiency versus outphasing angle is shown in Fig. 4.78d–f.

One can see from Fig. 4.78b that at 2.1 GHz, the upper and lower branch impedances are symmetric around real axis of the Smith chart and need equal amount of compensation to cancel the reactance at $\psi_c = 65°$. However, at 1.65 GHz and 2.35 GHz, asymmetric cancellation is required for upper and lower branch of outphasing. The compensating angle $\psi_c = 65°$ corresponds to 7.48 dB OPBO as calculated from Eq. (4.95). One can observe in Fig. 4.78d–f that the efficiency enhancement at $\psi_c = 65°$ occurs after applying proper compensation. Figure 4.78d–f also shows the efficiency of outphasing PA without applying any reactance compensation. One can depict an efficiency enhancement at OPBO after applying proper compensation in all the three graphs in comparison to the case in which there is no compensation applied. In order to achieve the compensation at multiple operating frequencies as discussed above, the reactance compensation network must be reconfigurable. In order to introduce this reconfigurability, reactance compensation network consists of tunable capacitance and tunable inductor. These tunable reactance compensating elements are realized using SKYWORK's varactor diode "SMV1430." Figure 4.79a shows the realization of tunable capacitor where varactors are connected in anti-series configuration. Two varactors connected in anti-series varactor (ASV) configuration has less distortions at high-power levels as compared to a single varactor [75]. Moreover, three varactor diodes are connected back to back to replace one varactor diode in ASV configuration to handle high output voltage swing at the output of PA. This tunable capacitance as shown in Fig. 4.79a is evaluated by sweeping bias voltage V_r over the frequency in ADS simulations. Fig. 4.79b shows the variation of capacitance with frequency for various values of V_r. One can see that at 1.65 GHz, a capacitance of 0.37 pF is realized at $V_r = 45$ V which is used for reactance compensation at upper branch of the outphasing PA. Similarly, at 2.1 GHz, the circuit is configured to present 0.5 pF capacitance when V_r is set to 10 V. When the bias voltage is set to 35 V, this circuit presents 0.32 pF capacitance at 2.35 GHz.

Realization of tunable inductor using ASV configuration is shown in Fig. 4.80a. A fixed inductor (L_f) is connected in parallel with the capacitor circuit as shown in Fig. 4.80a. The value of inductor should be selected such that the circuit presents

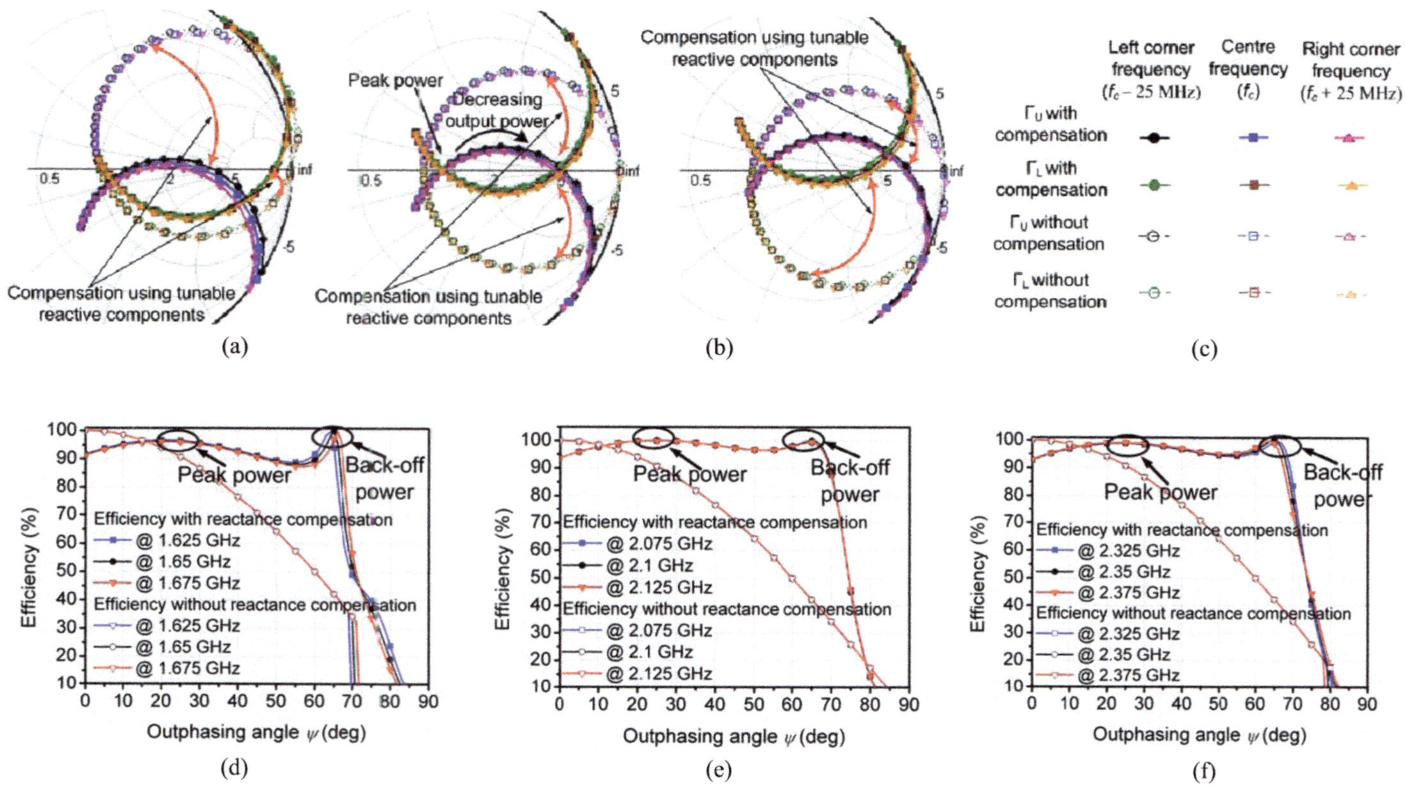

Fig. 4.78 Loads presented by Chireix combiner over ±25 MHz band centered at (**a**) 1.65 GHz, (**b**) 2.1 GHz, and (**c**) 2.35 GHz. Estimated efficiency of Chireix outphasing combiner over ±25 MHz band centered at (**d**) 1.65 GHz, (**e**) 2.1 GHz and (**f**) 2.35 GHz ([15], reprinted with permission from IEEE)

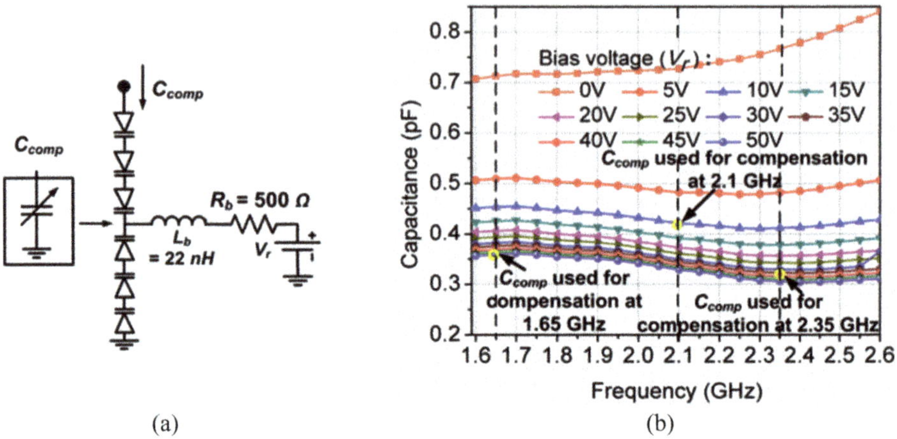

Fig. 4.79 (**a**) Tunable capacitance circuit and (**b**) capacitance versus frequency ([15], reprinted with permission from IEEE)

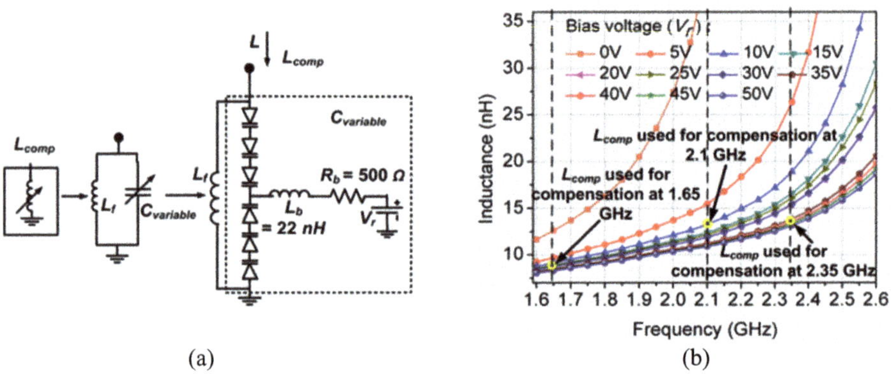

Fig. 4.80 (**a**) Tunable inductance circuit and (**b**) inductance versus frequency ([15], reprinted with permission from IEEE)

inductive reactance over the desired range of frequency. Fig. 4.80b shows the graph of inductance versus frequency over different bias voltages.

To compensate the reactance at lower PA branch of the outphasing system shown in Fig. 4.77 at multiple frequencies, this tunable inductor is used. At 1.65 GHz, the tunable inductor is biased at 30 V to present an inductance of 8.53 nH. Similarly, at 2.1 GHz, the circuit is biased at 10 V to realize 13.3 nH inductance and at 2.35 GHz, circuit is biased at 45 V to realize an inductance of 13.9 nH. Such tunable elements can be used to compensate the imaginary part of the load at multiple frequencies. Therefore, the efficiency enhancement at back-off power over expanded frequency range can be achieved.

Figure 4.81 shows the hardware implementation of reconfigurable outphasing transmitter with reconfigurable Chireix outphasing PA prototype.

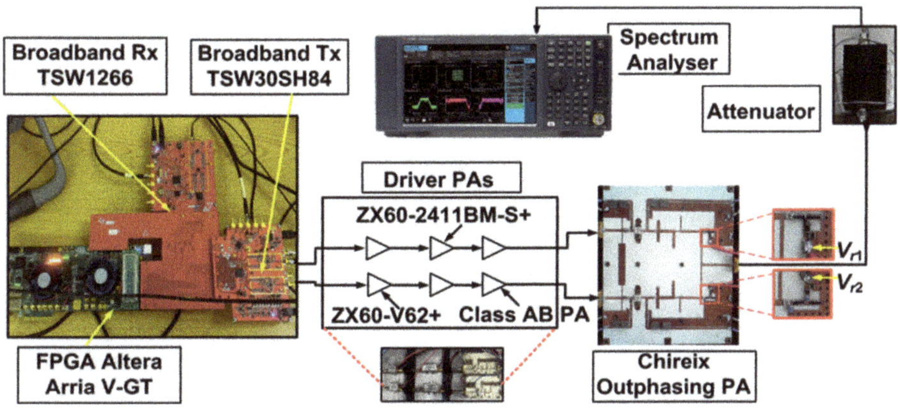

Fig. 4.81 Hardware setup for reconfigurable outphasing transmitter ([15], reprinted with permission from IEEE)

The hardware setup comprises a two-channel transmitter (TSW30SH84) from Texas Instruments. The baseband signals are generated in MATLAB and uploaded using Quartus to the RAM of Arria V (GT series) FPGA platform in I/Q format as shown in Fig. 4.81. This baseband data is then converted to analog signals using 16-bit DACs sharing the same clocks. The quadrature modulators then upconvert the baseband analog signal to RF signals in each transmitter channel. Since these two channels are synchronized with the same baseband clock of FPGA, one can set an accurate phase offset between the two channels at baseband level. This will provide an accurate outphasing between the RF outputs of the two channels. Later, the driver PAs are used to amplify power in each channel to drive the outphasing PA as shown in Fig. 4.81. V_{r1} and V_{r2} are the bias voltages applied to reactance compensation network at upper and lower PA branches. This reconfigurable outphasing system is tested with CW signal at three operating frequencies with ±25 MHz band around each frequency. The simulated and measured DE versus output power at three operating frequencies and their corner frequencies with ±25 MHz offset are shown in Fig. 4.82a–c.

One can see from Fig. 4.82a that the measured efficiency is varying between 51.5% and 62.7% for 6–7 dB OPBO at 1.65 GHz and its corner frequencies. At these frequencies, V_{r1} and V_{r2} are fixed at 45 and 30 V, respectively. Similarly, in case of reconfigured frequency of 2.1 GHz, measured efficiency varies between 40% and 48.1% for 6–7 dB OPBO as shown in Fig. 4.82b. When the system is reconfigured to the third frequency of 2.35 GHz, the measured efficiency degrades at 6–7 dB back-off in comparison to the simulated results. However, the efficiency enhancement is observed up to 5 dB OPBO at this frequency as shown in Fig. 4.82c. This may happen due to improper load modulation in hardware measurement at this frequency.

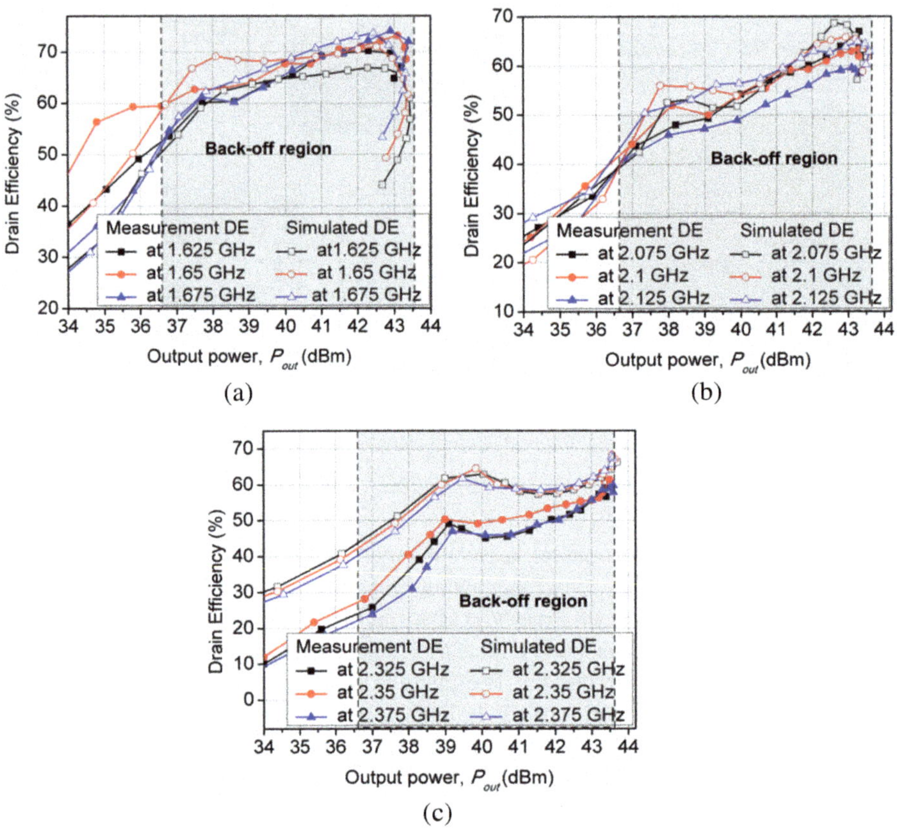

Fig. 4.82 Measured and simulated DE versus output power curves of outphasing PA over ±25 MHz band centered at (**a**) 1.65 GHz, (**b**) 2.1 GHz, and (**c**) 2.35 GHz ([15], reprinted with permission from IEEE)

4.5 Distributed Power Amplifiers

For an ultra-wideband PA the key limitation is device parasitic which must be absorbed or compensated with suitable broadband matching network. However, the gain bandwidth product of any amplifier is limited and both cannot be increased simultaneously beyond a certain limit. Efforts toward bandwidth expansion may result in a reduction in gain. A lower gain results in an inefficient architecture when several amplifier stages are cascaded to recover it. On the contrary, combining the outputs from various devices in parallel will enhance the overall output power but does not improve gain bandwidth product. Wheeler, while working on tubes [76], showed that this is proportional to the ratio of tube transconductance g_{m} to the square root of the product of input and output plate capacitances. Later, this concept was extended to the transistors using a simple transistor combined with coupling circuit in [77]. In such a case, the gain bandwidth product is inversely proportional to capacitance C which comprises device as well as coupling/combiner parasitic. One should seek for a solution where devices are combined with their output currents being added up without accumulating the effects of these shunt capacitances. The distributed amplifier is one of the schemes, where several devices are combined such that their input and output parasitic capacitances do not add up but the output currents are combined constructively. W.S. Percival was the first to present this concept in 1937, where the electrodes of the tubes were designed in a helical coil form, which combined with the inter electrode capacitors to form an artificial transmission line [78]. Later the idea was presented by Ginzton et al. in their article [2], which was followed by several researchers further. The development of distributed amplifier was initially carried out using vacuum tubes and later shifted to the transistors. Moser investigated GaAs MESFET-based distributed PA in 1967 [79] and was later followed by Jutzi in 1969 [80].

Figure 4.83 shows a generic schematic of this distributed amplifier using FET devices. The signal is coupled from the gate to the drain through the transistor. An RF signal when applied at the input of the gate line travels down the gate and will be absorbed at the termination Z_g. The gate of each transistor connected to this line is excited by the traveling voltages and will be coupled into the drain line through the trans-conductance of each transistor. The drain current at each node in the drain line generates traveling waves which travel toward right and appear as useful output. This is possible if all the drain currents add in phase as the signal propagates along the drain line and therefore one should keep the phase shifts per section provided by the gate and drain line same. Using the same transistors can further help in meeting this condition. Moreover, some additional capacitance can be added to compensate for these phase shifts. The signals traveling to the left will be absorbed by the terminating impedance Z_0.

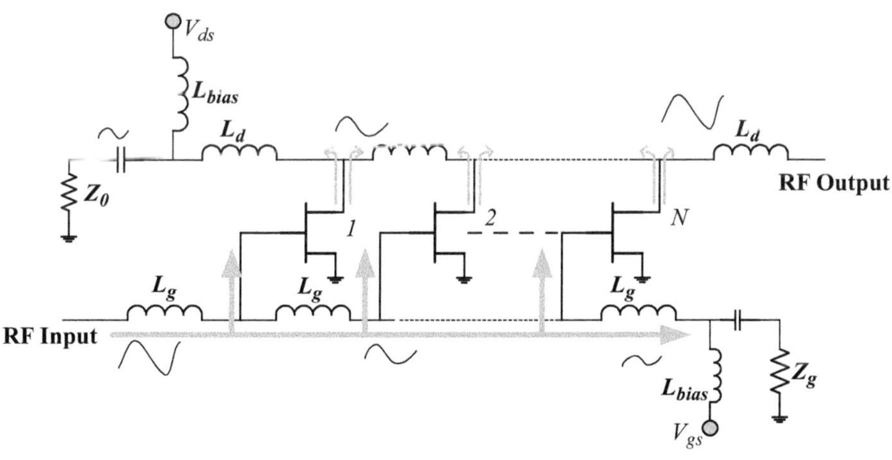

Fig. 4.83 Generic schematic of distributed amplifier using FET devices

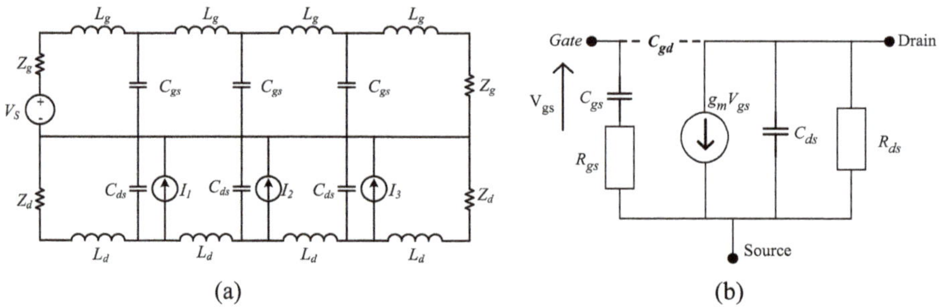

Fig. 4.84 (**a**) Equivalent circuit of distributed amplifier using FET devices, (**b**) equivalent circuit of unilateral FET

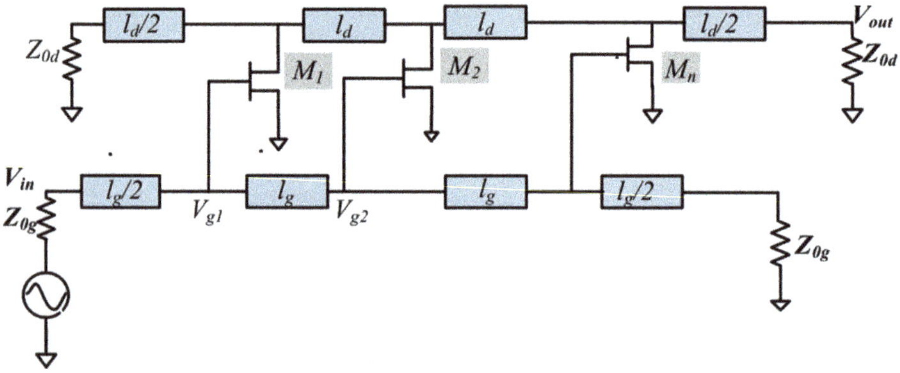

Fig. 4.85 Schematic representation of distributed amplifier using FETs

Figure 4.84a shows the equivalent circuit of the distributed amplifier. Here the simplified equivalent circuit of unilateral FET as shown in Fig. 4.84b is used where C_{gd} is neglected. One can see from Fig. 4.84a that the input and output capacitances of the transistor along with inductors at gate (L_{g}) and drain (L_{d}) represent an *LC* low-pass ladder network. This network can be visualized as an artificial transmission line with a series inductance connecting adjacent drains and gates. These artificial transmission lines are terminated with the appropriate impedances Z_{g} and Z_{d} in the respective gate and drain side in Fig. 4.84a.

In order to extend the operation of distributed amplifier to higher frequency, it is advantageous to replace inductors in gate and drain line with transmission line. This was proposed by Ayasli et al., who developed monolithic GaAs distributed amplifier and tested it successfully in 1981 [81] and 1982 [83]. Since the voltages in gate and drain are represented by traveling wave in these transmission lines, this approach is named as traveling wave amplifier or traveling wave distributed amplifier (TWDA). Ayasli et al., also presented a theoretical analysis for the TWDA constructed with periodically loaded transmission lines [81]. Figure 4.85 shows the topology of TWDA.

The voltage across the gate of M_1 transistor is given by.

$$V_{\mathrm{g1}} = V_{\mathrm{in}} \mathrm{e}^{-\gamma_{\mathrm{g}} l_{\mathrm{g}}/2} \tag{4.96}$$

$$V_{\mathrm{gk}} = V_{\mathrm{G1}} \mathrm{e}^{-(k-1)\gamma_{\mathrm{g}} l_{\mathrm{g}}} \tag{4.97}$$

Using Eqs. (4.96 and 4.97), one can write

$$V_{\mathrm{gk}} = V_{\mathrm{in}} \mathrm{e}^{-\left(k-\frac{1}{2}\right)\gamma_{\mathrm{g}} l_{\mathrm{g}}} \tag{4.98}$$

Similarly, it is assumed that the current flows at each node connected to the drain of each transistor. This current is divided into two parts moving in opposite direction. The current flowing out of kth node and heading toward the load $Z_{0\mathrm{d}}$ is given by,

$$I_{\mathrm{out}} = -\frac{1}{2} \sum_{k=1}^{n} I_{\mathrm{dk}} \mathrm{e}^{-\left(n-k+\frac{1}{2}\right)\gamma_{\mathrm{d}} l_{\mathrm{d}}} \tag{4.99}$$

The drain current of kth transistor depends on $g_{\mathrm{mk}} V_{\mathrm{gk}}$ which when applied to Eq. (4.98) results in,

$$I_{\mathrm{out}} = -\frac{1}{2} g_{\mathrm{m}} V_{\mathrm{in}} \mathrm{e}^{-n\gamma_{\mathrm{d}} l_{\mathrm{d}}} \mathrm{e}^{(\gamma_{\mathrm{g}} l_{\mathrm{g}} - \gamma_{\mathrm{d}} l_{\mathrm{d}})/2} \sum_{k=1}^{n} I_{\mathrm{dk}} \mathrm{e}^{-k(\gamma_{\mathrm{g}} l_{\mathrm{g}} - \gamma_{\mathrm{d}} l_{\mathrm{d}})} \tag{4.100}$$

The above equation assumes that each device is same and has same transconductance, that is, $g_{\mathrm{mk}} = g_{\mathrm{m}}$. Moreover, the term under summation can be further simplified by calculating sum of geometric progression as,

$$I_{\mathrm{out}} = -\frac{1}{2} g_{\mathrm{m}} V_{\mathrm{in}} \mathrm{e}^{-n\gamma_{\mathrm{d}} l_{\mathrm{d}}} \mathrm{e}^{(\gamma_{\mathrm{g}} l_{\mathrm{g}} - \gamma_{\mathrm{d}} l_{\mathrm{d}})/2} \frac{1 - \mathrm{e}^{-n(\gamma_{\mathrm{g}} l_{\mathrm{g}} - \gamma_{\mathrm{d}} l_{\mathrm{d}})}}{1 - \mathrm{e}^{-(\gamma_{\mathrm{g}} l_{\mathrm{g}} - \gamma_{\mathrm{d}} l_{\mathrm{d}})}} \tag{4.101}$$

For a lossless transmission line one can further modify Eq. (4.101) as,

$$I_{\mathrm{out}} = -\frac{1}{2} g_{\mathrm{m}} V_{\mathrm{in}} \frac{\sin\left\{\frac{n}{2}\left(\beta_{\mathrm{g}} l_{\mathrm{g}} - \beta_{\mathrm{d}} l_{\mathrm{d}}\right)\right\}}{\sin\left\{\frac{1}{2}\left(\beta_{\mathrm{g}} l_{\mathrm{g}} - \beta_{\mathrm{d}} l_{\mathrm{d}}\right)\right\}} \tag{4.102}$$

The available power gain is given by,

$$G_{\mathrm{p}} = \frac{\frac{1}{2} Z_{0\mathrm{d}} |I_{\mathrm{out}}|^2}{V_{\mathrm{in}}^2 / 2Z_{0\mathrm{g}}} = \frac{1}{4} g_{\mathrm{m}}^2 Z_{0\mathrm{g}} Z_{0\mathrm{d}} \left| \frac{\sin\left\{\frac{n}{2}\left(\beta_{\mathrm{g}} l_{\mathrm{g}} - \beta_{\mathrm{d}} l_{\mathrm{d}}\right)\right\}}{\sin\left\{\frac{1}{2}\left(\beta_{\mathrm{g}} l_{\mathrm{g}} - \beta_{\mathrm{d}} l_{\mathrm{d}}\right)\right\}} \right|^2 \tag{4.103}$$

For the case, where $\beta_g l_g = \beta_d l_d$, that is, the phase shift per section of the gate line is equal to the phase shift per section of the drain line, the available power gain expression is further simplified to

$$G_p = \frac{1}{4} n^2 g_m^2 Z_{0g} Z_{0d} \tag{4.104}$$

Like conventional cascaded stage, the available power gain increases with increase in number of stages, that is, n. However, one can see from Eq. (4.104), the gain in case of distributed amplifier with lossless assumption increases as n^2 whereas in a conventional cascaded stage, gain increases with $(G)^n$ [44]. Also, if lossy transmission line is considered, the traveling wave in gate and drain line decays exponentially and increasing number of sections n cannot compensate for an exponential decay of signals. In the presence of lossy transmission line, if one further assumes that $|\gamma_g| = |\beta_g|$ and $|\gamma_d| = |\beta_d|$, the available power gain G_P can be modified using Eq. (4.101) as,

$$G_P = \frac{1}{4} g_m^2 Z_{0g} Z_{0d} \left(\frac{e^{-n\alpha_d l_d} - e^{-n\alpha_g l_g}}{e^{\frac{(\alpha_g l_g - \alpha_d l_d)}{2}} - e^{-\frac{(\alpha_g l_g - \alpha_d l_d)}{2}}} \right)^2 \tag{4.105}$$

One can see that as n approaches to infinity the gain will reduce to zero. Therefore, one must find the optimum value of n that maximizes the available power gain G_P which is given by,

$$n_{opt} = \frac{\ln\left(\alpha_g l_g / \alpha_d l_d\right)}{\alpha_g l_g - \alpha_d l_d} \tag{4.106}$$

where α_g and α_d are the gate and drain line attenuation constants and can be derived from the theory of constant k line. Irrespective of the fact that TWDA looks like an attractive approach to realize wideband amplifier, its application is restricted to the gain block or driver amplifiers. The architecture wastes up to 50% of the output current available from various FET devices by allowing the current to flow in reverse direction and dissipates in Z_0 as shown in Fig. 4.85. This reduces the overall efficiency and output power per transistor in comparison to the case where all the current was tapped at desired output load. Moreover, the transistors do not contribute equal powers in the forward direction. This is only possible if each device sees its optimal load impedance R_{opt} guaranteeing best output power and efficiency out of each device. Therefore, the TWDA topology must provide these optimum loads to each PA along with key requirement of equalizing the phase shift per section of the gate and drain. In lieu of these requirements, modified TWDA architecture with tapered drain line is proposed for PA applications [83–88]. Figure 4.86a, b shows this modified TWDA architecture combining four FET devices with tapered drain line. One can see that there is no drain line termination Z_0 unlike in Fig. 4.85. Each transistor is connected with a drain line having different characteristic impedances.

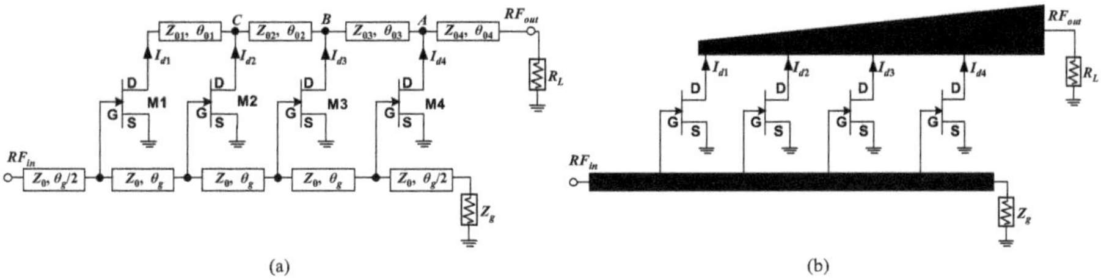

Fig. 4.86 TWDA with tapered drain line: (a) schematic and (b) layout

These characteristic impedances reduce gradually toward right, eventually having the lowest value for Z_{04} resulting into tapered line in the drain side.

In order to analyze the tapered drain line, Fig. 4.87a–c shows the drain side circuit broken into different fragments at different nodes where transistors are connected in drain line.

One can see from Figs. 4.86 and 4.87 that no current is flowing to the left side of the drain line assuming the device drain terminal presenting high impedance. The impedance seen by transistor M4, that is, Z_{in4}, can be calculated from Fig. 4.87a. One can calculate impedance seen by M4 transistor as,

$$Z_{\text{in4}} = \frac{V_{\text{A}}}{I_{\text{d4}}} = \left(1 + \frac{I_{\text{d1}} + I_{\text{d2}} + I_{\text{d3}}}{I_{\text{d4}}}\right) Z_{\text{L4}} \tag{4.107}$$

where it has been assumed that there is no current leakage toward drain terminal of the transistor. Similarly, the impedance seen by the added currents from M_1, M_2, and M_3 (i.e., $I_{\text{d1}} + I_{\text{d2}} + I_{\text{d3}}$) is given by,

$$Z_{\text{C4}} = \frac{V_{\text{A}}}{I_{\text{d1}} + I_{\text{d2}} + I_{\text{d3}}} = \left(1 + \frac{I_{\text{d4}}}{I_{\text{d1}} + I_{\text{d2}} + I_{\text{d3}}}\right) Z_{\text{L4}} \tag{4.108}$$

Similarly, one can use Fig. 4.87b to calculate impedance seen by M3 transistor as,

$$Z_{\text{in3}} = \frac{V_{\text{B}}}{I_{\text{d3}}} = \left(1 + \frac{I_{\text{d1}} + I_{\text{d2}}}{I_{\text{d3}}}\right) Z_{\text{L3}} \tag{4.109}$$

The impedance seen by the added currents from M_1, M_2 (i.e., $I_{\text{d1}} + I_{\text{d2}}$) in Fig. 4.87b is given by,

$$Z_{\text{C3}} = \frac{V_{\text{B}}}{I_{\text{d1}} + I_{\text{d2}}} = \left(1 + \frac{I_{\text{d3}}}{I_{\text{d1}} + I_{\text{d2}}}\right) Z_{\text{L3}} \tag{4.110}$$

where Z_{C4} is already calculated in Eq. (4.108). From Fig. 4.87c, one can obtain impedance seen by M_2 transistor as,

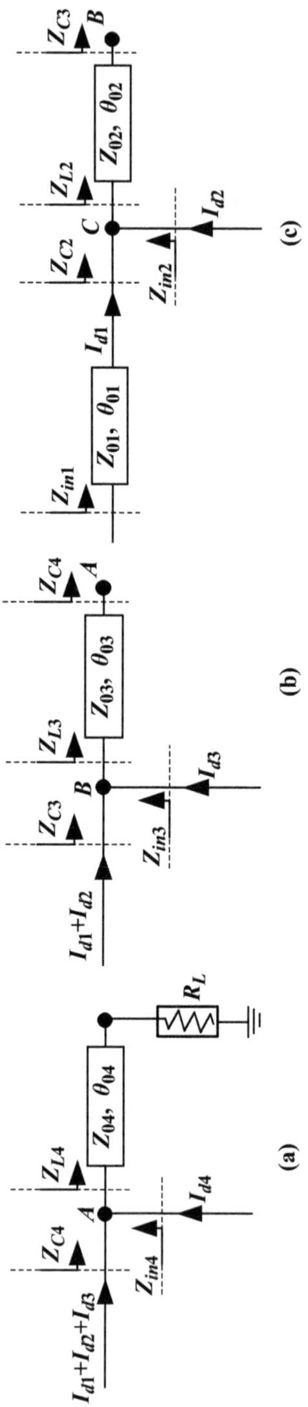

Fig. 4.87 Tapered drain line impedances seen at different nodes: (**a**) node A, (**b**) node B, and (**c**) node C

$$Z_{\text{in2}} = \frac{V_C}{I_{d2}} = \left(1 + \frac{I_{d1}}{I_{d2}}\right) Z_{L2} \tag{4.111}$$

The impedance seen by the current from M1 (i.e., I_{d1}) at node C in Fig. 4.87c is given by,

$$Z_{C2} = \frac{V_C}{I_{d1}} = \left(1 + \frac{I_{d2}}{I_{d1}}\right) Z_{L2} \tag{4.112}$$

In case of all the transistors being identical and operating in saturation, their drain currents are equal. Now, if one chooses R_L and Z_{04} as $R_{opt}/4$, Eq. (4.107) results in impedance Z_{in4} as R_{opt} seen by the transistor M4 irrespective of the electrical length θ_{04} chosen for the transmission line with characteristic impedance Z_{C4}. This gives additional freedom to choose the electric length of each section of the drain line in order to equalize the phase shift per section of the gate and drain line. The R_{opt} is the optimum impedance required by the transistor to operate in a desired class of operation. This results in Z_{C4} as $R_{opt}/3$ from Eq. (4.108). Now, if the characteristic impedance Z_{03} is selected as $R_{opt}/3$, the Z_{L3} is also $R_{opt}/3$ since the transmission line terminated with its characteristic impedance will present the input impedance as same as its characteristic impedance. This results in impedance Z_{in3} as R_{opt} seen by the transistor M3 from Eq. (4.109). One can also calculate Z_{C3} as $R_{opt}/2$ from Eq. (4.110). Similarly, the characteristic impedance Z_{02} is selected as $R_{opt}/2$ such that Z_{L2} is also $R_{opt}/2$ and is independent of the value of the chosen electrical length θ_{02}. This results in impedance Z_{in2} as R_{opt} seen by the transistor M2 from Eq. (4.111). The value of Z_{C3} is calculated as R_{opt} from Eq. (4.110). If the characteristic impedance Z_{01} is selected as R_{opt}, the impedance seen by the transistor M_1, that is Z_{in1} as R_{opt} irrespective of the electrical length θ_{01} chosen. Therefore, the tapered drain line has characteristic impedance $Z_{01} = R_{opt}$, $Z_{02} = R_{opt}/2$, $Z_{03} = R_{opt}/3$, $Z_{04} = R_{opt}/4$, and $R_L = R_{opt}/4$. The R_L can further be transformed to 50 Ω using any multisection impedance transformer.

In addition to the tapered drain line, the device scaling will also help in further improving the efficiency and the output power. This is due to the fact that in case of similar devices, the input signal starts from the device at right side and is amplified in each stage. At each stage, the output amplified signal adds up with the signal coming from the left in the drain line. This results in the largest voltage swing across the final stage at saturated output, whereas the preceding amplifiers are unable to attain their maximum voltage swing. Thus, in case of similar devices, the devices operating toward left of the final stage will operate in limited voltage swing and operate inefficiently. However, if one increases the voltage swing in preceding stages, less current will be required to produce the same output power. This is possible by scaling the transistor size from left to right and using the tapered drain line where the drain line impedance reduces from left to right. By adopting this approach, a wideband distributed PA with 77 GHz bandwidth is designed with improved PAE and output power compared to conventional distributed PA [89].

There are several other modifications done on TWDA architecture for various applications. In dual fed distributed PA, input signal is fed at both sides of the gate line [90]. In such a case, the output signals at the drain line has constructive contribution from signal moving from left to right in gate line as well as vice versa resulting in further amplified output signal. This topology has also better noise figure than conventional TWDA [90]. An alternate topology for dual-fed TWDA is single ended dual-fed TWDA [91]. In this arrangement reactive termination is used at the input and output line idle ports. This significantly improves gain and PAE in comparison to the conventional dual-fed TWDA [91].

Matrix PA is another variation of TWDA. In this architecture, various TWDA are stacked onto each other forming a matrix of active elements which leads to additive and multiplying gains at the same time [92]. Figure 4.88 shows a matrix PA using $m \times n$ active elements. A composite right/left-handed (CRLH) transmission line is used here [92]. The CRLH transmission line is special in the sense that it behaves as right-handed transmission line (positive permittivity) below a certain frequency and left-handed transmission line (negative permittivity) beyond that. The mathematical analysis performed in [92] reveals that the zero-phase condition of CRLH transmission line enables maximum output power with the reduction in overall circuit size. The open circuit CRLH transmission line used in this case reflects the forward traveling signal. The forward and reflected signals add up and forms the necessary gate signal for the TWDA section above it. The advantage of such architecture is better gain and power performance compared to conventional TWDA [92].

There are many variations in the topology of TWDA, yet the aim is to increase the gain, PAE, and output power. The key point in any TWDA is to obtain

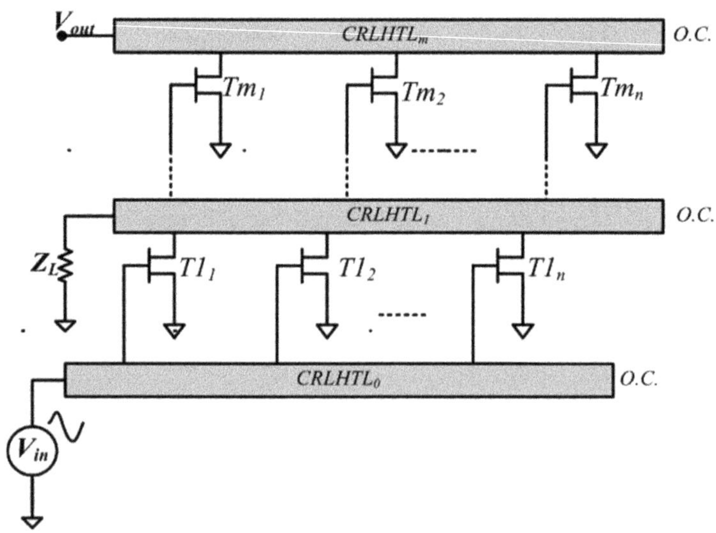

Fig. 4.88 Architecture of matrix PA using a composite right/left-handed (CRLH) transmission line

equalization of phase shift per section in gate and drain line since the output current is dependent on phase coherence of individual currents supplied from each device. This equalization may require extra capacitance to the drain line which sometimes adds more complexity in realizing these amplifiers. It is very important to present optimum load to each device such that it will give optimum output power and efficiency. Therefore, the gate and drain line impedances must be chosen judiciously to get the best performance out of TWDA.

4.6 Broadband Power Combining Techniques

Power combining is required to enhance the RF output power by combining several devices or PAs. High power at RF frequencies is sometimes difficult to obtain with the stand-alone device due to limited voltage handling and low gain pertaining to large parasitic in case of high current handling. Especially in GaAs-based transistors which are popular in RF applications, the low energy band-gap ($\approx$1.43 eV) limits the high-power application resulting in power combining which is essential for high-power output. With the advent of high-energy band-gap devices such as GaN ($\approx$3.4 eV), SiC ($\approx$3.05 eV), etc., these constraints are relaxed. Yet the power combining is still important for getting high power at high frequency and over larger bandwidths since the device scaling increases device parasitic and reduces operating frequency. In such a case, sometimes it is better to combine multiple PAs using a suitable power combiner to deliver a further amplified signal to the load. Combining power of more than one PAs can be accomplished at the device as well as at the circuit level [93, 94]. In the former case a number of devices are clustered in a region which is small compared to the wavelength. For circuit level power combining no such constraints are placed and the power is combined with the help of power combiners. Power combiner/distributor constitutes an integral part of high-power RF amplifier design, as well as of some special type of PAs. Being a reciprocal device, the same circuit can work both as a power combiner and power splitter. The key characteristics of a power combiner are its return loss, insertion loss, isolation, bandwidth, power handling capacity, integrating capability, combining efficiency, amplitude, and phase imbalance. Depending upon the application, a particular characteristic of the combiner will be emphasized. It is noteworthy that the splitter/combiner must be arranged such that if any branch of the splitter/combiner fails to operate, the total output power must not vary significantly [93]. The combining efficiency in this case is defined as,

$$\eta_{\text{comb}} = \frac{P_{\text{out,comb}}}{P_{\text{in,comb}}} \tag{4.113}$$

where $P_{\text{out,comb}}$ and $P_{\text{in,comb}}$ are the combined power at the output and input, respectively. The most common architectures of power combiner are corporate and chain combiners. In such a case, the power combining can be obtained in two

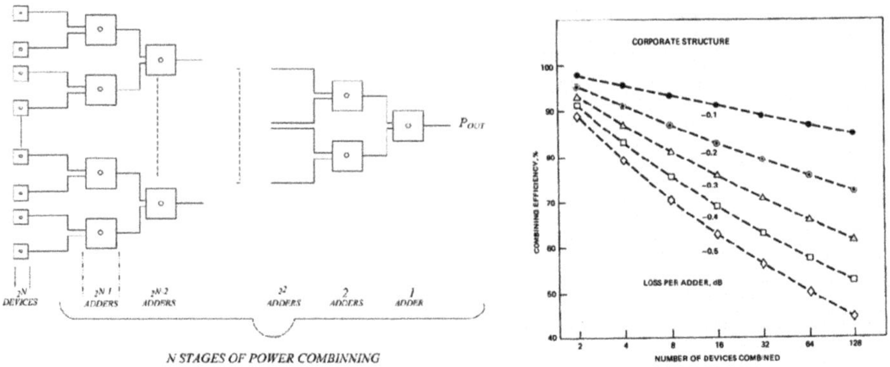

(a) (b)

Fig. 4.89 Corporate/tree power combining structure: (**a**) architecture and (**b**) combiner efficiency [95], reprinted with permission from IEEE)

different ways. In the first method, N power inputs are combined in a single stage and the second method contains a chain (serial) and tree (corporate) combining structures [95]. A general corporate/tree power combining structure is shown in Fig. 4.89a. It has N power combining stages to combine the power of 2^N identical devices. A nominal isolation is maintained between each output port (when used as a power divider) owing to isolation between the output ports of each combining element.

Although ideally any number of devices can be combined with this approach, in practice the losses incurred by each individual combiner limits the overall combining efficiency. A graph showing the combiner efficiency versus the number of devices combined is illustrated in Fig. 4.89b. It shows that as the number of combining devices is increased, the combiner efficiency deteriorates. Also, for higher loss per adder circuit, the rate of degradation in combiner efficiency is faster with increase in the combining device. One can also see that the corporate/tree-combiner is a combination of several binary combiners, where each combining element combines two inputs into a single output. In practice, these can be realized using conventional WPD/combiners [44]. The WPD/combiners offer good isolation between the two input paths providing good performance even in combining power devices which are not perfectly identical. However, in the case of chip design, the resistor as well as long quarter-wave lines may present wastage in terms of the effective chip area. Therefore, non-isolated combiner as shown in Fig. 4.90 can be used. In this topology, the two devices operating at saturation are directly combined through two transmission lines. An analysis similar to Fig. 4.87 results in the impedance seen by active devices as,

$$Z_1 = \frac{V_A}{I_1} = \left(1 + \frac{I_2}{I_1}\right) R_L \tag{4.114}$$

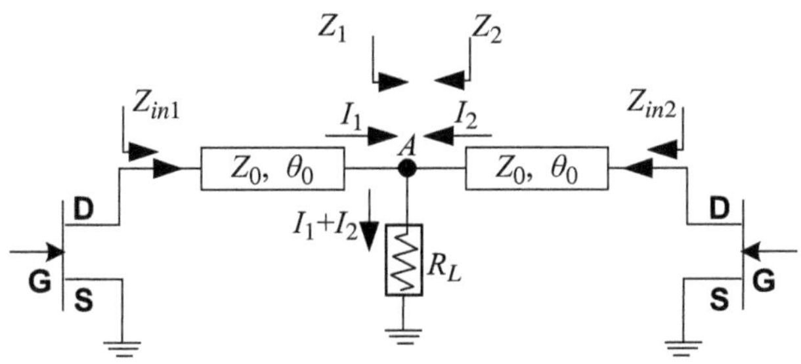

Fig. 4.90 Architecture of a non-isolated combiner

$$Z_2 = \frac{V_A}{I_2} = \left(1 + \frac{I_1}{I_2}\right) R_L \tag{4.115}$$

At saturation, if the two transistors are considered identical, one can write $I_1 = I_2$, which results in $Z_1 = Z_2 = 2R_L$. Choosing, $R_L = R_{opt}/2$ will provide $Z_1 = Z_2 = R_{opt}$, where R_{opt} is the optimum load seen by the device to obtain best performance in terms of power and efficiency. Later, one can choose the characteristic impedance of connecting lines Z_0 as R_{opt} such that both the devices will see $Z_{in1} = Z_{in2} = R_{opt}$. The θ_0 can be set to 90° in both the branches.

Figure 4.91a shows an N stage chain/serial combiner, where each coupler contributes (1/N)th of the total output power to the overall output. The coupling power of each coupler decides the number of couplers to be used for certain output power, that is, the power budget.

Unlike the corporate combiner, here the number of input signals can also be odd. One can easily add another chain combiner by adding a new source to the line through a coupler with coupling coefficient $10\log_{10}(N + 1)$ [95]. The losses of the coupling structure degrade the combining efficiency with increase in input devices as shown in Fig. 4.91b.

4.6.1 *Transformer-Type Power Combiner*

In an alternate approach, transformer-type combiners are used at the output stage in [43, 95–100]. They are of two types: series combining transformers (SCTs) and parallel combining transformers (PCTs). The advantages of a power combining transformer are impedance matching by controlling the turn ratio, DC isolation due to separated primary and secondary winding, and conversion of differential signal to single-ended signal [96]. Requirement of a transformer with high turn ratio in order to handle Watt level power has the drawbacks of low Q and low magnetic coupling. Instead, multiple transformers with reduced turn ratio are preferred. The transformer can be easily realized using microstrip line-based inductors.

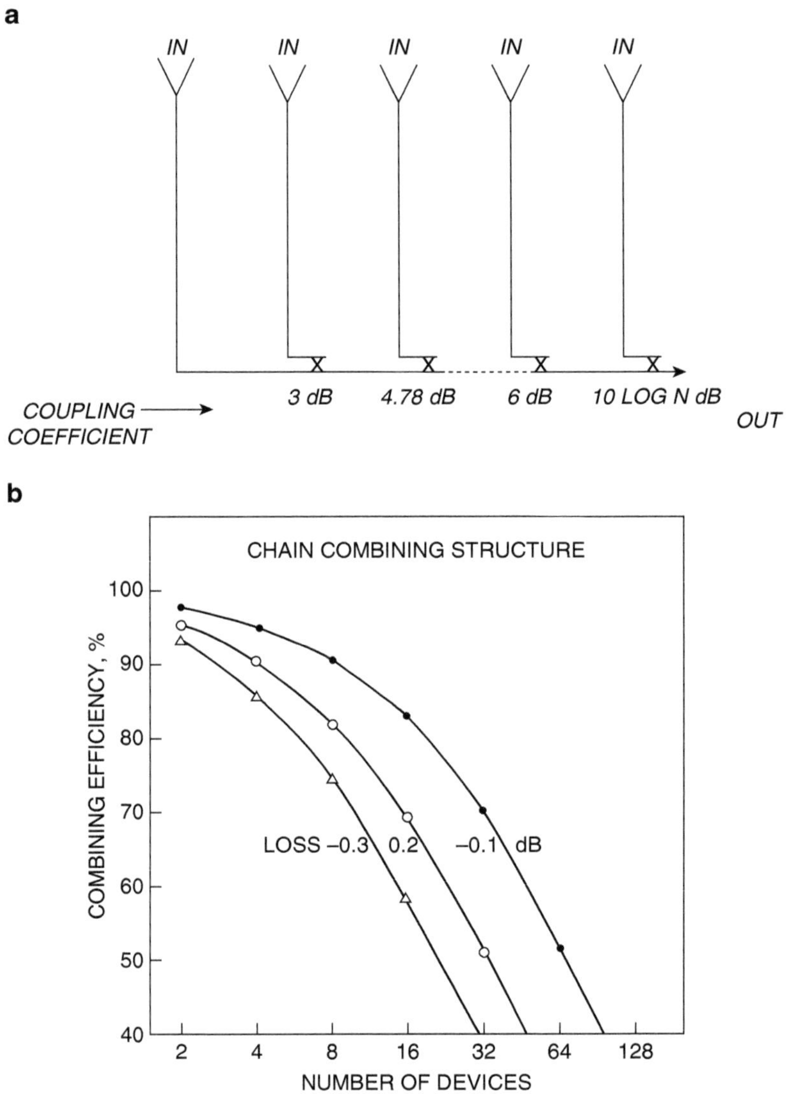

Fig. 4.91 Chain/serial power combining structure: (**a**) architecture and (**b**) combiner efficiency ([95], reprinted with permission from IEEE)

As stated earlier the two transformer combiner architectures, namely SCT and PCT are depicted in Fig. 4.92a, b. In both the cases M number of device outputs are combined to deliver the final output to load R_{load}. N_1 and N_2 are the number of primary and secondary turns of the transformer, respectively.

Recalling the current and voltage relation for any transformer in each case can be given as,

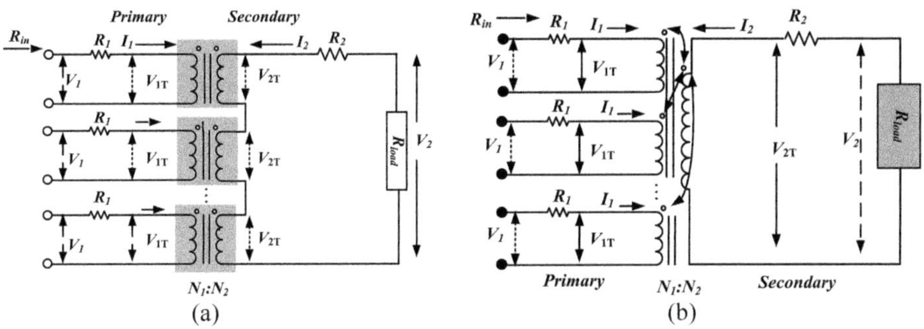

Fig. 4.92 Power combining transformers: (**a**) series combining transformer, (**b**) parallel combining transformer

$$\frac{I_2}{I_1} = \frac{V_{1T}}{V_{2T}} = \frac{N_1}{N_2} \tag{4.116}$$

where V_{1T} and V_{2T} are the voltages across primary and secondary windings of each transformer in both SCT and PCT configuration. Similarly, I_1 and I_2 are currents flowing through primary and secondary windings of each transformer configuration.

In case of SCT depicted in Fig. 4.92a, the secondary winding of all the transformers are connected in series. As a result, same current I_2 flows through each secondary winding. The coupled output signal at each secondary winding gets added at each stage, such that the final output signal is further amplified. The voltage across each transformer winding is $1/M$ times the overall voltage at the node just before R_2 which can be obtained using current I_2 and the effective load $R_2 + R_{\text{load}}$. Therefore using Eq. (4.116), the voltage at the primary winding of each transformer in SCT configuration can be written as,

$$V_{1T} = \frac{N_1}{N_2}\frac{1}{M}\left(R_2 + R_{\text{load}}\right) I_2 \tag{4.117}$$

Current I_1 can be expressed in terms of I_2 using transformer turn ratios N_1 and N_2 in Eq. (4.116). Later, this will be used in calculating input impedance looking into the transformer as,

$$R_{\text{in}} = \frac{V_1}{I_1} = R_1 + \frac{V_{1T}}{I_1} = R_1 + \left(\frac{N_1}{N_2}\right)^2 \frac{1}{M}\left(R_2 + R_{\text{load}}\right) \tag{4.118}$$

The power-combiner ratio for SCT topology is given by,

$$PCR_{\text{SCT}} = \frac{P_2}{P_1} = \frac{V_2 I_2}{V_1 I_1} = \frac{\left(\frac{N_1}{N_2}\right)^2 R_{\text{load}}}{R_1 + \left(\frac{N_1}{N_2}\right)^2 \frac{1}{M}\left(R_2 + R_{\text{load}}\right)} \tag{4.119}$$

In case of PCT topology as shown in Fig. 4.92b, a common secondary winding is used for M number of primary windings. In such a case, the current in the secondary winding is equally attributed from each primary winding. The input resistance R_{in} can be calculated as,

$$R_{\text{in}} = R_1 + M\left(\frac{N_1}{N_2}\right)^2 (R_2 + R_{\text{load}}) \tag{4.120}$$

The corresponding power combining ratio is given as,

$$\text{PCR}_{\text{PCT}} = \frac{P_2}{P_1} = \frac{V_2 I_2}{V_1 I_1} = \frac{M^2\left(\frac{N_1}{N_2}\right)^2 R_{\text{load}}}{R_1 + M\left(\frac{N_1}{N_2}\right)^2 (R_2 + R_{\text{load}})} \tag{4.121}$$

The transformer efficiency, defined as the power delivered to the load by the total input power, is represented as,

$$\eta = \frac{1}{M}\frac{P_2}{P_1} = \frac{1}{M}\text{PCR} \tag{4.122}$$

where *PCR* is given by Eqs. (4.119 and 4.121) in case of SCT and PCT, respectively. M is the number of transformers combined in both topologies.

4.6.2 Reactively Matched Power Combiner

Another broadband combining technique is reactively matched power combining. A broadband matching is incorporated within the combiner which provides impedance transformation, power combing, and biasing simultaneously from the same circuit. Since wideband matching is not a completely analytical design, it requires computer-aided optimization to obtain best performance. However, an appropriate design scheme is required for implementing an optimal reactively matched power combining technique for MMIC [7–12, 101]. In this topology, powers from multiple transistors are combined in parallel. The power combining can be corporate type, where each combining element should incorporate broadband matching in order to match the transistors to its optimum load. One can use load-pull simulations to find this optimum load. Figure 4.93a shows the generic topology of matching network as proposed in [9]. This matching network presents optimum load of $Z_{\text{L2,opt}}$.

This topology apparently looks like a ladder network used in filter design [9]; however, the topology being generic, any section of this ladder network can be chosen as low-pass as well as high-pass. Therefore, this topology comprises series and shunt elements that can be realized with inductor and capacitors and defined as

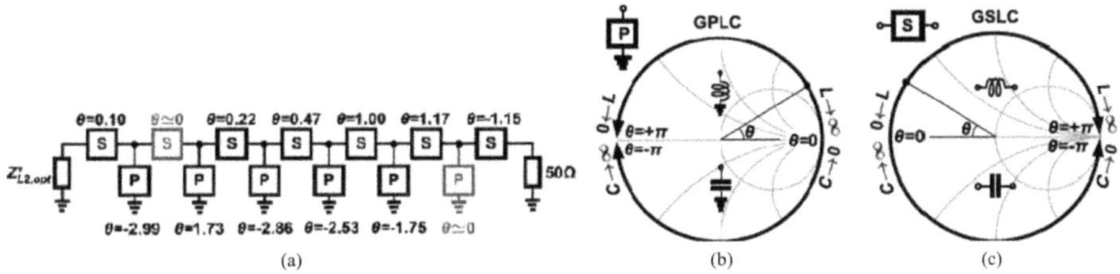

Fig. 4.93 Reactively matched circuit: (a) topology comprising of series and shunt sections of impedances Z_{GSLC} and Z_{GPLC}, (b) Smith chart showing inductor or capacitor realizing Z_{GPLC}, and (c) Smith chart showing inductor or capacitor realizing Z_{GSLC} ([9], reprinted with permission from IEEE)

generalized parallel LC (GPLC) and generalized series LC (GSLC). Figure 4.93b, c shows the behavior of GPLC and GSLC with the angle of the reflection coefficient θ of each cell at the center frequency of interest f_0. The impedance Z_{GPLC} and Z_{GSLC} corresponding to GPLC and GSLC sections can be mathematically represented as (Eqs. 4.123 and 4.124), respectively.

$$Z_{GPLC} = \begin{cases} Z_0 \frac{1+e^{j\theta}}{1-e^{j\theta}} \frac{f}{f_0} = j\left[Z_0 \cot(\theta/2)\right] \frac{f}{f_0} \text{ for } \theta \geq 0 \\ Z_0 \frac{1+e^{j\theta}}{1-e^{j\theta}} \frac{f_0}{f} = j\left[Z_0 \cot(\theta/2)\right] \frac{f_0}{f} \text{ for } \theta < 0 \end{cases} \tag{4.123}$$

$$Z_{GSLC} = \begin{cases} Z_0 \frac{1+e^{j(\pi-\theta)}}{1-e^{j(\pi-\theta)}} \frac{f}{f_0} = j\left[Z_0 \tan(\theta/2)\right] \frac{f}{f_0} \text{ for } \theta \geq 0 \\ Z_0 \frac{1+e^{j(\pi-\theta)}}{1-e^{j(\pi-\theta)}} \frac{f_0}{f} = j\left[Z_0 \tan(\theta/2)\right] \frac{f_0}{f} \text{ for } \theta < 0 \end{cases} \tag{4.124}$$

where Z_0 is the normalization impedance which can be chosen as 50 Ω, since optimum load is required to be transformed to the standard load of 50 Ω. The f is any frequency around f_0 where the desired performance is required.

From Fig. 4.93b, considering the Smith chart as impedance chart, one can see the right corner of the real axis where $\theta = 0$, the L approaches infinity, and C approaches 0 realizing an open circuit. Similarly, at the left corner of the real axis of Smith chart where $\theta = \pm\pi$, the L approaches 0, and C approaches infinity realizing a short circuit. Fig. 4.93c shows this trend for Z_{GSLC}. In this case, the left corner of the real axis of Smith chart where $\theta = 0$, the L approaches 0, and C approaches infinity realizing a short circuit. The right corner of the real axis where $\theta = \pm\,\pi$, the L approaches infinity, and C approaches 0 realizing an open circuit. From Eq. (4.123), it can be inferred that Z_{GPLC} is proportional to f for $\theta \geq 0$ and inversely proportional to f when $\theta < 0$ imitating a shunt combination of an inductor and capacitor. The values of the shunt inductor and capacitor can be calculated as,

$$L_{\text{shunt}}(\theta) = \frac{Z_{\text{GPLC}}}{j2\pi f_0} = \frac{Z_0 \cot(\theta/2)}{2\pi f_0} \quad \text{for } \theta \geq 0 \tag{4.125}$$

$$C_{\text{shunt}}(\theta) = \frac{1}{j2\pi f_0 Z_{\text{GPLC}}} = \frac{\tan(-\theta/2)}{2\pi f_0 Z_0} \quad \text{for } \theta < 0 \tag{4.126}$$

Similarly, from Eq. (4.124), one can see Z_{GSLC} is proportional to f for $\theta \geq 0$ and inversely proportional to f when $\theta < 0$ imitating series inductor and capacitor which are given by,

$$L_{\text{series}}(\theta) = \frac{Z_{\text{GSLC}}}{j2\pi f_0} = \frac{Z_0 \tan(\theta/2)}{2\pi f_0} \quad \text{for } \theta \geq 0 \tag{4.127}$$

$$C_{\text{shunt}}(\theta) = \frac{1}{j2\pi f_0 Z_{\text{GSLC}}} = \frac{\cot(-\theta/2)}{2\pi f_0 Z_0} \quad \text{for } \theta < 0 \tag{4.128}$$

In order to realize any optimum load, from the matching network of Fig. 4.93a, the values of θ for each series and shunt section will be optimized. The readily available optimization tools in CAD software such as random search and gradient optimization can be employed step by step to obtain the desired response. The cost function in this optimization can be reflection coefficient Γ_L corresponding to optimum load. For example, Fig. 4.93a shows an optimized section where θ parameter is altered using optimizer available in CAD [9]. Here, the center frequency f_0 and reference impedance Z_0 is chosen as 10 GHz and 50 Ω, respectively. Figure 4.94a shows the equivalent circuit comprising of L and C components until Γ_L is achieved better than −15 dB across a band of 6–18 GHz. The corresponding reflection coefficient achieved in each stage is shown in Fig. 4.94b plotted in Smith chart normalized with respect to 10 Ω [9].

The required order of matching network depends on bandwidth, impedance transformation ratio, and quality factor. One can start from lower order network and the number of GSLC and GPLC sections can be then progressively increased until the desired requirement is met. Using this technique, GaAs pHEMT MMIC was presented in [9] which operates at an ultra-wideband from 6 to 18 GHz with 10 W output power. In reactively matched topology, transmission lines and MIM capacitors can be used to obtain the minimum size of MMIC. Using reactively matched topology, MMIC PA up to 40 W was reported [7–12, 101], however at the cost of gain flatness. In order to optimize the gain performance up to Ku band, multiple gain stages can be used.

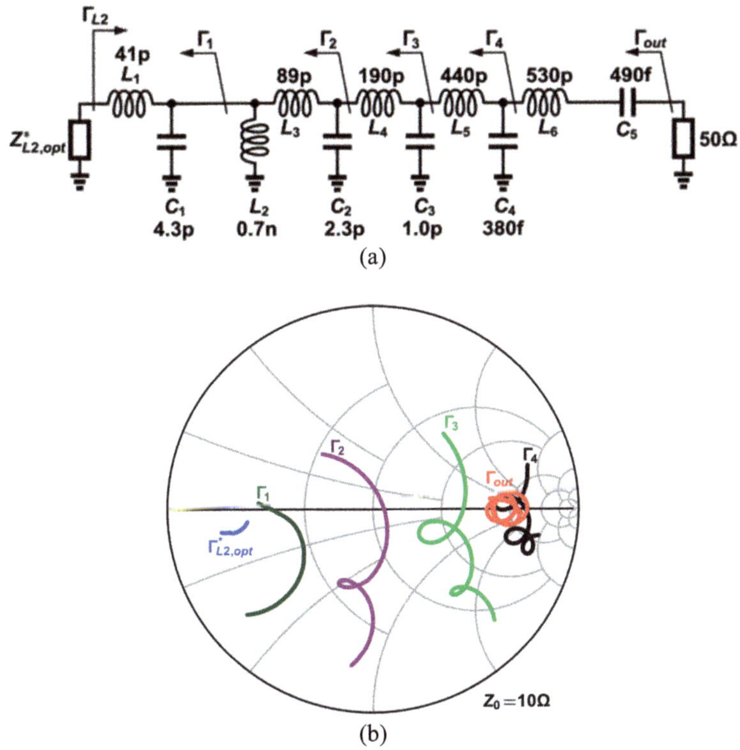

Fig. 4.94 Reactively matched topology using generalized parallel elements (GPLC) and generalized series elements (GSLC): (**a**) equivalent circuit obtained by replacing each GPLC and GSLC sections with corresponding inductors and capacitors, (**b**) Smith chart showing step-by-step transformation of 50 Ω load to optimum load at 6–18 GHz frequency band ([9], reprinted with permission from IEEE)

References

1. Ayasli, Y., Reynolds, L. D., Mozzi, R. L., & Hanes, L. K. (1984). 2-20 GHz GaAs traveling-wave power amplifier. *IEEE Transactions on Microwave Theory and Techniques, 32*(3), 290–295.
2. Ginzton, E. L., Hewlett, W. R., Jasberg, J. H., & Noe, J. D. (1948). Distributed amplification. *Proceedings of the IRE, 36*(8), 956–969.
3. Saxena, S., Rawat, K., & Roblin, P. (2017). Continuous Class-B/J power amplifier using nonlinear embedding technique. *IEEE Transactions on Circuits and Systems II: Express Briefs, 64*(7), 837–841.
4. Aggrawal, E., Rawat, K., & Roblin, P. (2017). Investigating continuous Class-F power amplifier using nonlinear embedding model. *IEEE Microwave and Wireless Components Letters, 27*(6), 593–595.
5. Sun, G., & Jansen, R. H. (2012). Broadband Doherty power amplifier via real frequency technique. *IEEE Transactions on Microwave Theory and Techniques, 60*(1), 99–111.

6. Sayed, A., von der Mark, S., & Boeck, G. (2004). An ultra wideband 5 W power amplifier using SiC MESFETs. In *34th European Microwave Conference, 2004, Amsterdam, The Netherlands* (pp. 57–60). Washington, DC: IEEE.
7. Kuwata, E., Yamanaka, K., Koyama, H., Kamo, Y., Kirikoshi, T., Nakayama, M., & Hirano, Y. (2011). C-Ku band ultra-broadband GaN MMIC amplifier with 20W output power. In *Proceedings of the Asia Pacific microwave conference* (pp. 1558–1561). Washington, DC: IEEE.
8. Litchfield, M., & Komiak, J. J. (2018). A 6–18 GHz 40 W reactively matched GaN MMIC power amplifier. In *Proceedings of of IEEE MTT-S International Microwave Symposium Digest*. Washington, DC: IEEE.
9. Meghdadi, M., & Medi, A. (2017). Design of 6–18-GHz high-power amplifier in GaAs pHEMT technology. *IEEE Transactions on Microwave Theory and Techniques, 65*(7), 2353–2360.
10. Mouginot, G., Ouarch, Z., Lefebvre, B., Heckmann, S., Lhortolary, J., Baglieri, D., Floriot, D., Camiade, M., Blanck, H., Le Pipec, M., Mesnager, D., & Le Helleye, P. (2010). Three stage 6–18 GHz high gain and high power amplifier based on GaN technology. In *Proceedings of IEEE MTT-S International Microwave Symposium Digest* (pp. 1392–1395). Washington, DC: IEEE.
11. Schmid, U., Sledzik, H., Schuh, P., Schroth, J., Oppermann, M., Brückner, P., Raay, F. V., Quay, R., & Seelmann-Eggebert, M. (2013). Ultra-wideband GaN MMIC chip set and high power amplifier module for multi-function defense AESA applications. *IEEE Transactions on Microwave Theory and Techniques, 61*(8), 3043–3051.
12. Tao, H. Q., Hong, W., & Zhang, B. (2017). 6-18 GHz 13W reactive matched GaN power amplifier MMIC. In *2017 12th European Microwave Integrated Circuits Conference* (pp. 357–360). Washington, DC: IEEE.
13. Lee, J.-W., & Webb, K. J. (2001). Broadband GaN HEMT push-pull microwave power amplifier. *IEEE Microwave and Wireless Components Letters, 11*(9), 367–369.
14. Hakala, I., Choi, D. K., Gharavi, L., Kajakine, N., Koskela, J., & Kaunisto, R. (2005). A 2.14-GHz Chireix outphasing transmitter. *IEEE transactions on Microwave Theory and Techniques, 53*(6), 2129–2138.
15. Aggrawal, E., & Rawat, K. (2019). Reconfigurable Chireix outphasing power amplifier over multiple frequency bands. *IEEE Transactions on Circuits and Systems II: Express Briefs*. https://doi.org/10.1109/TCSII.2019.2926529.
16. Carrubba, V. (2012). *Novel highly efficient broadband continuous power amplifier modes* (Doctoral dissertation, Cardiff University).
17. Wright, P., Lees, J., Benedikt, J., Tasker, P. J., & Cripps, S. C. (2009). A methodology for realizing high efficiency class-J in a linear and broadband PA. *IEEE Transactions on Microwave Theory and Techniques, 57*(12), 3196–3204.
18. Narendra, K., Prakash, C., Grebennikov, A., & Mediano, A. (2008). High-efficiency broadband parallel-circuit class E RF power amplifier with reactance-compensation technique. *IEEE Transactions on Microwave Theory and Techniques, 56*(3), 604–612.
19. Ortega-Gonzalez, F. J., Tena-Ramos, D., Patino-Gomez, M., Pardo-Martin, J. M., & Madueno-Pulido, D. (2013). High-power wideband L-band suboptimum class-E power amplifier. *IEEE Transactions on Microwave Theory and Techniques, 61*(10), 3712–3720.
20. Cripps, S. C., Tasker, P. J., Clarke, A. L., Lees, J., & Benedikt, J. (2009). On the continuity of high efficiency modes in linear RF power amplifiers. *IEEE Microwave and Wireless Components Letters, 19*(10), 665–667.
21. Du, X., You, C. J., Cai, J., Helaoui, M., Ghannouchi, F. M., Zhao, Y., & Li, X. (2018). Novel design space of load modulated continuous class-B/J power amplifier. *IEEE Microwave and Wireless Components Letters, 28*(2), 156–158.
22. Seo, M., Lee, H., Gu, J., Kim, H., Ham, J., Choi, W., Yun, Y., O, K. K., & Yang, Y. (2014). High-efficiency power amplifier using an active second harmonic injection technique under optimized third-harmonic termination. *IEEE Transactions on Circuits and Systems II: Express Briefs, 61*(8), 549–553.

23. Gowrish, B., Rawat, K., Basu, A., & Koul, S. K. (2013). Broadband matching network using band-pass filter with device parasitic absorption. In *Proc. 82nd ARFTG Microw. Meas. Conf., Columbus, OH, USA* (pp. 1–4). Piscataway: IEEE.
24. Carrubba, V., Clarke, A. L., Akmal, M., Lee, J., Benedikt, J., Tasker, P. J., & Cripps, S. C. (2011). On the extension of the continuous class-F mode power amplifier. *IEEE Transactions on Microwave Theory and Techniques, 59*(5), 1294–1303.
25. Jang, H., Roblin, P., & Xie, Z. (2014). Model-based nonlinear embedding for power-amplifier design. *IEEE Transactions on Microwave Theory and Techniques, 62*(9), 1986–2002.
26. Aggrawal, E., Saxena, S., & Rawat, K. Broadband power ampliifer design by exploring design space of continous class F mode. In *IEEE Asia Pacific Microwave Conference (APMC), 5th–9th December, 2016, New Delhi, India* (pp. 1–4). Piscataway: IEEE.
27. Raffo, A., Scappaviva, F., & Vannini, G. (2009). A new approach to microwave power amplifier design based on the experimental characterization of the intrinsic electron-device load line. *IEEE Transactions on Microwave Theory and Techniques, 57*(7), 1743–1752.
28. Carlin, H. J., & Komiak, J. J. (1979). A new method of broad-band equalization applied to microwave amplifiers. *IEEE Transactions on Microwave Theory and Techniques, 27*(2), 93–99.
29. Narendra, K., & Tee, Y. K. (2014). Optimized high-efficiency class E RF power amplifier for wide bandwidth and high harmonics suppression. *IET Circuits, Devices, System, 60*(10), 1–13.
30. Grebennikov, A. (2016). High-efficiency class-E power amplifier with shunt capacitance and shunt filter. *IEEE Transactions on Circuits and Systems I: Regular Papers, 63*(1), 12–22.
31. Zhou, J., Morris, K. A., Watkins, G. T., & Yamaguchi, K. (2015). Improved reactance-compensation technique for the Design of Wideband Suboptimum Class-E Power Amplifiers. *IEEE Transactions on Microwave Theory and Techniques, 63*(9), 2793–2801.
32. Chen, K., & Peroulis, D. (2011). Design of highly efficient broadband class-E power amplifier using synthesized low-pass matching networks. *IEEE Transactions on Microwave Theory and Techniques, 59*(12), 3162–3173.
33. Özen, M., Jos, R., & Fager, C. (2012). Continuous class-E power amplifier modes. *IEEE Transactions on Circuits and Systems II: Express Briefs, 59*(11), 731–735.
34. Kee, S. D., Aoki, I., Hajimiri, A., & Rutledge, D. (2003). The class-E/F family of ZVS switching amplifiers. *IEEE Transactions on Microwave Theory and Techniques, 51*(6), 1677–1690.
35. Asha Latha, Y. M., Rawat, K., & Roblin, P. (2019). Nonlinear embedding model-based continuous class E/F power amplifier. *IEEE Microwave and Wireless Components Letters, 29*(11), 714–717. https://doi.org/10.1109/LMWC.2019.2941919.
36. AlMuhaisen, A., Lees, J., Cripps, S. C., Tasker, P. J., & Benedikt, J. (2011). Wide band high-efficiency power amplifier design. In *2011 6th European Microwave Integrated Circuit Conference* (pp. 184–187). Piscataway: IEEE.
37. Latha, Y. M. A., Rawat, K., Helaoui, M., & Ghannouchi, F. M. (2019). Broadband Continuous Mode Power Amplifier with on-board Harmonic Injection. *IET Microwaves, Antenna and Propagation, 13*(9), 1402–1407.
38. Seo, M., Lee, H., Gu, J., Kim, H., Ham, J., Choi, W., & Yang, Y. (2014). High-efficiency power amplifier using an active second-harmonic injection technique under optimized third-harmonic termination. *IEEE Transactions on Circuits and Systems II: Express Briefs, 61*(8), 549–553.
39. Alizadeh, A., Yaghoobi, M., & Medi, A. (2017). Class-J2 power amplifiers. *IEEE Transactions on Circuits and Systems I: Regular Papers, 64*(8), 1989–2002.
40. Bathich, K., Markos, A. Z., & Boeck, G. (2011). Frequency response analysis and bandwidth extension of the Doherty amplifier. *IEEE Transactions on Microwave Theory and Techniques, 59*(4), 934–944.
41. Qureshi, J. H., Li, N., Neo, W. C. E., Rijs, F., Blednov, I., & de Vreede, L. C. N. (2010). A wideband 20 W LMOS Doherty power amplifier. In *2010 IEEE MTT-S International Microwave Symposium* (pp. 1504–1507). Piscataway: IEEE.

42. Wu, D. Y.-T., & Boumaiza, S. (2012). A modified Doherty configuration for broadband amplification using symmetrical devices. *IEEE Transactions on Microwave Theory and Techniques, 60*(10), 3201–3213.
43. Ghosh, S., & Rawat, K. (2018). A wideband two-stage Doherty power amplifier at high back-off by exploring feasible design space. *International Journal of RF and Microwave Computer-Aided Engineering, 28*(9), e21528.
44. Pozar, D. M. (2012). *Microwave engineering* (4th ed.). Hoboken, New Jersey: Wiley.
45. Belevitch, V. (1968). *Classical Network Theory*. San Francisco, (CA): Holden-Day.
46. Bode, H. W. (1959). *Network analysis and feedback amplifier design*. Boston, (MA): Tech. Composition Company.
47. Fano, F. M. (1960). Theoretical limitations on the broadband matching of arbitrary impedances. *Journal of the Franklin Institute, 249*(1), 57–83.
48. Ma, C., Pan, W., Shao, S., Qing, C., & Tang, Y. (2013). A wideband Doherty power amplifier with 100 MHz instantaneous bandwidth for LTE-advanced applications. *IEEE Microwave and Wireless Components Letters, 23*(11), 614–616.
49. Pang, J., He, S., Huang, C., Dai, Z., Peng, J., & You, F. (2015). A post-matching Doherty power amplifier employing low-order impedance inverters for broadband applications. *IEEE Transactions on Microwave Theory and Techniques, 63*(12), 4061–4071.
50. Giofre, R., Piazzon, L., Colantonio, P., & Giannini, F. (2013). A Doherty architecture with high feasibility and defined bandwidth behavior. *IEEE Transactions on Microwave Theory and Techniques, 61*(9), 3308–3317.
51. Rawat, K., Ajmera, G., Basu, A., & Koul, S. K. (2015). Design scheme for broadband D oherty power amplifier using broadband load combiner. *International Journal of RF and Microwave Computer-Aided Engineering, 25*(8), 655–674.
52. Giofrè, R., Colantonio, P., Giannini, F., & Piazzon, L. (2013). New output combiner for Doherty amplifiers. *IEEE Microwave and Wireless Components Letters, 23*(1), 31–33.
53. Piazzon, L., Giofrè, R., Colantonio, P., & Giannini, F. (2013). A wideband Doherty architecture with 36% of fractional bandwidth. *IEEE Microwave and Wireless Components Letters, 23*(11), 626–628.
54. Giofrè, R., Piazzon, L., Colantonio, P., & Giannini, F. (2014). An ultra-broadband GaN Doherty amplifier with 83% of fractional bandwidth. *IEEE Microwave and Wireless Components Letters, 24*(11), 775–777.
55. Ahn, H. R. (2006). *Asymmetric passive components in microwaveintegrated circuits* (1st ed.). New Jersey: Wiley.
56. Barthwal, A., Rawat, K., & Koul, S. K. (2017). A design strategy for bandwidth enhancement in three-stage Doherty power amplifier with extended dynamic range. *IEEE Transactions on Microwave Theory and Techniques, 66*(2), 1024–1033.
57. Pelk, M. J., Neo, W. C. E., Gajadharsing, J. R., Pengelly, R. S., de Vreede, L. C. N., & High-Efficiency, A. (2008). 100-W GaN three-way Doherty Amplifier for Base-Station applications. *IEEE Transactions on Microwave Theory and Techniques, 56*(7), 1582–1591.
58. Neo, W. C. E., Qureshi, J., Pelk, M. J., Gajadharsing, J. R., de Vreede, L. C. N., & Mixed-Signal Approach, A. (2007). Towards linear and efficient N-way Doherty amplifiers. *IEEE Transactions on Microwave Theory and Techniques, 55*(5), 866–879.
59. Barthwal, A., Rawat, K., & Koul, S. (2015). Bandwidth enhancement of three stage Doherty power amplifier using symmetric devices. *IEEE Transactions on Microwave Theory and Techniques, 63*(8), 2399–2410.
60. Srirattana, N., Raghavan, A., Heo, D., Allen, P. E., & Laskar, J. (2009). Analysis and design of a high-efficiency multistage Doherty power amplifier for wireless communications. *IEEE Transactions on Microwave Theory and Techniques, 53*(3), 852–859.
61. Golestaneh, H., Malekzadeh, F. A., & Boumaiza, S. (2013). An extended bandwidth three-way Doherty power amplifier. *IEEE Transactions on Microwave Theory and Techniques, 61*(9), 3318–3328.
62. Akbarpour, M., Helaoui, M., & Ghannouchi, F. M. (2015). Analytical design methodology for generic Doherty amplifier architectures using three-port input/output networks. *IEEE Transactions on Microwave Theory and Techniques, 63*(10), 3242–3253.

63. Gonzalez, G. (1997). *Microwave transistor amplifiers: Analysis and design* (Vol. 3, pp. 1–45). Upper Saddle River, NJ: Prentice Hall.
64. Rawat, K., Hashmi, M. S., & Ghannouchi, F. M. (2012). Double the band and optimize. *IEEE Microwave Magazine, 13*(2).
65. Ghosh, S., & Rawat, K. (2019). Hybrid analog/digital continuous class B/J mode for broadband Doherty power amplifiers. *IEEE Access, 7*, 74986–74995.
66. Darraji, R., Ghannouchi, F. M., & Hammi, O. (2011). A dual-input digitally driven Doherty amplifier architecture for performance enhancement of Doherty transmitters. *IEEE Transactions on Microwave Theory and Techniques, 59*(5), 1284–1293.
67. Rawat, M., Rawat, K., & Ghannouchi, F. M. (2010). Adaptive digital predistortion of wireless power amplifiers/transmitters using dynamic real-valued focused time-delay line neural networks. *IEEE Transactions on Microwave Theory and Techniques, 58*(1), 95–104.
68. Sáez, R. G., & Marqués, N. M. (2019). RF power amplier linearization in professional mobile radio communications using articial neural networks. *IEEE Transactions on Industrial Electronics, 66*(4), 3060–3070.
69. Chen, X., Chen, W., Ghannouchi, F. M., Feng, Z., & Liu, Y. (2016). A broadband doherty power amplier based on continuous-mode technology. *IEEE Transactions on Microwave Theory and Techniques, 64*(12), 4505–4517.
70. Zhang, R., Acar, M., van der Heijden, M. P., Apostolidou, M., & Leenaerts, D. M. W. (2014). Generalized semi-analytical design methodology of class-E outphasing power amplifier. *IEEE Transactions on Circuits and Systems I: Regular Papers, 61*(10), 2951–2960.
71. Ozen, M., Heijden, M., Acar, M., Jos, R., & Fager, C. (2017). A generalized combiner synthesis technique for class-E outphasing transmitters. *IEEE Transactions on Circuits and Systems I: Regular Papers, 64*(5), 1126–1139.
72. Roblin, P., Chang, H. C., Martinez-Rodriguez, F. J., Xie, C., & Martinez-Lopez, J. I. (2017). On the design of GaN Chireix power amplifiers using an embedding device model. *International Journal of Numerical Modelling: Electronic Networks, Devices and Fields, 30*(1), e2148.
73. Chang, H. C., Hahn, Y., Roblin, P., & Barton, T. W. (2018). New mixed-mode design methodology for high-efficiency outphasing Chireix amplifiers. *IEEE Transactions on Circuits and Systems I: Regular Papers, 66*(4), 1594–1607. https://doi.org/10.1109/TCSI.2018.2882770.
74. Roblin, P., Martinez-Rodriguez, F. J., Chang, H. C., Xie, C., & Martinez-Lopez, J. I. (2015). Transistor characterization and modeling and the use of embedding device models for the design of microwave power amplifiers. In *Proc. Integr. Nonlinear Microw. Millimetre-Wave CircuitsWorkshop (INMMiC)* (pp. 1–6). Washington, DC: IEEE.
75. Sasaki, T., & Hataoka, H. (1983). Investigation of intermodulation in a tuning varactor. *IEEE Transactions on Broadcasting, 29*(2), 77–81.
76. Wheeler, H. A. (1939). Wide-band amplifiers for television. *Proceedings of the IRE, 27*(7), 429–438.
77. Wong, T. T. Y. (1993). *Fundamentals of distributed amplification*. Norwood, MA: Artech House.
78. Percival, W. S. (1937), *Thermionic valve circuits*. British Patent, 460 56.
79. Moser, A. (1967). 140 MHz Kettenverstarker Mit Feldeffekttransistoren. *Inter. Elek Rundschau, 5*, 109–115.
80. Jutzi, W. (1969). A MESFET distributed amplifier with 2 GHz bandwidth. *Proceedings of the IEEE, 57*(6), 1195–1196.
81. Ayasli, Y., Vorhaus, J. L., Mozzi, R., & Reynolds, L. (1981). Monolithic GaAs travelling-wave amplifier. *Electronics Letters, 17*, 413–414.
82. Ayasli, Y., Mozzi, R. L., Vorhaus, J. L., Reynolds, L. D., Pucel, R. A., & Monolithic, A. (1982). GaAs 1-13-GHz traveling-wave amplifier. *IEEE Transactions on Microwave Theory and Techniques, 30*(7), 976–981.
83. Sewiolo, B., Fischer, G., & Weigel, R. (2009). A high-linearity tapered distributed amplifier for ultra-wideband-applications. In *2009 IEEE 10th Annual Wireless and Microwave Technology Conference*. Washington, DC: IEEE. https://doi.org/10.1109/GEMIC.2009.4815853.

84. Bae, K.-T., & Kim, D.-W. (2016). 2–6 GHz GaN distributed power amplifier MMIC with tapered gate-series/drain-shunt capacitors. *International Journal of RF and Microwave Computer-Aided Engineering, 26*(5), 456–465.
85. Prasad, S. N., & Ibrahim, A. S. (2006). Design guidelines for a novel tapered drain line distributed Power Amplifier. In *Proceedings of the 36th IEEE European Microwave Conference* (Vol. 10–15). Washington, DC: IEEE.
86. Xu, J., Wang, Z. G., Zhang, Y., & Ma, L. (2011). High-efficiency tapered distributed power amplifier with 2-μm GaAs HBT process. *Microwave and Optical Technology Letters, 53*(8), 1924–1927.
87. Campovecchio, M., Le Bras, B., Lajugie, M., & Obregon, J. (1994). Optimum design of distributed power-fet amplifiers. Application to a 2-18 GHz MMIC module exhibiting improved power performances. In *Proceedings of 1994 IEEE Microwave and Millimeter-Wave Monolithic Circuits Symposium*. Washington, DC: IEEE.
88. Duperrier, C., Campovecchio, M., Roussel, L., Lajugie, M., & Quéré, R. (2001). New design method of uniform and nonuniform distributed power amplifiers. *IEEE Transactions on Microwave Theory and Techniques, 49*(12), 2494.
89. Chen, J., Niknejad, A. M., & Stage-Scaled, A. (2010). Distributed Power Amplifier achieving 110 GHz bandwidth and 17.5 dBm peak output Power. In *Proceedings of IEEE Radio Frequency Integrated Circuits Symposium, 23–25 May 2010, Anaheim, CA, USA*. Washington, DC: IEEE.
90. Aitchison, C. S., Bukhari, N., Law, C., & Nazoa-Ruiz, N. (1988). The dual-fed distributed amplifier. In *IEEE MTT-S International Microwave Symposium Digest, New York, NY, USA* (Vol. 2, pp. 911–914). Washington, DC: IEEE.
91. Moazzam, M. R., & Aitchison, C. S. (1994). A high gain dual-fed single stage distributed amplifier. In *1994 IEEE MTT-S International Microwave Symposium Digest (Cat. No.94CH3389-4), San Diego, CA, USA* (Vol. 3, pp. 1409–1412).
92. Mazani, Z., Abdipour, A., & Afrooz, K. (2019). Matrix power amplifier with open-circuit composite right−/left-handed transmission line. *IEEE Microwave and Wireless Components Letters, 29*(3), 231–233.
93. Colantonio, P., Giannini, F., & Limiti, E. (2009). *High efficiency RF and microwave solid state Power amplifiers* (1st ed.). Hoboken, NJ: Wiley.
94. Bryerton, E. W., Weiss, M. D., & Popovic, Z. (1999). Efficiency of chip-level versus external power combining [microwave power amplifiers]. *IEEE Transactions on Microwave Theory and Techniques, 47*(8), 1482–1485.
95. Russell, K. J. (1979). Microwave power combining techniques. *IEEE Transactions on Microwave Theory and Techniques, 27*(5), 472–478.
96. An, K. H., Lee, O., Kim, H., Lee, D., Han, J., Yang, K. S., Kim, Y., Chng, J. J., Woo, W., C-H Lee, H., & Kim, J. (2008). Power-combining transformer techniques for fully-integrated CMOS power amplifiers. *IEEE Journal of Solid-State Circuits, 43*(5), 1064–1075.
97. Aoki, I., Kee, S. D., Rutledge, D. B., & Hajimiri, A. (2002). Distributed active transformer-a new power-combining and impedance-transformation technique. *IEEE Transactions on Microwave Theory and Techniques, 50*(1), 316–331.
98. Vasylyev, A., Weger, P., Bakalski, W., & Simburger, W. (2005). Fully-integrated 32 dBm, 1.5-2.9 GHz SiGe-bipolar power amplifier using power-combining transformer. *Electronics Letters, 41*(16), 908–909.
99. Javidan, J., Atarodi, M., & Luong, H. C. (2010). High power amplifier based on a transformer-type Power combiner in CMOS technology. *IEEE Transactions on Circuits and Systems II: Express Briefs, 57*(11), 838–842.
100. Zhao, Y., Long, J. R., & Spirito, M. (2010). Compact transformer power combiners for millimeter-wave wireless applications. In *2010 IEEE Radio Frequency Integrated Circuits Symposium, Anaheim, CA* (pp. 223–226). Washington, DC: IEEE.
101. Tao, H. Q., Hong, W., & Zhang, B. (2017). 6-18 GHz 13W reactive matched Ga*N power amplifier MMIC*. In *Proceedings of European Microwave Integrated Circuits Conference* (pp. 357–360). Washington, DC: IEEE.

Chapter 5
Digital Techniques for Broadband and Linearized Transmitters

5.1 Introduction

The radio frequency (RF) power amplifiers (PAs) are an essential component of the transmitter. However, in order to cater high-speed requirements of upcoming wireless communication, the broadband PAs are being employed in base stations. Such broadband operation in nonlinear PAs further adds to the distortion at the output. Moreover, broadband operation imposes restrictions on peripheral hardware to implement suitable linearization techniques for mitigation of such distortion terms. This chapter presents linearization techniques for multiband and broadband operations on system as well as algorithm levels. The chapter highlights the limitations of the established digital predistortion (DPD) linearization techniques for broadband transmission. To alleviate these limitations of conventional DPD schemes, the hybrid predistortion techniques are presented in detail, which take advantage of best features of "analog" as well as "digital" processing domains. Predistortion techniques rely on accurate characterization of intermodulation distortion (IMD) terms and their precise control can mitigate the nonlinearity in the PA. However, Delta-sigma technique is further presented for high-efficiency switched-mode PA applications, where amplitude is kept constant by converting the signal into pulses to drive the high efficiency switched-mode PAs. Keeping with the discussion of high efficiency, linearization challenges for multiband envelop tracking application is also presented and various digital techniques are compared in terms of linearization performance.

K. Rawat et al., *Bandwidth and Efficiency Enhancement in Radio Frequency Power Amplifiers for Wireless Transmitters*, Analog Circuits and Signal Processing,
https://doi.org/10.1007/978-3-030-38866-9_5

5.2 Nonlinear Behavior of Multiband RF Transmitters

Modern digital communication systems provide high data rate, flexibility in deployment, and better spectrum efficiency by adopting carrier aggregation (CA) to combine multiple RF bands effectively [1]. For example, noncontiguous CA allows the full utilization of the available spectrum resources, by including the scattered, unused RF bands along with the bands allocated for some legacy systems [1]. The long-term evolution (LTE)-advanced systems can theoretically provide transmission signal instantaneous bandwidth up to 100 MHz with downlink data rates of 1000 Mb/s and uplink data rates up to 500 Mb/s by aggregating five contiguous component carriers (CCs) [2]. Therefore, RF components of transmitter system are being upgraded to support CA operation. As a hardware requirement, dual-band [3–5] and tri-band PAs [6–8] are designed to be able to accommodate the signals in different RF bands concurrently [9], while using a single hardware. However, such multiband PAs have to meet strict linearity specifications to support data transmission over a broad frequency range and across multiple separated carriers while maintaining high efficiency [10]. Such restriction becomes even more stringent in case of noncontiguous CA when CCs from a single band or other bands have overlapping IMDs and harmonic terms [11]. As the distortion signals are overlapping with the desired signals, filtering may not be a feasible solution. Even standard preprocessing techniques may not be adequate as intermingling of distortion terms create new challenges in terms of modeling as well as model implementation. Therefore, the inter-band modulation, cross-modulation products, and harmonic content appearing within the range of each signal band need to be characterized and mitigated.

5.2.1 *Identification of IMD Terms for Multiband Operation*

In order to generate dual-band model for a PA, one can generalize the concept for two-tone transmission for the case of dual-band transmission. In such case, a two-tone input excitation $a_{\mathrm{in}}(\omega_1)$ and $a_{\mathrm{in}}(\omega_2)$ at frequency ω_1 and ω_2, respectively, is applied at the input port of the PA and various nonlinear terms are identified. As odd-order distortion are more prominent for PA modeling, the resulting odd-order intermodulation terms for the output b_{out} below ω_1 can be described with a Volterra system [12] as,

$$b_{\mathrm{out}}\left(\frac{p+1}{2}\omega_1 - \frac{p-1}{2}\omega_2\right) = V_p^- \cdot a_{\mathrm{in}}^{(p+1)/2}(\omega_1) \cdot a_{\mathrm{in}}^{*(p-1)/2}(\omega_2) \tag{5.1}$$

where p is an odd integer. Explicit equations for the above terms up to seventh order can be written as [12],

$$b_{\text{out}}(4\omega_1 - 3\omega_2) = V_7^- \cdot a_{\text{in}}^4(\omega_1) \cdot a_{\text{in}}^{*3}(\omega_2) \quad (5.2)$$

$$b_{\text{out}}(3\omega_1 - 2\omega_2) = V_5^- \cdot a_{\text{in}}^3(\omega_1) \cdot a_{\text{in}}^{*2}(\omega_2) \quad (5.3)$$

$$b_{\text{out}}(2\omega_1 - \omega_2) = V_3^- \cdot a_{\text{in}}^2(\omega_1) \cdot a_{\text{in}}^{*}(\omega_2) \quad (5.4)$$

$$b_{\text{out}}(\omega_1) = V_1^- \cdot a_{\text{in}}(\omega_1) \quad (5.5)$$

Similarly, the odd-order intermodulation terms of PA output b_{out} for the frequencies above ω_2 can be described with a Volterra system [12] as,

$$b_{\text{out}}\left(\frac{p+1}{2}\omega_2 - \frac{p-1}{2}\omega_1\right) = V_p^+ \cdot a_{\text{in}}^{(p+1)/2}(\omega_2) \cdot a_{\text{in}}^{*(p-1)/2}(\omega_1) \quad (5.6)$$

By expanding Eq. (5.6) for terms up to seventh order,

$$b_{\text{out}}(\omega_2) = V_1^+ \cdot a_{\text{in}}(\omega_2) \quad (5.7)$$

$$b_{\text{out}}(2\omega_2 - \omega_1) = V_3^+ \cdot a_{\text{in}}^2(\omega_2) \cdot a_{\text{in}}^{*}(\omega_1) \quad (5.8)$$

$$b_{\text{out}}(3\omega_2 - 2\omega_1) = V_5^+ \cdot a_{\text{in}}^3(\omega_2) \cdot a_{\text{in}}^{*2}(\omega_1) \quad (5.9)$$

$$b_{\text{out}}(4\omega_2 - 3\omega_1) = V_7^+ \cdot a_{\text{in}}^4(\omega_2) \cdot a_{\text{in}}^{*3}(\omega_1) \quad (5.10)$$

where the Volterra functions $V_k^{\pm} = f_k^{\pm}\left(|a_{\text{in}}(\omega_1)|^2, |a_{\text{in}}(\omega_2)|^2\right)$ are a function of the two envelopes. It can be noted from Eqs. (5.1) to (5.10) that the various output components, which generally take the form $a_{\text{in}}(\omega_1)^p.a_{\text{in}}^*(\omega_2)^q.|a_{\text{in}}(\omega_1)|^k.|a_{\text{in}}(\omega_2)|^m$, lie on the same frequency location as $a_{\text{in}}(\omega_1)^p \cdot a_{\text{in}}^*(\omega_2)^q$ and therefore a Volterra function comprised of absolute terms models the higher-order terms at each intermodulation frequency location. This two-tone analysis assumes that the separation $\delta = \omega_2 - \omega_1$ between the two tones verifies $\delta << \omega_1$. This allows for the capture and mitigation of each of the in-band and inter-band (intermodulation distortion (IMD)) signal components using a single receiver. It is also worth noticing that due to the small δ, only odd-order intermodulation terms (third-order intermodulation distortion (IMD3), fifth-order intermodulation distortion (IMD5), ...) are the near-band terms of interest, while even-order terms (second-order intermodulation distortion (IMD2), fourth-order intermodulation distortion (IMD4), ...) such as $((p+1)/(2)\omega_2 + (p-1)/(2)\omega_1)$ are far enough in the frequency domain to be easily filtered out.

Let us consider the case of two modulated input signals x_1 and x_2 at the frequencies $\omega_1 = \omega + \Delta$ and $\omega_2 = 2\omega$, respectively, where $\Delta << \omega$. It is convenient to analyze the case of a narrowband two-tone signal at these center frequencies to identify the output intermodulation terms near the two concurrent bands.

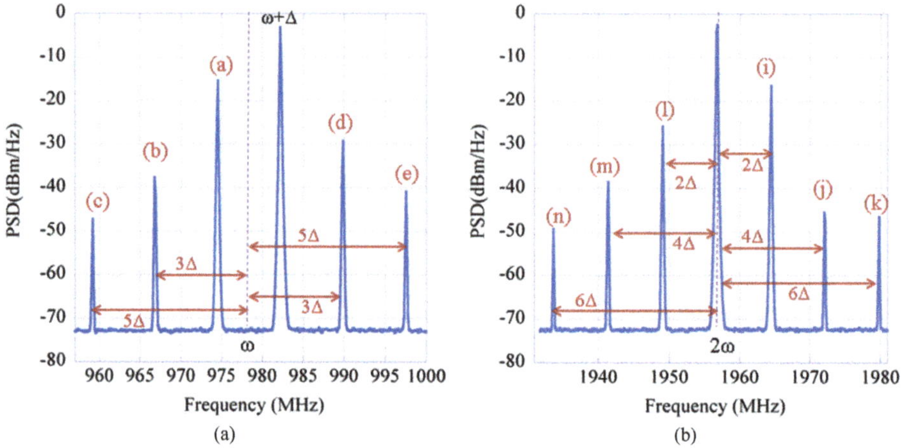

Fig. 5.1 Dual-band distortion component generation due to broadband power amplifier (PA). (**a**) Lower band ([12], reprinted with permission from IEEE), (**b**) upper band ([12], reprinted with permission from IEEE)

Table 5.1 Distortion terms in time and frequency domain ([12], reprinted with permission from IEEE)

Order	Lower band			Upper band		
	#	Frequency	IMD term	#	Frequency	IMD Term
2	(a)	$\omega_2 - \omega_1$	$x_2 x_1^* e^{j(\omega-\Delta)t}$	(i)	$2\omega_1$	$x_1^2 e^{j(2\omega+2\Delta)t}$
5	(b)	$2\omega_2 - 3\omega_1$	$x_2^2 x_1^{*3} e^{j(\omega-3\Delta)t}$	(j)	$4\omega_1 - \omega_2$	$x_1^4 x_2^* e^{j(2\omega+4\Delta)t}$
8	(c)	$3\omega_2 - 5\omega_1$	$x_2^3 x_1^{*5} e^{j(\omega-5\Delta)t}$	(k)	$6\omega_1 - 2\omega_2$	$x_1^6 x_2^{*2} e^{j(2\omega+6\Delta)t}$
4	(d)	$3\omega_1 - \omega_2$	$x_1^3 x_2^* e^{j(\omega+3\Delta)t}$	(l)	$2\omega_2 - 2\omega_1$	$x_2^2 x_1^{*2} e^{j(2\omega-2\Delta)t}$
7	(e)	$5\omega_1 - 2\omega_2$	$x_1^5 x_2^{*2} e^{j(\omega+5\Delta)t}$	(m)	$3\omega_2 - 4\omega_1$	$x_2^3 x_1^{*4} e^{j(2\omega-4\Delta)t}$

When one performs a Volterra model expansion of PA output to identify the near-band signal components, it can be observed that both even and odd intermodulation components fall near the first and second bands. These terms can also be practically observed in Fig. 5.1a, b, which show the intermodulation terms generated near the required bands at the output of an ultra-wideband (UWB) amplifier from Mini-Circuits (ZX60-14012L) when excited by a two-tone signal.

The lower-band (LB) single-tone signal $(\omega + \Delta)/2\pi$ is at 978.4 MHz + 3.84 MHz = 982.24 MHz, while upper-band (UB) single-tone $2\omega/2\pi$ is at 1956.8 MHz. The corresponding terms in the frequency and time domains, along with their respective calculated amplitudes, are given in Table 5.1 where the notation $a_{\text{in}}(\omega) = \exp(j\Delta t)$ and $a_{\text{in}}(2\omega) = x_2$ is used. From the intermodulation terms calculated at the output of the nonlinear system (as summarized in Table 5.1), a trend can be observed for even and odd intermodulation terms. These terms can be summarized in four groups containing both even- and odd-order intermodulation terms.

For $q = 1, 2, \ldots, n$, the lower-band lower frequency side components ((a), (b), and (c) in Fig. 5.1a) are given as,

$$b_{\text{out}}(\omega_1 - (2q-1)\Delta) = V_{12}^{(q)} \cdot x_2^q \cdot x_1^{*(2q-1)} \tag{5.11}$$

The LB upper frequency side components ((d) and (e) in Fig. 5.1a) are given as,

$$b_{\text{out}}(\omega_1 + (2q+1)\Delta) = V_{13}^{(q)} \cdot x_1^{2q+1} \cdot x_2^{*q} \tag{5.12}$$

The UB lower frequency side components ((l), (m), and (n) in Fig. 5.1b) are given as,

$$b_{\text{out}}(\omega_2 - 2q\Delta) = V_{22}^{(q)} \cdot x_2^{q+1} \cdot x_1^{*2q} \tag{5.13}$$

The UB upper frequency side components ((i), (j), and (k) in Fig. 5.1b) are given as,

$$b_{\text{out}}(\omega_2 + 2q\Delta) = V_{23}^{(q)} \cdot x_1^{2q} \cdot x_2^{*(q-1)} \tag{5.14}$$

Note that the Volterra functions are a function of the envelope of both the bands,

$$V_{b,p}^{(q)} = f_{b,p}^{(q)}\left[|a_{\text{in}}(\omega)|^2, |a_{\text{in}}(2\omega)|^2\right] \tag{5.15}$$

The above IMD2 and IMD3 components and harmonic of the LB signals can be differentiated in the frequency domain in a two-tone test as these are narrowband signals. However, when a modulated signal with a wider bandwidth is used, then some of these nonlinear components superimpose with the signal of interest, and therefore the effect of these terms has to be included in the model to correctly represent the PA behavioralmodel. Figure 5.2 shows the output of a broadband PA wherethree modulated complex input signals x_1, x_2, x_3 are transmitted at three different carrier frequencies $f_1 = 1.4$ GHz, $f_2 = 2.4$ GHz, and $f_3 = 3.7$ GHz. It can be observed that the higher frequency signals are very near to the harmonics of the signal at frequency f_1, that is $f_1 = f + \Delta_1 = 1.2 + 0.2 = 1.4$ GHz, $f_2 = 2f = 2.4$ GHz, and $f_3 = 3f + \Delta_2 = 3.6 + 0.1 = 3.7$ GHz, respectively. Due to the nonlinearity of the broadband PA, IMD terms are generated at 0.9, 1, 1.1, 1.3, 2.3, 2.6, 2.7, 2.8, 3.4, 3.5, 3.8, and 4.2 GHz frequencies.

In the practical scenarios of harmonic CA-modulated signals, some of the IMDs and cross-modulation distortion terms (CMDs) generated by the PA would fall within or nearby the bands of the carriers of the input signals. To illustrate the above, the IMD terms generated at angular frequencies $\omega_3 - 2\omega_1$, $\omega_2 - \omega_1$, $2\omega_2 - \omega_3$, and $\omega_3 - \omega_2$ are close to the signal at angular frequency ω_1. Similarly, these IMD terms generated at angular frequencies $\omega_3 - \omega_1$, $2\omega_3 - 2\omega_2$, $\omega_1 + \omega_3 - \omega_2$, and $2\omega_1$

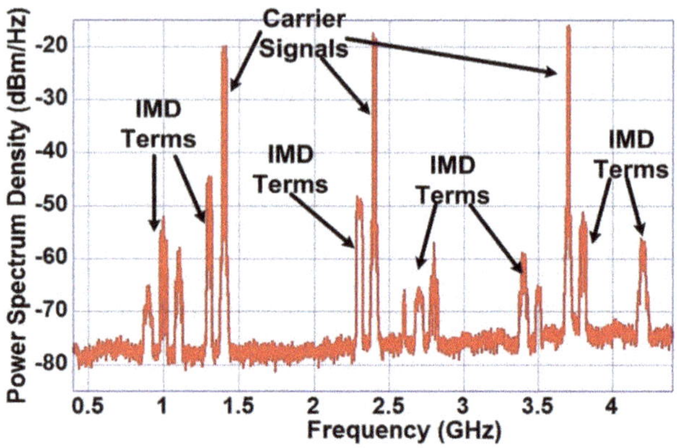

Fig. 5.2 Tri-band distortion component generation due to broadband PA ([13], reprinted with permission from IEEE)

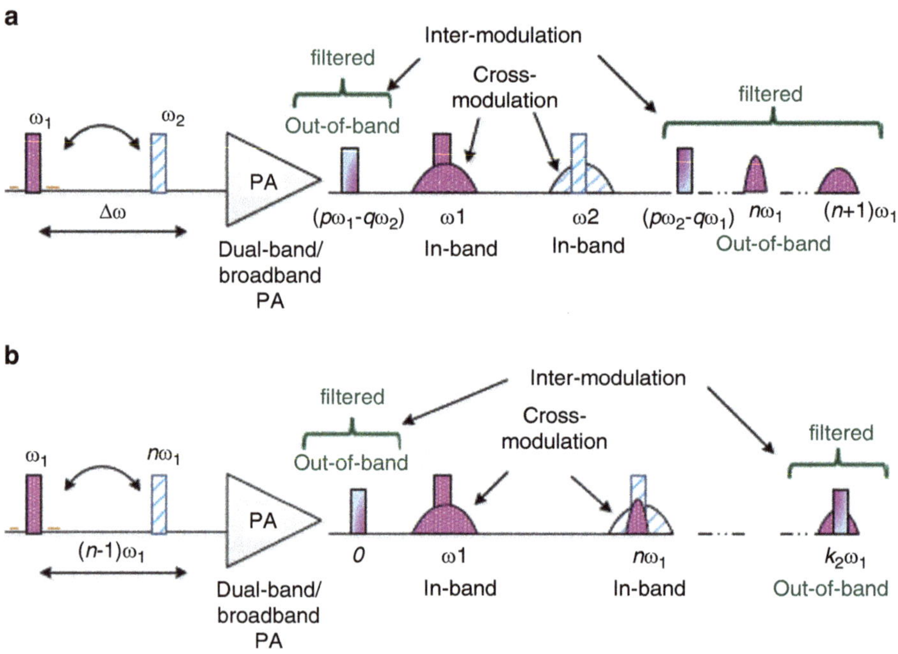

Fig. 5.3 (**a**) When two carrier frequencies are uncorrelated, that is not at harmonic frequencies and (**b**) scenario when two carrier frequencies are at harmonic frequencies ([12], reprinted with permission from IEEE)

fall close to the carrier signal at ω_2 and IMD terms $2\omega_2 - \omega_1$, $3\omega_2 - \omega_3$, $\omega_1 + \omega_2$, and $3\omega_1$ fall close by ω_3. There are two scenarios possible, based on the difference between frequency locations. In the first scenario, the frequencies of second and third carrier signals are far away from the first and third harmonic frequencies. In such case the IMD terms which are far away from the signal of interest can be filtered using multiband RF filter at the PA output. Figure 5.3a depicts the dual-band case where multiband carrier frequencies are unrelated. If the broadband PA is used for different set of multiband signals with different frequencies, corresponding RF filter should be either reconfigurable or an RF filter bank is required for each set of multiband frequencies. It is worth mentioning that the IMD terms due to in-band interference of each carrier is still required to be corrected using digital pre- or post-distortion techniques.

The second scenario has the frequencies of second and third carriers signals very close to the first and third harmonic, that is, $\omega_1 = \omega$, $\omega_2 = 2\omega$, and $\omega_3 = 3\omega$. In such case, IMD terms fall directly on ω, 2ω, and 3ω frequencies. Figure 5.3b depicts the dual-band case where multiband carrier frequencies are at harmonic frequencies. The IMD terms are very close to the carrier signals and hence cannot be filtered at the output of the PA using RF filters. Some digital processing pre- or post-solution is essential to remove IMD and harmonic distortions. In such case, hardware complexity is reduced; however, much higher digital processing resources are required.

It is worth mentioning that if the multiple bands are treated as a composite signal, the single-band DPD can be applied. However, for applying the single-band DPD, transmitter and receiver should be able to support such wide bandwidth to be able to train as well as apply the DPD model. As a rule of thumb, the analog-to-digital converters (ADCs) and digital-to-analog converters (DACs) in receivers and transmitters should support the sampling more than the N times Nyquist sampling rate of the composite signal. N represents the nonlinearity order of the DPD [13]. For an example signal with 100 MHz CA bandwidth indicated for LTE signals [10] and PA nonlinearity order of 7, the ADC/DAC sampling rate should be more than 1400 MHz. Utilizing multiple ADCs, DACs allow separate processing of each band with a much lower sampling rate. Besides, all the bands share the dynamic range of the composite signal, which places higher demand on the dynamic ranges of the ADCs and DACs. Therefore, each band is processed separately as shown in Fig. 5.4 for an example of tri-band DPD system.

As shown in Fig. 5.4, the DPD system using indirect learning scheme (refer Sect. 1.5) is composed of three major modules: (a) PA characteristic extraction module, (b) DPD coefficients estimation module, and (c) DPD application module. As discussed in Sect. 1.5, both the techniques, "direct learning architecture" and "indirect learning architecture," utilize the behavioral models as either "PA model" or "inverse model." Therefore it is essential to create an accurate model to capture all cross-modulation, intermodulation, and/or harmonic distortion terms. Behavioral models for the above state multiband transmission cases shown in Fig. 5.3a, b are discussed in detail in the following sections.

5.2.2 *Multiband Signals at Unrelated Frequencies*

When multiband signals are at unrelated frequencies, their harmonic signals are far apart in frequency domain and do not overlap with each other. The RF filter can

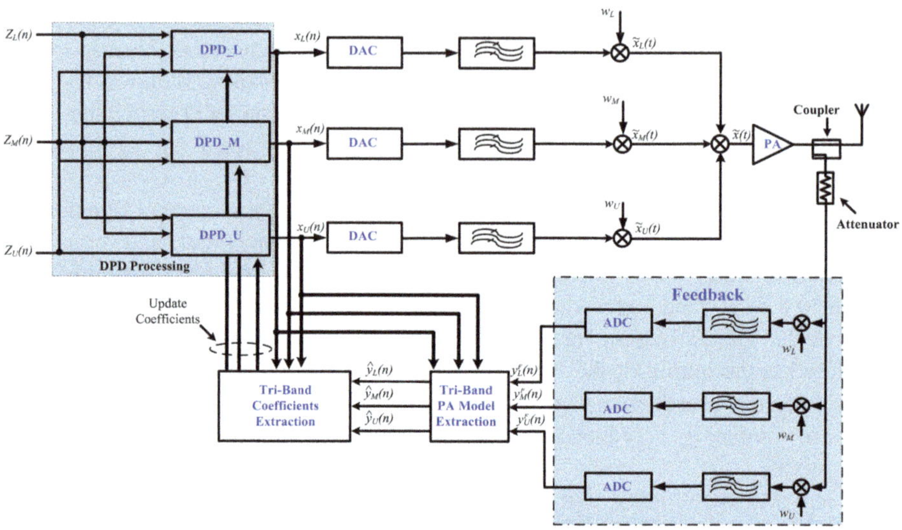

Fig. 5.4 Implementation of a tri-band digital predistortion (DPD) scheme using separated digital-to-analog converters (DACs) and analog-to-digital converters (ADCs)

easily filter such harmonics at the output of the PA. However, cross-modulation terms (CMD) and intermodulation terms (IMD) may interfere with the signal of interest for the scenario of noncontiguous CA, having closely spaced multiple bands. The multiband linearization technique combining the direct PA model extraction and the indirect DPD learning strategies is proposed to counter the spectral overlaps. A multidimensional memory polynomial (MD-MP) model can be utilized to extract IMDs as well as CMDs. The motivation for such a model derives from the observation of multiple bands of interest while achieving improved time as well as frequency selective modeling. Once the coefficients of MD-MP models are extracted for the inverse modeling, the inverse model is applied according to the indirect learning structure. For a tri-band transmission, using 3D-memory polynomial (3D-MP) model [9], the output of a LB is given as,

$$y_1(n) = \sum_{m=0}^{M-1} \sum_{k=0}^{K-1} \sum_{j=0}^{k} \sum_{i=0}^{j} a_{m,k,j,i} x_1(n-m) |x_1(n-m)|^{k-j} \times |x_2(n-m)|^{j-i} |x_3(n-m)|^{i} \tag{5.16}$$

$$y_2(n) = \sum_{m=0}^{M-1} \sum_{k=0}^{K-1} \sum_{j=0}^{k} \sum_{i=0}^{j} b_{m,k,j,i} x_2(n-m) |x_1(n-m)|^{k-j} \times |x_2(n-m)|^{j-i} |x_3(n-m)|^{i} \tag{5.17}$$

$$y_3(n) = \sum_{m=0}^{M-1} \sum_{k=0}^{K-1} \sum_{j=0}^{k} \sum_{i=0}^{j} c_{m,k,j,i} x_3(n-m) |x_1(n-m)|^{k-j} \times |x_2(n-m)|^{j-i} |x_3(n-m)|^{i} \tag{5.18}$$

where M is the memory depth, $a_{m,k,j,i}$, $b_{m,k,j,i}$, and $c_{m,k,j,i}$ are the coefficients, $x_1(n)$, $x_2(n)$, and $x_3(n)$ are the baseband-modulated input signals. The $y_1(n)$, $y_2(n)$, and $y_3(n)$ are the baseband-modulated output of the LB, middle-band (MB), and UB, respectively, using 3D-MP model, and K is the nonlinearity order. For the simplicity of expression, all memory depths and nonlinearity orders are assumed to be equal. Practically PA displays different nonlinearity profiles at different frequencies and selecting different memory depths and different nonlinearity orders for each band will allow to keep only essential monomial terms.

In the 3-dimentional phase-aligned (3D-PAV) model [8, 9], the baseband tridimensional model is constructed considering only the input carrier signals and the IMD components that fall within the three fundamental frequency bands. The output of PA for LB is given as,

$$\begin{aligned} y_1(n) &= \sum_{k=0}^{K-1}\sum_{j=0}^{k}\sum_{i=0}^{j} a_{k,j,i,0}x_1(n)|x_1(n)|^{k-j}|x_2(n)|^{j-i}|x_3(n)|^{i} \\ &+ \sum_{k=0}^{K-3}\sum_{j=0}^{k}\sum_{i=0}^{j}\sum_{\tau_1=0}^{M_1-1}\sum_{\tau_2=\tau_1+1}^{M_2-1} a_{k,j,i,1}(\tau_1,\tau_2)\,x_1^2(n-\tau_1)\,x_1^*(n-\tau_2) \\ &\times|x_1(n-\tau_1)|^{k-j}|x_2(n-\tau_1)|^{j-i}|x_3(n-\tau_1)|^{i} \\ &+ \sum_{k=0}^{K-3}\sum_{j=0}^{k}\sum_{i=0}^{j}\sum_{\tau_1=0}^{M_1-1}\sum_{\tau_2=\tau_1+1}^{M_2-1} a_{k,j,i,2}(\tau_1,\tau_2)\,x_1(n-\tau_1)\,x_2^*(n-\tau_2)\,x_2(n-\tau_1) \\ &\times|x_1(n-\tau_1)|^{k-j}|x_2(n-\tau_1)|^{j-i}|x_3(n-\tau_1)|^{i} \\ &+ \sum_{k=0}^{K-3}\sum_{j=0}^{k}\sum_{i=0}^{j}\sum_{\tau_1=0}^{M_1-1}\sum_{\tau_2=\tau_1+1}^{M_2-1} a_{k,j,i,3}(\tau_1,\tau_2)\,x_1(n-\tau_1)\,x_3^*(n-\tau_2)\,x_3(n-\tau_1) \\ &\times|x_1(n-\tau_1)|^{k-j}|x_2(n-\tau_1)|^{j-i}|x_3(n-\tau_1)|^{i} \end{aligned} \tag{5.19}$$

where M_1 and M_2 are the memory depths, and $a_{k,j,i,0}$, $a_{k,j,i,1}(\tau_1,\tau_2)$, $a_{k,j,i,2}(\tau_1,\tau_2)$ and $a_{k,j,i,3}(\tau_1,\tau_2)$ are the coefficients of the 3D-PAV model $x_1(n)$, $x_2(n)$ and $x_3(n)$ are the baseband-modulated input signals at different bands. Similar expressions can be written for second and third frequency bands.

5.2.3 Multiband Signals at Harmonically Related Frequencies

Let us consider two modulated complex baseband input signals x_1 and x_2 being transmitted at carrier frequencies $\omega_1 = \alpha\omega$ and $\omega_2 = \beta\omega$, respectively, where α, $\beta \in I$ and $\alpha \neq \beta$. By observing the distortion terms shown in Eqs. (5.7)–(5.15), the harmonic two-dimensional memory polynomial (H2D-MP) model is proposed to capture the harmonic interferences and model the PA output. The outputs of this model are given as [12, 13],

$$
\begin{aligned}
y_1(n) = & \sum_{m=0}^{M} V_{m,0}^{1,1}(n-m)\cdot x_1(n-m) \\
& + \sum_{m=0}^{M_l}\sum_{q=1}^{Q_l} V_{m,q}^{1,2}(n-m)\cdot x_2^{\alpha q}(n-m)\cdot x_1^{*(\beta q-1)}(n-m) \\
& + \sum_{m=0}^{M_u}\sum_{q=1}^{Q_u} V_{m,q}^{1,3}(n-m)\cdot x_1^{\beta q+1}(n-m)\cdot x_2^{*\alpha q}(n-m)
\end{aligned} \tag{5.20}
$$

and

$$
\begin{aligned}
y_2(n) = & \sum_{m=0}^{M} V_{m,0}^{2,1}(n-m)\cdot x_2(n-m) \\
& + \sum_{m=0}^{M_l}\sum_{q=1}^{Q_l} V_{m,q}^{2,2}(n-m)\cdot x_2^{\alpha q+1}(n-m)\cdot x_1^{*\beta q}(n-m) \\
& + \sum_{m=0}^{M_u}\sum_{q=1}^{Q_u} V_{m,q}^{2,3}(n-m)\cdot x_1^{\beta q}(n-m)\cdot x_2^{*(\alpha q-1)}(n-m)
\end{aligned} \tag{5.21}
$$

where

$$
V_{m,q}^{r,p}(n) = V_{m,q}^{r,p}\left(|x_r(n)|, |x_s(n)|\right), \quad r,s \in \{1,2\} \quad \text{and} \quad r \neq s, \tag{5.22}
$$

and $V_{m,q}^{b,p}\left(|x_r(n)|, |x_s(n)|\right)$ can be expressed as,

$$
V_{m,q}^{r,p}\left(|x_r(n)|, |x_s(n)|\right) = \sum_{k=0}^{K-1}\sum_{i=0}^{k} c_{m,q,k,i}^{r,p}\cdot |x_r(n)|^{k-i}\cdot |x_s(n)|^{i} \tag{5.23}
$$

where K is the nonlinearity order M, M_l and M_u are the memory depths, Q_l and Q_u are the intermodulation product orders, $c_{m,q,k,i}^{r,p}$ are the coefficients of the lower ($r = 1$) and upper band ($r = 2$). The number of coefficients in each band of the 2D-HMP model is $[(M + 1) + (M_l + 1)Q_l + (M_u + 1)Q_u]\cdot K(K + 1)/2$. All the equations from (5.16)–(5.19) and (5.20)–(5.21) are polynomial-based models, and for each band, output vector can be written in the matrix form as,

$$
\mathbf{Y} = \mathbf{UA} \tag{5.24}
$$

where $\mathbf{A}$ is a vector of coefficients to be extracted, $\mathbf{U}$ is the observation matrix for each monomial as column vector for an input signal vector containing training data, and $\mathbf{Y}$ is an output signal vector of the PA. Then using the least-squares (LS) solution, coefficient vector $\mathbf{A}$ is extracted as,

$$
\hat{\mathbf{A}} = \left(\mathbf{U}^H\mathbf{U}\right)^{-1}\mathbf{U}^H\mathbf{Y} = \mathbf{U}^{+}\mathbf{Y} \tag{5.25}
$$

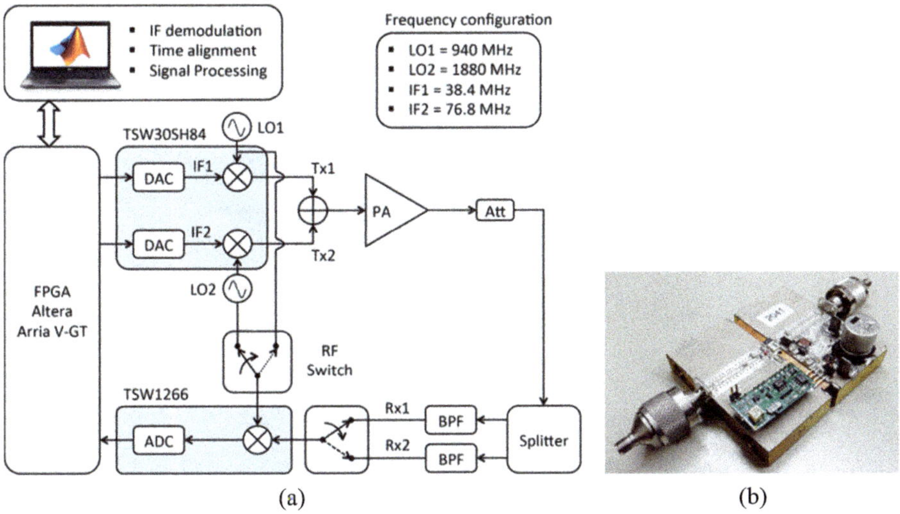

Fig. 5.5 (**a**) Experimental setup to implement dual-band concurrent signal at harmonic distance ([12], reprinted with permission from IEEE). (**b**) Broadband 10 W peak power PA from NXP ([12], reprinted with permission from IEEE)

where $\mathbf{U}^H$ denotes Hermitian transpose of $\mathbf{U}$ and $\mathbf{U}^+$ and is called the pseudo-inverse of the observation matrix.

Figure 5.5a, b shows dual-band DPD implementation when one of the bands is at the harmonic frequency of the other band. A dual-channel transmitter (TSW30SH84) from Texas Instruments incorporated is used for the concurrent dual-band transmission. The RF frequency range of each of the transmission channels is from 300 MHz to 4 GHz. The dual-channel transmitter incorporates two 16-bit DACs, which share the same clock. The input data is stored in pre-allocated field-programmable gate array (FPGA) memories using MATLAB and Quartus software. The data from the Arria-V (GT series) FPGA used can be sent to TSW30SH84 board at sampling rate of 307.2 MHz. The DAC in the transmit path is programmed with a sampling frequency of 1228.8 MHz, that is, with an interpolation factor of 4. The signal is up-converted using a TRF3705 quadrature modulator within the TSW30SH84 board and sent to an UWB PA. The output of the PA is down-converted and demodulated using a 500 MHz bandwidth TSW1266 receiver board from Texas Instruments incorporated with a sampling rate of 614.4 MHz. The clock shared by the two DACs is also provided to the receiverADCs.

A 10 MHz reference clock (Ref. clk) is used for the synchronization between the receiver and transmitter local oscillators (LOs). The signal is down-converted, digitized, and stored in the FPGA memory. The receiver board accepts an RF signal in the frequency range of 1880–2390 MHz; therefore, the second harmonic signal can be captured directly at the receiver. An input matching network was developed to extend the receiver operation in the fundamental band. An RF switch matrix from

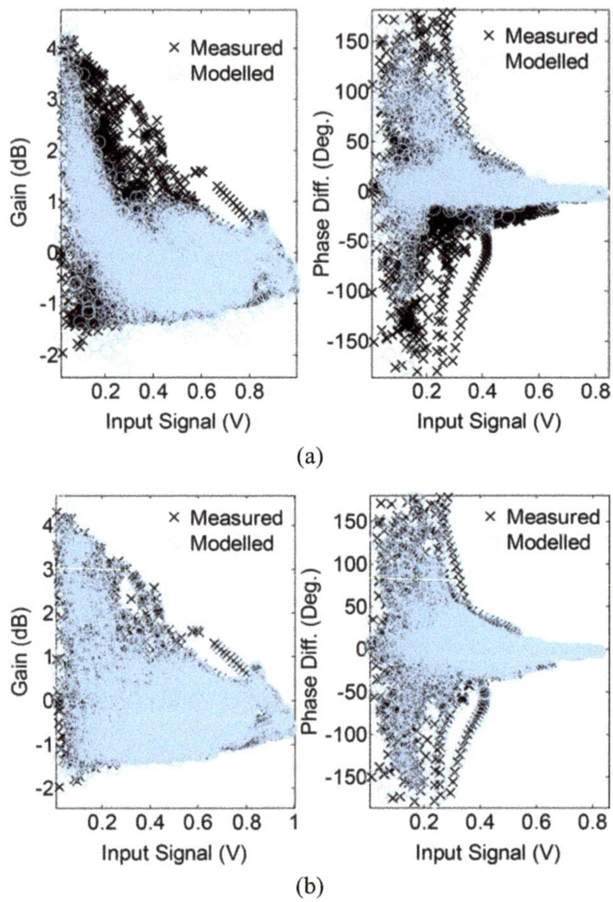

Fig. 5.6 Inverse modeling performance shown for 10 W peak-power PA from NXP at lower band (978.4 MHz): (**a**) two-dimensional memory polynomial (2D-MP) model ([12], reprinted with permission from IEEE) and (**b**) harmonic two-dimensional memory polynomial (H2D-HMP) model ([12], reprinted with permission from IEEE)

Mini Circuits is used before the receiver to select either the harmonic or fundamental frequency signal. The receiver of the feedback observation path captures one band at a time during the concurrent operation using a switch to select the proper LO and the RF input matching networks for the in-phase/quadrature (IQ) demodulator.

Figure 5.6a, b shows the LB inverse modeling performance for indirect learning architecture-based DPD operation when single-carrier WCDMA signal is applied in the lower band (ω) at 978.4 MHz and a 10 MHz LTE signal at the upper band (2ω) at 1916.8 MHz, respectively. Figure 5.6a shows the modeling performance when the lower-band data is modeled using 2D-MP model, which is dual-band version of Eqs. (5.16)–(5.18). It is evident that 2D-MP model is not able to model the PA output completely.This is expected as the 2D-MP model is appropriate for IMD

Table 5.2 Impact of higher-order terms, shown in Table 5.1, on modeling performance

Lower-band (WCDMA signal)			Upper-band (LTE signal)		
IMD term included	PA model NMSE (dB)	Inverse model NMSE (dB)	IMD term included	PA model NMSE (dB)	Inverse model NMSE (dB)
2D-MP	−28.9	−27.5	2D-MP	−28.1	−26.9
2D-MP + (a)	−38.2	−34.0	2D-MP + (i)	−34.4	−33.08
2D-MP + (a) + (d)	−39.3	−35.0	2D-MP + (i) + (l)	−35.4	−34.1
2D-MP + (a) + (d) + (b)	−39.39	−35.1	2D-MP + (i) + (l) + (m)	−35.6	−34.1
2D-MP +(a) + (d) + (b) + (c)	−39.45	−35.2	2D-MP + (i) + (l) + (m) + (j)	−35.6	−34.1

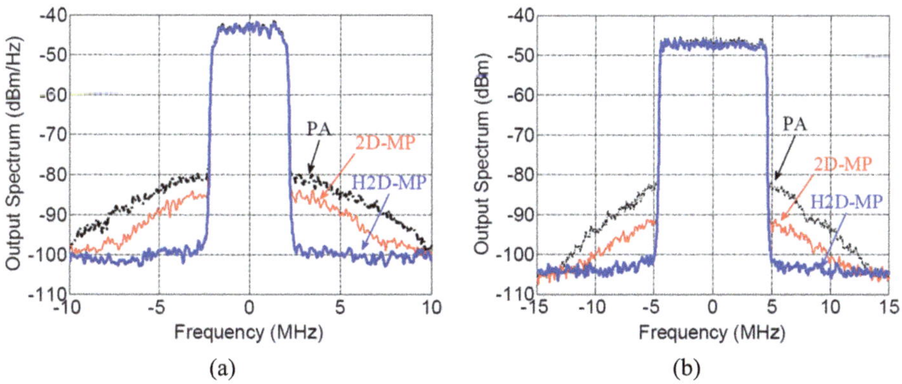

Fig. 5.7 DPD performance comparison for (**a**) lower band (978.4 MHz) and (**b**) upper band (1916.8 MHz) ([12], reprinted with permission from IEEE)

and CMD terms only, and is not capable of modeling harmonic terms. Figure 5.6b shows that with the inclusion of harmonic distortion terms and harmonically relevant higher-order CMD terms, the inverse model captures the characteristics more accurately. The benefit of inclusion of higher-order terms can be observed in terms of normalized mean square error (NMSE), as shown in Table 5.2.

The inclusion of first-order terms results in significant enhancement over the performance of the 2D-MP model. The MD-PAV model as given by Eq. (5.19) can be considered a subset of HMD-MP model, which contains only first-order terms.

Figure 5.7a, b shows the DPD results for 2D-MP and H2D-MP models for LB as well as UB. A significant improvement is observed in the ACPR performance of H2D-MP model. As mentioned in Chap. 1, adjacent channel power ratio (ACPR) provides out-of-band measure of signal quality. The transmission error for in-band is quantified in terms of NMSE between transmitted and received data. We can observe from Table 5.3 that H2D-MP model provides more than 10 dB improvement in terms of transmitted signal NMSE as well as ACPR for LB as well as UB signal.

Table 5.3 Transmission error reduction performance

	LB(NMSE(dB))	UB(NMSE(dB))	LB(ACPR(dB))	UB(ACPR(dB))
w/o DPD	−22.6	−20.6	−35.2	−33.8
2DMP DPD	−33.1	−33.9	−39.1	−41.7
H2DMP DPD	−44.3	−44.7	−53.7	−51.5

5.3 Nonlinear Harmonic Injection Scheme for Filterless Transmitter Architecture

The upcoming fifth-generation (5G) standard requirements are pushing conventional systems in terms of data rates and capacity to fulfill the demands of applications such as the Internet of Things, autonomous technologies, and machine-to-machine communication [14]. One of the propositions of the 5G system is newer cloud radio access network (C-RAN) architecture [15, 16]. C-RAN architecture contains a standard baseband unit (BBU) pool. This pool allows the connection of remote radio units (RRUs) with geographically distributed locations. Transceiver for such applications should be able to support different frequencies. Therefore, the most challenging component of such transceiver (in RRU) is an UWB/broadband RF PA [17, 18]. However, due to the nonlinear characteristics of a PA, harmonics of one signal may interfere with the other signals in the working frequency range of the UWB PA. The conventional method to suppress these harmonics is by using RF bandpass filter bank at desired bands. If such signals are not used concurrently, an RF switch is used to configure to the desired frequency. Figure 5.8 depicts a commercial [19], multiband transceiver with the filter bank for every desired band (dotted line in red). This solution is rigid, and in the case of the dynamic spectral access system, it can switch to only predefined bands. Besides, the filter banks add significantly to the transmitter size and introduce losses leading to reduced output power and sensitivity at the transmitter. Due to the inclusion of filter bank, the purpose of designing a compact multiband/UWB band PA is defeated.

A digitally programmable harmonic cancellation technique can provide the required flexibility, accuracy, and compactness of the solution. The digital cancellation technique is a two-step approach. During the first step, the harmonic signal is characterized and modeled using a digital model. The PA output consists of the fundamental signal and its harmonic signals. If the n_{th} order harmonic signal is to be suppressed, the digital model of nth harmonic is injected in the input of the PA. The first step to characterize the PA is the accurate delay estimation and compensation for the harmonic signal. The time alignment can be performed using cross-correlation either in time domain or in frequency domain. If y_n denotes the nth harmonic of the input signal x, the y_n should be correlated to x^n. The cross-correlation between y_n and x^n can be written in the frequency domain as:

$$F\left(x^n * y^n\right) = F\left(x^n\right) \cdot F\left(x^n\right) \tag{5.26}$$

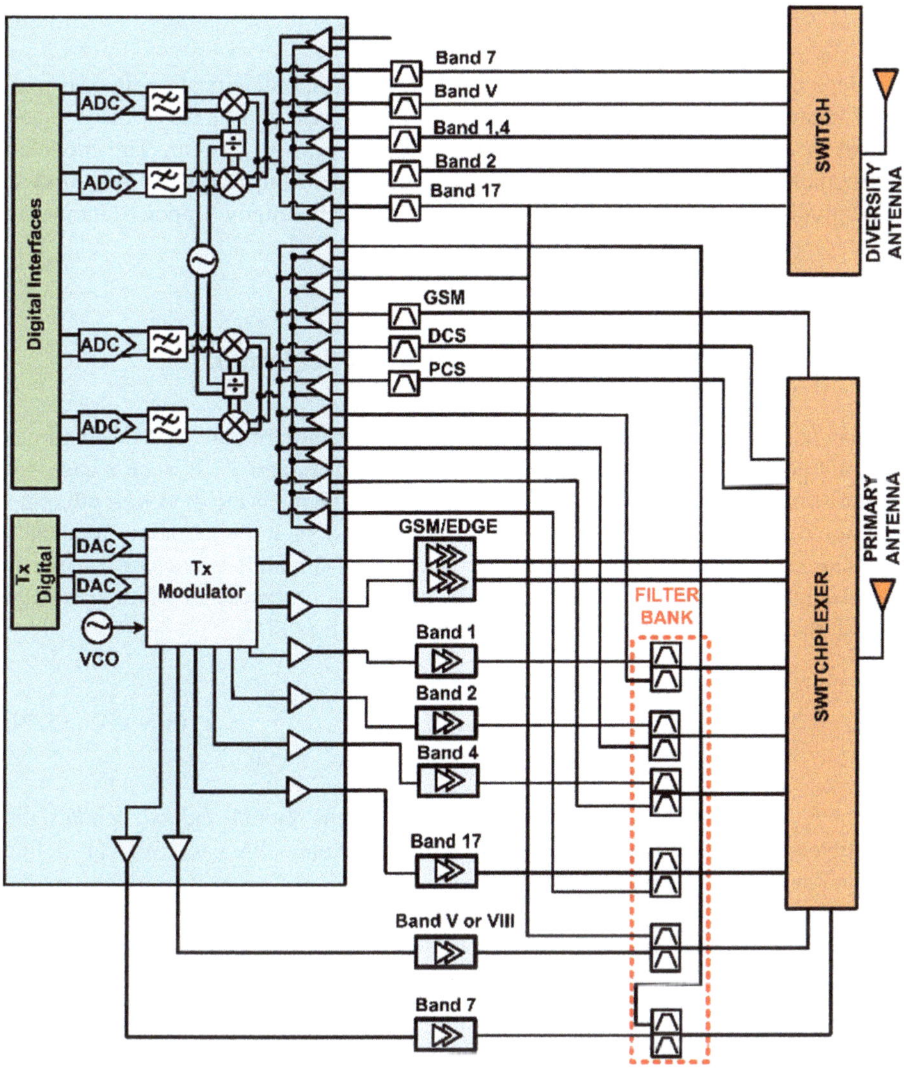

Fig. 5.8 A multimode, multiband transceiver module ([20], reprinted with permission from IEEE)

where F represents the Fourier transform; Eq. (5.26) can be rewritten as,

$$F\left(x^n * y_n\right) = K \cdot F\left(x^n\right) \cdot F\left(x^n\right) \mathrm{e}^{j(2\pi f\tau+\theta)} \tag{5.27}$$

where K is the gain, τ is the time delay, θ is the phase rotation of harmonic with respect to x^n. The value of cross-correlation in (5.27) is maximum when $(2\pi f\tau + \theta = 0)$ and therefore the time-adjusted harmonic signal is given as,

$$y_{\text{adj}} = y_n e^{-j(2\pi f\tau+\theta)} \tag{5.28}$$

Once the harmonic signal is delay aligned, the harmonic signal is injected with the fundamental signal at the input of the PA. The gain and phase of the injected signal is controlled utilizing the FPGA-based processing system. The modeled harmonics can be utilized for filterless architecture in the following two modes: (a) digitally supported feed-forward technique and (b) digitally supported harmonic injection technique.

5.3.1 Digitally Supported Feed-Forward Technique

As any harmonic signal may be impacted by the higher harmonic terms, sometimes it is not possible to find a linear correlation between y_n and x^n. In such a case, for accurate harmonic characterization, an uncorrelated signal is injected with nth order harmonic signal. The output of the feed-forward system at the harmonic frequency contains injected signal and harmonic of the required signal. If x_2 is a signal which is uncorrelated to the desired (fundamental) signal and x_1 represents the fundamental signal, the feed-forward system output can be written as [21],

$$y_2 = \sum_{i=0}^{k_p} \sum_{m_p=0}^{M_p} b_{m_p,i} \cdot x_1^p\left(n - m_p\right)\left|x_1\left(n - m_p\right)\right|^i + \sum_{m=0}^{M} c_m x_2\left(n - m\right) \tag{5.29}$$

where $b_{p,i}$ denotes the coefficients for the harmonic model and c_m denotes the coefficients associated with injected signal in the linear PA path. We can define the vector notation with N samples of input signal as,

$$\vec{x}_1(n) = \left[x_1(n), x_1\left(n+1\right), \ldots, x_1\left(n+N\right)\right] \tag{5.30}$$

$$\vec{x}_2(n) = \left[x_2(n), x_2\left(n+1\right), \ldots, x_2\left(n+N\right)\right] \tag{5.31}$$

Observation matrix for harmonic signal is,

$$\gamma_i(n) = \left[\vec{x}_1^{\,p}(n)\left|\vec{x}_1(n)\right|^i, \vec{x}_1^{\,p}(n)\left|\vec{x}_1\left(n-1\right)\right|^i, \ldots, \vec{x}_1^{\,p}\left(n-m_p\right)\left|\vec{x}_1\left(n-m_p\right)\right|^i\right] \tag{5.32}$$

and $\vec{\Gamma} = [\gamma_0, \gamma_1, \ldots, \gamma]$. Similarly, for x_2 signal, observation matrix can be given by,

$$\vec{\beta} = \left[\vec{x}_2(n), \vec{x}_2\left(n-1\right), \ldots, \vec{x}_2\left(n-M\right)\right] \tag{5.33}$$

The observation matrix for Eq. (5.32) can be given by $\mathbf{H} = [\Gamma \ \beta]$ and Eq. (5.32) can be written as,

$$\vec{y}_2 = \vec{\mathbf{H}}\vec{\delta} \tag{5.34}$$

where $\vec{\delta}$ is given by,

$$\vec{\delta} = \left[b_{0,0}, \ldots, b_{M_p,k_p}, c_0, \ldots, c_M\right] \tag{5.35}$$

Coefficient vector $\vec{\delta}$ can be calculated using LS solution.

$b_{m,i}$ can be identified as initial $(M_p + 1) \times (k_p + 1)$ coefficients of the vector δ. As x_2 and x_1 are uncorrelated signals, the contribution of pth harmonic can be separated as

$$y_{hrm} = \sum_{i=0}^{k_p} \sum_{m_p=0}^{M_p} b_{m_p,i} \cdot x_1^p\left(n - m_p\right)\left|x_1\left(n - m_p\right)\right|^i \tag{5.36}$$

Equation (5.36) shows the digital representation of the harmonic signal. Once the harmonic signal is accurately modeled, the x_2 is replaced by digitally reconstructed harmonic signal in baseband which eventually cancels the harmonic at the PA output. The amplitude of the signal is varied by digitally controlled attenuator while its phase is accurately tuned in baseband/digital domain. The output of the PA after feed-forward implementation is monitored at the feedback receiver. Using feedback receiver, amplitude, and phase are tuned in the digital domain (baseband level) by sensing the power/amplitude of the remaining harmonic content. The scheme for the described digitally supported feed-forward technique is shown in Fig. 5.9. The setup used is similar to the one described in Fig. 5.5a.

Figure 5.10a–d shows the characteristics of 1 W UWB PA from Mini-Circuits (ZX60-14012L), which is used for measurement at fundamental and second harmonic frequencies. The PA has approximately 4 dB gain compression at saturation for fundamental frequency ω (Fig. 5.10a). Due to the nonlinear operation, the second harmonic of the signal appears at 2ω. As per the characterization process discussed in the previous section, signal x_2 is injected via linear PA at 2ω from the lower transmitter channel in Fig. 5.9. The signal x_2 is injected with low powers and does not compress PA at the frequency 2ω. From Fig. 5.10b, d, one can observe that the phase characteristics are linear; however, dispersion is observed due to harmonic interference. The PA is driven at the frequency of 958.4 MHz (920 MHz + 38.4 MHz IF shift) with 10 MHz WCDMA signal and at frequency 1916.8 MHz ((1840 MHz + 76.8 IF shift) with uncorrelated 5 MHz LTE signal. The harmonic or fundamental frequency signal is selected at the receiver using an RF switch.

The PA output at 2ω frequency contains the second harmonic of 10 MHz WCDMA signal as well as injected 5 MHz LTE signal. It is anticipated that the bandwidth of the second harmonic is twice to that of the fundamental signal band-

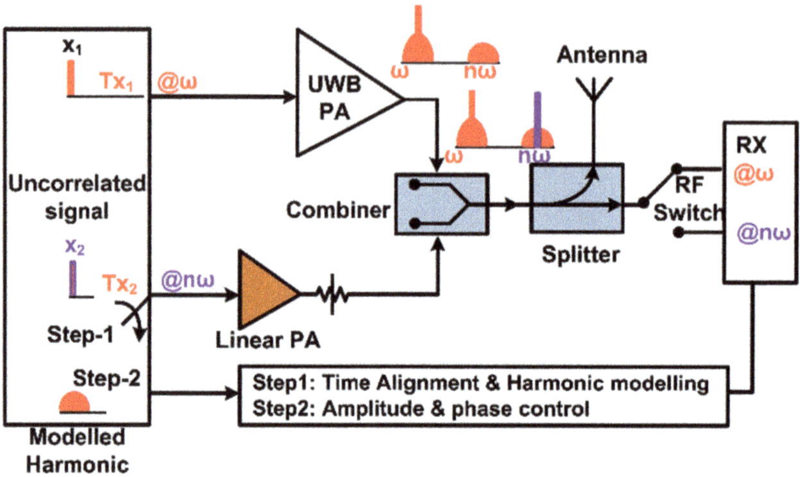

Fig. 5.9 Digitally supported feed-forward technique for filterless transmitter ([21], reprinted with permission from IEEE)

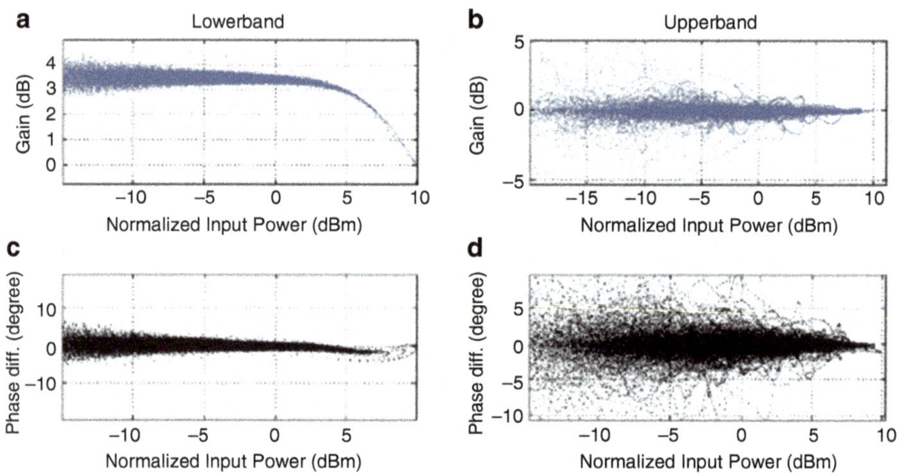

Fig. 5.10 Measured PA characteristics: (**a**) gain at frequency ω, (**b**) phase difference at frequency ω, (**c**) gain at frequency 2ω, and (**d**) phase difference at frequency 2ω ([21], reprinted with permission from IEEE)

width. We should note that the output is a linear combination of two uncorrelated signals, and one might assume that by merely subtracting x_2 from PA output, the harmonic signal can be easily separated. However, even with linear PA, x_2 encounters distortion and noise due to measurement path. The digital model captures the characteristics of the harmonic signal but is unable to capture the randomness of the noise. As a result, the digital technique is able to separate pure harmonic from the composite signal while rejecting the measurement noise in the captured harmonic

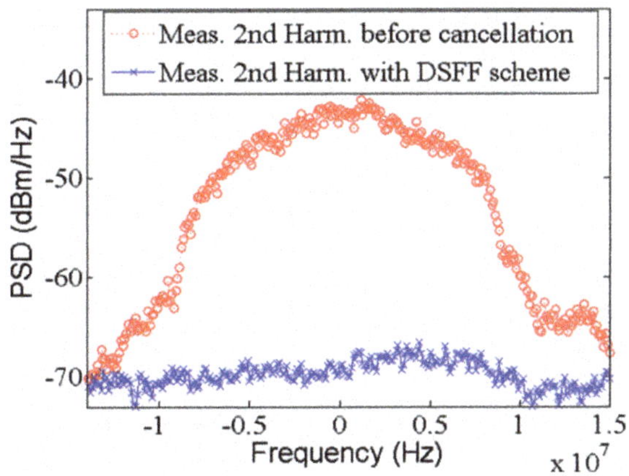

Fig. 5.11 Harmonic cancellation performance using digitally supported feed-forward cancellation scheme ([21], reprinted with permission from IEEE)

signal. Such digitally modeled harmonic signal is transmitted via the second channel to cancel the harmonic.

The phase compensation of 179.2° and gain factor of 0.998 were provided digitally to the baseband harmonic modeled signal when complete harmonic cancellation was observed. Figure 5.11 shows the optimized feedforward system output in the frequency domain. The harmonic cancellation of 25 dBc is observed. As the fundamental and harmonic paths are independent, the DPD can be applied for fundamental signal and harmonic cancellation can be applied for harmonic signal to provide a UWB linear system [21].

This method requires accurate modeling of the captured harmonic as well as additional transmission of an uncorrelated signal at the lower channel. Being a post-cancellation method, there is a requirement of additional UWB linear PAs for the second channel to cover the harmonic band of the targeted signal. Moreover, as the method is applied as a post-cancellation scheme, there is an additional requirement of a high-power RF combiner at PA output, leading to the higher hardware cost.

5.3.2 Digitally Supported Harmonic Injection Technique

Digitally supported harmonic injection technique is shown in Fig. 5.12. The fundamental frequency of 978.40 MHz (940 MHz RF frequency + 38.4 MHz IF frequency) is chosen; therefore, the second harmonic frequency is 1956.8 MHz.

The power spectral density (PSD) of PA output at the fundamental and the second harmonic frequency is shown in Fig. 5.13a, b for two different fundamental frequencies. One can see that there is a second harmonic cancellation of approximately 29 dBc. In both cases, the harmonic signal is 55 dBc below the fundamental frequency signal. By shifting the fundamental frequency from (900 MHz RF signal + 38.4 MHz IF-shift signal) to 1138.4 MHz (1100 + 38.4 MHz IF shift), the configurability of the system is justified.

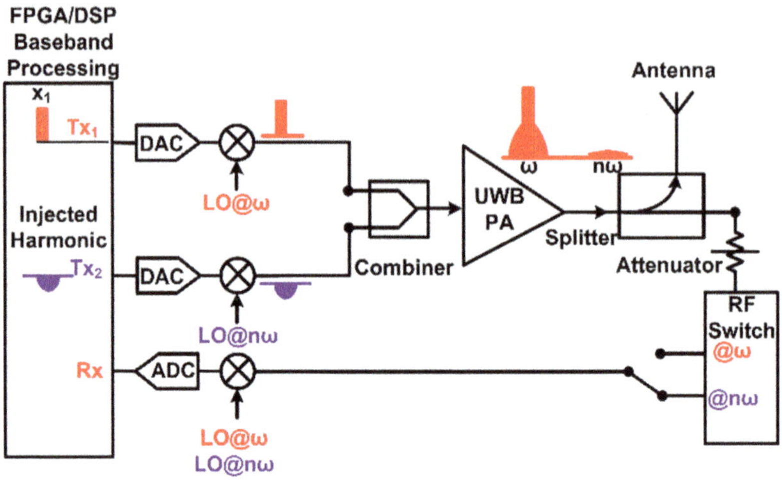

Fig. 5.12 Digitally supported harmonic injection technique for filterless transmitter ([20], reprinted with permission from IEEE)

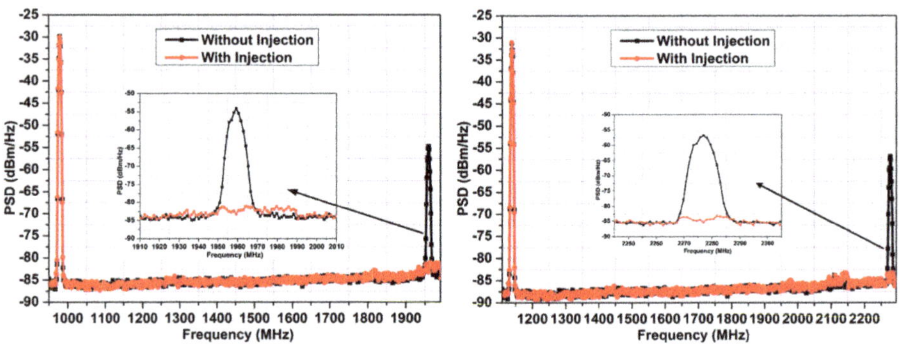

Fig. 5.13 Measured output spectrum of the amplifier with and without injection for a carrier frequency of (a) 978.40 MHz ([20], reprinted with permission from IEEE) and (b) 1138.4 MHz ([20], reprinted with permission from IEEE)

The characteristics of the PA change with the frequency; therefore, when fundamental frequency is changed, the coefficients of the harmonic model must be updated. Besides, a slight adaption of baseband harmonic model signal amplitude and phase is required. For nonconcurrent transmission case, when the frequency changes to a new frequency, the model adapts its coefficients according to the new frequency and additional filters are not required. Therefore, this method may enhance the compactness of the transmitter by replacing bulky filter banks with a parallel transmitter system. On the flip side, this method requires multichannel signal processing. The trade-off between hardware and software processing methods is required.

5.4 Delta Sigma-Based Transmitters

Chapter 1 discusses basic theory of DSM transmitters along with its various architecture depending on the DSM topology being low-pass, high-pass, or band-pass. In general, low-pass DSM is preferred for single-band transmission whereas band-pass DSM is preferred for multiband transmission. The DSM can be reconfigured to low-pass DSM and band-pass DSM by changing the position of poles and zeros of noise transfer function (NTF). Figure 5.14a shows the position of poles and zeros in low-pass and band-pass configuration. One can see that all zeros are in unity circle. Moreover, for low-pass configurations both poles and zeros are near positive real axis. However, in case of band-pass configuration, a pair of poles and zeros are near positive or negative imaginary axis. Figure 5.14b shows the magnitude response of NTF and signal transfer function (STF) of low-pass and band-pass configuration with normalized frequency. One can see from Fig. 5.14b that the magnitude plot of NTF has notches at the particular frequency where the positions of zeros are located.

5.4.1 *Multilevel Architectures*

Bi-level DSM has constant envelope and thus gives very high efficiency with switched-mode power amplifiers (SMPAs) [23–40]. However, the high quantization noise is the main bottleneck in case of bi-level DSM. This degrades the signal quality in terms of coding efficiency (CE) and signal-to-noise distortion ratio (SNDR) [30, 32, 33]. Due to this large quantization, the performance of transmitter is affected in two ways. In the first case, quantization noise is also amplified along

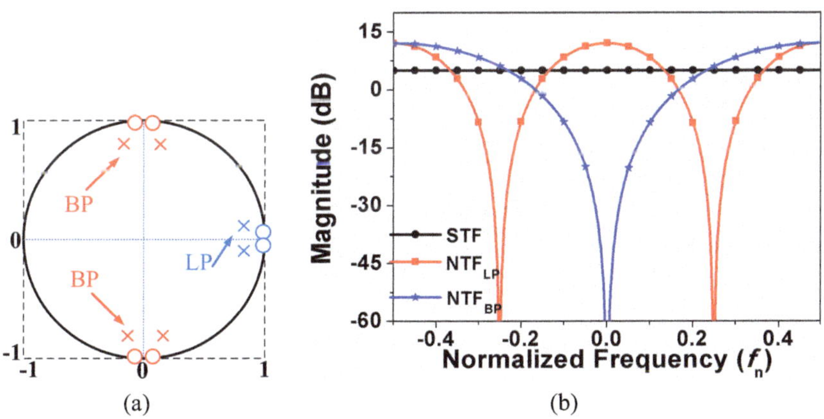

Fig. 5.14 Characteristics of low-pass and band-pass delta sigma modulation (DSM): (**a**) pole-zero plots, (**b**) magnitude response of noise transfer function (NTF) and signal transfer function (STF) with normalized frequency

with the required signal band and due to this most of the power at amplification stage is wasted in amplifying the quantization noise, thereby degrading the efficiency at amplification stage. Secondly, this out-of-band quantization noise must be filtered out before transmission; therefore, it will increase the complexity of subsequent filter stages. To solve the out-of-band quantization problem in case of bi-level DSM, researchers have recently proposed several architectures of multilevel DSM to decrease the generated quantization noise [41–46]. In order to improve the signal quality, multilevel DSM is considered as an efficient technique. If the number of quantization level increases at the output of DSM stage, then this will reduce the in-band and out-of-band quantization noise and consequently increase the CE and SNDR of DSM stage. However, as the number of quantization level increases, the envelope variation of DSM-modulated signal also increases and this decreases the efficiency of the PA [39]. Ultimately, this will reduce the overall transmitter efficiency, which is determined using CE and PA drain efficiency [39], [41]. To analyze the usefulness of multilevel DSM transmitter architecture the overall transmitter efficiency must be calculated [39]. The overall transmitter efficiency is equal to the product of PA efficiency and CE of DSM [44, 46].

$$\eta_T = \text{CE} \times \eta_{\text{PA}} \tag{5.37}$$

where CE is coding efficiency of DSM and η_{PA} is peak efficiency of PA when operated by continuous wave (CW) signal. By using multilevel DSM, the performance of DSM is improved in terms of CE and SNDR. The threshold levels selected for multiband DSM is given in Table 5.4 [46].

The multilevel DSMs are tested with 5 MHz LTE signals. The spectrum of output signal for two, three, four, and five-level DSM is shown in Fig. 5.15. One can see that the level of quantization noise decreases as number of quantization level increases.

The performance of multilevel DSM in terms of CE and SNDR for LTE 5 MHz signal are summarized in Table 5.5 [44]. One can see that the CE in case of five-level DSM is 96%, whereas for bi-level DSM CE is only 76%. Therefore, by using multilevel DSM the CE of DSM is increased by 20%. Similarly, SNDR for five-level DSM is 50 dB and for bi-level DSM it is only 40.5 dB implying an improvement of 9.5 dB [44]. Therefore, one can state that as the number of quantization levels increases, the CE and SNDR of DSM improves significantly.

Figure 5.16a, b show the detail of the scheme for selection among various quantizer level in case of multilevel DSM. Generally, the spacing between the level

Table 5.4 Quantization level for multi-level DSM [46]

Signal	Threshold levels (Cartesian DSM)
Level-2	–1,1
Level-3	–1,0,1
Level-4	–1,–0.33,0.33,1
Level-5	–1,–0.5,0,0.5,1

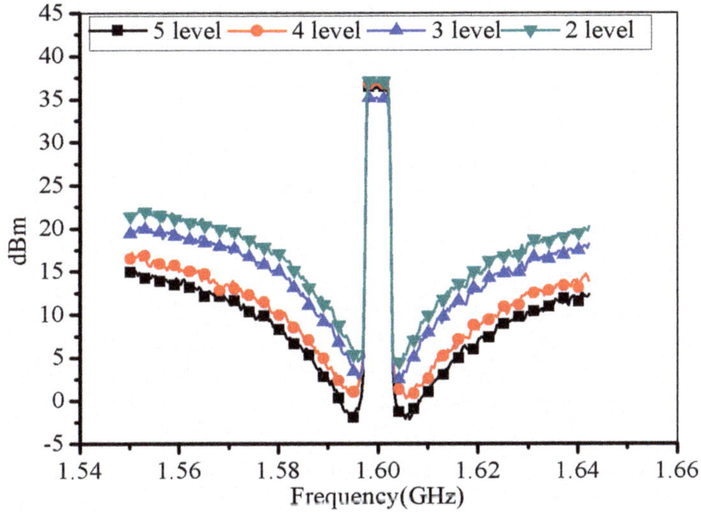

Fig. 5.15 Output signal spectrum of two-, three-, four-, and five-level DSM for 5 MHz LTE excitation [44]

Table 5.5 Signal-to-noise distortion ratio (SNDR) and coding efficiency (CE) for various levels of DSM signal

Signal type		2 levels	3 levels	4 levels	5 levels
LTE	CE (%)	76	83	94	96
	SNDR (dB)	40.5	41.9	48.4	50

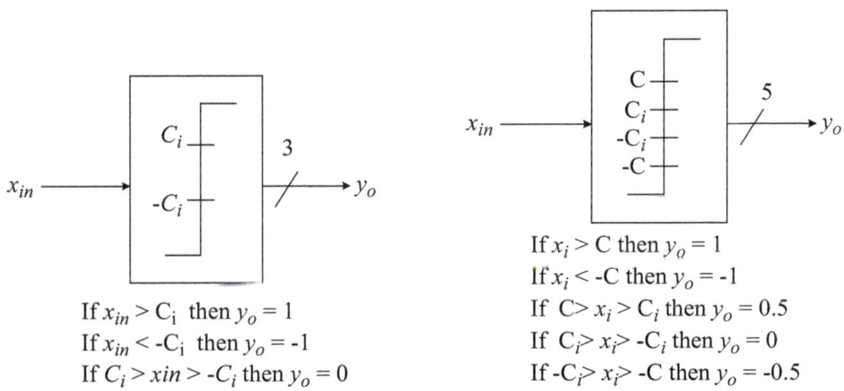

Fig. 5.16 Quantizer performance in multilevel DSM modulators: (**a**) 3-level, (**b**) 5-level

is selected uniform in case of multilevel DSM [46]. In this case, the threshold values are in the middle of quantizer levels.

If the output of DSM is limited to the value $\pm v$, the following equations relates the quantization level and spacing between threshold values [46],

$$\Delta = \frac{v - (-v)}{M - 1} \tag{5.38}$$

$$[L_k] = v + k^{*}\Delta \quad \text{where, } k = 0, \ldots, M \tag{5.39}$$

$$[C_i] = -v + \left(i + \frac{1}{2}\Delta\right) \quad \text{where, } i = 0, \ldots, M - 1 \tag{5.40}$$

where M is number of quantization level, Δ is spacing between threshold values, L_k is different levels of DSM output, and C_k is threshold values as shown in Fig. 5.16a, b. For example, in case of five-level DSM the quantization levels are -1, -0.5, 0, 0.5, 1 and the threshold values are -0.75, -0.25, 0.25, 0.75. One way to increase the CE and SNDR is by increasing the number of quantization level. However, by increasing the number of quantization levels, the peak-to-average power ratio (PAPR) of modulated signal increases and force the PA to work in power back-offs and lowers the PA efficiency. By adjusting the threshold values of quantizer in multilevel DSM the CE can be significantly improved for lower number of quantitation levels. By changing the quantizer threshold values for three- and five-level DSM, the CE and SNDR of DSM can be improved. An optimum value of threshold can be obtained to get the best CE and SNDR. Figure 5.17 shows the change in CE and SNDR of a 3 MHz LTE signal, when modulated by DSM for different threshold values [46]. The threshold value is varied from 0.01 to 0.99. The best SNDR is obtained at threshold value 0.46 and best CE is achieved at 0.99 threshold value.

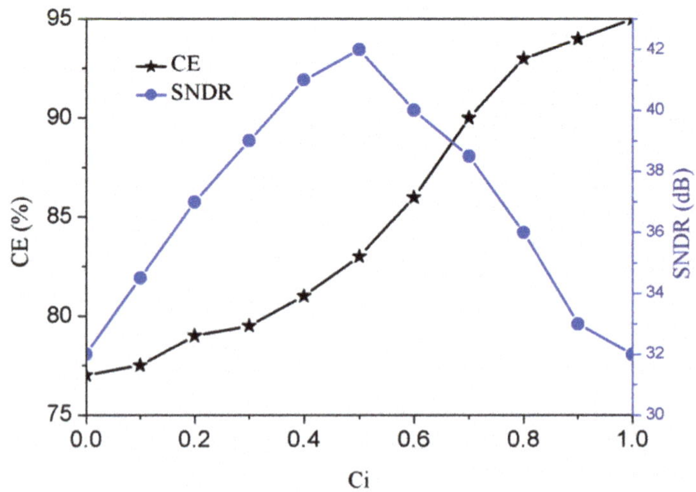

Fig. 5.17 Performance of three-level delta sigma modulators with threshold value in terms of signal-to-noise distortion ratio (SNDR) and coding efficiency (CE)

One can see that at a point where best CE is obtained, SNDR is not good and vice versa. Therefore, there is a trade-off between CE and SNDR performances by changing the quantization threshold value. One can also see from Fig. 5.17 that by compromising the SNDR by 5 dB the CE can be improved from 12 to 24% [46].

5.4.2 *Time Interleaving and Digital Sequencing in Multilevel Architecture*

The performance of multilevel DSM deteriorates at the RF amplification stage due to increase in PAPR of quantized signal as the number of quantization level increases. In [44], a new DSM-based transmitter architecture using level transformation is proposed. This architecture is shown in Fig. 5.18a, b.

The level transformation stage is added after DSM stage to maintain the efficiency of PA at higher quantization levels as shown in Fig. 5.18a, b. The input baseband signal is converted into *I* and *Q* components. These *I* and *Q* components are modulated by DSM blocks. DSM-modulated *I* and *Q* components are then passed through level transformation stage which is used to transform higher-level

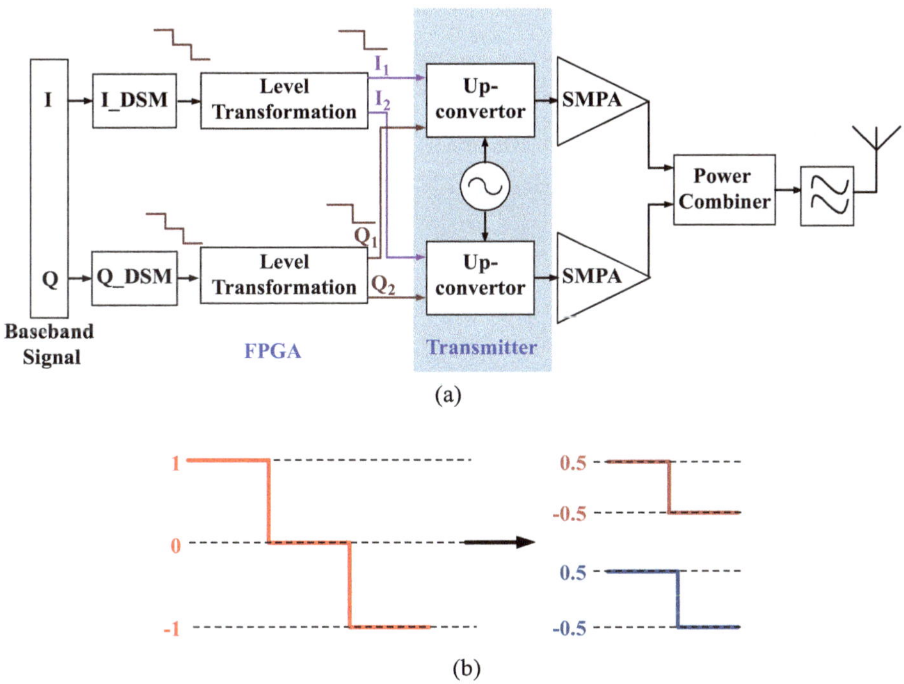

Fig. 5.18 DSM-based transmitter using level transformation: (**a**) architecture ([44], reprinted with permission from IEEE) and (**b**) level transformation scheme ([44], reprinted with permission from IEEE)

signal to lower-level signal. The level transformed $I-Q$ signals are then up-converted and amplified with separate PAs. For example, one can see in Fig. 5.18b that a three-level DSM is level transformed to two time-interleaved two-level DSM signals which are individually amplified by the two PAs. It is worth mentioning that the level conversion in Fig. 5.18b is intelligently performed such that a simple addition can recover the original levels. One can see that two-level signals of Fig. 5.18b when added results into three-level DSM signals of Fig. 5.18b.

Figure 5.19 shows the hardware setup for characterizing the performance of DSM transmitter with level transformation. The setup comprises Arria V (GT series) FPGA platform, where the output from FPGA is fed to two-channel transmitter (TSW30SH84) from Texas Instruments.

One can see that the signals are up-converted to RF signals using *IQ* modulators at each path in transmitter. The up-converted signals are fed to driver PAs which amplifies RF signal at each path to drive continuous class-F PAs. The decomposed amplified signals are combined with power combiner as shown in Fig. 5.19. Output is received and captured using receiver (TSW1266EVM) from Texas Instruments. The system is used to modulate and transmit 5 MHz LTE signal at 1.6 GHz carrier frequency. Figure 5.20 shows the output spectrum of DSM transmitter with level

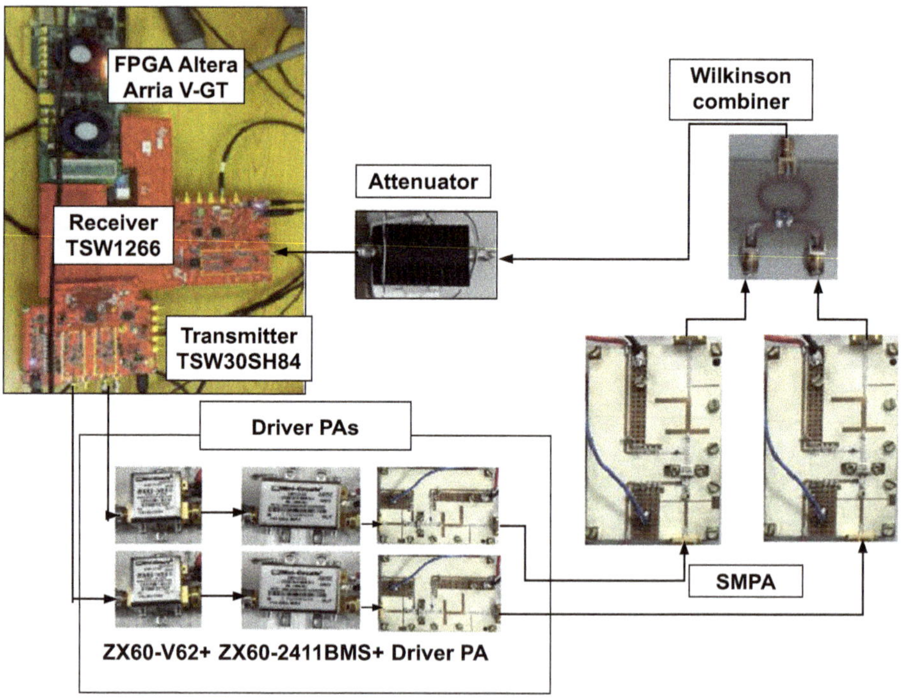

Fig. 5.19 Measurement test bench for DSM-based transmitter ([44], reprinted with permission from IEEE)

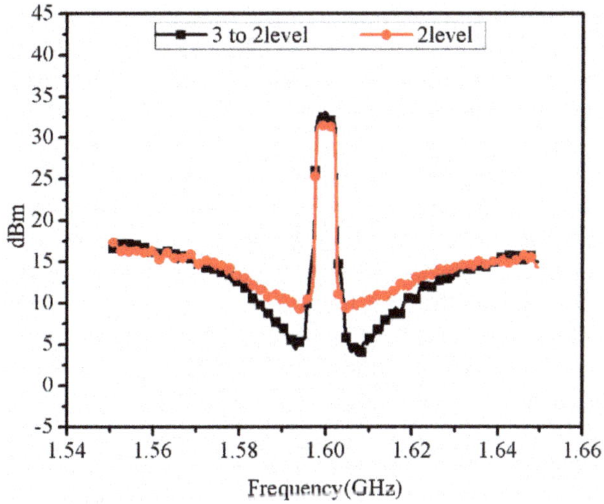

Fig. 5.20 Output spectrum of DSM transmitter with three- to two-level transformation ([44], reprinted with permission from IEEE)

Table 5.6 Performance evaluation of different quantizer level [44]

Level	CE (%)	SNDR (dB)	Back-off power (dB)
Two-level	76	40.5	0
Three-level	83	41.9	4.65
Three to two-level	80.25	41.2	0

transformation from three-level DSM to two-level DSM. For comparison purpose, output of two-level conventional DSM is also plotted in Fig. 5.20. One can see from this figure that the quantization noise power reduces in level transformation case in comparison to two-level conventional DSM.

The performance of level transformation-based DSM is also listed in Table 5.6. The SNDR and CE values are calculated for different levels using Eqs. (1.282) and (1.283) respectively. Drain efficiency values are taken at corresponding back-off where PA is operating as per the PAPR of the signal. In case of proposed three- to two-level DSM, the CE and SNDR obtained are 80.25% and 41.2 dB, respectively, which are close to three-level DSM values, while the PA amplifier works at peak power (0 dB back-off) which is equivalent to response of two-level DSM.

5.5 Hybrid Predistortion Scheme for Broadband Linearization

In recent years, DPD has established as a popular scheme for the linearization of PA. The advances in the DSP hardware and software-defined radios also support the use of DPD. However, with the advent of the fifth generation of communication systems, the instantaneous bandwidth of signal has increased five to ten times to that of fourth generation communication signals. It is anticipated that for time division duplex systems, LTE-CA will have signal bandwidth up to 100 MHz [47]. Inter-band CA utilizes several independent bands for the transmission of spectrum-efficient LTE signal. The transmitter will require several PAs or single concurrent multiband PA or a single concurrent UWB PA to support inter-band CA. Utilizing individual PA for each independent band will increase the cost of the transmitter system. Alternatively, a single wideband PA can be utilized that covers all the bands as shown in Sect. 5.3. However, multiband DPD has several limitations. First, the model complexity in case of multiband DPD becomes much higher as compared to the case of single PA as the number of bands keeps increasing. Consequently, the bulky models with polynomial basis tend to become numerically unstable. Second, in case of DPD implementation, the transmitter, as well as feedback receiver, should have minimum instantaneous bandwidth support up to five times to that of signal bandwidth. This allows to capture the IMD terms up to fifth order. As a result, the sampling rate required for the DPD technique is usually ten times the modulation bandwidth of the signal [47]. This requires high-speed analog-to-digital converters (ADC) and digital-to-analog converters (DAC) which may not be practically feasible in case of wide-band signals. Third, the dynamic range of DACs and ADCs is affected due to IF filter bandwidth. Besides, the DAC and ADC operates with limited dynamic range when the bandwidth of the signal increases. Due to limited dynamic range, the correction applied by DPD is restricted [48]. Figure 5.22 shows the decrease in the dynamic range or ACPR when the signal bandwidth increases. The power is concentrated for a single carrier at a given frequency, which provides the maximum dynamic range. The theoretical dynamic range of ADC is $6.02n + 1.76$ dB, where n is the number of bits. When the bandwidth increases, the same power distributes among different frequencies. This leads to a reduction in the signal dynamic range. Alternatively, at the cost of higher hardware complexity, several ADCs and DACs can be used in parallel to achieve a higher dynamic range for the broadband signal. Fourth, the DPD scheme requires access to baseband signal. This requirement is a hindrance for PA used in RF repeater systems, where incoming signal is RF in nature.

Due to above limitations, analog predistortion technique has some potential for the wideband operation in the order of 500 MHz and above. Figure 5.21 shows one preferred implementation strategy for analog predistortion, where incoming signal is RF in nature. The incoming RF signal is split into two paths: (a) linear path and (b) IMD generation path.

Linear path applies the RF signal to PA which amplifies the RF signal and IMD terms also appear due to PA nonlinearity. As the name suggests, second path is used to generate IMD terms by using some nonlinear RF component such as antiparallel diode pair.

A two-tone signal w applied at the input of the second path is given as,

$$x(t) = A\left[\cos\left(\omega_1 t + \varphi\right) + \cos\left(\omega_2 t + \varphi\right)\right] \tag{5.41}$$

where ω_1 and ω_2 are the carrier frequencies of the two tones. φ denotes the phase variation due to the delay in the lower branch composed of the two channels. The

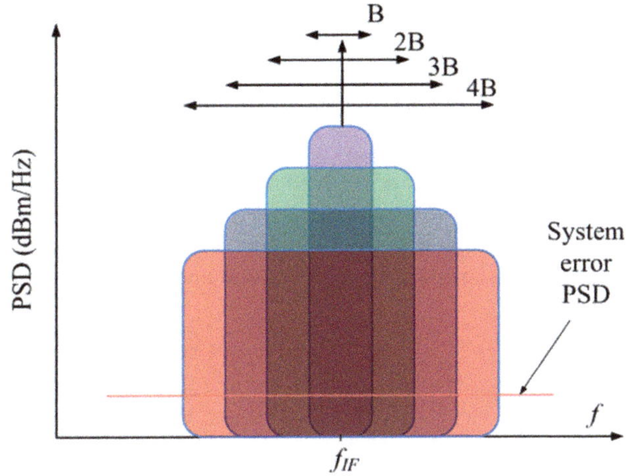

Fig. 5.21 Decrease in signal dynamic range (signal-to-noise ratio) with the increased bandwidth

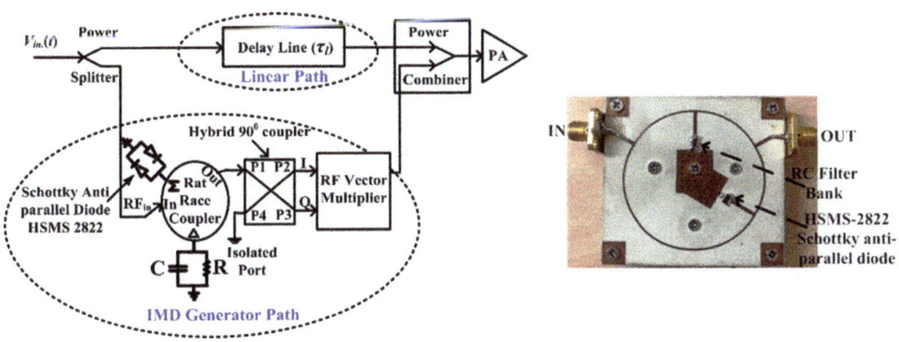

(a) (b)

Fig. 5.22 Implementation example of (**a**) cubic-analog predistorter, (**b**) fabricated intermodulation generator equipped with RC filter bank and HSMS 2822 Schottky antiparallel diode

signal at lower branch is 180° out of phase with respect to the signal at upper branch.

$$x_{\text{inv.}}(t) = -A\left[\cos\left(\omega_1 t + \varphi\right) + \cos\left(\omega_2 t + \varphi\right)\right] \tag{5.42}$$

Output of lower branch, that is, $x_{\text{inv.}}(t)$ is applied at the input of modified rat race coupler (RRC). The sigma port of RRC is mounted with antiparallel HSMS2822 Schottky diodes as shown in Fig. 5.22a, b. The antiparallel diodes produce the odd-order nonlinearities. IMD generation components are mostly analog nonlinear components such as diodes. As odd-order nonlinearity order has more impact, an antiparallel diode pair is commonly used component. In addition to IMD generation, it is also required to suppress the main signal. When this signal containing only IMD terms is added to the input signal at 180°, it facilitates cancellation of IMDs generated at PA output. A four-port RRC with sigma and delta ports can be used.

The output of antiparallel diode in RRC for the third-order nonlinearity is given as,

$$x(t) = A\left[\cos\left(\omega_1 t + \varphi\right) + \cos\left(\omega_2 t + \varphi\right)\right] \tag{5.43}$$

$$\begin{aligned} V_{\mathrm{HSMS}}(t) = &-Aa_1\left[\cos\omega_1 t + \cos\omega_2 t\right] - A^3 a_3\left[\cos^3\omega_1 t + \cos^3\omega_2 t\right] \\ &- 3A^3 a_3\left[\cos^2\omega_1 t\cos\omega_2 t + \cos\omega_1 t\cos^2\omega_2 t\right] \end{aligned} \tag{5.44}$$

Delta port of RRC is mounted with a resistor and capacitor. RRC is used to combine the antiparallel diode circuit and RC circuit. The output of RRC is given as,

$$\begin{aligned} V_{\mathrm{RRC}}(t) = &\tfrac{a_3 A^3}{4}\left(\cos 3\omega_1 t + \cos 3\omega_2 t\right) - \tfrac{3a_3 A^3}{4}\left[\cos\left(2\omega_1 t + \omega_2 t\right)\right. \\ &\left.+\cos\left(2\omega_1 t - \omega_2 t\right) + \cos\left(2\omega_2 t + \omega_1 t\right) + \cos\left(2\omega_2 t - \omega_1 t\right)\right] \end{aligned} \tag{5.45}$$

First term in Eq. (5.45) represents the third-order harmonics at $3\omega_1$ and $3\omega_2$ that can be easily filtered out. Next four terms are third-order IMDs. Filters can be used to remove the third-order IMD components that lie at frequency $2\omega_2 + \omega_1$, $2\omega_1 + \omega_2$. However, the third-order IMD terms ($2\omega_2 - \omega_1$, $2\omega_1 - \omega_2$) will be very close to the fundamental frequencies and therefore cannot be easily filtered out. For the input signal with the wider bandwidth, these third-order IMDs are of greater concern [48]. In order to remove the third-order IMD components that lie at frequency $2\omega_2 - \omega_1$, $2\omega_1 - \omega_2$, the output of RRC is combined with the original signal (at out-of-phase and provided to the PA).

Figure 5.23 shows the RRC output when input signal is a two-tone signal. Ideally, fundamental signal should be completely removed; however, practically fundamental signal is suppressed by 25 dB. It can also be observed that third-order IMDs are much higher than fifth- and higher-order IMDs.

The output of RRC is fed to the port 1 (P1) of quadrature 90° hybrid coupler or branch line coupler. Branch line coupler consists of symmetrical four ports with a 90° phase difference between two coupled ports. A signal coming from port 1 (P1) is split into two quadrature signals to the port 2 (P2) and port 3 (P3), with the remaining port 4 (P4) fully isolated from input port (P1) using 50 Ω load impedance.

The output from P2 and P3 of branch line coupler is fed at the input pins (RFIN_I and RFIN_Q) of vector multiplier which is often referred as vector modulator (VM). The branch line coupler is used to provide in-phase and quadrature-phase input to VM. ADL5390 VM consists of a matched pair of continuous variable gain amplifiers which has been used to vary the gain and phase of the input signal. Each amplifier has its separate control unit which is linear in magnitude.

When the two outputs from branch line coupler are in quadrature phase relation, the VM can be configured as variable gain and phase shifter by using its gain control pins. For the selected VM, phase can be shifted to the entire 360° and amplitude can be adjusted from −30 to 5 dB. The output of VM is combined with upper path signal using power combiner.

Ignoring the harmonics and filterable third-order IMDs, combined output is given as input to RF-PA, which is given as,

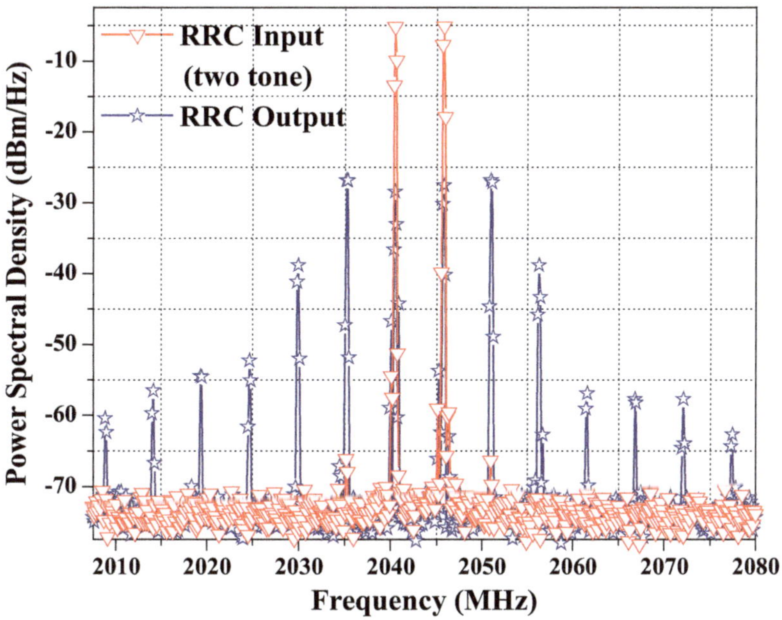

Fig. 5.23 Fundamental signal cancellation and IMD generation by the rat race coupler (RRC) for different frequencies ([48], reprinted with permission from IEEE)

$$y_{\mathrm{PD}}(t) = x(t) + V_{\mathrm{RRC}}(t) \tag{5.46}$$

$$\begin{aligned} y_{\mathrm{PD}}(t) = & A\left[\cos\left(\omega_1 t + \varphi\right) + \cos\left(\omega_2 t + \varphi\right)\right] \\ & - \delta_0\left[\cos\left(2\omega_1 t - \omega_2 t\right) + \cos\left(2\omega_2 t - \omega_1 t\right)\right] \end{aligned} \tag{5.47}$$

Here, the coefficient δ_0 affects third-order IMD, φ is the phase delay in terms of delay that occurs in the lower branch. The overall reduction in IMD is achieved by optimizing cubic predistorter linearizer parameters R, C, and φ.

Analog predistortion is more useful for the wideband operation in the order of 500 MHz and above; however, due to imperfections in individual analog components, IMD cancellation is not as good as its digital counterpart. Therefore, hybrid predistortion methods have been investigated to retain best of the "analog" and "digital" signal-processing techniques.

As discussed in Chap. 1, predistortion scheme can be divided into two main operations: (a) inverse nonlinearity generator (ING), which provides IMD components to counter the IMD component of the PA; (b) linear control elements such as attenuation, phase shift, and time-delay shift to apply the generated IMDs of ING with the suitable gain/phase to cancel the IMD components of the PA. Both of these operations can be implemented in either digital (baseband) domain, analog/RF domain, or hybrid domain.

DPD techniques incorporate ING as well as linear control elements in digital domain. APD technique incorporates both ING and linear control elements in analog domain. Hybrid methods can be differentiated into two categories based on the domain in which ING and linear control aspect are implemented. First category is digital vector modulation-based architecture. ING is implemented in analog domain while linear control is applied in digital domain. Another is analog vector modulator-based architecture. In this, ING is implemented in digital domain while linear control is applied in analog domain.

5.5.1 Digital Vector Modulator-Based Architecture

This architecture enhances the performance of APD by providing precise linear control in digital domain. Figure 5.24 shows the architecture of such hybrid predistorter [48]. It can be observed that the VM, hybrid coupler, and delay lines are not required, which are an essential part of the conventional APD scheme. The proposed scheme generates the input baseband signal and its out-of-phase version using two synchronized modulated sources. This is implemented using two-channel transmitter (TSW30SH84) from Texas instruments. The baseband part of the modulated sources is implemented in Arria V (GT series) FPGA platform as shown in Fig. 5.24. Baseband *I*/*Q* data are fed to random access memory (RAM) of FPGA allocated to each channel using MATLAB and Quartus software. This baseband data is then converted to analog signals using 16-bit DACs sharing the same clocks. These signals are further modulated and up-converted to RF signals using quadrature modulators in each path.

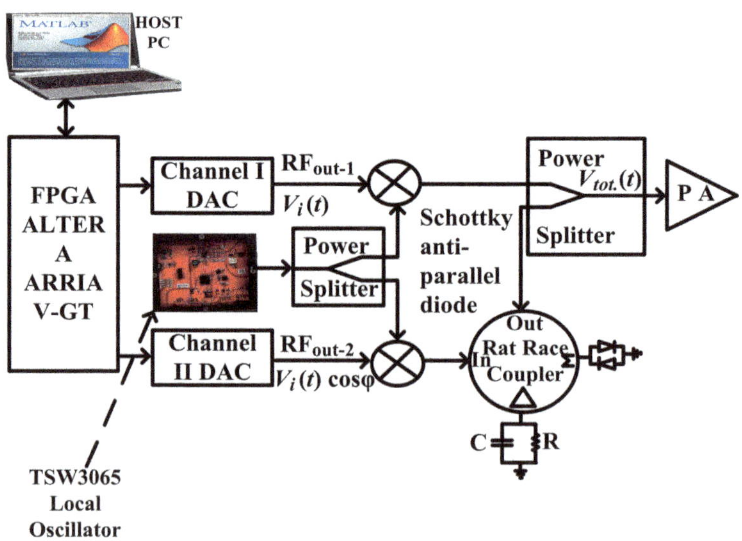

Fig. 5.24 Setup for digitally supported analog predistorter

Since the two channels are synchronized with the same baseband clock of FPGA, an accurate phase between the two channels can be set at baseband level. This provides required phase control between the two paths accurately as compared to APD. If the upper path described as channel 1 in Fig. 5.24 has an input signal $V_i(t)$, the signal in lower path depicted as channel 2 is represented by Re $(V_i(t)e^{-j\phi})$. The phase difference of 180° is required to generate inverted signal, whereas ϕ is required to compensate for any additional phase errors due to analog components in upper and lower path. The input to PA can be given as,

$$V_{\text{tot.}}(t) = V_i(t) + E_n\,(V_i(t)\cos\phi) \tag{5.48}$$

where $E_n(.)$ is the nonlinear function implemented by IMD generator in the lower path. Using the value of $\phi = 180° + \theta$, Eq. (5.48) can be rewritten as,

$$V_{\text{tot.}}(t) = V_i(t) + E_n\,(-V_i(t)\cos(\theta)) \tag{5.49}$$

The above Eq. (5.49) has two terms, first representing the signal and the second representing an error signal (E_n) from the output of a nonlinear component generated from the input signal with an appropriate phase shift and main signal removed. Phase delay that arises in lower branch is compensated digitally by introducing a phase difference in the digital domain. As phase is tuned in digital domain, very precise value can be provided for cancellation performance.

Figure 5.25 shows the automatic phase control by observing the ACPR at transmitter output. The power at the output of transmitter can be observed using power meter at adjacent band or by using feedback receiver, where ACPR can be measured in baseband and applied for phase correction.

With the aim to investigate broadband linearization, Fig. 5.26a, b shows the measured PSD for the contiguous and noncontiguous 8CC 160 MHz LTE signal with APD and hybrid RF-DPD. When RF-in RF-out APD is applied to linearize ZX60-V63+PA using contiguous 160 MHz LTE signal, an ACPR of −43.9 dBc is achieved which is an improvement of over 20.5 dB as compared to PA without linearization.

Figure 5.27a, b shows that the PA has almost 2 dB compression and 5° phase difference.

Similarly, when proposed linearization scheme is applied to noncontiguous 10001001 where 1 indicates on and 0 indicates off with total instantaneous bandwidth of 160 MHz LTE signal, it provides a correction of 9.2 dB and delivers an ACPR of −45.1 dBc. Its performance is further enhanced using HRF-DPD. As shown in Fig. 5.26b, with HRF-DPD, an ACPR of −56.1 dBc, which is achieved, which corresponds to an improvement of around 20.2 dB. With precise digital control, this technique is able to clearly enhance linearization performance significantly.

In the context of broadband linearization, Fig. 5.28 shows the linearization performance for LTE signal with 100 MHz bandwidth whose carrier frequencies are located at 1985 MHz and 2080 MHz. It can be appreciated from Fig. 5.28

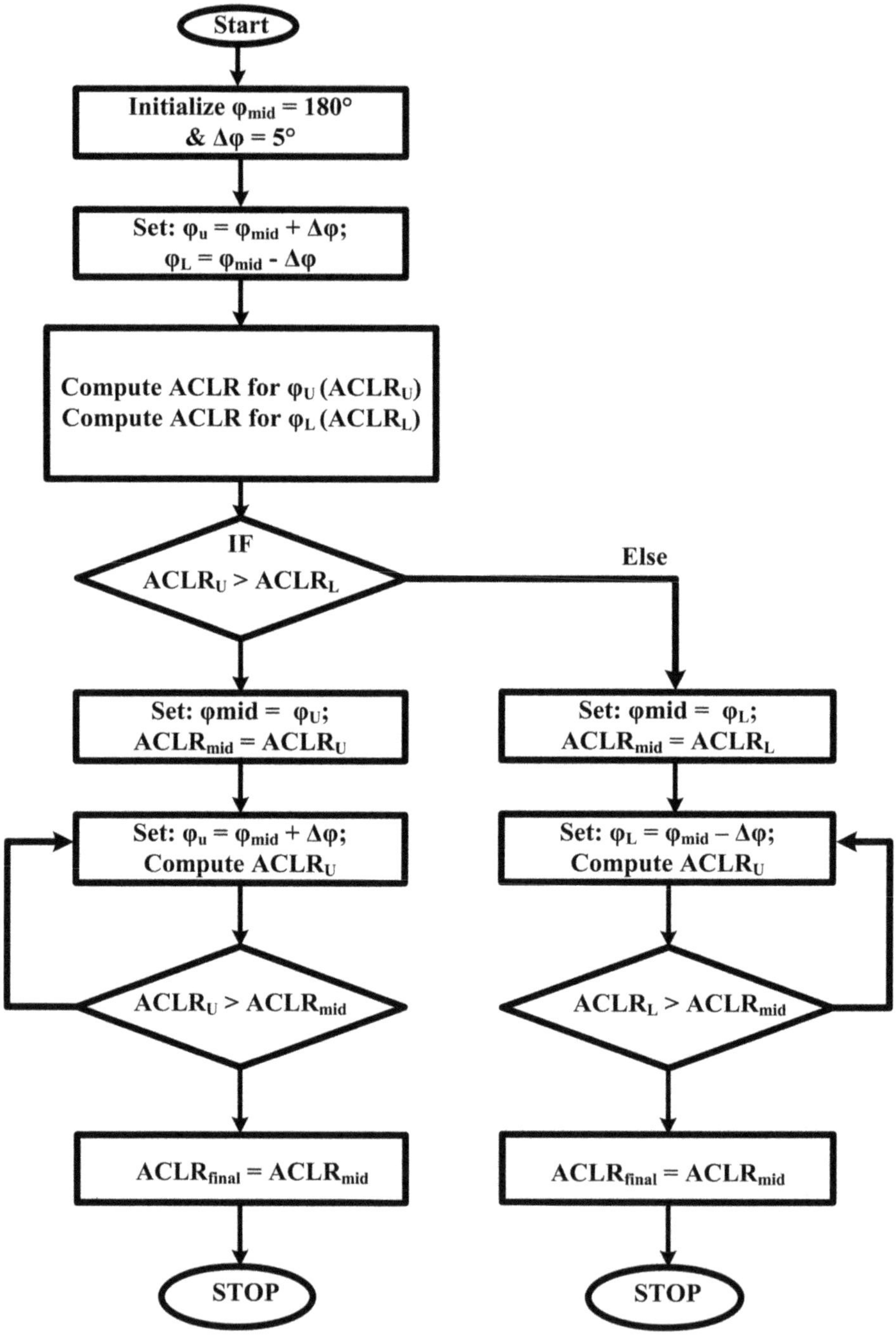

Fig. 5.25 Automatic phase delay compensation scheme by observing ACPR at transmitter output ([48], reprinted with permission from IEEE)

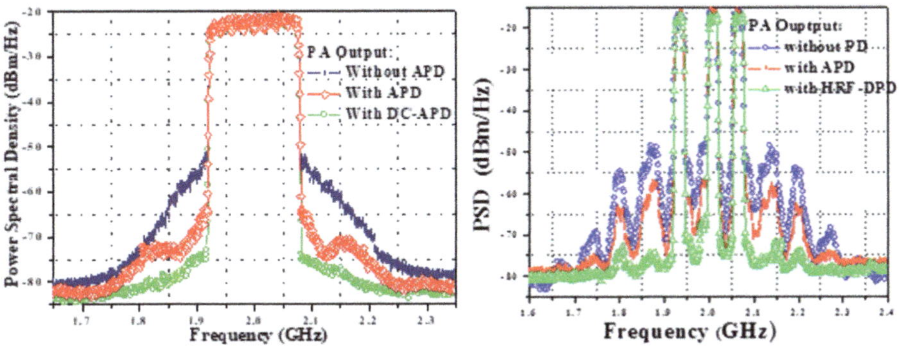

Fig. 5.26 Measured power spectrum densities of PA with and without the proposed RF-in RF-out APD and HRF-DPD excited by (**a**) contiguous 8CC 160 MHz LTE signal ([49], reprinted with permission from IEEE) and (**b**) noncontiguous 8CC 160 MHz LTE signal ([49], reprinted with permission from IEEE)

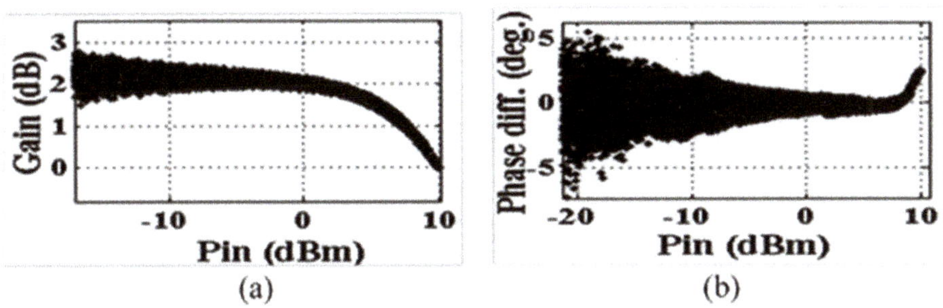

Fig. 5.27 ZX60V-63+ class AB PA (**a**) AM/AM characteristics, (**b**) AM/PM characteristics

that with precise digital control, HRF-DPD linearization not only provides in-band ACPR correction near the band, but it is also able to correct out-of-band IMD significantly [49].

5.5.2 Analog Vector Modulator-Based Architecture

The hybrid RF-DPD technique where required PD function is implemented as an LUT in an FPGA and controls the gain and phase of an analog vector modulator is discussed in section "Hybrid Predistortion" of Chap. 1. The architecture is also shown in Fig. 1.62, where the VM is referred as vector multiplier. The bits corresponding to the rectangular components G_I and G_Q of the PD's complex gain are obtained from Eqs. (1.272) and (1.273) after appropriate calibration. However, this assumes that the control voltages V_I and V_Q change the gain and phase of vector modulator in perfect quadrature relation as shown in Fig. 1.63a. In practice, the

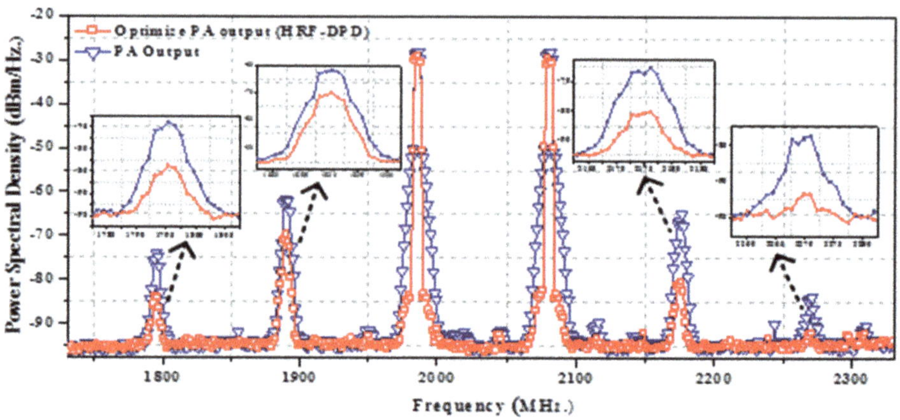

Fig. 5.28 Power spectrum of HRF-DPD showing cancellation of out-of-band IMD3 and IMD5 of LTE 100 MHz signal ([48], reprinted with permission from IEEE)

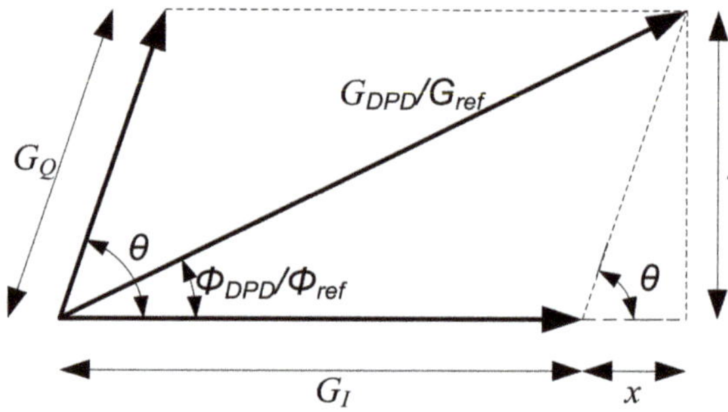

Fig. 5.29 Vector decomposition of the predistorter complex gain characteristic for generating LUT entries corresponding to the control voltages in case I–Q imperfection ([50], reprinted with permission from IEEE)

vector multiplier has quadrature imbalance and the control inputs change its gain and phase in non-quadrature fashion as shown in Fig. 5.29. This inhibits the vector multiplier to apply correct gain and phase values required for the predistortion.

One way to compensate for this imperfection is to convert the complex gain of PD function into rectangular components, i.e., G_I and G_Q, according to the non-quadrature relationship of Fig. 5.29. Accordingly, the mapping coefficients K_I and K_Q are generated to compensate for this quadrature imbalance. The values of x and y in Fig. 5.29 are given by:

$$y = G_{\text{DPD}} \sin(\Phi_{\text{DPD}}) \tag{5.50}$$

$$x = y \cot(\theta) \tag{5.51}$$

where θ is the quadrature imbalance in vector multiplier and represents a non-quadrature angle with which the G_I and G_Q vectors combine to give the desired complex gain. These rectangular components are now computed using Eqs. (5.50) and (5.51) and according to Fig. 5.29 as:

$$G_I = G_{\mathrm{DPD}} \left(\cos\left(\Phi_{\mathrm{DPD}}\right) - \sin\left(\Phi_{\mathrm{DPD}}\right) \cot\theta\right) \tag{5.52}$$

$$G_Q = G_{\mathrm{DPD}} \sin\left(\Phi_{\mathrm{DPD}}\right) \operatorname{cosec}\left(\theta\right) \tag{5.53}$$

Similar to the discussion in section "Hybrid Predistortion" of Chap. 1, the proportionality constants, K_I and K_Q, are obtained and substituted in Eqs. (5.52) and (5.53) to give:

$$\mathrm{Bit}_I = \frac{\mathrm{Bit}_{\mathrm{ref}} G_I}{\cos\left(\Phi_{\mathrm{ref}}\right) - \sin\left(\Phi_{\mathrm{ref}}\right) \cot\theta} \tag{5.54}$$

$$\mathrm{Bit}_Q = \frac{\mathrm{Bit}_{\mathrm{ref}} G_Q}{\sin\left(\Phi_{\mathrm{ref}}\right) \operatorname{cosec}\left(\theta\right)} \tag{5.55}$$

Here, the $\mathrm{Bit}_{\mathrm{ref}}$ is the bit value that corresponds to the unity gain in the RF path when all the entries of both the RAMs corresponding to V_I and V_Q in Fig. 1.62b are set to this bit value. The corresponding phase of the system obtained at this setting is Φ_{ref} which is relative phase with respect to the phase of the system when only Bit_I is applied, with Bit_Q set to zero. In case of perfect quadrature relation, θ is 90° and Φ_{ref} is exactly half of this value of θ, i.e., 45°. In such case, Eqs. (5.54) and (5.55) will reduce to Eqs. (1.272) and (1.273) which are used to map the G_I and G_Q to Bit_I and Bit_Q which are eventually corresponding to V_I and V_Q. Thus, Φ_{ref} quantify quadrature imbalance in the system and using it in Eqs. (5.54) and (5.55) for mapping results in unequal values of K_I and K_Q which compensates for the quadrature imbalance in the system. Figure 5.30 describes the measurement setup for characterizing this $I-Q$ imperfection in the hybrid RF-DPD system.

In this characterization, the VNA is set to power sweep mode to measure AM-PM characteristic of the system. The purpose is to find the non-quadrature relation between I and Q components in vector addition that happens in the RF vector multiplier. This is obtained by varying only one of the control inputs of the RF vector multiplier, while keeping the other at a nominal voltage value. Since, the control voltage changes according to the entries in the corresponding RAM I and RAM Q, the different entries are filled into these tables in order to inhibit variation in one control input while enabling the other. If VNA source is set to linear power sweep mode and the RAM entries are the bits arranged in their ascending order, the corresponding control voltage will also sweep linearly with the input power sweep. This condition is termed as THRU, where the entry in RAM is same as the address bit where it is stored. On contrary, the condition where all the entries of RAM are initialized to zero hexadecimal values is termed as ZERO. In such case, the corresponding control voltage to the vector multiplier is fixed to the nominal value and does not vary with the input RF power. Now, different combination of these conditions applied to the RAM I and RAM Q will retrieve the relative phase between I and Q components in vector addition. The bit values are loaded in the RAM by initializing it using the corresponding memory initialization file (mif). For example, if RAM I is set to THRU condition and RAM Q is set to ZERO condition,

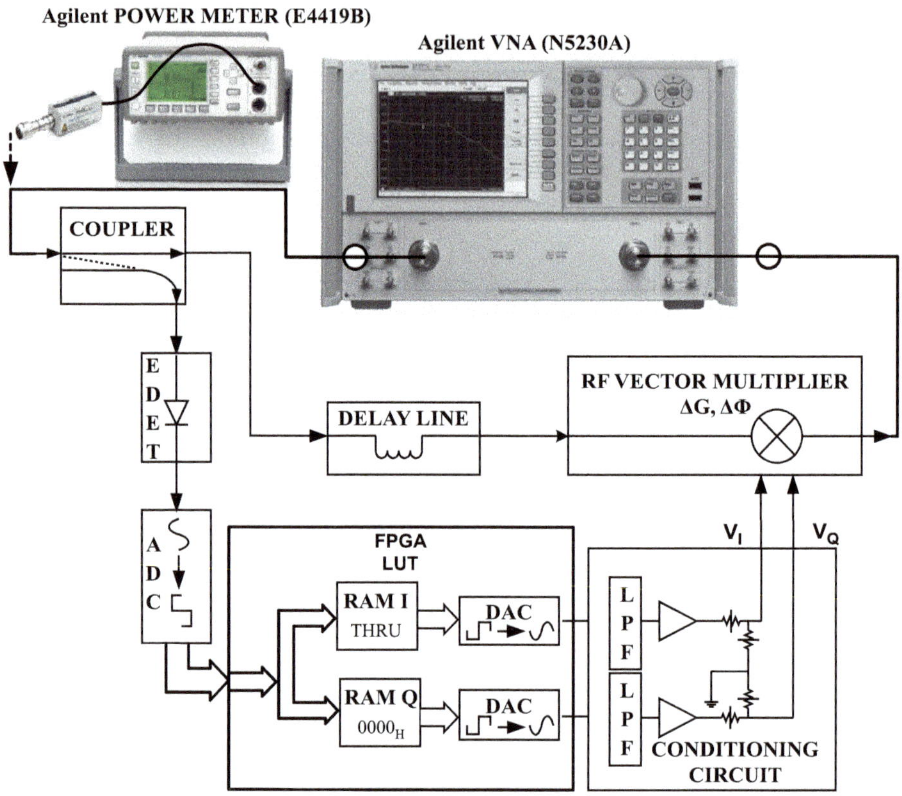

Fig. 5.30 Setup for characterizing the RF vector multiplier for $I-Q$ imperfection in the hybrid RF-DPD system ([50], reprinted with permission from IEEE)

the phase shift by the vector multiplier represents the *I*-axis as shown in Fig. 5.31. Similarly, Fig. 5.31 shows the change in vector multiplier phase representing the *Q*-axis. This is obtained by setting RAM *Q* and RAM *I* to THRU and ZERO conditions, respectively. The phase shift presented by the vector multiplier in case both RAM *I* and RAM *Q* are set to THRU is shown in Fig. 5.31 as *IQ* trace. The Φ_{ref} is obtained as 43.3° which is an angle between *I*-axis and *IQ* axis. The angle θ is measured as approximately 84.6° which is an angle between *I*-axis and *Q*-axis in Fig. 5.31. Now the LUT entries are generated using these values in Eqs. (5.54) and (5.55).

For proof of concept, a class AB PA is linearized with the hybrid RF-DPD system developed on a Stratix EP1S80 DSP/FPGA platform. The scheme for *I*–*Q* imperfection correction is evaluated in terms of ACLR improvement with respect to the case where LUT entries are generated using Eqs. (1.270) and (1.271). Both single carrier and multi-carrier 3G Wideband Code Division Multiple Access (WCDMA) signals with PAPR of 10.5 dB are used to characterize the PA which has peak output power of 40 dBm.

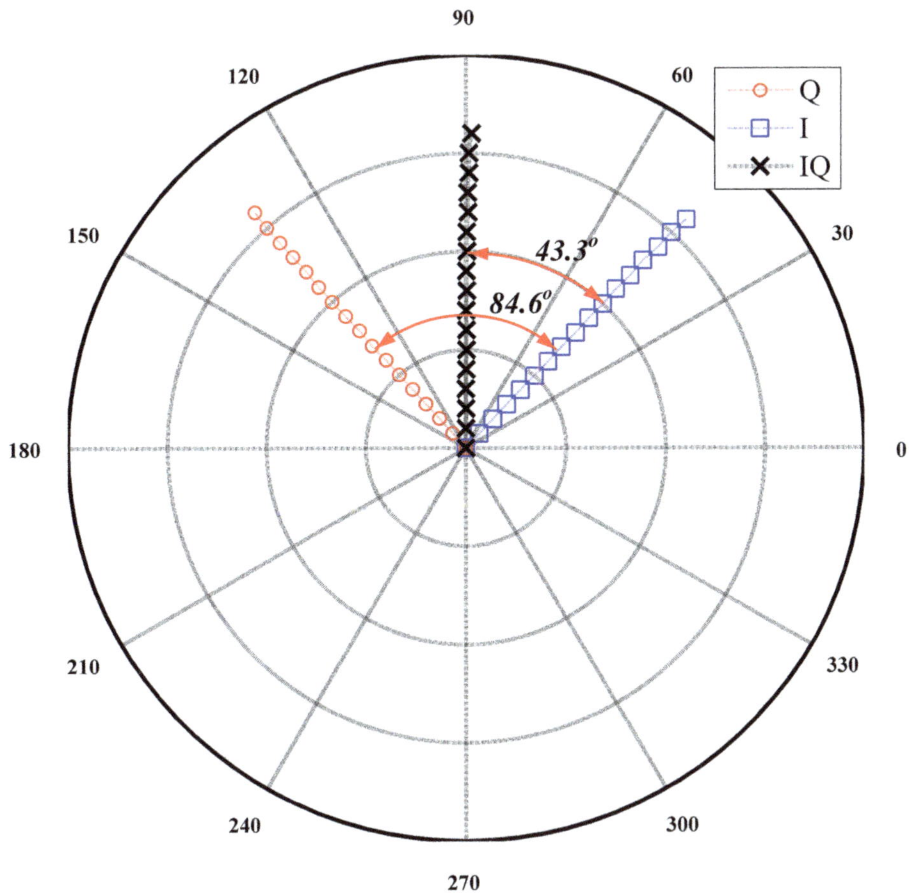

Fig. 5.31 Absolute phase measurement of different cases in RF vector multiplier characterization ([50], reprinted with permission from IEEE)

Figure 5.32a, b shows the results of the PA linearization for one-carrier WCDMA signals and three-carrier WCDMA signals, respectively. One can see that 2–3 dB improvement in terms of ACLR is achieved in both the cases with respect to the case where *I–Q* imbalances are not corrected.

This hybrid RF-DPD system is RF-in-RF-out systems [48], where the predistortion is applied in RF domain alleviating the restrictions on DAC sampling speed requirement. However, the nonlinear characterization setup is same as DPD as shown in Fig. 1.54; therefore, the ADC bandwidth requirements here is three to five times of the signal bandwidth. Table 5.7 summarizes the features of different predistortion linearization techniques.

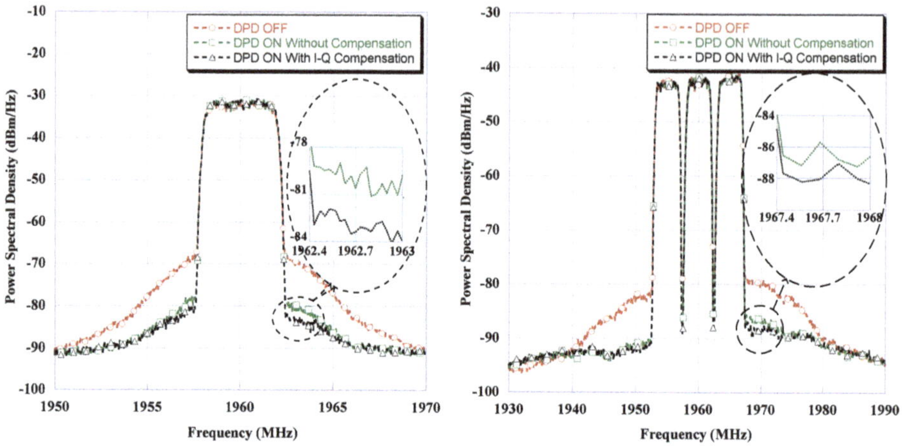

Fig. 5.32 Measured output spectrum of the PA (**a**) one-carrier WCDMA signal ([50], reprinted with permission from IEEE) and (**b**) three-carrier WCDMA signal ([50], reprinted with permission from IEEE)

Table 5.7 Feature comparison of different predistortion methods

Specifications	APD	DPD	Hybrid RF-DPD: Analog vector modulation	Hybrid RF-DPD: Digital vector modulation
Nonlinearity inducing element	Analog	Digital	Digital	Analog
Bandwidth of ADC	Not applicable	5× of the input signal	5× of the input signal	Not applicable
Bandwidth of DAC	Not applicable	5× of the input signal	1× of the input signal	1× of the input signal
Bandwidth of VM	5× of the input signal	Not applicable	5× of the input signal	Not applicable
Delay compensation	RF delay lines	Digital delay compensation	RF delay lines	Digital delay compensation
Compensation bandwidth	Ultra-wide	Medium	Wide	Ultra-wide
ACPR correction	Moderate 10–15 dB	High 20–30 dB	Moderate 10–18 dB	High 20–25 dB

5.6 Linearization Example for Bias Modulation and Envelope Tracking

To enhance spectral efficiency, complex modulation schemes like orthogonal frequency-division multiplexing (OFDM) is used in LTE-A. These spectral-efficient modulation schemes have high PAPR which possesses adverse effects on efficiency and linearity of RF PAs. Efficiency and linearization improvement techniques allow PAs to work in the saturation region. The efficiency of PA can be improved either by using load modulation techniques as used in Doherty PAs or by using dynamic supply voltage modulation as used in envelope tracking (ET) PAs. DPD is the most popular technique to linearize PA and dual-band DPD models have been proposed to linearize dual-band Doherty PAs with reduced computational complexity. However, there are few DPD models available for linearization of dual-band ET PA and their computational complexities are still high.

5.6.1 Envelope Tracking PA

Figure 5.33 shows the block diagram of concurrent dual-band DPD for ET PA. The bandwidth of the envelope RF signal can be several times higher than carrier frequency separation of dual-bands [51–52]. This can result in speed mismatch between envelope signal and carrier signals in ET PA. In order to fix this problem, envelope shaping function *M(n)* is used to generate slow envelope. This is given as [53],

$$M(n) = \left((V_{\mathrm{Th}})^6 + (E(n))^6\right)^{1/6} \tag{5.56}$$

where V_{Th} is the minimum envelope provided to dynamic supply voltage modulator. *E*(*n*) is the generalized mean of the envelope of input signals [52], which is defined as,

$$E(n) = \left(\frac{1}{2}\left(|z_1(n)|^2 + |z_2(n)|^2\right)\right)^{1/2} \tag{5.57}$$

5.6.2 Conventional Dual-Band ET PA Behavioral Models

The output of ET PA contains intermodulation distortions (IMD), cross-modulation distortions, and distortions generated due to envelope supply modulator (EM). Thus, the output of ET PA is a function of IMDs, cross-modulation distortions of input signals, and distortions of their combined envelope *E*(*n*). In [54], the output of 3D-MP behavioral model is given as,

$$\begin{aligned}\hat{y}_{1-\mathrm{3D-MP}}(n) = & \sum_{m=0}^{M-1}\sum_{k=0}^{R-1}\sum_{p=0}^{N-1}\sum_{q=0}^{p}\sum_{r=0}^{q} a^1_{m,k,p,q,r}\cdot z_1\left(n-\tau_m\right) \\ & \times \left|z_1\left(n-\tau_m\right)\right|^{p-q}\left|z_2\left(n-\tau_m\right)\right|^{q-r}E\left(n-\tau_k\right)^r\end{aligned} \tag{5.58}$$

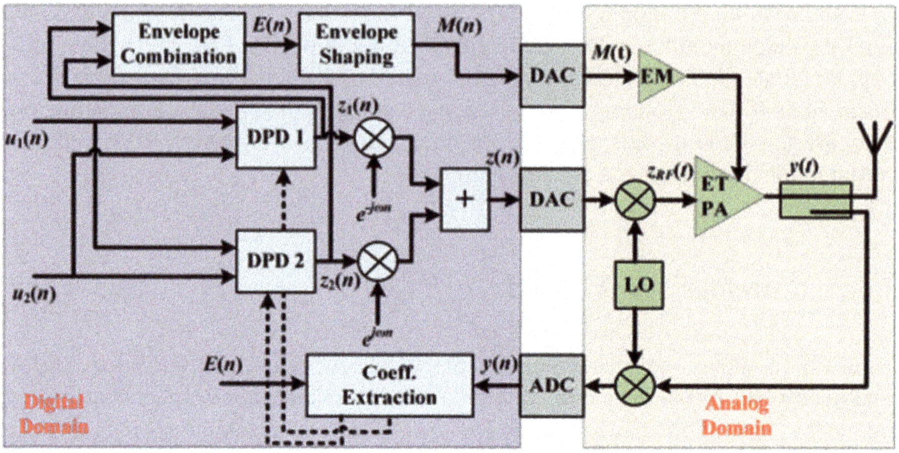

Fig. 5.33 Block diagram for DPD implementation in envelope tracking PAs ([51], reprinted with permission from IEEE)

where $Z_1(n)$ and $Z_2(n)$ are the complex baseband input signals. $a^1_{m,k,p,q,r}$ are the complex coefficients. M and R are the memory depths and N is the nonlinearity order. The number of coefficients is very high which results in an increase in computational complexity while using LS method to extract these coefficients.

In [55], the output of dual-band ET PA is characterized using 2D P-A-V-based model. The output of this model is given as,

$$\begin{aligned}\hat{y}_{1-2\mathrm{D}-\mathrm{PA}-V}(n) = &\sum_{k=0}^{N-1}\sum_{i=0}^{k} a^1_{k,i,1}\cdot z_1(n)|z_1(n)|^{k-i}|z_2(n)|^i\\ &+\sum_{k=0}^{N-1}\sum_{i=0}^{k}\sum_{m=1}^{M_1} a^1_{k,i,m,2}\cdot z_1(n-\tau_m)|z_1(n-\tau_m)|^{k-i}|z_2(n-\tau_m)|^i\\ &+\sum_{k=0}^{N-1}\sum_{i=0}^{k}\sum_{m=1}^{M_1} a^1_{k,i,m,3}\\ &\cdot z_1^2(n)z_1^*(n-\tau_m)|z_1(n-\tau_m)|^{k-i}|z_2(n-\tau_m)|^i\\ &+\sum_{k=0}^{N-1}\sum_{i=0}^{k}\sum_{m=1}^{M_1} a^1_{k,i,m,4}\cdot z_1(n)z_2^*(n-\tau_m)z_2(n)|z_1(n-\tau_m)|^{k-i}.\\ &\times|z_2(n-\tau_m)|^i\end{aligned} \tag{5.59}$$

where M_1 and M_2 are the memory depths of this model. N is the nonlinearity order. In [52], the 3-D distributed memory polynomial (3D-DMP) model is proposed which has output given as,

$$
\begin{aligned}
\hat{y}_{1-\mathrm{3D-DMP}}(n) = & \sum_{i=0}^{N_1-1}\sum_{p=0}^{P_1-1} a_{ip}^{1}\cdot z_1\left(n-\tau_i\right)\left|z_1\left(n-\tau_i\right)\right|^{p} \\
& + \sum_{i=0}^{N_2-1}\sum_{p=0}^{P_2-1}\sum_{j=1}^{M_2-1}\sum_{q=0}^{Q_2-1} b_{ipjq}^{1}\cdot z_1(n)\left|z_1\left(n-\tau_i\right)\right|^{p}\left|z_2\left(n-\tau_j\right)\right|^{q} \\
& + \sum_{i=0}^{N_3-1}\sum_{p=0}^{P_3-1}\sum_{k=1}^{K_3-1}\sum_{r=0}^{R_3-1} c_{ipkr}^{1}\cdot z_1(n)\mid z_1\left(n-\tau_i\right)\mid^{p}E\left(n-\tau_k\right)\mid^{r}
\end{aligned} \tag{5.60}
$$

where P_1, Q_2, P_2, R_3, and P_3 are the nonlinearity order. N_1, M_2, N_2, K_3, and N_3 are the memory depth, respectively. The number of coefficients required in this model is very less when compared to 3D-MP model but still it is comparable with 2D PAV model.

5.6.3 Behavioral Model of ET-PA

Equation (5.60) can be simplified to yield less complex behavioral model. During time alignment, the delay is adjusted between input and output baseband signals. Therefore, two memory summation is not required in the second term of Eq. (5.60). Thus, first and second term of Eq. (5.60) can be simplified, combined together as in [56], and written as,

$$
\begin{aligned}
\hat{y}_1(n) = & \sum_{k=0}^{N-1}\sum_{i=0}^{k}\sum_{m=0}^{M} a_{kim}^{1}\cdot z_1\left(n-\tau_m\right)\left|z_1\left(n-\tau_m\right)\right|^{k-i}\left|z_2\left(n-\tau_m\right)\right|^{i} \\
& + \sum_{k=0}^{P-1}\sum_{i=0}^{k}\sum_{l=0}^{R}\sum_{m=0}^{Q} c_{kilm}^{1}\cdot z_1\left(n-\tau_m\right)\left|z_1\left(n-\tau_m\right)\right|^{k-i}E\left(n-\tau_l\right)^{i}
\end{aligned} \tag{5.61}
$$

Now the second term of Eq. (5.61) contains four summations, that is, two summations for memory depth and two for nonlinearity order. The two summations of nonlinearity order can be further simplified. This is simplified in [57, 58] by taking an example of odd-order memoryless nonlinearity terms. By using the same approach, we got the proposed model output,

$$
\begin{aligned}
\hat{y}_1(n) = & \sum_{k=0}^{N-1}\sum_{i=0}^{k}\sum_{m=0}^{M} a_{kim}^{1}\cdot z_1\left(n-\tau_m\right)\left|z_1\left(n-\tau_m\right)\right|^{k-i}\left|z_2\left(n-\tau_m\right)\right|^{i} \\
& + \sum_{k=0}^{P-1}\sum_{i=0}^{k}\sum_{m=0}^{Q} c_{kilm}^{1}\cdot z_1\left(n-\tau_m\right)\cdot h_k^{1}\left(\left|z_1\left(n-\tau_m\right)\right|, E\left(n-\tau_l\right)\right)
\end{aligned}
$$

where M, Q, and R are the memory depths. N and P are the nonlinearity orders defined as,

$$h_k^1\left(|n-\tau_m)\,|, E\left(n-\tau_l\right)\right.$$
$$=\left\{\begin{array}{ll}\sum\limits_{i=0}^{k/2}\binom{k/2+1}{i}\binom{k/2}{i}|z_1\left(n-\tau_m\right)|^{k-2i}\cdot E(n-\tau_i)^{2i}, & k\in 2\mathbb{N}\\ \left(\sum\limits_{\omega}^{(k-1)/2}\binom{(k-1)/2+1}{i}\binom{(k-1)/2}{i}\right. & \\ \left.|z_1\left(n-\tau_m\right)|^{k+2}\cdot E(n-\tau_l)^{2i}\right), & k\in 2\mathbb{N}+1\\ \quad\times\left(|je_g^1|z_1\left(n-\tau_m\right)|+E\left(n-\tau_l\right)|\right) & \end{array}\right. \tag{5.62}$$

where $\binom{n}{k}=\frac{n!}{(n-k)!\times k!}$, e_g^r $(r=1,2)$ is the real-valued generalized coefficients for LB and UB. Here its range is from 0 to 2.

Table 5.8 shows the comparison of the number of coefficients in a band (LB or UB), where the proposed model has complexity as compared to that of the 3D-DMP model in terms of coefficients.

Figure 5.34 shows the test bench used for ET PA characterization and linearization. The measurement setup consists of vector signal generator (VSG), signal and spectrum analyzer (SSA), EM, ET PA, and attenuators. Two OFDM signals of bandwidth 10 MHz and 5 MHz, PAPR of 9.5 dB and 9.77 dB, respectively, are used. These signals are 80 MHz apart and transmitted from "SMW200A" VSG of R&S at a carrier frequency of 1950 MHz and a sampling rate of 122.88 MHz. The LM3290-91 EVM is used as EM to efficiently provide a dynamic supply voltage

Table 5.8 Comparison of coefficient in a band (LB or UB) ([51], reprinted with permission from IEEE)

Model	Number of Coefficients
3D-MP	$M\times R\times N\times(N+1)(N+2)/6$
2D P-A-V	$(1+M_1+2M_2)\times N\times(N+1)/2$
3D-DMP	$N_1\cdot P_1+(N_2-1)\cdot P_2\cdot(M_2-1)\cdot Q_2+(N_3-1)\cdot P_3\cdot(K_3-1)\cdot R_3$
Proposed model	$N\cdot(N+1)\cdot(M+1)/2+P\cdot(Q+1)\cdot(R+1)$

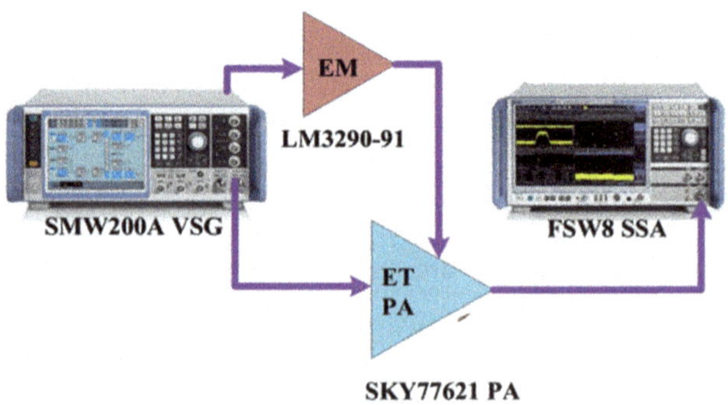

Fig. 5.34 Measurement test bench ([51], reprinted with permission from IEEE)

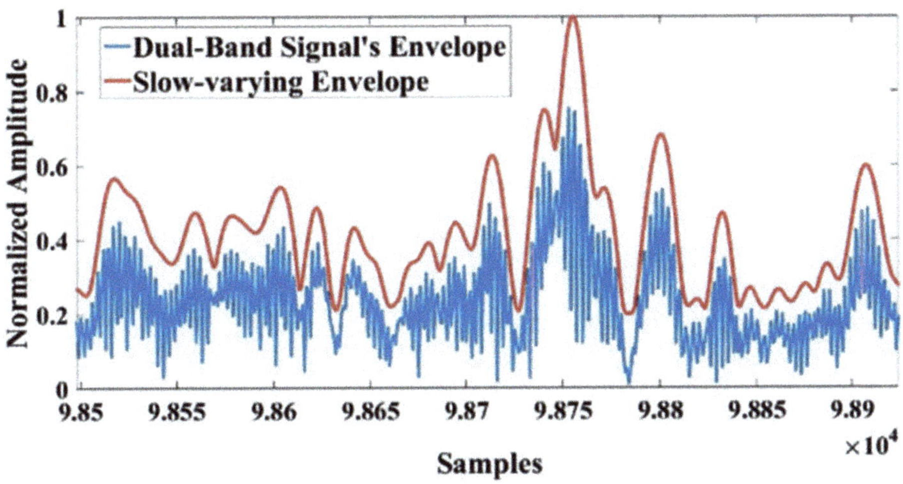

Fig. 5.35 Time domain waveforms of envelopes for envelope tracking (ET) PA ([51], reprinted with permission from IEEE)

for the PA to maximize overall efficiency of EM and PA together. The output of LM3290-91 is provided to Skyworks SKY77621 Multimode Multiband (MMMB) PA. The PA output is passed through attenuators and then fed to "FSW8" SSA of R&S. The output $y(t)$ is converted into baseband signal $y(n)$ in SSA. The baseband output signal $y(n)$ is processed using MATLAB. The outputs are obtained using digital band-pass filtering of $y(n)$ in MATLAB. After that, they are time-aligned using frequency domain method as discussed in Chap. 1. Figure 5.35 shows the time-domain instantaneous dual-band signal's envelope and slow-varying signal envelope $M(n)$ [53].

Indirect learning approach is a well-established method for the DPD. The inverse modeling of PA is required in the indirect learning architecture (ILA). In inverse modeling, the output signal with appropriate small-signal (linear) gain normalization and input signal are swapped with each other. Then coefficients for different models are extracted using LS method. Table 5.9 shows the inverse modeling results in terms of Adjacent Channel Error Power Ratio (ACEPR) and NMSE. As from Table 5.9, the proposed model gives good performance with least number of coefficients as compared with other models. The predistorted signals are generated using the coefficients extracted earlier. These predistorted signals are passed through ET PA to produce linearized outputs. Table 5.10 shows the performance of DPD in terms of ACPR and NMSE. In DPD results, NMSE is the normalized mean square error between the transmitted and received signals. As from Table 5.10, the proposed DPD model gives good linearization performance with least number of coefficients as compared with other models.

Figure 5.36 shows the frequency spectra of output signal without DPD and with DPD of various behavioral models. As seen from the graph, 2D PAV, 3D-MP, and the proposed DPD model output signal have nearly same ACPR.

Table 5.9 Inverse modeling results for different models ([51], reprinted with permission from IEEE)

Dual-band ET behavioral models	NMSE (dB)		ACEPR (dBc)		No. of coeff. (LB or LB)
	LB	UB	LB	UB	
2D P-A-V ($N = 7, M_1 = M_2 = 5$)	−43.84	−42.09	−49.96	−50.07	448
3D-MP ($N = 7, M = 5, R = 4$)	−42.96	−41.81	−49.24	−51.08	1680
3D-DMP ($P_1 = 7$, $N_1 = P_2 = N_2 = Q_2 = P_3 = N_3 = R_3 = 5$, $M_2 = K_3 = 4$)	−40.81	−35.36	−46.58	−43.81	635
Proposed model ($N = 7, P = 5, M = Q = 4$, $R = 3, e_g^1 = 0.25, e_g^2 = 1.75$)	−43.19	−42.98	−49.64	−51.24	240

Table 5.10 Digital predistortion (DPD) results for different models ([51], reprinted with permission from IEEE)

Dual-band ET behavioral models	NMSE (dB)		ACPR (dBc)		No. of coeff. (LB or UB)
	LB	UB	LB	UB	
Without DPD	−19.33	−18.17	−28.80	−24.94	N/A
2D P-A-V ($N = 7, M_1 = M_2 = 5$) DPD	−41.11	−40.33	−48.55	−47.85	448
3D MP (N = 7, M = 5, R = 4) DPD	−40.38	−40.10	−48.25	−47.97	1680
3D DMP ($P_1 = 7$, $N_1 = P_2 = N_2 = Q_2 = P_3 = N_3 = R_3 = 5$, $M_2 = K_3 = 4$) DPD	−38.24	−36.42	−45.95	−44.62	635
Proposed Model ($N = 7, P = 5$, $M = Q = 4, R = 3, e_g^1 = 0.25, e_g^2 = 1.75$) DPD	−40.97	−40.55	−48.39	−48.10	240

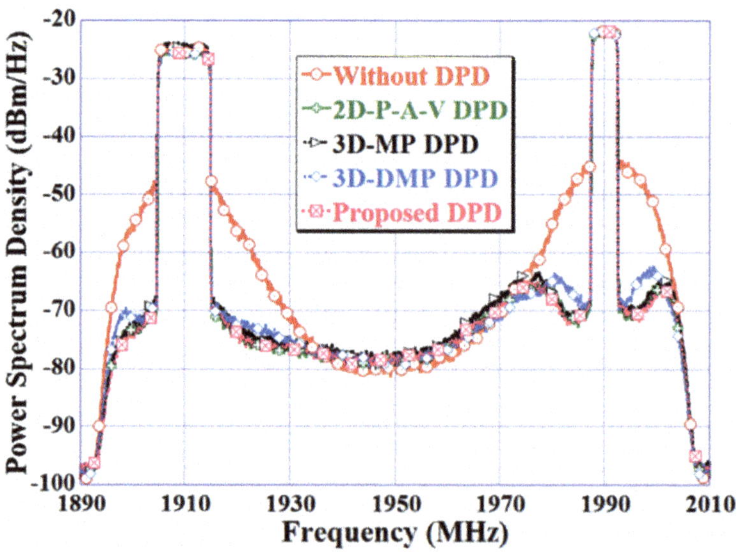

Fig. 5.36 Spectra of various signals ([51], reprinted with permission from IEEE)

References

1. Shen, Z., Papasakellariou, A., Montojo, J., Gerstenberger, D., & Xu, F. (2012). Overview of 3GPP LTE-advanced carrier aggregation for 4G wireless communications. *IEEE Communications Magazine, 50*(2), 122–130.
2. Bassam, S. A., Chen, W., Helaoui, M., & Ghannouchi, F. M. (2013). Transmitter architecture for CA: Carrier aggregation in LTE-advanced systems. *IEEE Microwave Magazine, 14*(5), 78–86.
3. Saad, P., Colantonio, P., Piazzon, L., Giannini, F., Andersson, K., & Fager, C. (2012). Design of a concurrent dual-band 1.8–2.4-GHz GaN-HEMT Doherty power amplifier. *IEEE Transactions on Microwave Theory and Techniques, 60*(6), 1840–1849.
4. Rawat, K., & Ghannouchi, F. M. (2012). Design methodology for dual-band Doherty power amplifier with performance enhancement using dual-band offset lines. *IEEE Transactions on Industrial Electronics, 59*(12), 4831–4842.
5. Liu, R., Schreurs, D., De Raedt, W., Vanaverbeke, F., & Mertens, R. (2011). *Concurrent dual-band power amplifier with different operation modes*. IEEE MTT-S International Microwave Symposium Digest, pp. 1–4.
6. Wang, Z., & Park, C.-W. (2012). *Concurrent tri-band GaN HEMT power amplifier using resonators in both input and output matching networks*. Proceedings of IEEE Wireless Microwave Technology Conference (WAMICON), pp. 1–4.
7. Nghiem, X. A., & Negra, R. (2012). *Novel design of a concurrent tri-band GaN-HEMT Doherty power amplifier*. Asia Pacific Microwave Conference Proceedings (APMC), pp. 364–366.
8. Kwan, A. K., et al. (2014). Concurrent multi-band envelope modulated power amplifier linearized using extended phase-aligned DPD. *IEEE Transactions on Microwave Theory and Techniques, 62*(12), 3298–3308.
9. Younes, M., Kwan, A., Rawat, M., & Ghannouchi, F. M. (2013). Linearization of concurrent tri-band transmitters using 3-D phase-aligned pruned Volterra model. *IEEE Transactions on Microwave Theory and Techniques, 61*(12), 4569–4578.
10. LTE. (2013). Evolved Universal Terrestrial Radio Access (E-UTRA); Base Station (BS) radio transmission and reception (release 11).
11. Liu, Y., Roblin, P., Yu, H., Shao, S., & Tang, Y. (2015). *Novel multiband linearization technique for closely-spaced dual-band signals of wide bandwidth*. IEEE MTT-S International Microwave Symposium Digest, pp. 1–4.
12. Rawat, M., Roblin, P., Quindroit, C., Salam, K., & Xie, C. (2015). Concurrent dual-band modeling and digital predistortion in the presence of unfilterable harmonic signal interference. *IEEE Transactions on Microwave Theory and Techniques, 63*(2), 95–104.
13. Jaraut, P., Rawat, M., & Ghannouchi, F. M. (2018). Harmonically related concurrent tri-band behavioral modeling and digital predistortion. *IEEE Transactions on Circuits and Systems II: Express Briefs, 66*(6), 1073–1077.
14. Mehmood, Y., Haider, N., Imran, M., Timm-Giel, A., & Guizani, M. (2017). M2M communications in 5G: State-of-the-art architecture, recent advances, and research challenges. *IEEE Communications Magazine, 55*(9), 194–201.
15. Checko, A., et al. (2015). Cloud RAN for mobile networks-a technology overview. *IEEE Communication Surveys and Tutorials, 17*(1), 405–426.
16. Alimi, I. A., Teixeira, A. L., & Monteiro, P. P. (2018). Toward an efficient C-RAN optical Fronthaul for the future networks: A tutorial on technologies, requirements, challenges, and solutions. *IEEE Communication Surveys and Tutorials, 20*(1), 708–769.
17. Qian, H. J., Liang, J. O., & Luo, X. (2016). Wideband digital power amplifiers with efficiency improvement using 40-nm LP CMOS technology. *IEEE Transactions on Microwave Theory and Techniques, 64*(3), 675–687.
18. Park, J. S., Wang, Y., Pellerano, S., Hull, C., & Wang, H. (2018). A CMOS wideband current-mode digital polar power amplifier with built-in AM–PM distortion self-compensation. *IEEE Journal of Solid-State Circuits, 53*(2), 340–356.

19. MB86L11A 2G/3G/4G LTE Transceiver, fujitsu. (2018). Retrieved from https://www.fujitsu.com/downloads/MICRO/fswp/pdf/products/FSWP_RFT_MB86L11A_FS.pdf.
20. Tripathi, G. C., Rawat, M., & Roblin, P. (2019). *Harmonic cancellation technique for ultra-wideband filter-less 5G*. 93rd ARFTG microwave measurement conference, Boston, MA.
21. Rawat, M., Roblin, P., Quindroit, C., Salam, K., & Xie, C. (2014). *Digitally supported feed-forward harmonic cancellation for filter-less ultra-wideband transmitters*. IEEE International Microwave and RF Conference, Bangalore, India, pp. 84–87.
22. Yu, H., Ratnasamy, V., Roblin, P., Rawat, M., & Xie, C. (2015). *Automatic feed-forward cancelation of modulated harmonic*. IEEE 86th ARFTG microwave measurement conference, pp. 1–3.
23. Samulak, A., Fischer, G., &Weigel, R. (2008). *Demonstrator of Class-S Power Amplifier based on GaN transistors*. IEEE German Microwaves Conference (GeMIC).
24. Morgan, D. R. (2013). A three-state signal coding scheme for high efficiency class-S amplifiers. *IEEE Transactions on Circuits and Systems-I, 60*(7), 1681–1691.
25. Aggrawal, E., Rawat, K., & Roblin, P. (2017). Investigating continuous class-F power amplifier using nonlinear embedding model. *Microwave and Wireless Components Letters, 27*, 593–595.
26. Li, Y., Montgomery, B. J., & Neihart, N. M.. (2016). *Development of a concurrent dual-band switch-mode power amplifier based on current-switching class-D configuration*. IEEE Wireless and Microwave Technology Conference (WAMICON).
27. Kumar, N., Datt Poonia, J., & Rawat, K. (2017). *Class S power amplifier based on CSCD with delta-sigma modulation*. 2017 IEEE applied electromagnetics conference, AEMC, Aurangabad, pp. 1–2.
28. Ebrahimi, M. M., Helaoui, M., & Ghannouchi, F. M. (2009). *Efficiency enhancement of a WiMAX switching mode GaN power amplifier through layout optimization of distributed harmonic matching networks*. Proc. IEEE European Microwave Conference, pp. 1732–1735.
29. Johnson, T., & Stapleton, S. P. (2006). RF class-D amplification with bandpass sigma–delta modulator drive signals. *IEEE Transactions on Circuits and Systems I: Regular Papers, 53*(12), 2507–2520.
30. Ebrahimi, M. M., Helaoui, M., & Ghannouchi, F. M. (2013). Delta-sigma-based transmitters: Advantages and disadvantages. *IEEE Microwave Magazine, 14*(1), 68–78.
31. Nielsen, M., & Larsen, T. (2007). A transmitter architecture based on delta–sigma modulation and switch-mode power amplification. *IEEE Transactions on Circuits and Systems II: Express Briefs, 54*(8), 735–739.
32. Ebrahimi, M. M., Helaoui, M., & Ghannouchi, F. M. (2011). Time-interleaved delta-sigma modulator for wideband digital GHz transmitters design and SDR applications. *Journal of Progress Electromagnetics Research B, 34*, 263–281.
33. Ghannouchi, F. M., Hatami, S., Aflaki, P., Helaoui, M., & Negra, R. (2010). Accurate power efficiency estimation of GHz wireless delta sigma transmitters for different classes of switching mode power amplifiers. *IEEE Transactions on Microwave Theory and Techniques, 58*(11), 2812–2819.
34. Kumar, N., & Rawat, K. (2018). *Delta sigma modulation based digital transmitter for single and dual band transmission*. IEEE MTT-S International Microwave and RF Conference (IMaRC), Kolkata, pp. 1–4.
35. Schreier, R., & Temes, G. C. (2005). *Understanding delta-sigma data converters*. Piscataway, NJ: IEEE Press.
36. Aziz, P. M., Sorensen, H. V., & Spiegel, J. (1996). An overview of sigma-delta converters. *IEEE Signal Processing Magazine, 13*(1), 61–84.
37. Jouzdani, M., Ebrahimi, M. M., Helaoui, M., & Ghannouchi, F. M. (2017). Complex delta–sigma-based transmitter with enhanced linearity performance using pulsed load modulation power amplifier. *IEEE Transactions on Microwave Theory and Techniques, 65*(9), 3324–3335.
38. Cordeiro, R. F., Prata, A., Oliveira, A. S. R., Vieira, J. M. N., & De Carvalho, N. B. (2017). Agile all-digital RF transceiver implemented in FPGA. *IEEE Transactions on Microwave Theory and Techniques, 65*(11), 4229–4240.

39. Ebrahimi, M. M., & Helaoui, M. (2013). Reducing quantization noise to boost efficiency and signal bandwidth in delta–sigma-based transmitters. *IEEE Transactions on Microwave Theory and Techniques, 61*(12), 4245–4251.
40. Grebennikov, A. (2011). *RF and microwave transmitter design* (1st ed.). New York: Wiley.
41. Zhu, Q., Ma, R., Duan, C., Teo, K. H., & Parsons, K. (2014). *A 5-level discrete-time power encoder with measured coding efficiency of 70% for 20-MHz LTE digital transmitter.* IEEE MTT-S International Microwave Symposium.
42. Podsiadlik, T., Dooley, J., & Farrell, R. (2018). Analysis of 3-level bandpass sigma-delta modulators with 2-level output. *IEEE Transactions on Circuits and Systems II: Express Briefs, 65*(4), 506–510.
43. Elsayed, F., & Helaoui, M. (2013). Linearized multi-level $\Delta\ \Sigma$ modulated wireless transmitters for SDR applications using simple DLGA algorithm. *IEEE Journal on Emerging and Selected Topics in Circuits and Systems, 3*(4), 594–601.
44. Kumar, N., & Rawat, K. (2017). *Efficiency enhancement in delta-sigma modulator based transmitter using level transformation.* IEEE MTT-S International Microwave and RF Conference (IMaRC), Ahmedabad, pp. 1–5.
45. Singh, R., Tripathi, G. C., & Rawat, M. (2015). *Performance analysis of multilevel delta sigma modulators for 3G/4G communication.* 2015 IEEE UP Section Conference on Electrical Computer and Electronics (UPCON), Allahabad, pp. 1–5.
46. Ebrahimi, M. M., Helaoui, M., & Ghannouchi, F. M. (2012). *Improving coding efficiency by compromising linearity in delta-sigma based transmitters.* IEEE Radio and Wireless Symposium, pp. 411–414.
47. Hu, Y., & Boumaiza, S. (2016). Power scalable wideband linearization of power amplifier. *IEEE Transactions on Microwave Theory and Techniques, 64*(5), 1456–1464.
48. Gumber, K., & Rawat, M. (2017). A modified hybrid RF Predistorter linearizer for ultra wideband 5G systems. *IEEE Journal on Emerging and Selected Topics in Circuits and Systems, 7*(4), 547–557.
49. Gumber, K., & Rawat, M. (2018). Low cost RFin-RFout predistorter linearizer for high power amplifier and ultra-wideband signals. *IEEE Transactions on Instrumentation and Measurement, 67*(9), 2069–2081.
50. Rawat, K., Rawat, M., & Ghannouchi, F. M. (2010). Compensating I-Q imperfections in hybrid RF/ digital predistortion with adapted look up table implemented in FPGA. *IEEE Transactions on Circuits and Systems II: Express Briefs, 57*(5), 389–393.
51. Jaraut, P., & Rawat, M. (2017). *3D generalized coefficient supported model for concurrent dual-band digital predistortion of envelope tracking power amplifier.* IEEE MTT-S International Microwave & RF Conference, Ahmedabad, India, pp. 1–4.
52. Gilabert, P., & Montoro, G. (2015). 3-D distributed memory polynomial behavioral model for concurrent dual-band envelope tracking power amplifier linearization. *IEEE Transactions on Microwave Theory and Techniques, 63*(2), 638–648.
53. Cidronali, A., Giovannelli, N., Mercanti, M., Maddio, S., & Manes, G. (2013). Concurrent dual-band envelope tracking GaN PA design and its 2d shaping function characterization. *International Journal of Microwave and Wireless Technology, 5*, 669–681.
54. Gilabert, P., Montoro, G., Lopez, D., & Garcia, J. (2013). *3D digital predistortion for dual-band envelope tracking power amplifiers.* Proceedings of Asia–Pacific Microwave Conference, pp. 734–736.
55. Kwan, A., Younes, M., Zhang, S., Chen, W., Darraji, R., Helaoui, M., & Ghannouchi, F. M. (2014). *Dual-band predistortion linearization of an envelope modulated power amplifier operated in concurrent multi-standard mode.* IEEE MTT–S International Microwave Symposium Digest, Tampa Bay, FL, pp. 1–4.
56. Bassam, S., Helaoui, M., & Ghannouchi, F. (2011). 2-D digital predistortion (2-D-DPD) architecture for concurrent dual-band transmitters. *IEEE Transactions on Microwave Theory and Techniques, 59*(10), 2547–2553.

57. Liu, Y. J., Zhou, J., Chen, W., Zhou, B., & Ghannouchi, F. M. (2012). Low complexity 2D behavioural model for concurrent dual-band power amplifiers. *Electronics Letters, 48*(11), 620–621.
58. Ding, L., Yang, Z., & Gandhi, H., *Concurrent dual-band digital predistortion*. IEEE MTT-S International Microwave Symposium Digest, Montreal, Canada, pp. 1–3.

Chapter 6
Advance Materials for Power Amplifier Design and Packaging

6.1 Introduction

This chapter describes recent trends related to advancement in the area of PA design. The PA design based on complementary metal-oxide-semiconductor (CMOS) technology has been discussed with their limitations in terms of handling high power. The circuit topology to extract more power out of metal-oxide-semiconductor (MOS) transistors while withstanding high voltage swings is presented in Sect. 6.2.1. Later the design based on III–V semiconductor technology such as GaN (gallium nitride) and GaAs (gallium arsenide) are presented with their advantages and limitations. The microelectromechanical system (MEMS)-based reconfigurable PAs are also discussed to cater the need of upcoming trend in the development of multiband/multi-standard transmitters. The challenges in extending PA design at millimeter wave are discussed in Sect. 6.4. This chapter also discusses various trends in high-power device packaging and some recent trends in high-power device integration and assembly.

6.2 On-Chip Power Amplifiers

This section describes on-chip PA which includes device as well its biasing and matching scheme which is realized in a single chip. Such circuits are traditionally called monolithic microwave integrated circuits (MMIC). Traditionally GaAs materials are considered as one of the popular choices for MMIC design. Recently, GaN MMIC design which has high power density in comparison to GaAs technology is gaining interest. On the other hand, silicon-based complementary metal-oxide-semiconductor (CMOS) technology is being used widely for integrated circuits (IC) since it is cost effective, good in performance, and has high level of integration

K. Rawat et al., *Bandwidth and Efficiency Enhancement in Radio Frequency Power Amplifiers for Wireless Transmitters*, Analog Circuits and Signal Processing,
https://doi.org/10.1007/978-3-030-38866-9_6

[1–8]. This section describes both CMOS and MMIC technology used for developing RF PAs.

6.2.1 MOS-Based Design

The significant scaling of MOS transistor has enhanced the transition frequency to around 100 GHz, which makes CMOS technology more attractive for RF applications. In spite of low cost, high-cutoff frequency, and high level of integration, CMOS devices have been scarcely favored for PA applications [9–11]. This is because of the CMOS PA design issues such as low breakdown and high knee voltage which limits the drain-to-source voltage swing and hence the output power and efficiency. In submicron CMOS technologies, the breakdown and degradation mechanism are due to gate oxide breakdown, hot carrier degradation, punch through, and drain bulk breakdown [12, 13]. To avoid these effects, drain-to-gate, gate-to-source, and drain-to-source voltages should not exceed the prespecified values. The typical approach to obtain high output power is to increase the device size which increases the drain current swing. But this method reduces the input impedance and hence designing an input matching network becomes difficult. Also, in Watt-level PAs, the optimum load (R_{opt}) seen by the transistor becomes small as R_{opt} is inversely proportional to output power (P_{out}). Designing the output matching network to match low R_{opt} to 50 Ω load becomes difficult since transformation ratio becomes large resulting in large chip area, narrow bandwidth, high loss, and thus low efficiency. To overcome these limitations, multiple devices can be combined in series or parallel so that the power of individual devices can be combined to achieve high output power. The parallel field effect transistor (FET) combining PAs are based on transformer techniques and hence occupy large die area due to on-chip transformers [13–15].

An alternative approach is to combine the devices in series so that the output voltage swing increases while maintaining device operation within it's voltage breakdown limit. The basic series topology is cascode configuration of common source (CS) and common gate (CG) stages to improve gain performance in low-frequency designs as shown in Fig. 6.1.

The RF signal is applied to the gate of CS transistor while the gate of CG transistor is RF grounded. Under large signal operation, the voltage will not be distributed equally across CS and CG transistors. Since there is no RF swing at the gate of CG transistor, drain-to-gate voltage (V_{dg}) swing would be RF swing at the drain of CG transistor. The CS transistor has smaller V_{dg} swing compared to CG transistor since the gate of CS transistor has RF swing due to the RF input signal. Hence, this configuration has a problem of unequal voltage division between the transistors, and unfortunately the CG transistor will be stressed more across drain and gate terminals which may operate the transistor beyond the breakdown voltage.

In order to resolve the problem of uneven stress, a self-biased cascode structure was introduced as shown in Fig. 6.2 [12]. With the aid of feedback resistor R_{b} and

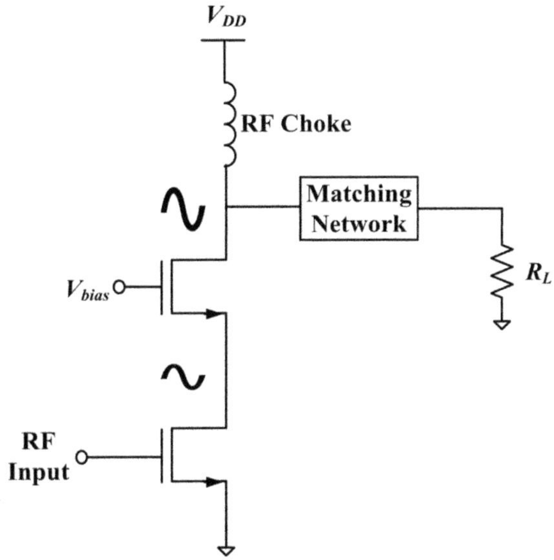

Fig. 6.1 Power amplifier (PA) in cascode configuration

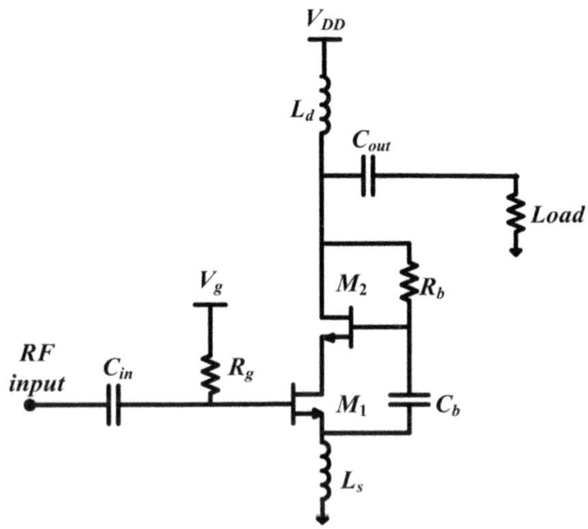

Fig. 6.2 A self-biased cascode PA

capacitor C_b, an attenuated RF swing can be provided at the gate of CG transistor. The RF swing at the gate reduces the drain-to-gate voltage (V_{dg}) and helps to operate below breakdown. Another approach was used in high-voltage/high-power PAs to achieve high output voltage swing using series connection of multiple transistors as shown in Fig. 6.3 [16, 17]. This configuration uses feedback resistors along with a capacitor to provide RF swing at gates of stacked devices. The feedback resistors also ensure that the DC drain-to-source voltage (V_{ds}) is equal across all the transistors and prevents from early breakdown unlike cascode configuration.

The stacked topology is a popular technique among CMOS PAs because of its superior performance [18–23]. As the number of transistors increases in stacked

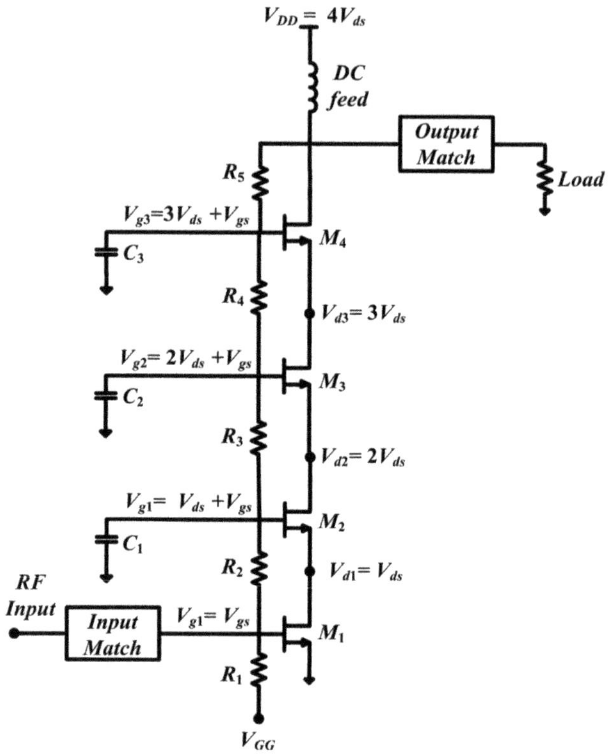

Fig. 6.3 A high-voltage/high-power (HiVP) PA

Table 6.1 Comparison of single, parallel, and stacked topologies

Parameter	Single FET	*n* Parallel topology	*n* Stacked topology
Peak drain voltage	V_m	V_m	nV_m
Peak drain current	I_m	nI_m	I_m
Input capacitance	C_g	nC_g	C_g
Output impedance	R_{opt}	R_{opt}/n	nR_{opt}
Output power	$\frac{1}{8}V_m I_m$	$\frac{1}{8}V_m \times nI_m$	$\frac{1}{8}nV_m \times I_m$

topology, the optimum load impedance increases compared to single transistor or parallel configuration. The comparison of single, parallel, and stacked topologies is shown in Table 6.1. In parallel combining topology, *n* devices are combined in parallel such that the current swing becomes *n* times of the single FET device, thereby increasing the output power by *n* times. Although output power is increased in parallel topology by the virtue of increased current swing, the output impedance is reduced by *n* times since the voltage swing is same as that of the single FET topology. This makes it difficult to match small impedance (R_{opt}/n) to 50 Ω. Unlike parallel topology, increasing the number of transistors in the stacked topology makes the output impedance closer to 50 Ω and thereby making the matching network simpler.

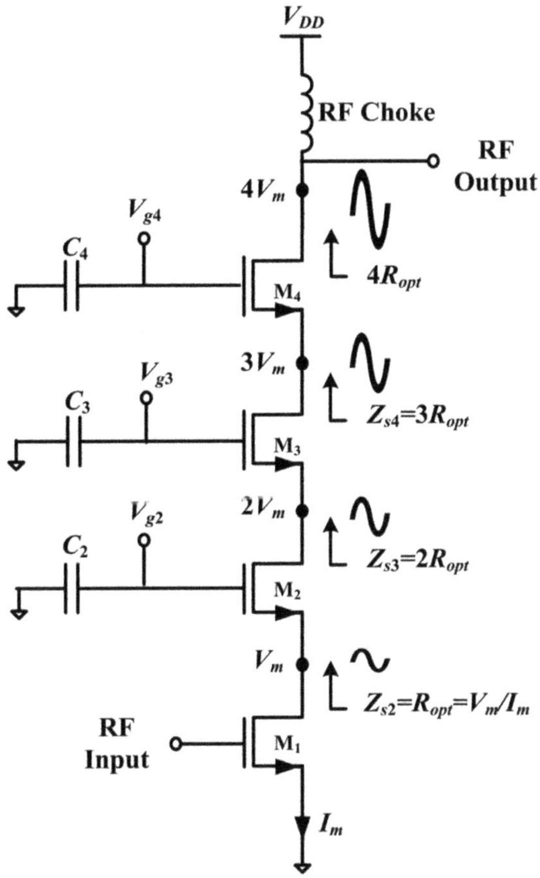

Fig. 6.4 Stacked field-effect transistor (FET) topology

Stacked Topology for RF CMOS Power Amplifier

In stacked topology, the CS input transistor is connected in series with CG transistors for in-phase addition of voltage swings at the output while maintaining the drain current constant. The basic structure of stacked topology is shown in Fig. 6.4 [18].

Unlike conventional cascode topology, the gate of CG transistors has RF swing which is achieved by adding small capacitors (C_2, C_3, and C_4) at the gate terminal. Equal voltage swings are obtained by properly choosing the values of external capacitors C_2, C_3, and C_4. These external capacitors appear in parallel to the gate-to-source capacitance (C_{gs}) of the corresponding transistors. The effect of external gate capacitor on the stacked FETs input impedance is shown in Fig. 6.5.

The input impedance Z_{sk} of the kth transistor can be written from Fig. 6.5 as [18],

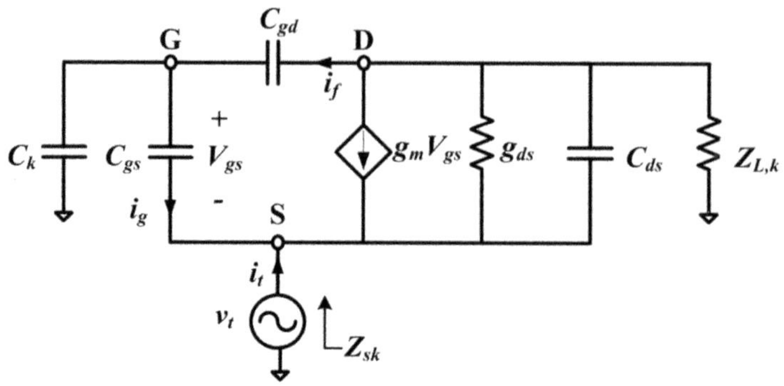

Fig. 6.5 Effect of external gate capacitor on the stacked topology

$$Z_{sk} = \frac{C_{gs} + C_k + C_{gd}\left(1 + g_m Z_{L,k} + sC_{gs}Z_{L,k} + sC_k Z_{L,k}\right)}{\left(g_m + sC_{gs}\right)\left(C_{gd} + C_k + sC_{gd}C_k Z_{Lk}\right)} \tag{6.1}$$

The load impedance of $(k - 1)$th transistor is the input impedance of kth transistor, that is, $Z_{sk} = Z_{L,\,k-1}$.

If the operating frequency f_0 is much smaller than the cutoff frequency f_T, that is, $f_0 \ll f_T$ of the device, then Eq. (6.1) can be simplified as [18],

$$Z_{sk} \approx \frac{1}{g_m}\left(1 + \frac{C_{gs}}{C_k}\right) \tag{6.2}$$

In order to obtain $Z_{L,k}$ equal to kR_{opt}, the gate capacitance C_k can be determined by equating $Z_{sk} = Z_{L,\,k-1}$ to $(k-1)R_{opt}$. Therefore, C_k can be given as [18],

$$C_k = \frac{C_{gs}}{(k-1)\,g_m R_{opt} - 1} \tag{6.3}$$

By using Eq. (6.3), the value of gate capacitances can be determined such that the kth transistor sees kR_{opt}, which ensures equal V_{ds} swing across all the stacked transistors unlike conventional cascode configuration. The impedance $Z_{L,k-1}$ (Z_{sk}) seen by the drain terminal of transistor M_{k-1} is real as long as $f_0 \ll f_T$. At high RF frequencies beyond Ku band, the standard stacked topology is not valid because of the parasitic effects of the transistor [21, 22]. Now the direct cascoding of transistors may not ensure in-phase voltage combining due to effect of parasitic elements. Especially at millimeter-wave frequencies, intermediate nodes impedance would have reactance part and hence reduces the efficiency because RF current flows through the parasitic capacitances [20]. In order to avoid the reduction in efficiency and output power, intermediate nodes need to be reactively tuned. To achieve the proper complex impedance matching between the transistors, additional tuning elements need to be used for optimal performance. The complex intermediate

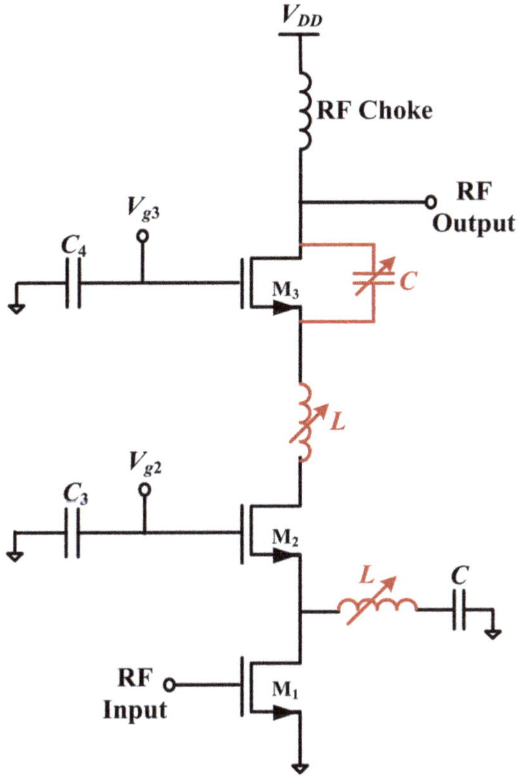

Fig. 6.6 Intermediate node-tuning techniques

node can be matched using shunt inductance between transistors, shunt inductor-tuning technique, or by shunt feedback drain-source capacitor-tuning technique as shown in Fig. 6.6 [20]. This ensures the in-phase voltage combining at the output node and ensures increase in output power and hence efficiency.

6.2.2 GaAs and GaN MMIC Designs

Although CMOS is an inexpensive and well-developed technology, it is difficult to simultaneously obtain high output power and high gain due to the limited breakdown voltage, low transconductance, and lossy silicon substrate [24]. Since the development of liquid-encapsulated Czochralski (LEC) method, GaAs ingots were grown which are suitable for high-frequency transistors and low loss passive components [25]. GaAs-based MMIC are well known over many years for high-power and high-frequency applications. But high electron mobility transistor (HEMT) using III–V GaN compounds have achieved great importance among designers in recent years. GaN MMIC-based technology has several advantages in terms of high bandwidth, high power density, high breakdown electric field, and high electron saturation

Table 6.2 Comparison between GaAs and GaN MMIC technologies

Parameter	GaAs	GaN
Maximum current	~0.5 A/mm	~1 A/mm
Breakdown voltage	20–40 V	>100 V
Thermal conductivity	47 W/m K	360 W/m K
Output power density	0.5–1.5 W/mm	3–8 W/mm
Operating Voltage	5–20 V	18–48 V

carrier velocity in comparison with other competing technologies such as GaAs and InP devices [26]. Due to high power density, the size requirements of the MMICs can be reduced. The breakdown field of GaN is five times of GaAs due its wide bandgap. The power density is also 10–20 times of GaAs-based devices [27]. This particularly helps in amplifier design, where GaN-based devices are smaller in size and have a lower capacitance for the same operating power. This enables GaN-based amplifiers to operate over a wide bandwidth making input and output matching simpler. These features lead to higher drain–source bias voltage and low drain currents which also gives the facility to design a simple and low-cost bias voltage circuits. The comparison between GaAs and GaN MMIC technologies are shown in Table 6.2 [34].

Recently, the gain and drain efficiency of GaN HEMT-based PAs are sharply improved at X and Ku-band by advanced device process [28, 29]. However, at Ku-band, high power is achieved by combining several smaller devices or low-power PAs [30, 31]. The power-combining techniques discussed in Sect. 4.6 can be used in device and circuit level to implement high-power/broadband MMIC PA. The distributed and reactively matched techniques are popular for ultra-wideband PAs which can operate up to multi-octave frequencies. A non-uniformly distributed PA can also provide multi-octave operation from C to Ku band [32]. The chip photograph of a 10 W non-uniformly distributed GaN MMIC PA is shown in Fig. 6.7 where two non-uniformly distributed PAs are combined using Lange couplers. The chip size is 5.2 mm × 3.6 mm. This MMIC PA was fabricated using 0.25 μm GaN HEMT technology from Fujitsu. The measurement results over the operating frequency of 6–18 GHz are shown in Fig. 6.8.

At 18 GHz, the MMIC PA achieved a linear gain of 10.4 dB and a maximum output power of 40.9 dBm (12.3 W) with an associated power-added efficiency (PAE) of 18%. One can see that maximum output power achieved over the band from measurement is 13.5 W. Such ultra-wide bandwidth can also be achieved using reactively matched topology. Ref. [33] presents the design of three-stage 6–18 GHz GaN MMIC which uses reactively matched topology to achieve multi-octave operating frequency range. The chip photograph of 6–18 GHz wideband high-power MMIC PA is shown in Fig. 6.9.

The MMIC is fabricated with 0.25 μm AlGaN/GaN on SiC substrate process from UMS. Three stages are used to achieve small signal gain more than 18 dB. In each of these stages, transistors of size 8 × 75 μm are used as the basic cell. The final power stage combines the power from four basic transistor cells to achieve high output power. In order to drive the power stage, a driver stage is implemented

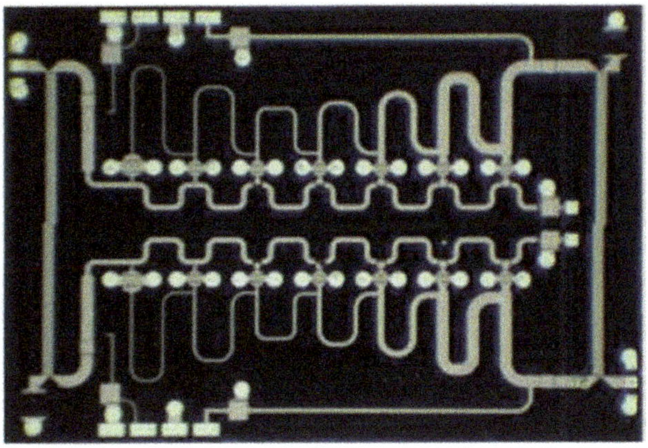

Fig. 6.7 Chip photograph of non-uniformly distributed PA ([32], reprinted with permission from IEEE)

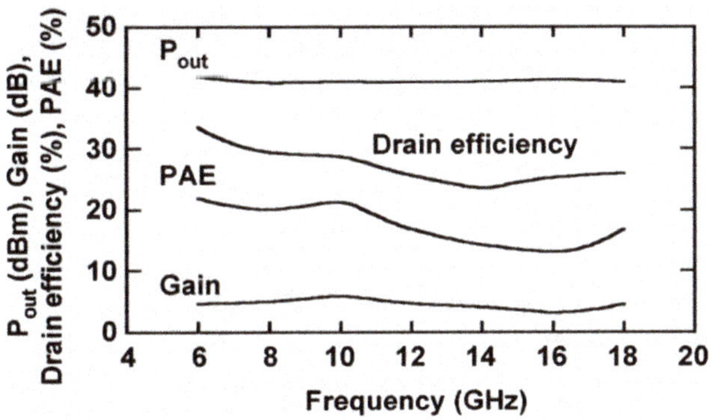

Fig. 6.8 Measurement results of 6–18 GHz non-uniformly distributed PA ([32], reprinted with permission from IEEE)

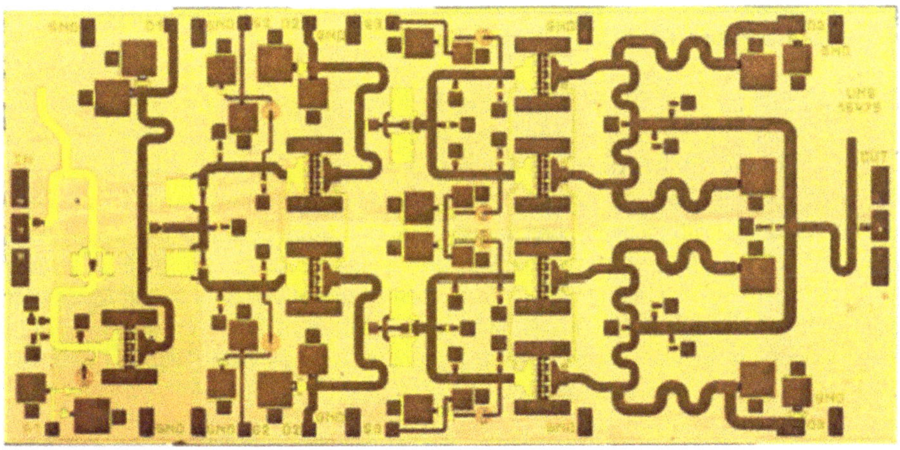

Fig. 6.9 Chip photograph of three-stage 6–18 GHz GaN MMIC PA ([33], reprinted with permission from IEEE)

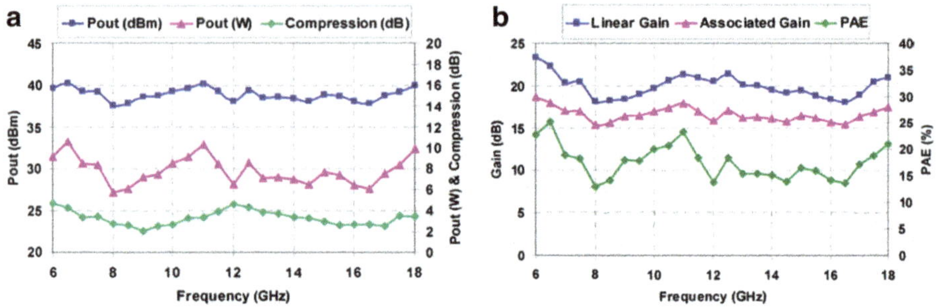

Fig. 6.10 (**a**) Measured output power and gain compression (**b**) small signal gain, power-added efficiency (PAE), and associated gain of three-stage 6–18 GHz GaN MMIC PA over frequency ([33], reprinted with permission from IEEE)

by combining two basic cells which is further driven using a single transistor. The chip size of the GaN MMIC is 6.43 mm × 3.08 mm. The measurement results of three-stage 6–18 GHz GaN MMIC PA is shown in Fig. 6.10a, b.

One can see from Fig. 6.10a the output performance in dBm and Watt and the corresponding gain compression over frequency when input power is 22 dBm. The measured output power is from 6 to 10 W over the frequency from 6 to 18 GHz. The linear gain, associated gain, and PAE over the frequency are shown in Fig. 6.10b. The linear gain varies from 18 to 23 dB over the band. The associated gain from measurement results is more than 15 dB in the entire bandwidth. The PAE of GaN MMIC varies from 13 to 25% over the band of 6–18 GHz.

Advancements in GaN MMIC Technology

Although silicon carbide (SiC) is an excellent thermal conductor, GaN transistors are thermally limited due to reliability issues. Diamond which has the highest thermal conductivity has excellent microwave properties. Research has been going on in developing GaN on diamond [34]. Raytheon has demonstrated that GaN on diamond transistor can deliver output power comparable to GaN on Si with reduced size due to the improvement in power density up to 3.6 times. It is also observed that GaN is chemically stable over most of the temperature range used in Si CMOS processing [34]. This shows the possibility of low cost and high yield GaN processing of GaN on Si. The development of GaN on Si also helps in the integration of GaN with Si CMOS on a single chip.

6.3 Reconfigurability and Tuning in Power Amplifiers Using Microelectromechanical Systems

The upcoming radios are based on software-defined platform transparent to any standards and applications. This is possible if the RF front-end is multiband and multi-standard. Therefore, reconfigurability and tuning are now introduced in several RF circuits to make them suitable for such software defined radio (SDR) platforms. The reconfigurability and tuning have been introduced in PA such that it can provide optimal performance at multiple bands and for various standards. For example, reconfiguring Doherty PA high-efficiency operation to various back-off levels can handle signals belonging to different standards and crest factors [35–37]. In addition, the operating frequencies of such PAs can also be tuned to various frequency bands providing multiband features. The reconfigurability and tunability can be introduced either using varactors or variometer (variable inductance) or switches. With switches, one can provide digital control over the reconfigurability and this is more useful in case of SDR platform with programmable feature. One classical way of realizing switches is using p-doped, intrinsic, n-doped (PIN) diode [38, 39]. The PIN diode can handle high power, however, does not provide inherent isolation between control and RF signal circuit. On the contrary, the RF-MEMS has inherent decoupled circuitry for control and RF signal. These devices provide inherent low-pass filter characteristic at their control line and hence avoiding any pickup from control line to pass to RF [38]. This is advantageous in case of radio applications where the interaction between control and RF circuit may lead to cross talk. In addition, RF MEMS devices are superior in terms of reduced parasitic and losses [38]. For example, MEMS-based varactor does not present any flicker noise due to its inherent passive nature [38]. Some key challenges in using RF-MEMS device is related to their low reliability and low power-handling which is now being answered with technology advancement [35–37, 40–42]. This section describes some reconfigurable and tunable RF PAs using RF-MEMS technology.

Referring to Fig. 1.12 and discussion on class E PA shunt capacitor topology in section "Class E Power Amplifier", the reconfigurability can be introduced in class E by varying shunt capacitance and matching network [43]. The shunt capacitor can be realized using inherent device drain source capacitance C_{ds} and an additional capacitor. This additional capacitor will increase the overall shunt capacitance in class E configuration of Fig. 1.12. One can use a switch with this additional capacitor, which allows to increase the capacitance at low frequency when turned *on*. However, at higher frequency, this switch can be kept *off* to remove the effect of additional capacitor resulting in only device inherent capacitance C_{ds} appearing as shunt capacitor in class E topology. Figure 6.11 shows one such circuit topology of reconfigurable quad-band MEMS-based class E PA.

Figure 6.11 shows transistors M_1 and M_2 constituting two cascaded stages of PA. These transistors are metal-oxide-semiconductor field effect transistors (MOSFETs) with high breakdown voltage (7.5 V). In order to reconfigure the PA to two frequencies of operation, a switch is implemented using MOSFET M_3 as shown in

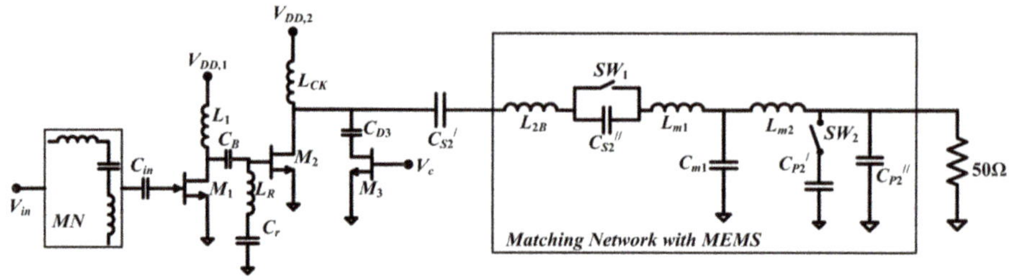

Fig. 6.11 Circuit topology for reconfigurable class E PA using microelectromechanical system (MEMS)-based load matching network

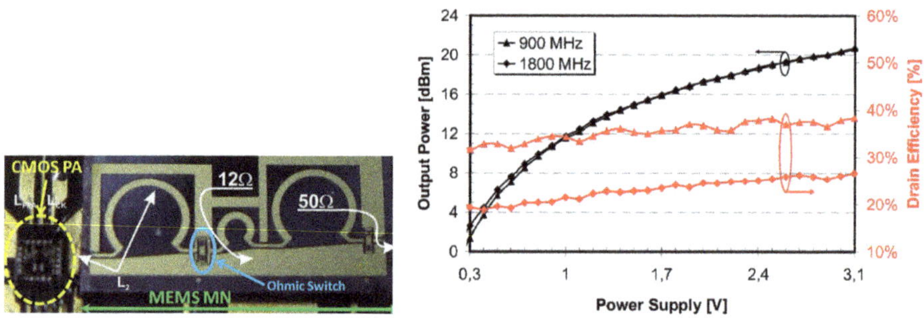

(a) (b)

Fig. 6.12 Reconfigurable class E PA, (**a**) circuit photograph and (**b**) measured results ([43], reprinted with permission from IEEE)

Fig. 6.11. At lower operating frequency, M_3 is switched *on* in order to increase the shunt capacitance by adding capacitance C_{D1} in shunt to the device as shown in Fig. 6.11. At higher operating frequency, MOSFET M_3 is switched *off* and capacitance in shunt is reduced. Later, MEMS switches are used to reconfigure a capacitance appearing in series with inductor L_{2B} in Fig. 6.11 which are considered as part of class E shunt capacitor topology [43]. The capacitor C_{S2} in Fig. 6.11 comprises $C_{S2}{}^{/}$ and $C_{S2}{}^{//}$, where C_{S2} is reconfigured by actuating MEMS switch SW_1. The MEMS switch SW_1 is *on* at lower operating frequency, implementing the series capacitance $C_{S2} = C_{S2}{}^{/}$. At higher operating frequency, SW_1 is *off* and the capacitor C_{S2} is lowered to the value $C_{S2}{}^{/} + C_{S2}{}^{//}$ as shown in Fig. 6.11 [43]. The output matching network comprises two cascaded L-match sections. The second section comprises capacitors $C_{P2}{}^{/}$ and $C_{P2}{}^{//}$. When MEMS switch SW_2 is *on*, a high value of capacitance $C_{P2} = C_{P2}{}^{/} + C_{P2}{}^{//}$ will constitute the second section of L-match. This condition is used while operating at lower operating frequency. The circuit topology of Fig. 6.11 and its operation described above is used to realize reconfigurable class E PA while operating at 900 and 1800 MHz operating frequencies [43]. Figure 6.12a, b shows the photograph and measurement results of fabricated prototype respectively.

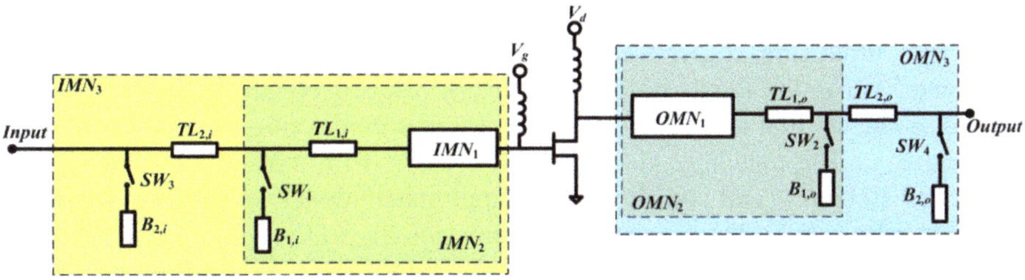

Fig. 6.13 Circuit topology of reconfigurable tri-band PA with MEMS switches

Figure 6.12a shows two different dies fabricated using AMS 0.35 μm CMOS process and FBK RF-MEMS are assembled in an FR4 substrate. These dies are connected using bond wires. The output bond wire also realizes series inductance of class E circuit as shown in Fig. 6.11. The CMOS die area is 1.44 mm^2, whereas RF-MEMS circuit occupies 29.88 mm^2. Figure 6.12b demonstrates the performance of the fabricated reconfigurable class E PA with drain bias. One can see from Fig. 6.12b that at drain supply of 3 V, the output power is 20 dBm at both the operating frequencies. The drain efficiencies are 38% and 26% at 900 and 1800 MHz, respectively.

Another MEMS-based high-efficiency PA with band-reconfigurability and up to three different frequency bands was introduced using MEMS switch [42]. The scheme was demonstrated for triple-band operation with 1 W PA design achieving the PAE of 60% in three different frequency bands, 900, 1500, and 2000 MHz. The methodology is generic and can be extended to a larger number of frequency bands. Figure 6.13 shows the circuit topology of this reconfigurable triple band PA. This comprises first input matching network (IMN_1) and first output matching network (OMN_1) terminating the FET device as shown in Fig. 6.13.

When switches SW_1 and SW_2 at input and output side, respectively, are *off*, the impedance terminating IMN_1 and OMN_1 is reference impedance Z_0 Ω if the characteristic impedances of transmission lines $TL_{1,i}$ and $TL_{2,i}$ at the input side and $TL_{1,o}$ and $TL_{2,o}$ in the output side are Z_0 Ω. The IMN_1 and OMN_1 are then designed to transform this Z_0 Ω to optimum source and load impedances at the device input and output terminals, respectively, at first operating frequency f_1. In case of frequency f_2, switch SW_1 at the input side is *on* and first stubs $B_{1,i}$ is connected between lines $TL_{1,i}$ and $TL_{2,i}$ in the input matching network. Similarly, switch SW_2 in the output side is also *on* connecting stub $B_{1,o}$ between the transmission lines $TL_{1,o}$ and $TL_{2,o}$ in the output matching network. The stub lengths are selected such that they add appropriate reactance to the IMN_1 and OMN_1 in shunt, providing transformation of reference impedance Z_0 Ω to optimum source and load impedances at the device input and output terminals, respectively, at second operating frequency f_2. Since the characteristic impedance of transmission line $TL_{2,i}$ and $TL_{2,o}$ is Z_0 Ω, it is transparent at second operating frequency f_2. Therefore, stubs $B_{1,i}$ with IMN_1 and $B_{1,o}$ with OMN_1 will constitute second input and output matching networks, respectively, responsible for matching at second operating frequency f_2 as shown in Fig. 6.13. At frequency f_3, SW_1 and SW_2 at input and output side, respectively, are set to the *off* states. Whereas, the switches SW_3 and SW_4 in the input and output network respectively are set to the *on* state. This disconnects first stubs $B_{1,i}$ and $B_{1,o}$, whereas connecting second set of stubs $B_{2,i}$ and $B_{2,o}$. The matching networks IMN_1 along with transmission lines $TL_{1,i}$, $TL_{2,i}$, and stubs $B_{2,i}$ will constitute third input matching network IMN_3. Similarly, matching networks OMN_1 along with transmission lines $TL_{1,o}$, $TL_{2,o}$, and stubs $B_{2,o}$ will constitute third output matching network OMN_3. Therefore, IMN_3 and OMN_3 are used to transform reference impedances Z_0 Ω to optimum source and load impedances at the device input and output terminals, respectively, at third operating frequency f_3. Using the same scheme for extending operating frequencies of PA to n-bands, $2(n-1)$ switches are required, where both input and output matching have $n-1$ switches. Moreover, no switches are inserted in series; therefore, this configuration has less effect on output power and PAE. The frequency f_1, f_2, and f_3 are selected in an order $f_1 > f_2 > f_3$, where f_1 is the highest and f_3 is the lowest frequency.

Figure 6.14 shows photograph of tri-band reconfigurable PA operating at 875/1500/2000 MHz [44]. The authors in [44] represent SW1 and SW2 of Fig. 6.13 with same name SW1 in Fig. 6.14. Similarly, SW3 and SW4 of Fig. 6.13 are represented as SW2 in Fig. 6.14.

The circuit uses GaAs FET as active device. The substrate used is alumina with dielectric constant of 9.2. Each MEMS switch has 0.4 dB loss and isolation greater than 40 dB over DC-2000 MHz [45, 46]. The switch configurations are given in Table 6.3 [44]. Since, the authors in [44] represents all the switches with only two names SW1 and SW2 in Fig. 6.14, the various switch configurations in Table 6.3 are listed according to Fig. 6.14.

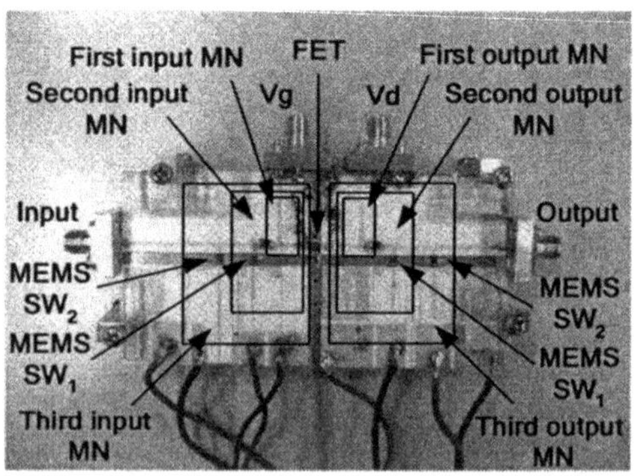

Fig. 6.14 Photograph of tri-band reconfigurable PA ([44], reprinted with permission from IEEE)

Table 6.3 Switch configurations for different frequencies of operation

Switch #	875 MHz	1500 MHz	2000 MHz
SW_1	OFF	ON	OFF
SW_2	ON	OFF	OFF

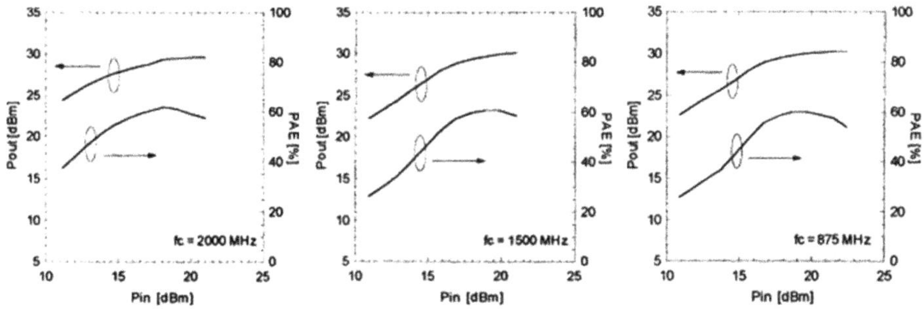

(a) (b) (c)

Fig. 6.15 Measured performance of tri-band reconfigurable PA (**a**) at 2000 MHz, (**b**) 1500 MHz, (**c**) 875 MHz ([44], reprinted with permission from IEEE)

Figure 6.15a–c shows measured performance of the designed PA in terms of output power and PAE at each operating frequency.

One can see from Fig. 6.15a–c that the maximum output power is 30 dBm or more at each frequency of operation. The maximum PAE is better than 60% in each case. The respective gain is 11.1, 10.7, and 11.2 dB at 2000, 1500, and 875 MHz.

Since modern wireless communication requires PA that can handle envelope varying signals, the reconfigurability is also introduced in Doherty PA and Chireix outphasing PAs [35–37], [39], [47]. In case of Doherty PA, both back-off region and frequencies are reconfigured using MEMS switches in order to provide multiband/multi-mode features [35–37]. For example, packaged MEMS single-pole double-throw (SPDT) RMSW221 from Radant MEMS which has high power-handling capability is used to reconfigure operating frequency and back-off [36, 37]. Later, same MEMS switches are used to simultaneously reconfigure both frequency and back-off region in multiband/multi-mode Doherty PA [35]. This architecture utilizes two tuners in main and auxiliary path as a part of load combiner. These tuners will tune the load combiner parameters to obtain various load modulation factor β as described in Chap. 1. At the same time, the tuner will also change effective electrical length to obtain quarter-wave length at different operating frequencies. In addition, the input and output matching network both are also designed as reconfigurable option using MEMS SPDT switches in order to optimize the performance at each frequency as well as back-off. At any particular operating frequency, one can draw generic configuration of Doherty PA as shown in Fig. 6.16.

This configuration is similar to Fig. 1.35 except that it utilizes one more transformer Z_K. This quarter-wave transformer now transforms load R_K to common load Z_K^2/R_K. In addition, one can see the offset line is absorbed in the input matching network of auxiliary PA. Referring to Eqs. (1.175) and (1.176), if the optimum load of the transistor is R_{opt}, one can choose the load combiner parameters of Fig. 6.16 as,

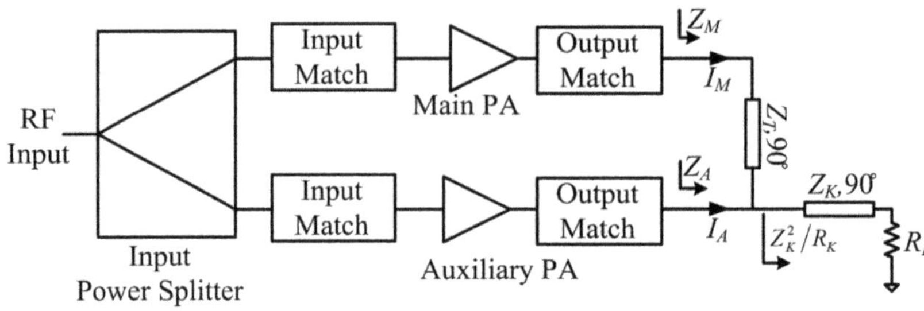

Fig. 6.16 Architecture of reconfigurable two-stage Doherty PA

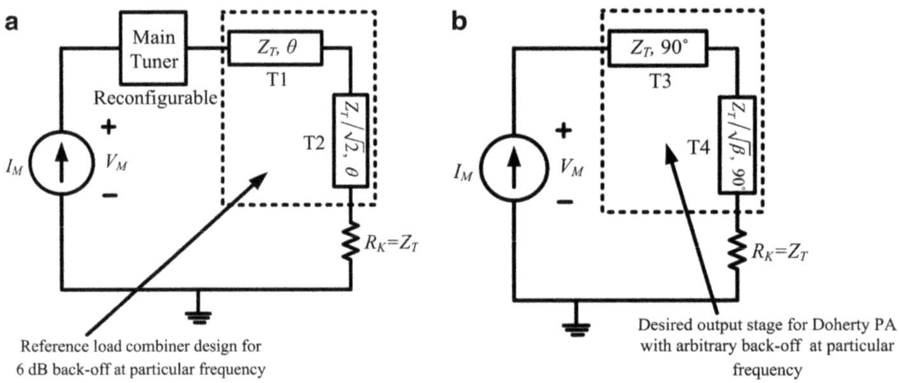

Fig. 6.17 Equivalent circuits of Doherty PA at low-power operation: (**a**) proposed scheme with reconfigurable tuner option, (**b**) desired load combiner

$$R_K = R_{\text{opt}}; \quad Z_T = R_K = R_{\text{opt}} \text{ and } Z_K = \frac{R_K}{\sqrt{\beta}} = \frac{R_{\text{opt}}}{\sqrt{\beta}} \tag{6.4}$$

Using the load combiner parameters as given in Eq. (6.4) in Eqs. (1.175) and (1.176), one can obtain the desired load modulation at arbitrary X dB back-off related to β given in Eq. (1.174).

At low-power levels where the auxiliary PA is *off*, equivalent circuit of the Doherty PA as shown in Fig. 6.16 is shown in Fig. 6.17a, b.

The key idea is to convert a reference load combiner design set for 6 dB back-off to any arbitrary back-off using a reconfigurable tuner described as main tuner as shown in Fig. 6.17a [35]. The electrical length of the transformers T_1 and T_2 are kept as a free design parameter to help in obtaining the optimum response at each operating condition. Figure 6.17b shows the desired output load combiner at low-power region for a particular frequency of operation. Now the *ABCD* parameters of load combiner cascaded with main tuner as shown in Fig. 6.17a are compared with *ABCD* parameters of desired load combiner as shown in Fig. 6.17b. The following expression is obtained from this comparison [35].

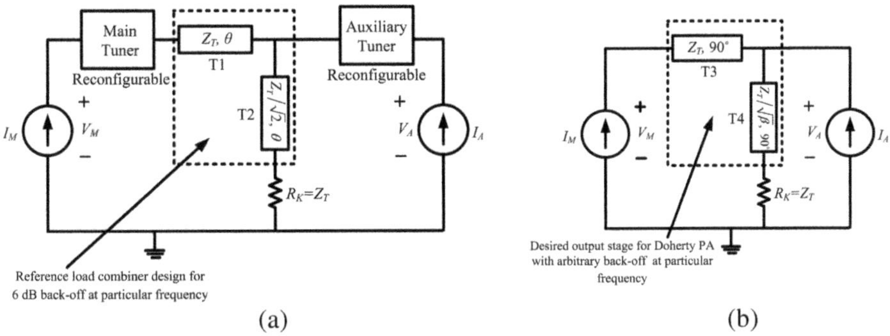

Fig. 6.18 Equivalent circuits of Doherty PA at high-power operation: (**a**) proposed scheme with reconfigurable tuners option, (**b**) desired load combiner

$$\begin{bmatrix} -\sqrt{\beta} & 0 \\ 0 & -\frac{1}{\sqrt{\beta}} \end{bmatrix} = \begin{bmatrix} A & B \\ C & D \end{bmatrix}_{\text{Main Tuner}} \times \begin{bmatrix} \cos^2\theta - \sqrt{2}\sin^2\theta & jZ_T\left(1+\frac{1}{\sqrt{2}}\right)\sin\theta\cos\theta \\ j\frac{1}{Z_T}\left(1+\sqrt{2}\right)\sin\theta\cos\theta & \cos^2\theta - \frac{1}{\sqrt{2}}\sin^2\theta \end{bmatrix} \times \begin{bmatrix} -\sqrt{\frac{\beta}{2}} & 0 \\ 0 & -\sqrt{\frac{2}{\beta}} \end{bmatrix} \tag{6.5}$$

Similarly, at high-power region, when auxiliary PA is *on*, Fig. 6.18a, b shows the equivalent circuit of the proposed scheme with main as well as auxiliary tuner and a desired load combiner configuration at a particular frequency.

The *ABCD* parameters of main tuner is already obtained from Eq. (6.5) and therefore considered known parameter in Fig. 6.18. The *ABCD* parameters of Fig. 6.18a are then compared with the *ABCD* parameters of Fig. 6.18b to obtain the *ABCD* parameters of auxiliary tuner [35]. The following expression as obtained from this comparison can be used to obtain *ABCD* parameter of auxiliary tuner [35].

$$\begin{bmatrix} j\beta & jZ_T \\ \frac{j}{Z_T} & 0 \end{bmatrix} = \begin{bmatrix} A & B \\ C & D \end{bmatrix}_{\text{Main Tuner}} \times \begin{bmatrix} \dfrac{\cos^2\theta - \sqrt{2}\sin^2\theta + j\left(1+\frac{1}{\sqrt{2}}\right)\sin\theta\cos\theta}{\cos\theta + j\frac{1}{\sqrt{2}}\sin\theta} & jZ_T\sin\theta \\ \dfrac{1}{Z_T}\dfrac{\cos^2\theta - \frac{1}{\sqrt{2}}\sin^2\theta + j\left(1+\sqrt{2}\right)\sin\theta\cos\theta}{\cos\theta + j\frac{1}{\sqrt{2}}\sin\theta} & \cos\theta \end{bmatrix} \times \begin{bmatrix} A & B \\ C & D \end{bmatrix}_{\text{Auxiliary Tuner}} \times \begin{bmatrix} 1 & 0 \\ \frac{\beta}{Z_T} - \frac{Z_T}{Z_K^2} & 1 \end{bmatrix} \tag{6.6}$$

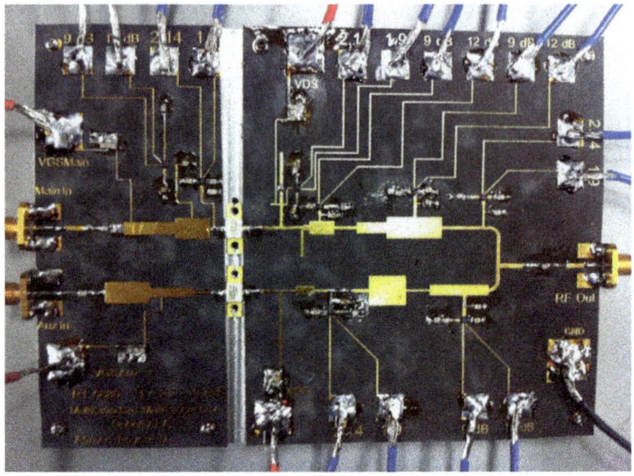

Fig. 6.19 Photograph of prototype of reconfigurable MEMS-based Doherty PA ([35], reprinted with permission from IEEE)

As a proof of concept, a multifrequency multi-standard reconfigurable Doherty PA is designed by considering different values of θ and β. The prototype is designed to handle signals with peak-to-average power ratio (PAPR) of 6, 9, and 12 dB with corresponding β values of 2, 2.82, and 4 operating at frequencies of 1.9, 2.14, and 2.6 GHz. At these frequencies, respective values of θ are chosen as 65.8°, 74.1°, and 90° [35]. The circuit utilizes overall eight tunable elements, two in each main and auxiliary tuner, two in each matching network used in the input and output of main PA [35]. Figure 6.19 shows the photograph of the prototype developed based on the above scheme [35]. The main PA is designed using 10 W GaN HEMT device CGH40010F from Wolfspeed, whereas auxiliary PA is designed using 25 W HEMT device CGH40025F of the same make [35]. Single-pole double-throw (SPDT) MEMS switch RMSW221 from RadantMEMS, which has actuation voltage of 110 V, is used, and therefore it cannot be actuated with high peaks of voltage swing in Doherty PA output. Figure 6.20a–c shows the measured efficiency characteristics of the prototype operating at various frequencies.

One can see from Fig. 6.20a that at 2.6 GHz, the efficiency at peak power is 67%. The corresponding efficiency values are 58.5, 54, and 42% at 6, 9, and 12 dB output power back-offs. Figure 6.20b shows the efficiency enhancement at 2.14 GHz, where efficiency at peak power is 66%, whereas efficiency measured at 6, 9, and 12 dB output power back-off is 58, 57, and 42%, respectively. Similar reconfigurability can be seen in Fig. 6.20c at 1.9 GHz.

6.4 Design and Challenges in High-Frequency and High-Power Packaging

The high-power device requires packaging which provides good thermal relief in order to keep the junction temperature of the device low and steady. This

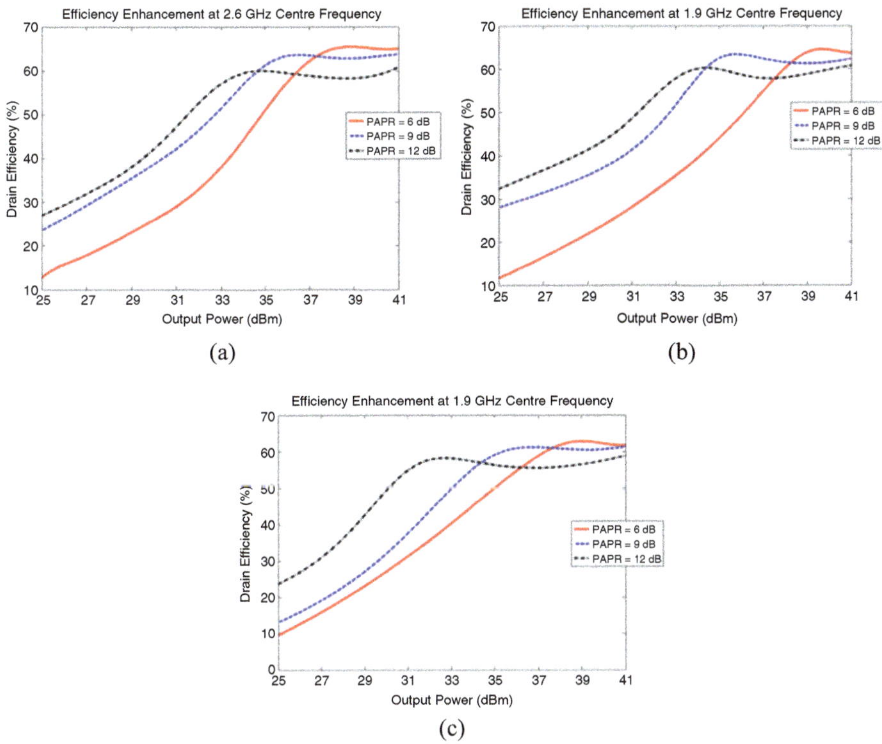

Fig. 6.20 Measured efficiency of reconfigurable MEMS-based Doherty PA prototype: (**a**) efficiency at 2.6 GHz, (**b**) efficiency enhancement at 2.14 GHz, and (**c**) efficiency enhancement at 1.9 GHz ([35], reprinted with permission from IEEE)

guarantees good device performance with high reliability and durability. Therefore, package for high-power devices utilizes advance materials for thermal interfacing, heat spreaders, and bonding materials. Figure 6.21 shows typical bare die device mounting scheme for GaN HEMT devices.

One can see that a direct connection of die with copper carrier is avoided and rather a molybdenum–copper (MoCu) shim is used in between. This avoids a large difference in terms of coefficient of thermal expansion (CTE) between GaN and metal heat spreader copper. The GaN device has low CTE as compared to copper. When high CTE thermal interfacing material such as solders are used to attach the die to the copper carrier, due to high mismatch in CTE, the die may break. Therefore, alloys of copper with CTE is often used as shim as shown in Fig. 6.21. Some of these alloys can be copper-tungsten (CuW), copper-molybdenum-copper (CuMoCu), and MoCu as shown in Fig. 6.21. Although the thermal conductivity (TC) of these alloys are inferior than copper as well as they are costly and heavy, yet they are useful in die attachment for relaxing the thermally induced stress. Table 6.4 lists some materials including semiconductors, die attach materials, and heat spreader

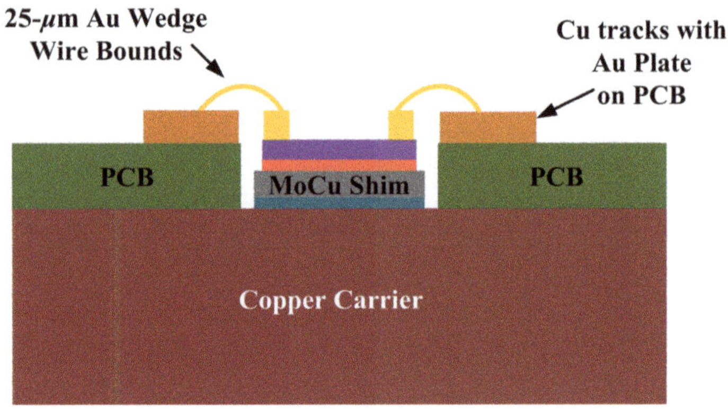

Fig. 6.21 High-power die device mounting scheme

Table 6.4 Thermal conductivity and coefficient of thermal expansion for materials used as semiconductors, die attach materials, and heat spreader [48–50]

Material	CTE (ppm/K)	Thermal conductivity (W/m K)
Aluminum	24	237
Alumina	7.3	30
Beryllium oxide	8.25	250
Cu	17	401
CuW (20/80)	8.3	230
Chemical-vapor-deposition (CVD) diamond	1.5	1000–1800
Gold-germanium	13	44
Gold-in-tin (AuSn)	16	57
GaN	4.3	130
GaAs	5.9	46
MoCu (30/70)	8.0	180
Si carbide	2.4	250

[48–50]. In addition to appropriate thermal management, the high-power device packages should present low parasitic which has minimal effect on RF performance especially in case of high-frequency or wide-band applications. For example, the MoCu shimmer used in Fig. 6.21 under the die device may also be extended below printed circuit board (PCB) [51]. This arrangement perhaps has reduced parasitic with low source-ground inductance presented to the die device. This will enhance the performance in terms of gain and power over the wide bandwidth [49].

There are two popular methods of GaN die attachment, namely eutectic and epoxy (specifically recently developed sintered silver epoxy) [48]. Epoxy in general requires dispensing along with pick-and-place unit. Therefore, this die-attachment process can be automated for high-throughput assembly process. In case of thermal epoxies, the deposition in thick layers have lower CTE than solders and therefore can directly be used to bond the device to higher CTE metal such as copper.

(a) (b) (c)

Fig. 6.22 Plastic overmolded surface mount device (SMD) packages: (**a**) quad-flat no-lead (QFN), (**b**) dual-flat no-lead (DFN), (**c**) small-outline integrated circuit (SOIC) [54, 55]

However, they may suffer with lower TC. On the contrary, thermal epoxies with high TC may require low temperature typically below 40 °C for reliable bonding [48]. This is a severe limitation in case of devices where junction temperature may rise beyond 150 °C such as GaN [52]. On the other hand, eutectic die attachment has best performance in terms of thermal as well as RF at very high-power levels. For example, gold-tin (AuSn) and gold-silicon (AuSi) eutectic solders can withstand temperatures up to 320 and 420 °C, respectively [48]. However, in comparison to epoxy, it is difficult to automate eutectic bonding for mass production [48, 53].

For low to medium power devices with output power typically up to 4 W, the plastic packages such as quad-flat no-lead (QFN), dual-flat no-lead (DFN), and small-outline integrated circuit (SOIC) are quite popular. These plastic overmolded surface mount device (SMD) packages are low-cost solutions and are shown in Fig. 6.22 [54–55]. For power device packaging, each of these packages may have exposed ground pad present at the bottom side for surface-mount assembly with appropriate RF grounding and thermal relief. The assembly of these devices over PCB require appropriate vias on PCB to appropriately dissipate heat generated by the devices.

The most common packaging for high-power devices are air cavity ceramic (ACC) packages [56]. Figure 6.23a shows these packages, where gate and drain access are provided using flat pins/leads. However, source is connected through the carrier. Figure 6.23b shows the package structure where most commonly used flange material is Cu/Mo70Cu/Cu, a laminate of copper and Cu-Mo. This material can provide low thermal resistance compared to the Cu-W in addition to good CTE match with the device material. Figure 6.23b also shows that the ACC package is made of three parts, that is, flange, ring frame, and lid. The flange and ring frame are brazed at high temperature and their joint assembly is called header as shown in Fig. 6.23b. The die is then attached (typically using eutectic soldering) to the flange and its corresponding gate and drain pads are wire bonded with the flat leads of the ACC package. Later the lid is glued on the top to close the package. The ACC superior performance comes with a price of having high handling time and also the brazing of the ring frame with the flange sometimes creates stress and distortions. Also, with each package variation a unique header must be created and therefore limiting the activity into small scale production.

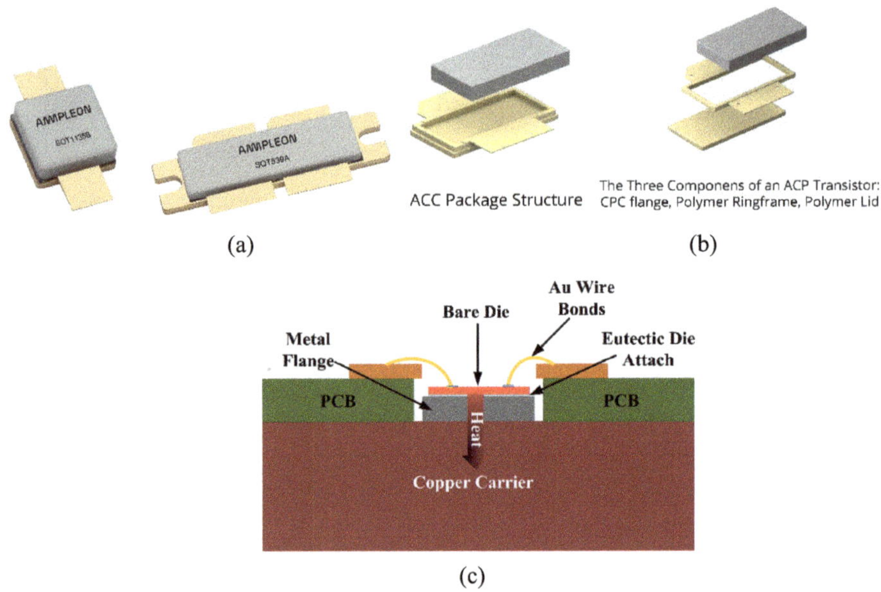

Fig. 6.23 Air cavity ceramic packages: (**a**) photographs of ACC packages [56], (**b**) ACC package structure with three components [56], (**c**) assembly scheme for ACC packages transistors in RF PA circuit

Figure 6.23c also shows the assembly scheme of these packages with flange mount assembly. One can see that the flange can be directly bolted or soldered over the heat sink by inserting through cut-out in PCB. The source of the device is therefore directly grounded through the flange and thereby providing good contact with low thermal resistivity. Sometimes a gap filler such as indium foils are inserted between flange and heat sink to reduce thermal resistance and improve physical contact between flange and heat sink [48].

The ACC packages can be used for high-power devices packages. On the contrary, the plastic overmold packages due to their higher thermal resistance results in high operating temperature and therefore degrading the performance and making it unsuitable for high-power operation. However, the cost of ACC packaging is around ten times higher than plastic packages.

The high temperature molded plastic packages are recently gaining popularity as they have good heat dissipation characteristics close to ACC and rather cheaper than ACC packages. Figure 6.24 shows photograph of high temperature molded plastic packages [58]. These packages have been in products delivering up to 200 W CW power without exceeding the operating temperature of the device beyond its recommended junction temperature [57].

The traditional plastic mold has a glass transition temperature (T_g) between 115 °C and 140 °C, above which it transforms from hard state to soft state [57]. This further results in change in coefficient of thermal expansion and flexural

Fig. 6.24 High temperature molded plastic packages [58]

modulus, eventually damaging device, wire bonding, etc. which are connected to the plastic overmold material. Therefore, for the GaN devices which may have junction temperature reaching above 200 °C, new overmold materials are required with higher T_g [57]. In such case the mold compounds with T_g ranging from 170 to 235 °C allows the device to operate at temperature close to its maximum junction temperature without creating stress on the device. Some of these mold compounds with high T_g have certain mechanical properties (adhesion strength, moisture absorption, etc.) and electrical properties (dielectric constant, loss tangent, etc.) which restricts its direct use in certain applications. Therefore, a buffer layer is used in such case to isolate sensitive component with mold compound. For example, silicon elastomer die coat can be used for this buffer layer to minimize stress on the device [57]. The plastic packaging generally utilizes die attachment using epoxy. This is perhaps due to room temperature dispensing and pick-and-place process which is more compatible with plastic packaging. The epoxies generally have inferior performance as compared to eutectic die attachment. However, recently new materials are developed containing silver, gold, or copper particles embedded in an epoxy/organic matrix. When properly cured, these will sinter together to form a strong bond with good thermal properties with TC up to 200 W/m K exceeding even eutectic die attach. Table 6.5 shows sintered epoxy materials with manufacturer part number, composition, and TC [57]. For comparison purpose, the TC of AuSn and AuSi is also listed in this table.

All these sintered epoxies are dispensed in a similar manner as traditional epoxy and can be automated for high throughput. These sintered epoxies have enabled GaN device packaging in a low-cost plastic package without compromising in thermal as well as RF performance. The plastic parts also use copper heat slug rather than Cu-Mo flange in ACC taking advantage of high TE of copper as compared to Cu-Mo. Table 6.6 compares the thermal resistance between junction and case (θ_{J-C}) for ACC and plastic overmold package using Au-Sn eutectic and silver-sintered epoxy die attachment, respectively [57].

Table 6.5 List of available sintered epoxy materials with thermal properties [57]

Product/manufacturer	Thermal conductivity (W/m K)	Technology
Agromax, Fortibond/Alpha	135–200	Ag nano-particle sintering
SSp 2020/Loctite Ablestik	>100	Ag sintering paste
mAgic Sintering/Heraeus	150–200	Ag sintering
XH9890 series/NAMICS	60–160	Ag nano-particles in organic matrix
Aurofuse TR-191R/Tanaka	150	Au sub-micron particle paste
Reference	190	AuSi
Reference	59	AuSn

Table 6.6 Comparison of plastic overmold using silver-sintered epoxy with ACC packaging using Au-Sn eutectic bonding [57]

Die/power level	Die attach material/thermal conductivity	$\theta_{J\text{–}C}$	Package type
24 mm GaN 100 W	AuSn preform 59 W/m K	1.75 °C/W	Air cavity
14 mm GaN 45 W	AuSn preform 59 W/m K	2.3 °C/W	Air cavity
14 mm GaN 45 W	Ag-sintered epoxy 160 W/m K	1.6 °C/W	Plastic overmold
24 mm GaN 100 W	Ag-sintered epoxy 160 W/m K	1.3 °C/W	Plastic overmold

The table also reports the TC of Au-Sn and silver-sintered epoxy. One can see there is reduction of more than 25% in $\theta_{J\text{–}C}$ in plastic overmold package as compared to ACC.

6.5 Techniques for Millimeter-Wave Power Amplifier Design

In order to handle very high speed in order of Gbps in 5G communication, the RF transmitter development is targeting RF PA design at millimeter-wave (mm-wave) frequencies. The PA designers have recently focused toward the design challenges at centimeter-wave (cm-wave) near 30 GHz and mm-wave frequencies beyond 30 GHz. The design challenges are widely analyzed in terms of device technology, circuit topologies, system architectures and in perspective of SOC integration [59]. A Ka-band PA was designed to operate around 28 GHz using 0.15 μm GaAs pHEMT technology [60]. The process has power density of 630 mW/mm and the gate periphery is set to achieve 26 dBm output power at around 1 dB gain compression. The PA has three common source stages with gate periphery of input, intermediate, and output stages of 160 μm, 300 μm and 750 μm, respectively. Each stage is biased in Class AB mode and the optimum impedance for output stage is obtained from the load-pull analysis [60]. The interstage matching performs conjugate match of the two devices for maximum gain [60]. The PAE reported for this work was 27–32% at 24–25 dBm output power with IMD3 below −30 dBc [60]. CMOS-based PAs are also reported at mm-wave frequencies [61–66]. Some design consideration in terms of layout as well as circuit topology suitable for mm-wave

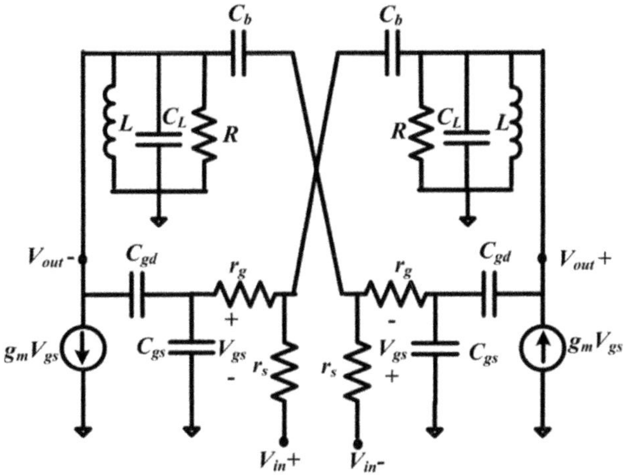

Fig. 6.25 Topology of pseudo-differential pair of common-source amplifiers with neutralization [66]

PA design is discussed in [63]. The major challenge in mm-wave frequency is to compensate for the device and interconnect parasitic which are the main limitations for obtaining high power and efficiency at mm-wave frequency. The parasitic such as gate resistance (r_g) and gate-to-drain capacitances (C_{gd}) can be minimized by layout optimization [64]. An optimized layout is proposed in 40 nm bulk CMOS technology for the design of 60 GHz dual-mode PA [63]. In addition to the layout optimization, a neutralization technique is also used to minimize the effect of parasitic negative feedback path presented by C_{gd} on power gain and stability [63]. A neutralization technique is easy to implement in differential amplifier by cross-connecting the capacitors between the gate and drain terminals of differential stage [63, 66]. Figure 6.25 shows an equivalent circuit of two common source amplifiers driving the tuned loads combined in pseudo-differential pair with neutralization [66].

One can see that two additional capacitors cross-connected between the gate and the drain terminals can provide cancellation to unwanted feedback via C_{gd}. A comparison between pseudo-differential common source and cascode amplifier is performed in simulation using 65-nm CMOS. Both the amplifiers drive 100 Ω loads [67]. Considering the layout parasitic at mm-wave frequencies, the gain advantage of cascode over the common source topology is less significant. Therefore, a differential common source stage with neutralizing capacitors may be considered as an alternative solution at mm-wave frequencies [66]. Figure 6.26 shows topology of three stages differential common-source amplifier with neutralization [66].

The PA is designed to operate at 60 GHz with integrated input and output baluns [66]. The design exploits neutralization and transformer coupling which helps in achieving higher gain with unconditional stability and good isolation between the input and output. The common-source amplifiers are used to operate below 1 V and each stage provides power gain of 12.4 dB in simulation which reduced to 9.5 dB

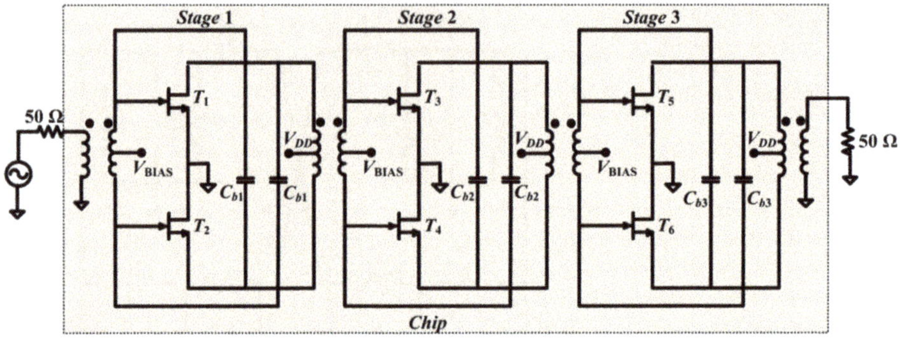

Fig. 6.26 Topology of the three-stage pseudo-differential pairs of common-source amplifiers with neutralization [66]

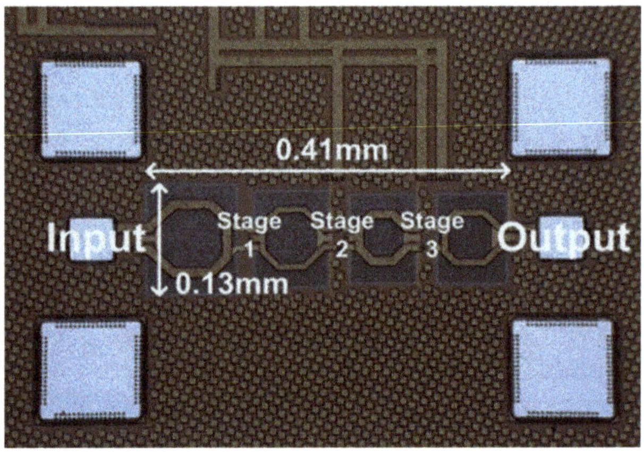

Fig. 6.27 Chip photograph of prototype of the three-stage pseudo-differential pairs of common-source amplifiers with neutralization ([66] reprinted with permission from IEEE)

with estimated parasitic of interconnects [66]. Figure 6.27 shows photograph of chip fabricated based on topology discussed above.

Total chip area is 0.87 × 0.8 mm^2 including bondpads. The PA and its input and output baluns occupies 0.41 × 0.13 mm^2. Figure 6.28 shows measurement results for this chip in terms of output power, gain, and PAE at 62 GHz.

One can see from Fig. 6.28 that the output power of 10.3 dBm is achieved with 7.1 dB gain compression. At this power PAE is 15.2%. The output power at 1 dB gain compression is about 5 dBm.

Similar efforts have been carried out in multiband Doherty PA design in 130 nm SiGe BiCMOS process by Global Foundries [68, 69]. This is fully integrated multiband Doherty PA operating at 28/37/39 GHz [68, 69]. Board level Doherty PA uses quarter-wave transformer as discussed in Chaps. 1 and 4. These quarter-wave transformers are unfeasible in CMOS and BiCMOS technology as it consumes large on-chip area. Therefore, transformer-based solution is proposed to replace quarter-wave transformer in Doherty PA design. In addition, the driver amplifiers are designed in such a manner that power-dependent adaptive uneven feeding replaces

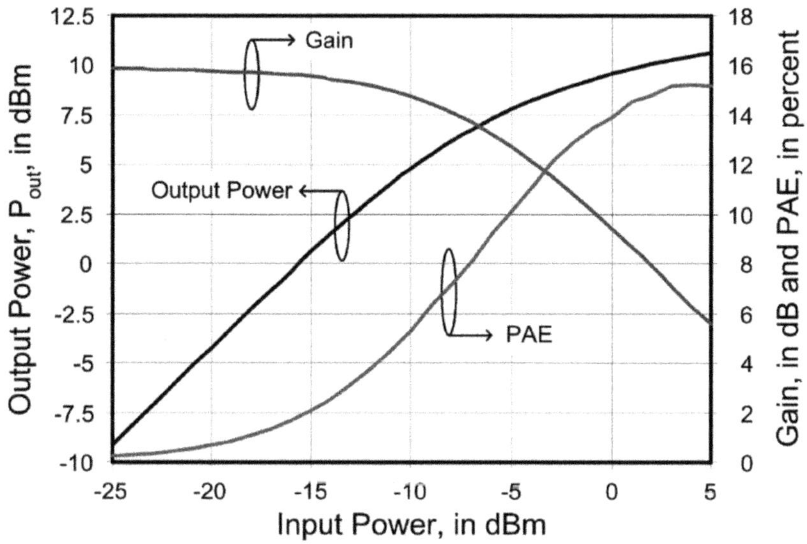

Fig. 6.28 Measurement results: output power, gain, PAE with input power drive ([66], reprinted with permission from IEEE)

the conventional uneven power division technique at the input of Doherty PA. As the input drive to Doherty PA topology increases, class C bias auxiliary PA is gradually turned *on* and its effective transconductance increases. This results into increase in input conductance, and therefore the input impedance of auxiliary PA increases. On contrary, the input impedance of main PA is comparatively consistent [68]. This increase in input impedance of auxiliary PA will act as load for the driver amplifier in the auxiliary path. Therefore, the driver amplifier for auxiliary PA is designed in such a manner that it shall deliver more power with this increase in its load impedance as auxiliary PA is turned *on* and pushed toward saturation [69]. Therefore, one can achieve power-dependent adaptive feeding which improves Doherty PA performance. Figure 6.29 shows the photograph of the chip fabricated. The overall size of the chip is 780 μm × 2250 μm.

One can see from the layout of Fig. 6.29 that this Doherty PA has five stages. The first stage is differential quadrature hybrid coupler for splitting of input drive in quadrature relationship among the main and auxiliary PA. Second stage is varactor loaded transmission line for adjusting relative phase between main and auxiliary path for 28/37/39 GHz. The third stage is driver stage which will drive the main and auxiliary PA along with a power-dependent adaptive uneven feeding. Fourth stage is the main and auxiliary PA followed by transformer-based load combiner. Figure 6.30 shows the measured results of this Doherty PA when tuned to three operating frequencies, i.e., 28 GHz, 37 GHz, and 39 GHz. At 28 GHz, the peak power is 16.8 dBm with 29.4% collector efficiency and PAE of 20.3% as shown in Fig. 6.30a. At 5.9 dB back-off the collector efficiency is around 25% which is 1.72 and 3.39 times better than class B and class A PA, respectively. The gain is

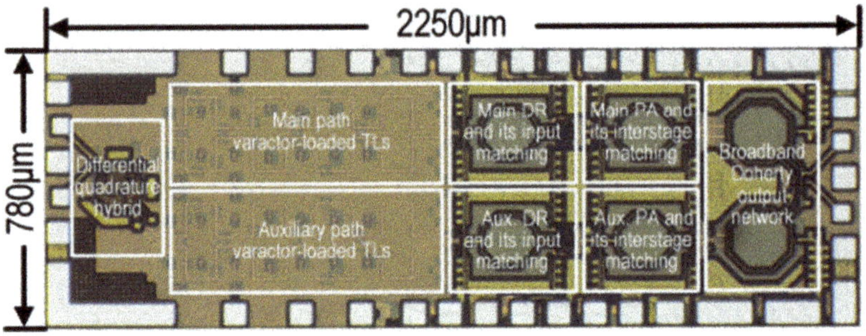

Fig. 6.29 Photograph of fabricated chip of 28/37/39 GHz Doherty PA ([69], reprinted with permission from IEEE)

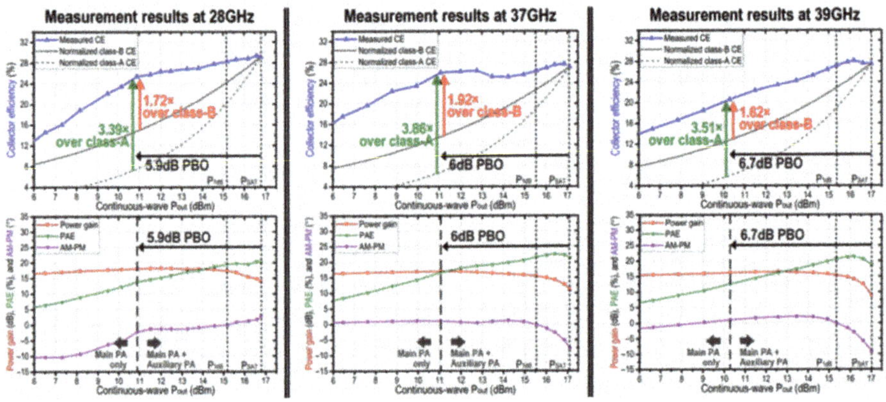

(a) (b) (c)

Fig. 6.30 Continuous wave measurement results of Doherty PA at 28 GHz, 37 GHz and 39 GHz ([69], reprinted with permission from IEEE)

around 14 dB at peak power, whereas the corresponding small signal gain is around 17 dB indicating around 3 dB compression at peak power. Figure 6.30b shows the measured performance at 37 GHz. The peak power is 17 dBm where the collector efficiency is around 28% and corresponding PAE is 21.4%. At 6 dB back-off, the collector efficiency is around 25% which is comparatively 1.92 times and 3.86 times the collector efficiency of class B and class A PA, respectively. The gain at saturation is around 11 dB with around 5 dB gain compression with respect to the small signal gain of around 16 dB. Figure 6.30c shows measured performance at 39 GHz. The peak power is measured as 17 dBm with collector efficiency reported around 28% and corresponding PAE is around 19%. The collector efficiency at 6.7 dB back-off is measured around 20% which is around 1.62 and 3.51 times the efficiency of class B and class A, respectively. The small signal gain is around 15 dB whereas the gain at peak power is around 10 dB implying the PA is operating at 5 dB gain compression at peak output power.

There are several other attempts in the design of mm-wave PA, yet researchers are exploring better device technology as well circuit topologies to reach to the optimum solution.

References

1. Issac, R. D. (2000). The future of CMOS technology. *IBM Journal of Research and Development, 44*(3), 369–378.
2. Huang, P., Tsai, Z., Lin, K., & Wang, H. (2010). A high-efficiency, broadband CMOS power amplifier for cognitive radio applications. *IEEE Transactions on Microwave Theory and Techniques, 58*(12), 3556–3565.
3. Lu, C., Pham, A.-V. H., Shaw, M., & Saint, C. (2007). Linearization of CMOS broadband power amplifiers through combined multigated transistors and capacitance compensation. *IEEE Transactions on Microwave Theory and Techniques, 55*(10), 2053–2058.
4. Joo, T., Koo, B., & Hong, S. (2013). A WLAN RF CMOS PA with large signal MGTR method. *IEEE Transactions on Microwave Theory and Techniques, 61*(3), 1272–1279.
5. Fritzin, J., Sundstrom, T., Johannson, T., & Alvandpour, A. (2010). Reliability study of a low-voltage class-E power amplifier in 130 nm CMOS. In *Proceedings of IEEE International Symposium on Circuits and Systems (ISCAS)*, Paris, France, pp. 1907–1910.
6. Fritzin, J., & Alvandpour, A. (2009). Low voltage class-E power amplifiers for DECT and Bluetooth in 130 nm CMOS. In *Proceedings of IEEE Topical Meeting on Silicon Monolithic Integrated Circuits in RF Systems*, pp. 57–60.
7. Thyagarajan, S. V., Niknejad, A. M., & Hull, C. D. (2014). A 60 GHz drain source neutralized wideband linear power amplifier in 28 nm CMOS. *IEEE Transactions on Circuits and Systems I: Regular Papers, 61*(8), 2253–2262.
8. Fritzin, J., Svensson, C., & Alvandpour, A. (2011). A +32 dBm 1.85 GHz class-D out-phasing RF PA in 130nm CMOS for WCDMA/LTE. In *Proceedings of ESSCIRC*, pp. 12–16.
9. Moazzami, R., & Hu, C. (1997). Projecting gate oxide reliability and optimizing reliability screens. *IEEE Transactions on Electron Devices, 37*(7), 1643–1650.
10. Taur, Y., Buchanan, D. A., Chen, W., Frank, D. J., Ismail, K. E., Lo, S. H., Sai-Halasz, G. A., Viswanathan, R. G., Wann, H. J. C., Wind, S. J., & Wong, H. S. (1997). CMOS scaling into the nanometer regime. *Proceedings of the IEEE, 85*(4), 486–504.
11. Kimura, M. (1999). Field and temperature acceleration model for time-dependent dielectric breakdown. *IEEE Transactions on Electron Devices, 46*(1), 220–229.
12. Sowlati, T., & Leenaerts, D. M. W. (2003). A 2.4-GHz 0.18-μm CMOS self-biased cascode power amplifier. *IEEE Journal of Solid-State Circuits, 38*(8), 1318–1324.
13. Aoki, I., Kee, S. D., Rutledge, D. B., & Hajimiri, A. (2002). Fully integrated CMOS power amplifier design using the distributed active-transformer architecture. *IEEE Journal of Solid-State Circuits, 37*(3), 371–383.
14. Aoki, I., Kee, S. D., Magoon, R., Aparicio, R., Bohn, F., Zachan, J., Hatcher, G., McClymont, D., & Hajimiri, A. (2008). A fully integrated quadband GSM/GPRS CMOS power amplifier. *IEEE Journal of Solid-State Circuits, 43*(12), 2747–2758.
15. Liu, G., Haldi, P., Liu, T. K., & Niknejad, A. M. (2008). Fully integrated CMOS power amplifier with efficiency enhancement at power back-off. *IEEE Journal of Solid State Circuits, 43*(3), 600–609.
16. Ezzeddine, A. K., & Huang, H. C. (2003). The high voltage/high power FET (HiVP). In *Proceedings of IEEE Radio Frequency Integrated Circuits Symposium*, pp. 215–218.
17. Ezzeddine, A. K., & Huang, H. C. (2006). Ultra-broadband GaAs HIFET MMIC PA. In *Proceedings of IEEE International Microwave Symposium*, pp. 1320–1323.

18. Pornpromlikit, S., Jeong, J., Presti, C. D., Scuderi, A., & Asbeck, P. M. (2010). A watt-level stacked-FET linear PA in silicon-on-insulator CMOS. *IEEE Transactions on Microwave Theory and Techniques, 58*(1), 57–64.
19. Fritsche, D., Wolf, R., & Ellinger, F. (2012). Analysis and design of a stacked power amplifier with very high bandwidth. *IEEE Transactions on Microwave Theory and Techniques, 60*(10), 3223–3231.
20. Dabag, H. T., Hanafi, B., Golcuk, F., Agah, A., Buckwalter, J. F., & Asbeck, P. M. (2013). Analysis and design of stacked-FET mm-wave power amplifiers. *IEEE Transactions on Microwave Theory and Techniques, 61*(4), 1543–1556.
21. Kim, Y., Koh, Y., Kim, J., Lee, S., Jeong, J., Seo, K., & Kwon, Y. (2011). A 60 GHz broadband stacked FET power amplifier using 130 nm metamorphic HEMTs. *IEEE Microwave and Wireless Components Letters, 21*(6), 323–325.
22. Kim, Y., & Kwon, Y. (2015). Analysis and design of millimeter-wave power amplifier using stacked-FET structure. *IEEE Transactions on Microwave Theory and Techniques, 63*(2), 691–702.
23. Wu, H. F., Cheng, Q. F., Li, X. G., & Fu, H.-P. (2016). Analysis and design of an ultra broadband stacked power amplifier in CMOS technology. *IEEE Transactions on Circuits and Systems II: Express Briefs, 63*(1), 49–53.
24. Kuo, C. C., Lin, Y. H., Lu, H. C., & Wang, H. (2013). A K-band compact fully integrated transformer power amplifier in 0.18-μm CMOS. In *Proceedings of IEEE Asia Pacific Microwave Conference*, pp. 597–599.
25. Metz, E. P. A., Miller, R. C., & Mazelsky, R. (1962). A technique for pulling single crystals of volatile materials. *Journal of Applied Physics, 33*(6), 2016–2017.
26. Liberati, R. M., & Calori, M. (2008). High performance future hybrid transceiver module using GaN power devices for seeker applications. In *Proceedings of IEEE Radar Conference*, pp. 1–4.
27. Masuda, S., Yamada, M., Ohki, T., Makiyama, K., Okamoto, N., Nakasha, Y., Imanishi, K., Kikkawa, T., & Shigematsu, H. (2011). C-Ku band GaN MMIC T/R frontend module using multilayer ceramics technology. In *Proceedings of IEEE International Microwave Symposium*, pp. 1–4.
28. Azam, S., & Wahab, Q. (2010). *The present and future trends in high power microwave and millimeter wave technologies*. London: INTECH Open Access Publisher.
29. Bettidi, A., Calori, M., Cetronio, A., Cicolani, M., Costrini, C., Lanzieri, C., Maccaroni, S., Marescialli, L., & Peroni, M. (2008). Innovative T/R module in state-of the-art GaN technology. In *Proceedings of IEEE Radar Conference*, pp. 1–5.
30. Sudow, M., Fagerlind, M., Thorsell, M., Andersson, K., Billstrom, N., Nilsson, P., & Rorsman, N. (2008). An AlGaN/GaN HEMT-based microstrip MMIC process for advanced transceiver design. *IEEE Transactions on Microwave Theory and Techniques, 56*(8), 1827–1833.
31. Yang, J. M., Aust, M., Chung, Y., Lee, W., Stokes, R., Dupery, R., Lai, R., Gorospe, B., & Kagiwada, R. (2004). Affordable high performance InP X-band transceiver module for large aperture phased array applications. In *Proceedings of IEEE International Microwave Symposium*, pp. 1761–1763.
32. Masuda, S., Akasegawa, A., Ohki, T., Makiyama, K., Okamoto, N., Imanishi, K., Kikkawa, T., & Shigematsu, H. Over 10W C-Ku band GaN MMIC non-uniform distributed power amplifier with broadband couplers. In *Proceedings of IEEE International Microwave Symposium*, pp. 1388–1391.
33. Mouginot, G., Ouarch, Z., Lefebvre, B., Heckmann, S., Lhortolary, J., Baglieri, D., Floriot, D., Camiade, M., Blanck, H., Le Pipec, M., Mesnager, D., & Le Helleye, P. (2010). Three stage 6–18 GHz high gain and high power amplifier based on GaN technology. In *Proceedings of IEEE International Microwave Symposium*, pp. 1392–1395.
34. Kolias, N. J. (2015). Recent advances in GaN MMIC technology. In *Proceedings of IEEE Custom Integrated Circuits Conference (CICC)*.

35. Mohamed, A. M. M., Boumaiza, S., & Mansour, R. R. (2013). Electronically tunable Doherty power amplifier for multi-mode multi-band base stations. *IEEE Transactions on Circuits and Systems I: Regular Papers, 61*(4), 1229–1240.
36. Mohamed, A. M. M., Boumaiza, S., & Mansour, R. R. (2013). Doherty power amplifier with enhanced efficiency at extended operating average power levels. *IEEE Transactions on Microwave Theory and Techniques, 61*(12), 4179–4187.
37. Mohamed, A. M. M., Boumaiza, S., & Mansour, R. R. (2013). Reconfigurable Doherty power amplifier for multifrequency wireless radio systems. *IEEE Transactions on Microwave Theory and Techniques, 61*(4), 1588–1598.
38. Fischer, G., Eckl, W., & Kaminski, G. (2002). RF-MEMS and SiC/GaN as enabling technologies for a reconfigurable multi-band/multi-standard radio. *Bell Labs Technical Journal, 7*(3), 169–189.
39. Kalyan, R., Rawat, K., & Koul, S. K. (2016). Reconfigurable and concurrent dual-band Doherty power amplifier for multiband and multistandard applications. *IEEE Transactions on Microwave Theory and Techniques, 65*(1), 198–208.
40. Koul, S. K., & Dey, S. (2018). K-band 4-bit phase shifter using two back to back MEMS SP16T switching networks. *Journal of Microelectromechanical Systems, 27*(4), 643–655.
41. Dey, S., & Koul, S. K. (2015). Reliability analysis of Ku-band 5-bit phase shifters using MEMS SP4T and SPDT switches. *IEEE Transactions on Microwave Theory and Techniques, 63*(12), 3997–4012.
42. Dey, S., Koul, S. K., Poddar, A. K., & Rohde, U. L. (2018). Reliable and compact 3-and 4-bit phase shifters using MEMS SP4T and SP8T switches. *Journal of Microelectromechanical Systems, 27*(1), 113–124.
43. Larcher, L., Brama, R., Ganzerli, M., Iannacci, J., Bedani, M., & Gnudi, A. (2009). A MEMS reconfigurable quad-band class-E power amplifier for GSM standard. In *Proceedings of IEEE Conference on Micro Electro Mechanical Systems*, pp. 364–368.
44. Okazaki, H., Narahashi, S., Fukuda, A., & Miki, T. (2005). Band-reconfigurable high-efficiency power amplifier using MEMS switches. In: *Proceedings of IEEE International Workshop on Radio-Frequency Integration Technology: Integrated Circuits for Wideband Communication & Wireless Sensor Networks*, pp. 83–86.
45. Fukuda, A., Okazaki, H., Hirota, T., & Yamao, Y. (2005). Novel band-reconfigurable high efficiency power amplifier employing RF-MEMS switches. *IEICE Transactions on Electronics, 88*(11), 2141–2149.
46. Seki, T. (2002). Development and packaging of RF MEMS series switch. In *Proceedings of IEEE Asia Pacific Microwave Conference*, pp. 266–272.
47. Aggrawal, E., & Rawat, K. (2019). Reconfigurable chireix outphasing power amplifier over multiple frequency bands. *IEEE Transactions on Circuits and Systems II: Express Briefs (Early Access)*. https://doi.org/10.1109/TCSII.2019.2926529.
48. Samanta, K. K. (2016). PA thermal management and packaging: Wideband PA and packaging, history, and recent advances: Part 2. *IEEE Microwave Magazine, 17*(11), 73–81.
49. Samanta, K. K. (2015). Advanced ultra-wide-band GaN power amplifiers. In *Proceedings of IEEE European Microwave Conference*, Paris.
50. Coppola, L., Huff, D., Wang, F., Burgos, R., & Boroyevich, D. (2007). Survey on high-temperature packaging materials for SiC-based power electronics modules. In *Proceedings of IEEE Power Electronics Specialists Conference*, pp. 2234–2240.
51. Samanta, K. K. (2016). Advanced broadband GaN power amplifiers with kW CW power. In *Proceedings of IEEE MTT-S International Microwave Symposium Workshop: Advances in High Power SSP Technology for kW Operation at Microwave Frequencies*.
52. Chin, H. S., Cheong, K. Y., & Ismail, A. B. (2010). A review on die attach materials for SiC-based high-temperature power devices. *Metallurgical and Materials Transactions B: Process Metallurgy and Materials Processing Science, 41*(4), 824–832.
53. Manikam, V. R., & Cheong, K. Y. (2011). Die attach materials for high temperature applications: A review. *IEEE Transactions on Components, Packaging and Manufacturing Technology, 1*(4), 457–478.

54. Qorvo Datasheet. (2018). QPM1002, Data Sheet Rev. E.
55. Allegro MicroSystems Datasheet, LLC. A6260, 6260-DS, Rev. 7.
56. Ampleon Package Overview. (2016). AMP LL20160802.
57. Martin, Q. (2015). High power plastic packaging with GaN. In *CS MANTECH Conference*, pp. 59–62.
58. Leth, D., Pelletier, L., & Shah, M. (2015). Designing with plastic RF power transistors, freescale semiconductor white paper, RFPLASTICWP, Rev. 2.
59. Lie, D. Y. C., Mayeda, J. C., Li, Y., & Lopez, J. (2018). A review of 5G power amplifier design at cm-wave and mm-wave frequencies. *Hindawi Wireless Communications and Mobile Computing, 2018*, 6793814. https://doi.org/10.1155/2018/6793814.
60. Curtis, J., Zhou, H., & Aryanfar, F. (2016). A fully integrated Ka-band front end for 5G transceiver. In *Proceedings of IEEE International Microwave Symposium*, San Francisco, CA.
61. Shakib, S., Park, H.-C., Dunworth, J., Aparin, V., & Entesari, K. (2016). A highly efficient and linear power amplifier for 28-GHz 5G phased array radios in 28-nm CMOS. *IEEE Journal of Solid-State Circuits, 51*(12), 3020–3036.
62. Shakib, S., Park, H., Dunworth, J., Aparin, V., & Entesari, K. (2016). 28 GHz efficient linear power amplifier for 5G phased arrays in 28nm bulk CMOS. In *Proceedings of IEEE International Solid-State Circuits Conference (ISSCC)*, San Francisco, CA, pp. 44–45.
63. Zhao, D., & Reynaert, P. (2013). A 60-GHz dual-mode class AB power amplifier in 40-nm CMOS. *IEEE Journal of Solid-State Circuits, 48*(10), 2323–2337.
64. Heydari, B., Bohsali, M., Adabi, E., & Niknejad, A. M. (2007). Millimeter-wave devices and circuit blocks up to 104 GHz in 90 nm CMOS. *IEEE Journal of Solid-State Circuits, 42*(12), 2893–2902.
65. Sadhu, B., Tousi, Y., Hallin, J., Sahl, S., Reynolds, S., Renström, O., Sjögren, K., Haapalahti, O., Mazor, N., Bokinge, B., & Weibull, G. (2017). A 28 GHz 32-element phased array transceiver IC with concurrent dual polarized beams and 1.4 degree beam-steering resolution for 5G communication. In *Proceedings of IEEE International Solid- State Circuits Conference (ISSCC)*, San Francisco, CA, pp. 128–129.
66. Chan, W. L., & Long, J. R. (2010). A 58–65 GHz neutralized CMOS power amplifier with PAE above 10% at 1-V supply. *IEEE Journal of Solid-State Circuits, 45*(3), 554–564.
67. Luo, Z., Steegen, A., Eller, M., Mann, R., Baiocco, C., Nguyen, P., Kim, L., Hoinkis, M., Ku, V., Klee, V., Jamin, F., Wrschka, P., Shafer, P., Lin, W., Fang, S., Ajmera, A., Tan, W., Park, D., Mo, R., Lian, J., Vietzke, D., Coppock, C., Vayshenker, A., Hook, T., Chan, V., Kim, K., Cowley, A., Kim, S., Kaltalioglu, E., Zhang, B., Marokkey, S., Lin, Y., Lee, K., Zhu, H., Weybright, M., Rengarajan, R., Ku, J., Schiml, T., Sudijono, J., Yang, I., & Wann, C. (2004). High performance and low power transistors integrated in 65 nm bulk CMOS technology. In *Proceedings of IEEE International Electron Devices Meeting (IEDM)*, pp. 661–666.
68. Hu, S., Wang, F., & Wang, H. (2019). A highly efficient and linear power amplifier for 28-GHz 5G phased array radios in 28-nm CMOS. *IEEE Journal of Solid-State Circuits, 54*(6), 1586–1599.
69. Hu, S., Wang, F., & Wang, H. (2017). A 28GHz/37GHz/39GHz multiband linear Doherty power amplifier for 5G massive MIMO applications. In *Proceedings of IEEE International Solid- State Circuits Conference (ISSCC)*, San Francisco, CA, pp. 32–33.

Index

K. Rawat et al., *Bandwidth and Efficiency Enhancement in Radio Frequency Power Amplifiers for Wireless Transmitters*, Analog Circuits and Signal Processing,
https://doi.org/10.1007/978-3-030-38866-9

Milton Keynes UK
Ingram Content Group UK Ltd.
UKHW052121200324
439639UK00001B/5

9 783030 388652